CRYSTALLIZATION OF POLYMERS, SECOND EDITION

In *Crystallization of Polymers*, second edition, Leo Mandelkern provides a self-contained, comprehensive, and up-to-date treatment of polymer crystallization. All classes of macromolecules are included and the approach is through the basic disciplines of chemistry and physics. The book discusses the thermodynamics and physical properties that accompany the morphological and structural changes that occur when a collection of molecules of very high molecular weight is transformed from one state to another. The first edition of *Crystallization of Polymers* was published in 1964. It was regarded as the most authoritative book in the field. The first edition was composed of three major portions. However, due to the huge amount of research activity in the field since publication of the first edition (involving new theoretical concepts and new experimental instrumentation), this second edition has grown to three volumes.

Volume 2 provides an authoritative account of the kinetics and mechanisms of polymer crystallization, building from the equilibrium concepts presented in Volume 1. As crystalline polymers rarely, if ever, achieve their equilibrium state, this book serves as a bridge between equilibrium concepts and the state that is finally achieved. A comprehensive treatment of surrounding theories is given, and experimental results for simple and complex polymer systems are described in detail.

This book will be an invaluable reference work for all chemists, physicists and materials scientists working in the area of polymer crystallization.

LEO MANDELKERN was born in New York City in 1922 and received his bachelors degree from Cornell University in 1942. After serving in the armed forces during World War II, he returned to Cornell, receiving his Ph.D. in 1949. He remained at Cornell in a post-doctoral capacity until 1952.

Professor Mandelkern was a staff member of the National Bureau of Standards from 1952 to 1962 where he conducted research in the physics and chemistry of polymers. During that time he received the Arthur S. Fleming Award from the Washington DC Junior Chamber of Commerce "As one of the outstanding ten young men in the Federal Service".

In January 1962 he was appointed Professor of Chemistry and Biophysics at the Florida State University, Tallahassee, Florida, where he is still in residence.

He is author of *Crystallization of Polymers*, first edition, published by McGraw-Hill in 1964. He is also author of *Introduction to Macromolecules*, first edition 1972, second edition 1983, published by Springer-Verlag.

Besides the Arthur S. Fleming Award he has been the recipient of many other awards from different scientific societies including the American Chemical Society and the Society of Polymer Science, Japan.

Professor Mandelkern is the author of over 300 papers in peer reviewed journals and has served on the editorial boards of many journals, including the *Journal of the American Chemical Society*, the *Journal of Polymer Science* and *Macromolecules*.

CRYSTALLIZATION OF POLYMERS

SECOND EDITION

Volume 2

Kinetics and mechanisms

LEO MANDELKERN

R. O. Lawton Distinguished Professor of Chemistry, Emeritus
Florida State University

PUBLISHED BY THE PRESS SYNDICATE OF THE UNIVERSITY OF CAMBRIDGE
The Pitt Building, Trumpington Street, Cambridge, United Kingdom

CAMBRIDGE UNIVERSITY PRESS
The Edinburgh Building, Cambridge CB2 2RU, UK
40 West 20th Street, New York, NY 10011-4211, USA
477 Williamstown Road, Port Melbourne, VIC 3207, Australia
Ruiz de Alarcón 13, 28014 Madrid, Spain
Dock House, The Waterfront, Cape Town 8001, South Africa

http://www.cambridge.org

First published 1964 by McGraw-Hill, New York
Second edition, Volume 1 2002, Volume 2 2004

Printed in the United Kingdom at the University Press, Cambridge

Typeface Times 11/14 pt   *System* LaTeX 2_ε [TB]

A catalog record for this book is available from the British Library

ISBN 0 521 81682 3 hardback

Contents

Preface

This volume is concerned with crystallization kinetics and mechanisms and serves a crucial role in understanding the actual crystalline state that develops in polymers under real conditions. Attention is focused on the nonequilibrium aspects of the crystalline state that actually forms. It serves as a bridge to Volume 3 and the discussion of morphology, structure and properties. As the reader will find, there are still some important problems in this area that are in need of resolution. In discussing these, the author has tried to be as objective as possible and has presented the diverse viewpoints. Professor R. G. Alamo and Dr. F. C. Stehling read portions of the manuscript and offered constructive criticisms. However, the contents of the volume are the sole responsibility of the author.

It is a pleasure once again to acknowledge a great debt to Mrs. Annette Franklin for her expert typing of the manuscript and preparing it in final format. The author also acknowledges the contribution of Dr. James A. Haigh for the original calculations that appear throughout the book. Most of the illustrations in final form are due to the efforts of Dick Roche and Steve Leukanech. Their contributions are gratefully acknowledged.

The permission granted by American Chemical Society, *Anales de Quimica, Chemical Reviews, Colloid and Polymer Science, Deutsche Bunsen-Gesellschaft für Physikalische Chemie*, Elsevier, *Industrial and Engineering Chemistry, Journal of Applied Physics, Journal of Chemical Physics, Journal of Macromolecular Science, Journal of Material Science, Journal of Physical Chemistry, Journal of Polymer Science, Journal of Thermal Analysis and Calorimetry, Macromolecules*, Materials Research Society, *Polymer, Polymer Engineering and Science, Polymer Journal, Physical Review Letters, Proceedings of the Royal Society of London*, Royal Society of Chemistry, and Springer-Verlag to reproduce material appearing in their publications is gratefully acknowledged.

Tallahassee, Florida
September 2003

Leo Mandelkern

9

Crystallization kinetics of homopolymers: bulk crystallization

9.1 Introduction

In discussing the crystalline state of polymers the concern up to now has been primarily with equilibrium concepts and expectations. Although the fusion of polymers is well established to be a problem in phase equilibrium, stringent experimental procedures must be employed to approach the necessary conditions. It is concluded from equilibrium theory that for homopolymers of regular structure high levels of crystallinity should be attained.(1) However, this conclusion is contrary to usual experience. Since the crystalline phase only develops at a reasonable rate at temperatures well below the equilibrium melting temperature, the state that is actually observed in a real system is a nonequilibrium one. The constitution and properties of the state that is achieved are the result of the competition between the kinetic factors involved in the transformation and the requirements for thermodynamic equilibrium. In a formal sense one is dealing with a metastable state. This generalization is not unique to polymer crystallization. It also holds for a vast majority of substances that undergo a liquid to crystal transformation. For polymers, however, kinetic control is the dominant factor. For example, the crystallization from the melt of an *n*-alkane of about one hundred carbon atoms occurs extremely rapidly at a temperature just infinitessimally below its equilibrium melting temperature. On the other hand, a comparable crystallization rate in a low molecular weight linear polyethylene fraction of the same molecular weight is only achieved at undercoolings of about 25 °C. Polymer crystallization only occurs at a reasonable rate at temperatures well below the equilibrium melting temperature. The reasons for this behavior will emerge from the discussion of crystallization kinetics that ensues.

Kinetic studies, by their nature, represent a general type of experimental observation. Therefore, caution must be exercised in deducing specific molecular mechanisms from kinetic data. Very often several different mechanisms can explain the same experimental results. Despite this restraint it is possible, with care,

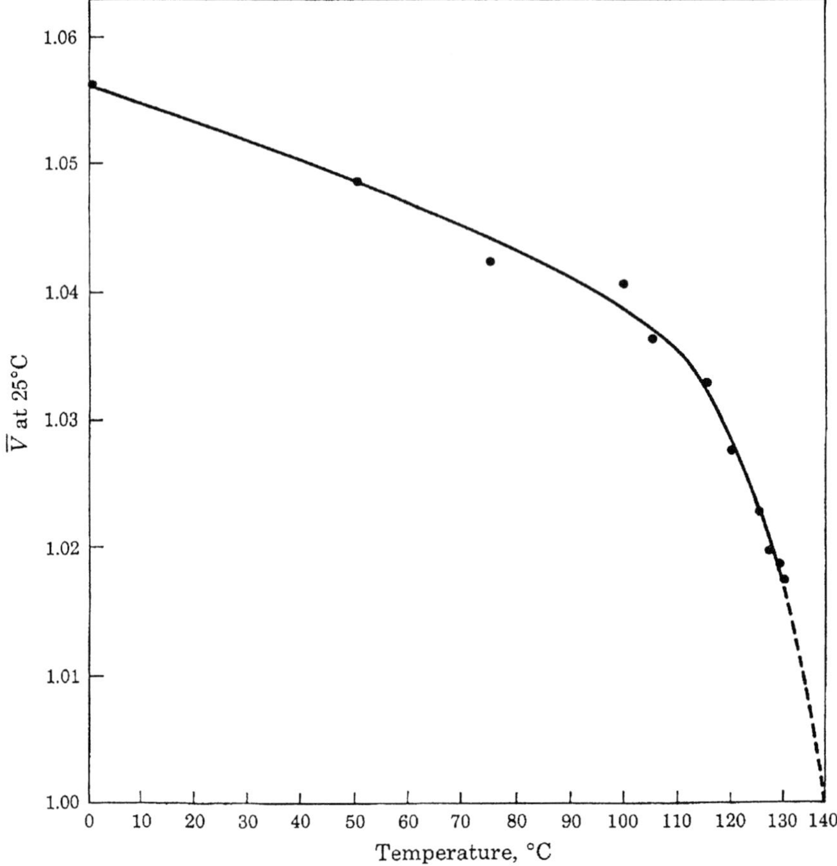

Fig. 9.1 Plot of specific volume at 25 °C for an unfractionated linear polyethylene (Marlex50) against the crystallization temperature. Time of crystallization is adjusted so that the crystallization is essentially complete at the specified temperature.(2)

to deduce some of the salient features that are important to the understanding of the crystallization process in polymers. Although the general results that are obtained are important, any specific deductions that are made are not necessarily unique. The possibility of different interpretations always needs to be recognized.

There are many examples that demonstrate that polymer properties depend on the crystallization conditions, and thus the crystallization kinetics. This generalization can be illustrated by examining some simple properties. In Fig. 9.1 the specific volume at 25 °C of an unfractionated linear polyethylene is plotted against the crystallization temperature.(2) In this example the crystallization was conducted isothermally from the pure melt. The times of crystallization were adjusted so that no significant crystallization occurred on cooling, prior to the specific volume measurement. For crystallizations carried out at relatively low temperatures, the specific volumes varied from 1.06 to 1.04 $cm^3 g^{-1}$. These values are in the range usually reported for this polymer. However, as the crystallization temperature is raised a sharp decrease in the specific volume occurs. This quantity is now sensitive to small changes in the crystallization temperature. The lower values are indicative of the high levels of crystallinity that are achieved as the crystallization is conducted at temperatures closer to the melting temperature. The extrapolation of the plot in Fig. 9.1, as indicated by the dashed curve, portends the distinct possibility that at still higher crystallization temperatures even lower specific volumes would be observed. For crystallization conducted at the melting temperature, a specific volume of about 1.00 $cm^3 g^{-1}$ is predicted. This value is very close to the density of the completely crystalline polymer as deduced by Bunn from the x-ray determination of the unit cell of polyethylene.(3) It implies the formation of a nearly perfect macroscopic single crystal under these crystallization conditions. Experimental confirmation of the extrapolated curve would involve crystallization for such intolerably long periods of time as to be impractical to carry out. As a general rule the transformation in polymers is rarely, if ever, complete. For a pure homopolymer of regular structure the fraction transformed can vary from about 0.90 to as low as 0.30, depending on molecular weight, molecular weight distribution and crystallization temperature.[1]

A wide range in densities can thus be obtained at 25 °C for the same crystalline homopolymer. The values depended directly on the crystallization temperature. Other thermodynamic, physical, and mechanical properties are also sensitive to the manner in which the crystallization is conducted. Another example of the importance of crystallization conditions in governing properties is illustrated in Fig. 9.2.(4) Here the crystallite thickness of a molecular weight fraction of linear polyethylene ($M_w = 1.89 \times 10^5$, $M_n = 1.79 \times 10^5$) is plotted as a function of the crystallization temperature. There is a dramatic increase in these values at about 125 °C. A profound effect is observed in the maximum crystallite thickness.

A striking example of the importance of crystallization kinetics and mechanisms on properties is seen in the influence of molecular weight. It was found in Chapter 2 (Volume 1) that above a very modest chain length the equilibrium melting temperature and equilibrium level of crystallinity only increase slightly with a further increase in chain length. Yet, because of kinetic factors the crystallinity level that can actually be obtained depends very strongly on molecular weight. This dependence is reflected in the major influence of molecular weight on all of the microscopic and macroscopic properties of crystalline polymers. In general, the morphology of polymers, as well as other key structural variables, can be expected to be governed by the

[1] A detailed discussion of the level of crystallinity that can be attained will be given in Volume 3. The role of molecular constitution and crystallization conditions will be examined in detail at that point.

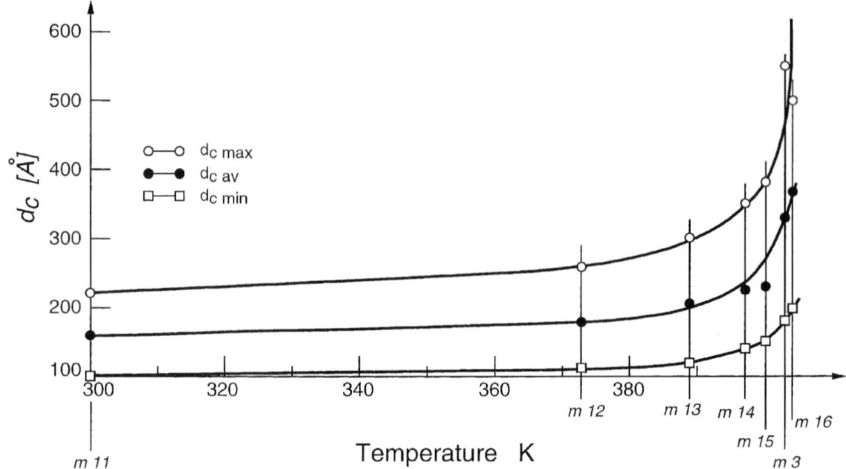

Fig. 9.2 Plot of crystallite thickness as a function of crystallization or quenching temperature for a molecular weight fraction of linear polyethylene, $M_w = 1.89 \times 10^5$, $M_n = 1.79 \times 10^5$.(4)

crystallization conditions. This behavior will be discussed in more detail in chapters concerned with morphology, structure and properties (Volume 3). A complete understanding of properties and behavior of crystalline polymeric systems requires knowledge and information of the mechanisms involved in the transformation, in addition to the equilibrium characterization. In principle, this information can be deduced from crystallization kinetic experiments.

There are several different kinds of experimental methods that are commonly used to observe the time development of crystallinity in polymers. One method is concerned with assessing the isothermal rate at which the total amount of crystallinity develops from the supercooled liquid. This procedure involves measuring the overall rate of crystallization. In carrying out these experiments it is necessary to follow a change in a property that is sensitive to crystallinity. Measurements of the changes in density or specific volume are commonly used for this method. However, other techniques have also been successfully used. These include wide- and small-angle x-ray scattering, vibrational spectroscopy, nuclear magnetic resonance, depolarized light microscopy and differential scanning calorimetry. Each method has a characteristic and different sensitivity to crystallinity.

In another widely used method, the isothermal rate of formation and subsequent growth of spherulites are studied by either polarized light microscopy or small-angle light scattering. Spherulites are spherical aggregations of individual crystallites that develop into supermolecular structures that are easily visible in the light microscope. Although spherulite structures are widely observed they are not, however, a universal mode of polymer crystallization. In particular, spherulites are not observed at the extremes of either high or low molecular weights and their formation is sensitive to polydispersity.(5–7) Thus, kinetic studies that are restricted to spherulite growth rates do not encompass the complete range of molecular weights. In favorable situations, particularly when the crystallization takes place from dilute solution, the growth rate of specific crystal faces can be followed using electron microscopy.(8a,b, 9)

9.2 General experimental observations

The isothermal rate at which the total amount of crystallinity develops in homopolymers follows a universal pattern that was first observed by Bekkedahl in his classic study of natural rubber.(10) Some typical isotherms that represent the course of crystallization in polymers are illustrated in Figs. 9.3 and 9.4 for natural rubber (10) and linear polyethylene (11) respectively. Changes with time of the specific volume were used to follow the crystallization in both of these examples. For polyethylene, the relative extent of the transformation is plotted against the time; for natural rubber the percentage decrease in specific volume is given. When the polymer sample is quickly transferred from a temperature above the melting temperature to the predetermined crystallization temperature, there is a well-defined time interval during which no crystallinity is observed. This time has often been termed the induction or incubation period. It was pointed out many years ago that this time interval must depend on the detector used to monitor the development of crystallinity.(12) Different experimental techniques vary in their sensitivity to detect crystallinity.(13) Within a given technique, the sensitivity will also depend on the particular instrument used. A quantitative investigation of these differences, involving several different methods, has been reported.(13) Imai, Kaji and collaborators (14,15) have shown, using small-angle x-ray scattering produced by synchrotron radiation, that density fluctuations occur within the so-called induction period. Subsequently, higher level fluctuations have also been observed.(13,16) An understanding of the molecular and structural basis for these fluctuations is evolving. It has been suggested that nucleation and growth processes take place during this induction period.(13,16,17) Serious consideration also needs to be given to transient (non-steady-state) nucleation that precedes the attainment of the steady state.(17a,b) The non-steady-state nucleation, with its associated embryonic nuclei, will occur within the induction time scale. Rheological and light scattering studies have shown that a network structure develops at the very early stages of crystallinity development.(17c,d) Presumably, the network structure is formed within the induction period by very small crystallites. A variety of possible structures can develop within the induction period and some have been observed.(17e) However, the induction time itself is a measure of the sensitivity of the detector being used. Spreading out of the time scale, by studies at low undercoolings, will enable a definitive analysis of the structures involved.

After the onset of observable crystallization, the process proceeds at an accelerating rate that is autocatalytic in character. Finally, a pseudo-equilibrium level of crystallinity is approached. A small, but finite, amount of crystallinity develops at very slow rates over many decades of time in this region. This flat portion of the isotherm is often referred to as the "tail". There are thus several regions in an isotherm where the rate of change of crystallinity with time is clearly different. However, the crystallization process is a continuous one. No sudden changes, or discontinuities, are discernible in the isotherms. The sigmoidal shaped isotherms, which are illustrated in Figs. 9.3 and 9.4, are characteristic of the overall crystallization of all homopolymers. Such isotherms are prima facie evidence of nucleation and growth processes.

When the change in the property being studied is plotted against the logarithm of time, as illustrated in Fig. 9.5 for a molecular weight fraction of linear polyethylene, important additional information can be obtained.(18) The major features of the crystallization isotherms are still maintained. However, in this plot the isotherms all have the same shape. With the exception of the tail portion in some cases, the isotherms can be superposed upon one another by merely shifting the curves along the horizontal axis. This implies that there is a reduced temperature–time variable. Consequently, from a purely experimental point of view a single isotherm, based on the reduced variable involving time and temperature, can be constructed. This isotherm is representative of the crystallization of a given polymer. The implications of this observation will be discussed subsequently.

An important feature of Figs. 9.3, 9.4 and 9.5 is the strong negative temperature coefficient displayed by the crystallization process. This phenomenon is observed in all polymer crystallization conducted in the vicinity of the melting temperature. For the example shown in Fig. 9.5 the rate decreases by more than four orders of magnitude over only a seven degree temperature interval. A temperature coefficient of this type and magnitude is quite different from a chemical process. It is a very strong indication of nucleation controlled crystallization. We shall find, in detailed discussion later in this chapter, that nucleation plays a central role in polymer crystallization.

When the crystallization is conducted over the complete accessible temperature range, i.e. from below the melting temperature to above the glass temperature, a more complex temperature dependence evolves. A typical example is given in Fig. 9.6 for natural rubber.(19) Here the time taken for half of the crystallization to develop is plotted as a function of the crystallization temperature. Many other homopolymers behave in a similar manner. This type of temperature dependence is not restricted to long chain polymers. It is also observed during the crystallization

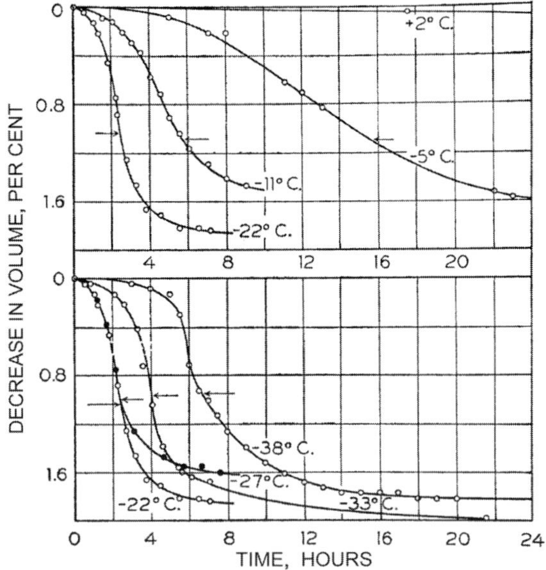

Fig. 9.3 Crystallization of natural rubber at various temperatures as measured by the decrease in specific volume. Temperature of crystallization is indicated for each isotherm. (From Bekkedahl (10))

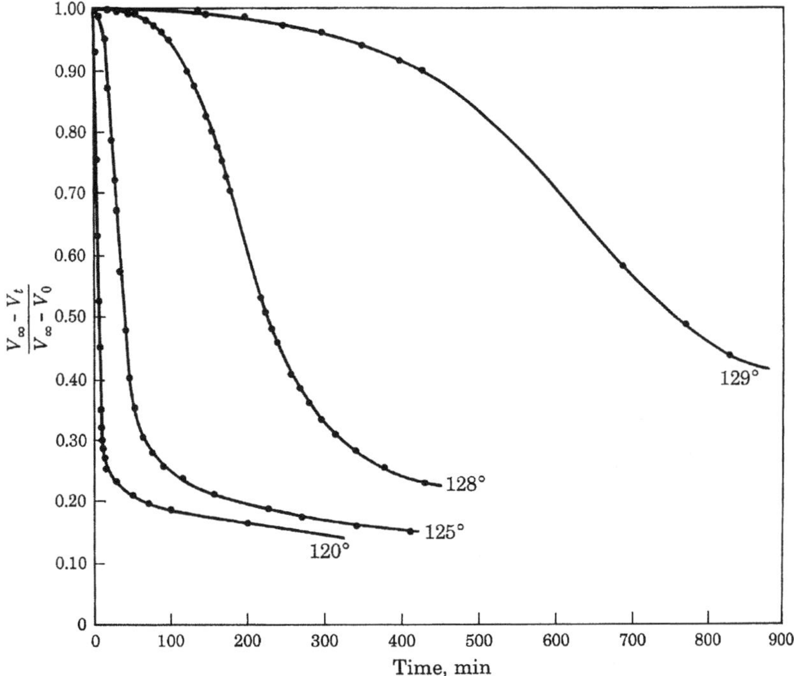

Fig. 9.4 Plot of quantity $(V_\infty - V_t)/(V_\infty - V_0)$ against the time for the crystallization from the melt of an unfractionated linear polyethylene. Temperature of crystallization is indicated for each isotherm.(11)

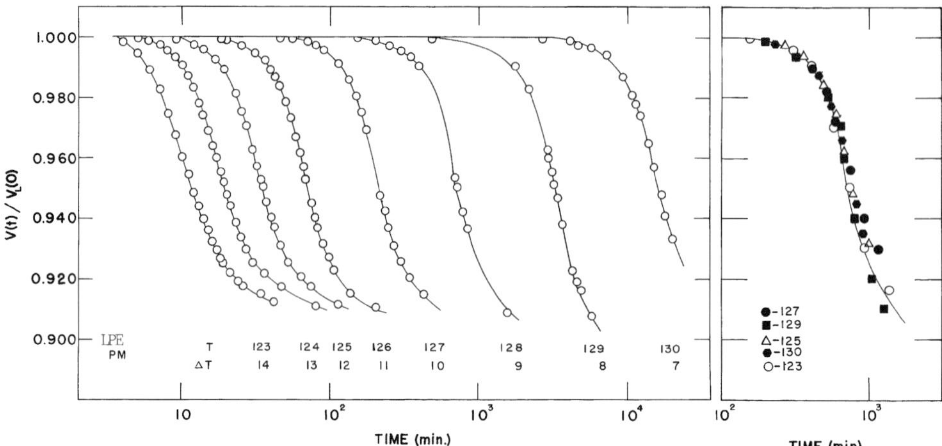

Fig. 9.5 Plot of $V(t)/V_L(0)$ against log time for an unfractionated very high molecular weight linear polyethylene at indicated crystallization temperatures. Insert at right demonstrates a common superposed isotherm.(18)

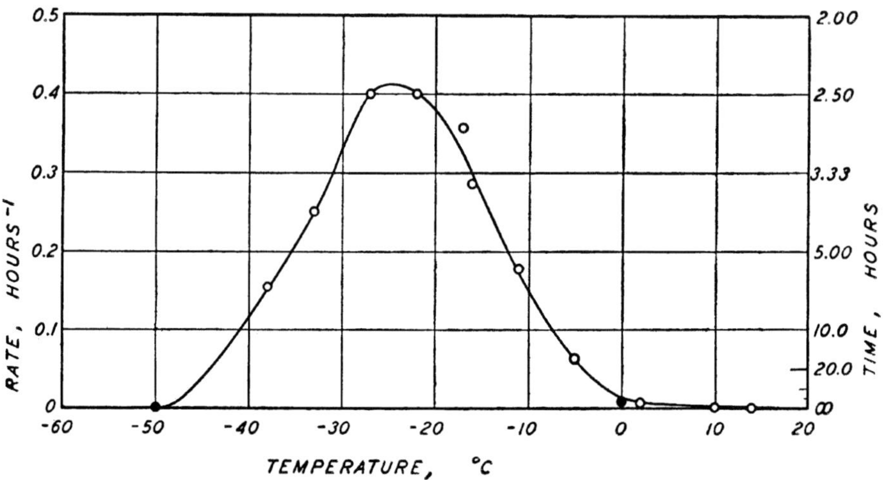

Fig. 9.6 Plot of the crystallization rate of natural rubber over an extended temperature range. The rate plotted is the reciprocal of the time requested for one-half the total volume change. (From Wood and Bekkedahl (19))

of monomeric substances.(20,21) At temperatures in the vicinity of T_m^0 the crystallization rate is very slow, so that for any reasonable time of measurement the appearance of crystallinity will not be detected. As the temperature is lowered, the rate progressively increases and eventually passes through a maximum. At crystallization temperatures below the maximum, the overall rate of crystallization is retarded once again. This interval coincides with the temperature range where the

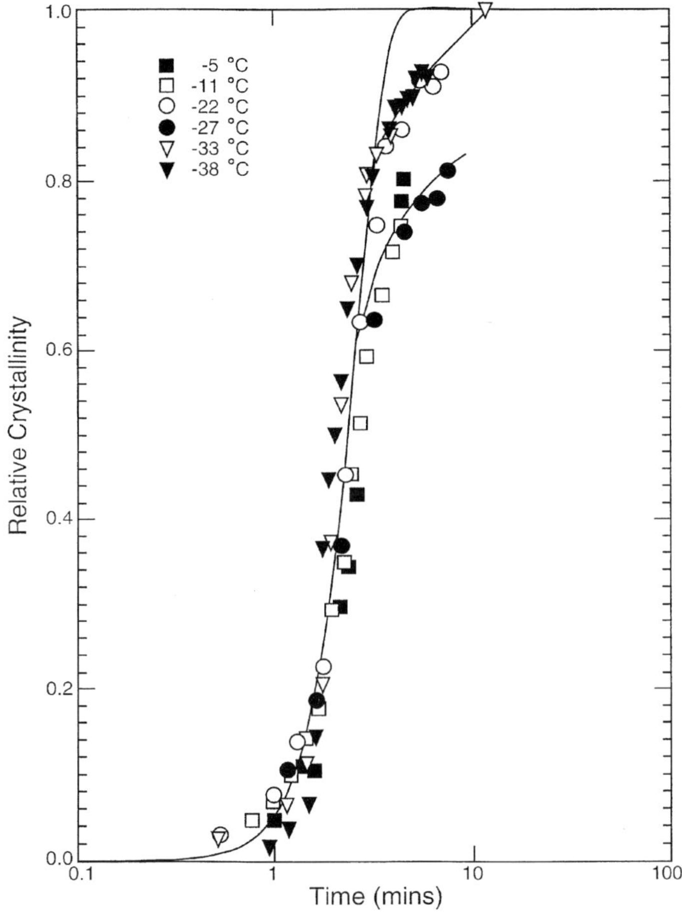

Fig. 9.7 Demonstration of superposition over extended temperature range for natural rubber. Plot of relative crystallinity against log time for crystallization at indicated temperatures. (Data from Fig. 9.3 and Bekkedahl (10))

glass temperature of the supercooled polymeric liquid is approached. For some polymers the crystallization rate becomes so rapid at temperatures below T_m^0 that it is extremely difficult to detect the temperature at which the rate is a maximum. As is illustrated in Fig. 9.7, the superposition of isotherms is also maintained for crystallization over this extended temperature range. The isotherms of other polymers that show a rate maximum can also be superposed over the complete range of crystallization temperatures.

Studies of the rates of spherulite formation and growth in thin films have provided important information about the crystallization kinetics of polymers. In the vicinity of the melting temperature, the rate at which spherulites are formed depends very strongly on the crystallization temperature and increases very rapidly

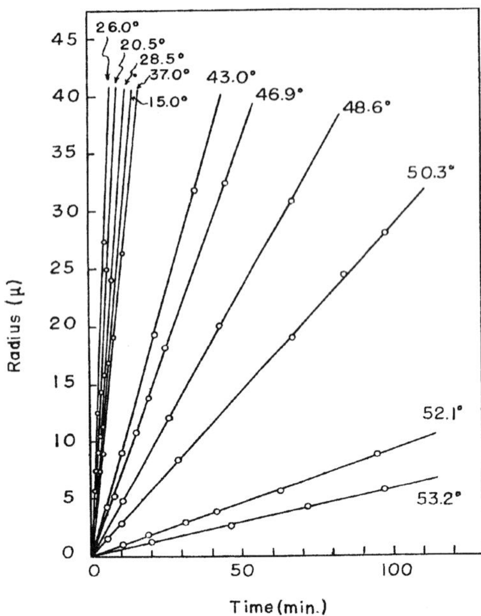

Fig. 9.8 Plot of spherulite growth rates against time for poly(ethylene adipate), $M = 9900$, at indicated crystallization temperatures. (From Takayanagi (23))

as the temperature is lowered. For example, for poly(decamethylene adipate) the rate at which spherulitic centers are generated deceases by a factor of 10^5 as the crystallization temperature is raised from 67 to 72 °C.(22) These results clearly indicate that the birth of spherulites is also governed by a nucleation process.

An impressive body of experimental evidence, involving many polymers, demonstrates that at a fixed temperature the radius of a growing spherulite of a pure homopolymer increases linearly with time. These observations are consistent with the autocatalytic nature of the isotherms illustrated in Figs. 9.3, 9.4 and 9.5. The linear growth rate is also sensitive to temperature as illustrated in Fig. 9.8 for poly(ethylene adipate).(23) The growth rates are clearly defined by the slopes of the straight lines. A marked negative temperature coefficient is again observed. Thus, both the birth of spherulites and their growth are nucleation controlled processs. As the level of crystallinity increases, problems of resolution caused by the overlapping of spherulites make counting and measuring their sizes difficult. Consequently, although a potent quantitative tool, this technique is restricted to relatively low levels of crystallinity and a restricted molecular weight range.

As the crystallization temperature is lowered the growth rate increases until a maximum is observed. With a further decrease in the temperature, the rate of growth diminishes. The temperature variation of the spherulitic growth rate is thus qualitatively similar to the temperature coefficient of the overall rate of

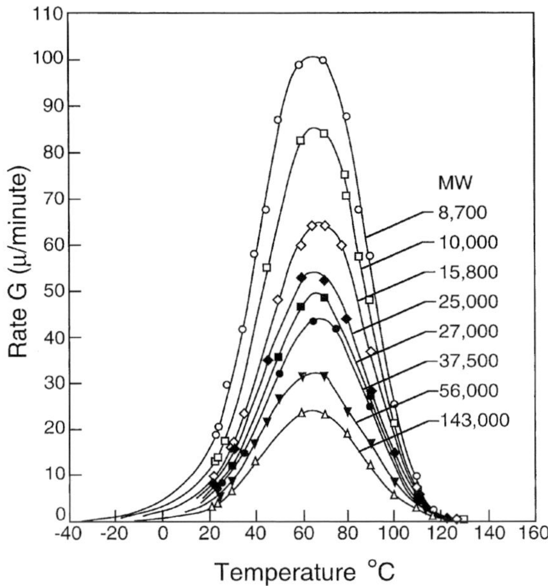

Fig. 9.9 Plot of spherulite growth rates against crystallization temperature for indicated molecular weights. (From Magill (24))

development of crystallinity. A typical example of the spherulite growth rate over an extended temperature range is shown in Fig. 9.9 for molecular weight fractions of poly(tetramethyl-*p*-silphenylene siloxane).(24) The equilibrium melting temperature of this polymer is 152 °C while the glass temperature is about −17 °C.

With this qualitative discussion of the major characteristics of polymer crystallization, we are in a position to develop the subject in a more quantitative manner. We consider first the basic theoretical developments. The experimental results will then be presented and the two compared. Based on this comparison, further modifications and refinements that are needed in the theory will be discussed.

9.3 Mathematical formulation

The development of a new phase within a mother phase, such as a crystal within a liquid, involves the birth of the phase and its subsequent development. The former process is termed nucleation and the latter, growth. It is also possible for growth to be nucleation controlled. In this case it is necessary to distinguish between the initiation or primary nucleation and growth or secondary nucleation. For most cases of interest, isothermal crystallization can be described in terms of the nucleation frequency N and the growth rates G_i of the different crystallographic planes designated by the subscript i. The amount of material transformed as a function of the time can be calculated, subject to the restraints that are imposed on the kinetic process.

The underlying theoretical basis for the overall crystallization kinetics of polymers is found in the theory developed for metals and other monomers.(25–29) In monomeric systems the fraction of the liquid transformed to crystal varies from 0 to 1 over the time course of the crystallization. As was pointed out, the transformation in polymers is rarely, if ever, complete. This well-established experimental fact, and the reasons for this restraint, present a major problem that needs to be resolved in extending the crystallization kinetics theories from monomers to polymers. However, before discussing polymers, it is instructive to examine the theoretical base that has been established for monomers.

A simple case to consider is that in which, once born, a given center grows unimpeded by the initiation and growth of other centers. This model, termed free growth, serves as a convenient frame of reference for further theoretical development. It was proposed by von Göler and Sachs.(25) We let N' be the steady-state nucleation frequency per unit of untransformed mass. The number of nuclei generated in a time interval $d\tau$, at time τ is given by

$$dn = N'(\tau)\,\lambda(\tau)\,d\tau \tag{9.1}$$

Here, $\lambda(\tau)$ is the fraction of untransformed material at time τ. If $w(t,\tau)$ is the mass of a given center at time t, that was initiated at time $\tau(\tau \leq t)$ and grew without restriction, then

$$1 - \lambda(t) = \int_0^t w(t,\tau)\,dn = \int_0^t w(t,\tau)\,N'(\tau)\,\lambda(\tau)\,d\tau \tag{9.2}$$

Alternatively, Eq. (9.2) can be expressed as

$$1 - \lambda(t) = \frac{\rho_c}{\rho_l}\int_0^t v(t,\tau)\,N(\tau)\,\lambda(\tau)\,d\tau \tag{9.3}$$

where $N(\tau)$ is now the nucleation frequency per unit of untransformed volume, $v(t,\tau)$ is the corresponding volume of the growing center, and ρ_c and ρ_l are the densities of the crystalline and liquid phases, respectively.

The evolution of an individual growing center is assumed to be completely independent of the mass or volume already transformed as well as the growth of other centers. In particular, the effect of the impingement of growing centers upon one another is neglected. Thus, no mechanism has been imposed for the cessation of the crystallization. Although these assumptions are unrealistic, the results turn out to be of practical importance.

In order to solve the integral equation, Eq. (9.3), it is necessary that the nucleation and growth processes, i.e. $v(t,\tau)$ and $N(\tau)$, be specified. A very simple nucleation mechanism is one where $N(\tau)$ is independent of the amount of material transformed, i.e. it is constant. This implies either steady-state homogeneous nucleation or a specific type of heterogeneous nucleation where the number of nuclei activated is constant with time. Another possibility involving heterogeneous nucleation is that all nuclei are activated at $t = 0$. The simple possibilities are not, however, necessarily the correct ones in terms of physical reality.

The growth rates can be either isotropic, i.e. the same in all directions, or anisotropic, where the rates in the different directions are not the same. Besides the geometry and dimensionality of the growth, one also has to consider its time dependence. Growth can be either interface controlled, so that the rate is linearly dependent on (t,τ) in one dimension, or diffusion controlled, so that the dimension will vary as $(t - \tau)^{1/2}$. Thus, for isotropic linear growth

$$v(t,\tau) = f_3 G_x G_y G_z (t - \tau)^3 \tag{9.4}$$

where G_x, G_y, and G_z are the linear growth rates in the x, y, z directions respectively. The growth vectors are assumed to be independent of time. The shape factor f is characteristic of the geometry of the growing center. The assumption of a linear rate of growth implies that the rate of volume change of a growing center is proportional to its surface area. Growth is therefore governed by processes that occur at the crystal–liquid interface. If, on the other hand, the diffusion of the crystallizing entity to the interface is rate controlling, then $v(t,\tau)$ will vary with $(t - \tau)^{3/2}$ for the isotropic case. For a linear growth, that is restricted to two dimensions

$$v(t,\tau) = f_2 G_x G_y \gamma_z (t - \tau)^2 \tag{9.5}$$

where γ_z represents the dimension held fixed. Similarly, for growth in one dimension,

$$v(t,\tau) = f_1 G_x \gamma_y \gamma_z (t - \tau) \tag{9.6}$$

Corresponding expressions can be written for diffusion controlled growth.

For the case where $N(\tau)$ is constant, equal to N, and growth is three dimensional, Eq. (9.3) becomes

$$1 - \lambda(t) = \frac{\rho_c}{\rho_l} N f_3 G_x G_y G_z \int_0^t (t - \tau)^3 \lambda(\tau)\, d\tau \tag{9.7}$$

Then, for isotropic growth

$$1 - \lambda(t) = \frac{\rho_c}{\rho_l} N f_3 G^3 \int_0^t (t - \tau)^3 \lambda(\tau)\, d\tau \tag{9.8}$$

The solution of this integral equation is (25)

$$\lambda(t) = 1 - \cosh k_3 t \cos k_3 t \tag{9.9}$$

where k_3, considered to be a rate constant, is given by $(\frac{3}{2}Nf_3 G_x G_y G_z)^{1/4}$. Isotropic growth in one and two dimensions, with $N(\tau)$ constant, gives

$$\lambda(t) = \frac{2}{3}\exp\left(\frac{k_2 t}{2}\right)\cos\left(\frac{\sqrt{3}}{2}k_2 t\right) + \frac{1}{3}\exp(-k_2 t) \tag{9.10}$$

and

$$\lambda(t) = \cos k_1 t \tag{9.11}$$

respectively.(29a) Here the subscripts 1, 2, 3 indicate the dimensionality of the growth. The above equations do not provide for a natural termination of the crystallization process, since real solutions exist for $1 - \lambda(t) > 1$, i.e. after the crystallization process is complete. In fact, the functions involved oscillate about $1 - \lambda(t) = 1$. Therefore, their use must be arbitrarily restricted to the physically sensible region $0 \leq 1 - \lambda(t) \leq 1$. The solutions for diffusion controlled growth lead to a similar dilemma. These results are a natural consequence of the free-growth approximation.

For the initial portions of the transformation, the rate equations can be expanded in series terms of $k_i t$. When the first two terms of the expansion are retained, Eqs. (9.9), (9.10) and (9.11) become

$$1 - \lambda(t) \cong (k_3 t)^4/4 \tag{9.12}$$

$$1 - \lambda(t) \cong (k_2 t)^3/6 \tag{9.13}$$

$$1 - \lambda(t) \cong (k_1 t)^2/2 \tag{9.14}$$

for the respective growth geometries, with interface control. These expressions are commonly called the free-growth approximations. They suggest that for the most rudimentary analysis of the kinetic data a plot of $\log(1 - \lambda(t))$ against time $\log t$ should yield a straight line.

A comparison of the complete solution of the Göler–Sachs equations, Eqs. (9.9), (9.10) and (9.11), with the approximate forms given by Eqs. (9.12), (9.13) and (9.14) indicates that the differences between them are not significant. For the cases of three- and two-dimensional growth the two resulting expressions are virtually identical up to $1 - \lambda(t) = 1$, the physically real region. It will be of interest to ascertain subsequently how this simple development actually describes the crystallization kinetics of polymers.

When the condition of a constant nucleation rate is relaxed, a variety of other possibilities exists. A similar statement can also be made if the growth rate constants are not held fixed. We focus attention on the consequences of removing certain of the restrictive conditions on the nucleation rate. If, for example, it is assumed that all the nuclei are activated at time $t = 0$ then the crystallization rates will be similar to those given by Eqs. (9.12), (9.13) and (9.14), with the exponent of t reduced by one integer. Another case that can be given a simple mathematical treatment is when the rate of nucleus formation follows a first-order rate law with the specific rate constant v. The case of a fixed number $\bar{N}$ of potential nuclei, or heterogeneities, being initially present and having the specific probability v of developing into crystallites is detailed in the following. The corresponding nucleation rate is then given by

$$N = \bar{N} \exp(-\tau) \tag{9.15}$$

where $\tau = vt$. For this type of nucleation, the Göler–Sachs model becomes

$$1 - \lambda(t) = \frac{C_3 G^3 \bar{N}}{v^3} \left[\exp(-vt) - 1 + vt - \frac{(vt)^2}{2!} + \frac{(vt)^3}{3!} \right] \equiv A_3 \tag{9.16}$$

$$1 - \lambda(t) = \frac{C_2 G^2 \bar{N}}{v^2} \left[2 - 2\exp(-vt) - 2vt + (vt)^2 \right] \equiv A_2 \tag{9.17}$$

for three- and two-dimensional growth, respectively. The constants C_3 and C_2 incorporate the geometric and density factors involved. When vt is large, corresponding to a high probability for the initial growth of all potential or nuclei sites, Eqs. (9.16) and (9.17) reduce to

$$1 - \lambda(t) = C_3' t^3 \tag{9.18}$$

and

$$1 - \lambda(t) = C_2' t^2 \tag{9.19}$$

These equations have the same form as Eqs. (9.13) and (9.14) that were obtained for the initial portion of a system nucleating at a constant rate that grows in either two or one dimension respectively. In the other extreme, if vt is small Eqs. (9.16) and (9.17) reduce to

$$1 - \lambda(t) = C_3'' v t^4 \tag{9.20}$$

and

$$1 - \lambda(t) = C_2'' v t^3 \tag{9.21}$$

respectively. In this extreme, not all the centers are initiated at $t = 0$. Rather they are activated at a constant rate. For an actual mechanism operating between these two extremes, the potential nuclei will become depleted at some intermediate stage of the transformation. The exponent of t will then vary between 3 and 4 for three-dimensional growth and 2 and 3 for two-dimensional growth. The exponent in this situation will not have a constant integral value over the complete transformation range.

The Göler–Sachs formulation has the obvious shortcoming that there is no mechanism for the termination of the transformation. This is of serious concern for the crystallization of metals and other low molecular weight substances where the actual transformation is complete. Serious discrepancies are observed between the free-growth theory and experiment in these systems, particularly toward the end of the transformation. It is clear that some mechanism for termination needs to be introduced into the analysis. The mutual interference of growing crystalline regions that originate from separate nuclei needs to be taken into account. In a liquid–crystal type transformation when two such regions impinge upon one another a common interface develops. All growth will cease along this interface. This type of impingement is thus a natural mechanism by which crystallite growth can cease. It is independent of the type of nucleation that is operative. This concept, primarily geometrical in nature, is an important one. The initial treatment of this problem was given by Johnson and Mehl (26), Avrami (27), Evans (28) and Kolmogorov.(29) We shall discuss the Avrami approach in some detail since it has been the one most widely applied to polymers.

In addressing the impingement problem Avrami introduced the concept of phantom nuclei and the "extended volume" of transformed material that results from such nuclei. Phantom nuclei are nuclei that are allowed to develop in the volume that has already been transformed. Thus their designation as phantom. The total number of nuclei, real and fictitious, that are generated in the time interval $d\tau$ is given by

$$dn' = N'\lambda'\,d\tau + N'(1-\lambda)'\,d\tau = N'\,d\tau \tag{9.22}$$

Here $N'(1-\lambda)'\,d\tau$ represents the number of nuclei that would have originated in the mass fraction $1-\lambda$, if it had not become transformed. The extended fraction transformed, $(1-\lambda)'$, includes the contribution from the phantom nuclei. It can be expressed as

$$(1-\lambda)' = \int_0^t w(t,\tau)\,dn' = \int_0^t w(t,\tau)\,N'(\tau)\,d\tau \tag{9.23}$$

The extended fraction transformed will be larger than the actual fraction transformed. It then remains to relate the actual fraction transformed, $1-\lambda$, to the extended one.(27) Consider a small region, selected at random, where a fraction $1-\lambda$ remains untransformed at time τ. During the interval τ to $\tau+d\tau$ the extended fraction transformed in this region will increase by $d(1-\lambda)'$ and the true fraction by $d(1-\lambda)$. In this new extended fraction, $d(1-\lambda)'$, a fraction λ, on the average, will lie in previously untransformed material and contribute to $d(1-\lambda)$, while the remainder will already be in transformed material. Consequently, the relation between $d(1-\lambda)$ and $d(1-\lambda)'$ can be written as

$$\frac{d(1-\lambda)}{d(1-\lambda)'} = \lambda(t) \tag{9.24}$$

This result is the reason that the concept of phantom nuclei was introduced and included in the definition of $(1-\lambda)$. These conclusions follow only if $d(1-\lambda)$ can be treated as a completely random volume element.

The integration of Eq. (9.24) yields

$$(1-\lambda(t)) = 1 - \exp[(1-\lambda)'] \tag{9.25}$$

When $(1-\lambda)'$ from Eq. (9.23) is substituted into Eq. (9.25)

$$(1-\lambda(t)) = 1 - \exp\left[-\int_0^t w(t,\tau)\,N'(\tau)\,d\tau\right] \tag{9.26}$$

or alternatively,

$$(1-\lambda(t)) = 1 - \exp\left[-\frac{\rho_c}{\rho_l}\int_0^t v(t,\tau)\,N(\tau)\,d\tau\right] \tag{9.27}$$

Either equation, (9.26) or (9.27), is the basic Avrami relation. They should not be identified with expressions that are derived from them. These equations describe the kinetics of phase transformations for a one-component monomeric system. Only an integral has to be evaluated, rather than the need to solve an integral equation as in the free-growth case. The integral can be evaluated by specifying the nucleation and growth laws that are operative. A quantitative description of the fraction transformed as a function of time is then obtained. There are obviously many possibilities that can be considered, with many different expressions resulting. Any one of these is a descendent of the basic Avrami equation (9.26 or 9.27). Strictly speaking, it is not proper to identify any of the derived ones as the Avrami equation. A computer simulation agrees with the Avrami result, provided the distribution of nuclei is random and the phantom nuclei are included in the calculation.(29b) An alternative derivation of the general Avrami expression, one that avoids the introduction of "phantom nuclei", has been given by Yu and Lai.(29c)

One possible set of conditions out of many, which has been popular, is for the steady-state nucleation rate to be achieved at $t = 0$ and to remain invariant with the fraction of material transformed. Thus $N(\tau)$ can be treated as a constant. Similarly, the crystallite growth rate is assumed to be linear and constant. With these simplifying assumptions analytical solutions of Eqs. (9.26) and (9.27) are obtained. For three-dimensional linear growth,

$$\ln(1/\lambda) = \frac{\pi}{3}\frac{\rho_c}{\rho_l} N G^3 t^4 \equiv k_s t^4 \tag{9.28}$$

For a disk of fixed thickness l_c developing radially according to a linear growth law,

$$\ln(1/\lambda) = \frac{\pi}{3} l_c \frac{\rho_c}{\rho_l} N G^2 t^3 \equiv k_d t^3 \tag{9.29}$$

Similarly, for a one-dimensional growing system,

$$\ln(1/\lambda) = k_r t^2 \tag{9.30}$$

With these specified restrictions on the nucleation and growth rates, the derived Avrami equation can be expressed as

$$\lambda(t) = \exp\{-kt^n\} \tag{9.31}$$

or

$$1 - \lambda(t) = 1 - \exp\{-kt^n\} \tag{9.31a}$$

Either equation, (9.31) or (9.31a), is commonly termed the Avrami equation. However, they are merely derived expressions that are based on a very specific set of assumptions. Alternatively Eq. (9.31) can be expressed as

$$\ln \ln \lambda = -\ln k - n \ln t \tag{9.32}$$

The exponent n is usually termed the Avrami exponent. The value of n appropriate to systems with invariant nucleation and growth rates is dependent on the geometry of the growth. The n values for specific geometries for either interface or diffusion controlled growth are summarized in Table 9.1. It is clear from this summary that even using the derived Avrami expression the exponent n does not define a unique nucleation and growth set.

In another approach to the problem, Evans examined the expansion of circles and spheres.(28) This analysis can be related to crystal growth in either two or three dimensions. An analog in two dimensions is raindrops falling randomly on a pond. In this case, the probability that the number of circles that pass over a representative point P, within a certain time t from the start of the rain, is exactly C is given by Poisson's relation $\exp(-E)E^C/C!$. Here E is the expected number of waves. The probability that the point P (selected at random) will escape being crossed by an expanding wave, i.e. $C = 0$, is given by $\exp(-E)$. Thus the fraction of the area of the pond that remains uncovered at time t is $\theta(t) = \exp(-E)$. In terms

Table 9.1. *Values of exponent n for various types of nucleation and growth acts*

Growth habit	Homogeneous nucleation				Heterogeneous nucleation[a]
	Linear growth		Diffusion controlled growth		Linear growth
	Steady state	$t = 0$[b]	Steady state	$t = 0$	
Sheaf-like[c]	6	5	7/2	5/2	$5 \leq n \leq 6$
Three-dimensional	4	3	5/2	3/2	$3 \leq n \leq 4$
Two-dimensional	3	2	2	1	$2 \leq n \leq 3$
One-dimensional	2	1	3/2	1/2	$1 \leq n \leq 2$

[a] According to Eq. (9.15).
[b] All nuclei activated at $t = 0$.
[c] From L. B. Morgan, *Phil. Trans. R. Soc.*, **247**, 13 (1954).

of crystallization, $\theta(t)$ can be identified with the fraction untransformed. Thus

$$\lambda(t) = \exp(-E) \tag{9.33}$$

It then remains to calculate E and to specify the rate of droplet formation, i.e. the nucleation rate.

To obtain E, the contribution dE due to an annulus of breadth dr situated around P at a radial distance r, is integrated over all values of r from zero to Gt. G is the lineal growth velocity and the annulus possesses an area equal to $2\pi r\, dr$. During a period equal to $(t - r/G)$, any point within the annulus will be capable of sending out circles that will reach P before time t. Thus

$$dE = N(t - r/G)2\pi r\, dr \tag{9.34}$$

where N is now the steady-state nucleation rate per unit area. Then

$$E = 2\pi N \int_0^{Gt} (tr - r^2/G)\, dr = \pi N G^3 t^3/3 \tag{9.35}$$

Substituting into Eq. (9.32)

$$\lambda(t) = \exp(-E) = \exp(-\pi N G^3 t^3/3) = \exp(-k_{\mathrm{d}} t^3) \tag{9.36}$$

Equation (9.36) is identical to Eq. (9.29) obtained by the Avrami analysis for the same types of nucleation and growth. Calculation of three-dimensional growth by this method yields Eq. (9.28). The Evans method can also be applied to other growth geometries with the corresponding Avrami expressions resulting.

Heterogeneous nucleation is also important in analyzing polymer crystallization. Certain types can be treated directly by the Avrami method. The results obtained, however, are specific to the details assumed. Many possibilities exist for activating heterogeneous nuclei. In analyzing heterogeneous nucleation by the Göler–Sachs method the specific case of the nuclei being initiated by a first-order rate law was treated. Under the same conditions the Avrami analysis gives

$$\ln\left(\frac{1}{\lambda}\right) = \frac{C_3 G_3 \bar{N}}{v^3}\left[\exp(-vt) - 1 + vt - \frac{(vt)^2}{2!} + \frac{(vt)^3}{3!}\right] = A_3 \tag{9.37}$$

for three-dimensional lineal growth, and

$$\ln\left(\frac{1}{\lambda}\right) = \frac{C_2 G^2 \bar{N}}{v^2}[2 - 2\exp(-vt) + (vt)^2] = A_2 \tag{9.38}$$

for the two-dimensional case.

The right-hand sides of Eqs. (9.37) and (9.38) are the same as were obtained by Göler and Sachs (Eqs. (9.16) and (9.17)). The right-hand sides thus reduce in the same way for large and small values of v. For three-dimensional growth n is in the range of 3 to 4. For two-dimensional growth, $2 \leq n \leq 3$, while for one-dimensional growth, $1 \leq n \leq 2$. These values are included in the tabulation of Table 9.1. It is apparent that even with the mild restrictions that have been imposed, details of the growth geometry and type of nucleation cannot be elucidated from the kinetic isotherms solely by specifying the value of the exponent n. Depending on the specifics of the nucleation and growth, the Avrami exponent does not need to be a constant integer over the complete course of the transformation. In fact, the expectation is that it would not be so when the myriad of available possibilities is taken into account. Put another way, the fact that n is not integral, or varies over the course of the transformation, is not by itself a shortcoming of the basic Avrami equation (Eq. (9.26) or (9.27)). This conclusion is true even for a lineal growth rate.

As has been indicated earlier, both Eq. (9.31) and Eq. (9.31a) have been commonly termed the Avrami relation. Either one is a very simple and convenient expression to use and hence their widespread adoption. However, it should be recognized that they are restricted in scope because of the limited nucleation and growth processes that have been considered. Thus, caution must be exercised when interpreting results using these equations. Theoretical isotherms based on the derived Avrami equations are plotted in Fig. 9.10 in accordance with Eq. (9.31a) for $n = 1$, 2, 3 and 4. The curves in Fig. 9.10 have several important features in common. Their general shapes are in qualitative accord with the experimental observations that were described previously. When examined in detail, differences exist that distinguish one curve from another. As n increases from 1 to 4, the time interval at which crystallinity becomes apparent becomes greater. However, once the transformation

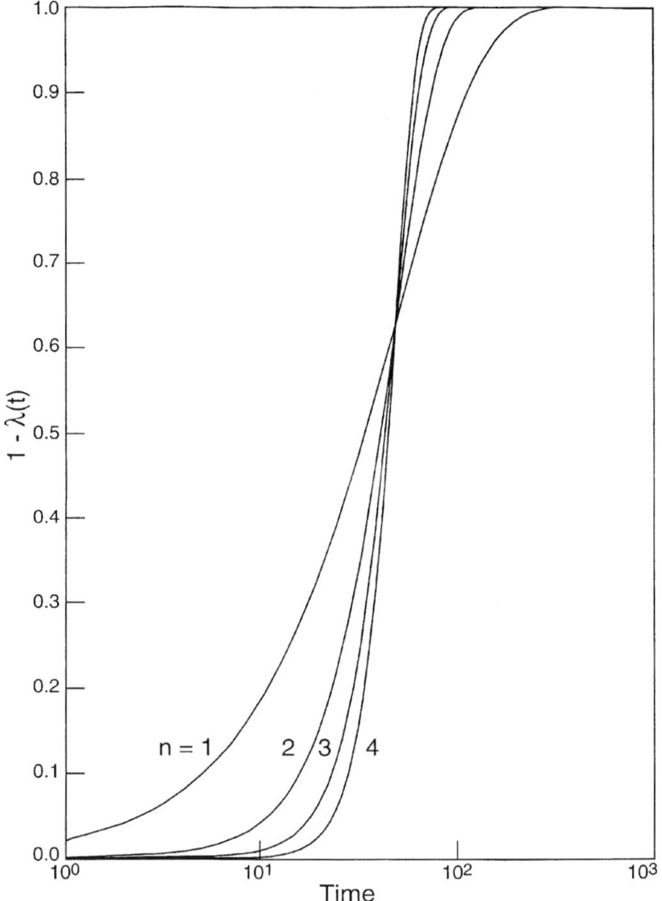

Fig. 9.10 Plot of theoretical derived Avrami isotherms according to Eq. (9.31a), for $n = 1, 2, 3$ and 4.

develops, the crystallization rate becomes greater the higher the value of the exponent n. All the curves have a common point of intersection at $1 - \lambda(t) = 0.6$. Beyond this point the approach to termination becomes more rapid the larger the value of n. Because of the similarity in shape of the theoretical curves it can be anticipated that in the analysis of experimental data, even when the assumptions made in the derivation are adhered to, it will be a difficult matter to decide between $n = 4$ or 3. This is particularly true for the early stages of the transformation where the most reliable experimental data are obtained. From purely a mathematical point of view, larger values of n can also be involved. However, the theoretical isotherms for n values of 5 or 6 can scarcely be distinguished from that for $n = 4$.

It is of interest at this point to compare the character of the isotherms that are predicted by the simplified Avrami form (Eq. 9.31a) and the free growth of

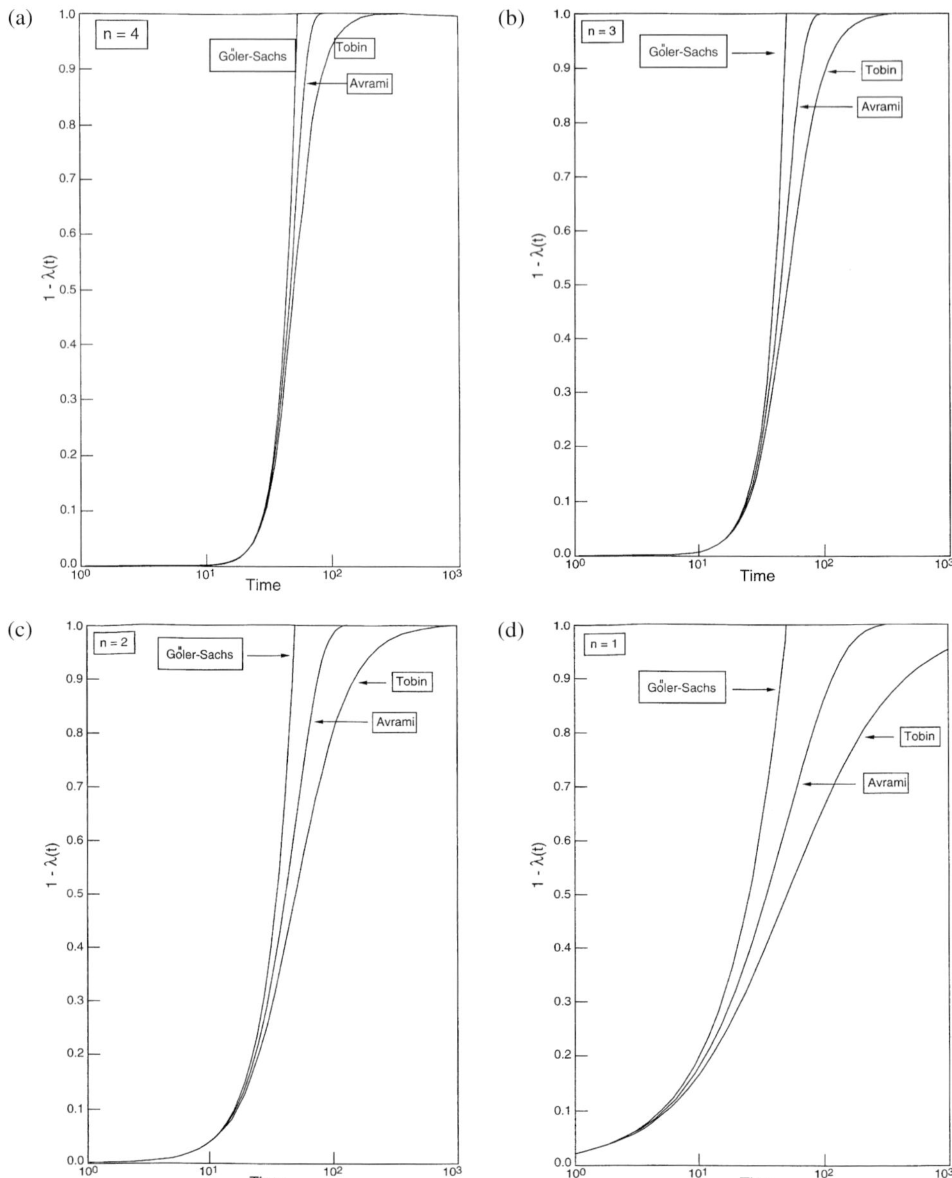

Fig. 9.11 Comparison of theoretical isotherms between Göler–Sachs, derived Avrami and Tobin. (a) $n = 4$, (b) $n = 3$, (c) $n = 2$, (d) $n = 1$.

Göler–Sachs (Eqs. 9.12–9.14). This comparison is given in Fig. 9.11 for $n = 1$, 2, 3 and 4. (The curves marked Tobin are given for later discussion.) The isotherms are very close to one another for small extents of the transformation for all values of n. However, as the transformation progresses the agreement between the two theories depends on the n value. For example, for $n = 4$ the isotherms are virtually identical to about 30% of the transformation. The difference between the two isotherms remains small up to about 70% of the transformation. At this point the difference, in terms of experiment, is well within any error. At the higher levels of the transformation the two isotherms diverge in a significant manner. The Avrami isotherm becomes more protracted reflecting the influence of the termination process. The agreement, and closeness of the isotherm, is reduced as n decreases. For example, when $n = 1$ agreement is only found for the first 10% of the transformation. In summary, the two theories give essentially the same results at low levels of absolute crystallinity. However, as the transformation progresses important differences are predicted particularly towards the end of the process where the Avrami termination step is important. The significance of these differences, in terms of polymer crystallization, can only be decided by analysis of experimental data.

The influence of heterogeneous nucleation on the crystallization kinetics is examined in Fig. 9.12 in terms of the Avrami formulation. In this figure the

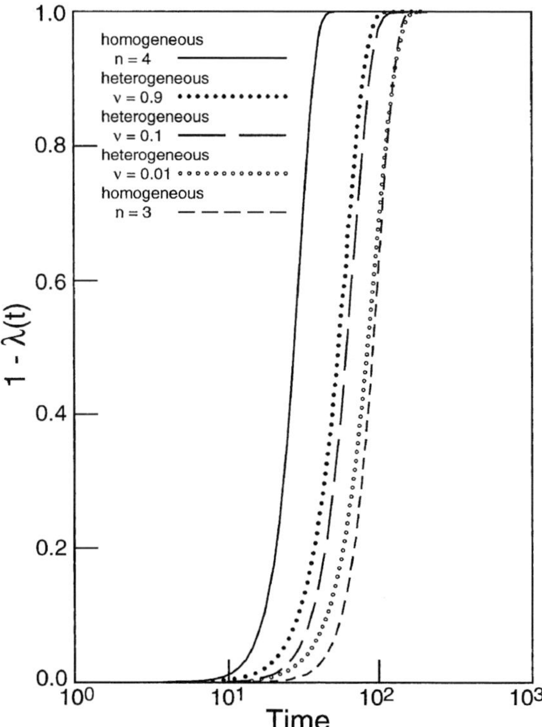

Fig. 9.12 Comparison of theoretical isotherms for homogeneous and heterogeneous nucleation. Plot of extent of transformation against log time for indicated situations. Homogeneous: derived Avrami with $n = 4$ and 3. Heterogeneous: first-order rate law $v = 0.9$, 0.01 and 0.001. (From Eqs. (9.37) and (9.38))

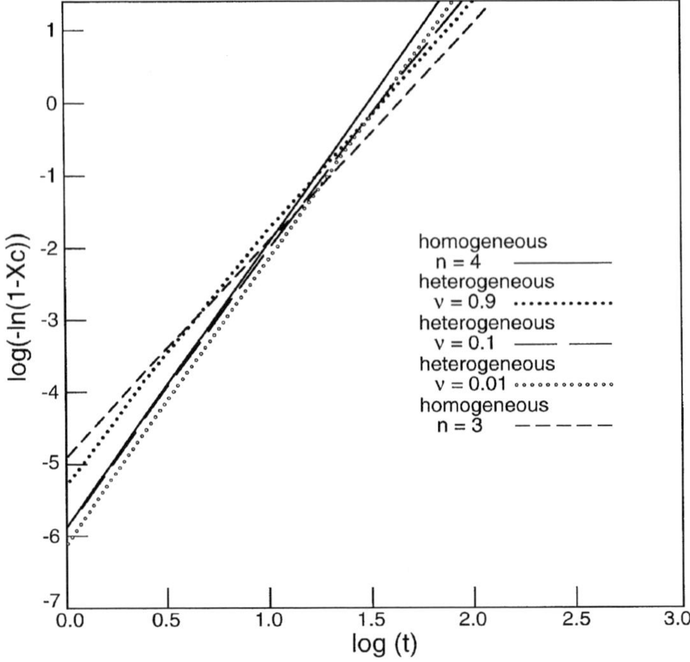

Fig. 9.13 Comparison of theoretical homogeneous and heterogeneous nucleation by plotting $\ln(-\ln \lambda)$ against log time. Situations correspond to those in Fig. 9.12.

isotherms for homogeneous steady-state nucleation and lineal growth (two- and three-dimensional) are compared with those where nuclei are activated according to a first-order law (Eqs. (9.37) and (9.38)), with different rate constants. The role of the heterogeneities in influencing the shape of the isotherms is not as great as might have been anticipated. There is just a slight difference between the isotherms at the early stages of the transformation. However, as the crystallization progresses and the concentration of available nuclei decreases, the rate of crystallization is reduced, particularly as $1 - \lambda(t)$ approaches unity. These isotherms are replotted in Fig. 9.13 as $\ln(-\ln \lambda)$ against $\log t$. This type of plot illustrates more vividly the changes that occur with this type of heterogeneous nucleation. When the isotherms are analyzed in the conventional manner, by using Eq. (9.32), the Avrami exponent is found to be $n = 4$ at the early stages of the transformation and approaches 3 towards the end of the crystallization. Thus, if the value of n were calculated in this manner, it would vary with the extent of the transformation for the same growth geometry. Over a limited range in experimental data a fractional value of n would be obtained. This type of result is often misinterpreted as reflecting a change in the growth geometry. A similar conclusion is reached for two-dimensional growth. In this case n varies from 3 to 2 over the range $(1 - \lambda) = 0$ to 1.

The isotherms illustrated in Fig. 9.12 only represent one type of nuclei activation. If, for example, all of the potential nuclei are activated at $t = 0$, equations of the form of Eqs. (9.28) to (9.31) result. There is obviously a variety of activation processes that can be postulated, each of which will result in its own unique isotherm. Hence, for a given growth geometry there can be many differently shaped isotherms that have n values that not only differ but apparently vary with the time course of the transformation. Hence, it should not be unexpected when considering the complete transformation to observe decreasing values of the Avrami exponent with time. This observation is not caused by a deficiency in the general Avrami formulation (Eqs. 9.26 and 9.27) but rather by the varying nucleation rate with time.

The two theories that have been outlined were developed for the crystallization of monomeric substances where the transformation is complete. The termination of the crystallization process in monomers has been attributed by the Avrami analysis to the mutual impingement of growing centers. The extent of the liquid–crystal transformation in homopolymers is never complete. It can vary from 30% to 90% depending on the circumstances. In considering a theoretical base for the crystallization kinetics of long chain molecules, the superposability of the isotherms gives strong support to the concurrence of nucleation and growth processes. The widespread observations that the radius of a growing spherulite increases linearly with time are consistent with the assumption that growth is controlled by processes occurring at the crystallite–melt interface. These assumptions are consistent with the monomeric theories that have been discussed above. Therefore, a similar approach is suggested in treating polymer crystallization kinetics.

In order to adapt the Avrami concept to polymers, and to account for the incomplete transformation, an arbitrary normalization procedure was introduced.(12) Following the Avrami framework, and analogous to Eq. (9.24), it was proposed that for polymeric systems

$$\frac{d(1-\lambda)}{d(1-\lambda)'} = 1 - U(t) \tag{9.39}$$

where $U(t)$ is the "effective fraction" of the mass transformed at time t. It is defined as that fraction of the total mass into which further crystal growth cannot occur. This quantity includes the actual mass transformed as well as the disordered chain units that are noncrystallizable at time t. The additional assumption is made that the "effective fraction" transformed is proportional to the actual mass fraction transformed, the proportionality factor being $1/[1-\lambda(\infty)]$. The quantity $1-\lambda(\infty)$ is the weight fraction of polymer that is crystalline at the termination of the process. The proportionality factor has been assumed to be independent of time. For small values of the amount transformed, $d(1-\lambda) \simeq d(1-\lambda)'$, this difficulty is alleviated.

With

$$U(t) = \frac{1 - \lambda(t)}{1 - \lambda(\infty)} \tag{9.40}$$

the integration of Eq. (9.39) yields

$$\ln \frac{1}{\lambda[1 - \lambda(\infty)]} = \frac{[1 - \lambda(t)]'}{[1 - \lambda(\infty)]} \tag{9.41}$$

Since $(1 - \lambda(t))'$ is the same as for monomeric substances

$$\ln\left[\frac{1 - \lambda(\infty)}{\lambda(t)}\right] = \frac{1}{1 - \lambda(\infty)} \frac{\rho_c}{\rho_l} \int_0^t v(t,\tau)\, N(\tau)\, d\tau \tag{9.42}$$

In order to express the time dependence of $1 - \lambda$ the details of the nucleation and growth processes need to be specified. If the integration of Eq. (9.42) is carried out under the same set of assumptions made previously for low molecular weight substances, an equation of the form

$$\ln\left[\frac{1 - \lambda(\infty)}{\lambda(t)}\right] = \frac{1}{1 - \lambda(\infty)} kt^n \tag{9.43}$$

results. The interpretation of the exponent n, as given in Table 9.1, is still valid. The results embodied in Eq. (9.43) contain the same restrictions as previously, since specific growth and nucleation rates have been assumed. Equation (9.43) is then a derived Avrami equation proposed for homopolymers.

In order to compare theory with experiment, the equations must be recast in terms of directly measured quantities. For example, if the specific volume is being measured, Eq. (9.43) can be rewritten as

$$\ln\left[\frac{V_\infty - V_t}{V_\infty - V_0}\right] = -\frac{1}{1 - \lambda(\infty)} kt^n \tag{9.44}$$

where V_0 is the initial volume (the volume of the melt at $t = 0$), V_∞ is the final volume, and V_t the volume at time t. Other properties that change with the amount of the transformation can be treated in a similar manner. If expressed in terms of the relative volume shrinkage, $a(t) = (V_0 - V_t)/V_0$, Eq. (9.44) becomes

$$\ln\left[\frac{1 - a(t)}{a(\infty)}\right] = -\frac{1}{1 - \lambda(\infty)} kt^n \tag{9.45}$$

If the degree of crystallinity $1 - \lambda(t)$ is defined as $(V_c - V_t)/(V_l - V_c)$,[2] then

$$\frac{1 - \lambda(t)}{1 - \lambda(\infty)} = 1 - \exp\left(-\frac{1}{1 - \lambda(\infty)} kt^n\right) \tag{9.46}$$

[2] This definition is predicated on the assumption that the specific volumes of the crystalline and amorphous phases are additive in a polymer that is only partly crystalline.

The normalization that has been proposed to the Avrami equation is a very formal one. The unique structural features of long chain molecules, the connectivity of chain units and the melt structure have not as yet been explicitly taken into account.

9.4 Comparison of theory with experiment: overall crystallization

There are some inherent problems in analyzing the experimental results of overall crystallization kinetics that need to be recognized. It is of paramount importance that the initial melt be chemically and structurally pure. Residual crystallites, chemical impurities, partial orientation of the chains and other inhomogeneities lead to misleading results.(30–34b) Very high molecular weight species often retain some orientation in the melt.(34) Structural irregularities in disordered chains are often subtle and are not easily detected. These factors can have a major influence on the crystallization kinetics. There can also be difficulties in reproducibility, and thus in interpretation, among polymers that have the same chemical and structural repeating units. A major reason for these discrepancies is that the molecular weight and polydispersity strongly influence the crystallization kinetics. It is also possible that the same polymer can apparently give different results because of impurities present in some cases and absent in others. In some instances unapparent to the investigator, a small amount of a comonomer is deliberately introduced into the chain by the manufacturer. Thus, it is possible that a polymer having the same repeating units but coming from different sources will not always yield the same results.(35) Residual catalyst fragments can also influence the co-unit of the crystallization.

Despite these concerns, when crystallization kinetics experiments are conducted so that prior to crystallization the sample is completely molten, the chains in random conformation and degradative processes avoided, the resulting isotherms are extremely reproducible. They are independent of the initial temperature of the melt at which the sample is held.(12,35,36) An important factor is that complete melting be ensured. There are some reports where the expected reproducibility is not achieved. In these cases, the aforementioned conditions may not be operative, or the rate of cooling of the sample is such that isothermal crystallization does not occur at the predetermined temperature.

The literature concerning the experimental results of the overall crystallization kinetics of homopolymers is quite vast. It encompasses virtually all types of repeating units of crystallizable polymers. From this abundance, a representative set of examples has been chosen that illustrates the basic experimental results. These data allow for a quantitative assessment of the theoretical developments that have been presented so far.

To compare experimental results with theory we start with the Göler–Sachs free-growth expressions, approximated by Eqs. (9.12) to (9.14). Accordingly, a plot

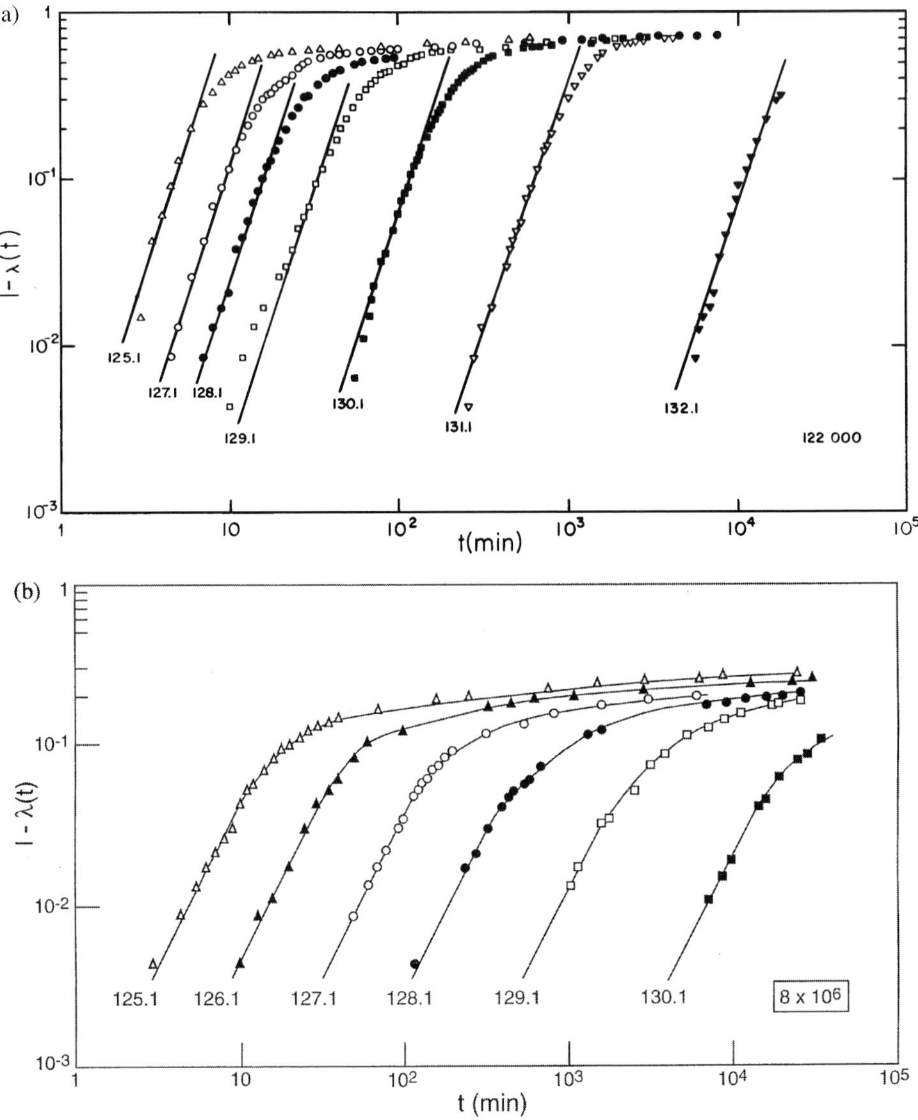

Fig. 9.14 Plot of log crystallinity, or relative crystallinity, against log time for representative polymers. (a) Linear polyethylene, $M = 122\,000$ (34); (b) linear polyethylene $M = 8 \times 10^6$ (34); (c) poly(1,3-dioxolane) (38); (d) poly(phenylene sulfide) (37).

of the log of either $(1 - \lambda(t))$ or the relative extent of the transformation against log time should be linear. The slope of the straight line should correspond to the Avrami exponent n. Examples of this analysis are given in Figs. 9.14a–d.(34,37,38) Despite the simplicity of this approach there is good adherence of the experimental data to this theory. The straight lines drawn correspond to the theory and the slopes

(c)
1-λ
10^0
10^{-1}
10^{-2}
10^{-3}
34
35
36
37
38
39
40
41
3
10^0
10^1
10^2
10^3
10^4
10^5
t (min.)
(d)
Relative Crystallinity
10^0
10^{-1}
10^{-2}
T °C
235.5
246.0
255.5
263.5
120.5
115.5
110.5
10
100
1000
10000
100000
Time (mins)

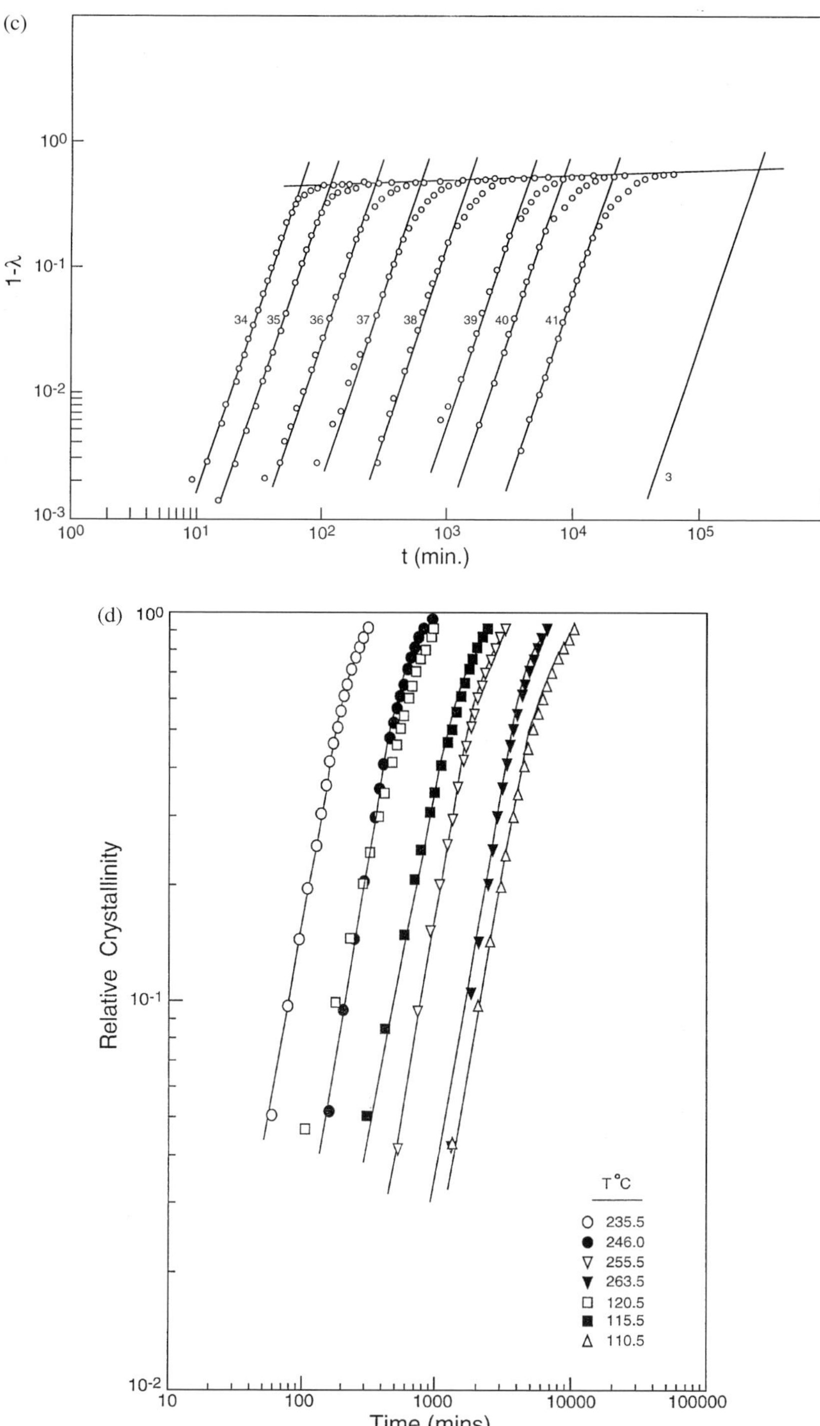

Fig. 9.14 (*cont.*)

represent the Avrami exponent. The linear polyethylene fraction, $M = 122\,000$, fits the Göler–Sachs theory to absolute crystallinity levels of about 45%. However, when the molecular weight is increased to 8×10^6, deviations occur in the range of 20–25% crystallinity.(34) On a relative basis, the data for poly(phenylene sulfide) fit the Göler–Sachs theory for about 90% of the transformation.(37) The results for poly(1,3-dioxolane) are similar.(38) In other examples, not shown, a poly(ethylene oxide) fraction ($M = 365\,000$) adheres to theory for 50–60% of the transformation;(39) while poly(ether ether ketone) follows the Göler–Sachs formulation for more than 80% of the transformation.(40) This kind of agreement has been found for all homopolymers. Based on the widespread experimental results we can conclude that the free-growth approach explains a significant portion of the transformation. When deviations from theory occur, they are such that the crystallization proceeds at a more protracted rate than is expected. It is noteworthy, and certainly surprising, that this simple theory can quantitatively explain a major portion of the transformation of such a complex crystallizing system.

The next step in analyzing experiment is to compare the results with one of the derived Avrami expressions. We recall that Avrami introduced a specific mode for termination of the transformation. There are essentially two ways in which the comparison can be made. One reliable, and informative, way is to superpose the experimental isotherms, based on either absolute or relative crystallinity, onto a master isotherm of a $1 - \lambda(t)$ against $\log t$ plot. This master experimental isotherm can then be compared with the theoretical derived Avrami type plots as illustrated in Fig. 9.10. In this way, both the best fit and the extent of the transformation at which deviation from theory occurs can be clearly and definitively established. Alternatively, taking advantage of Eq. (9.32), a plot of $\ln(-\ln\lambda)$ against $\ln t$ should give a straight line with slope equal to n. It is informative to examine both of these methods.

Figure 9.15 gives several examples of the fitting method.(19,34,41–43) Here, either absolute or relative crystallinity levels are plotted against $\log t$. The best fit with a derived Avrami is indicated by the solid curve. The corresponding n value is also given. In these examples, as well as many others, there is always a good fit between theory and experiment for a major portion of the transformation. For the high molecular weight linear polyethylene fraction (Fig. 9.15a) adherence to the derived Avrami is found for a level of crystallinity of about 15% or approximately half of the transformation.(34) For poly(oxetane) (Fig. 9.15b), good agreement is found for about 25% crystallinity,(41) while for natural rubber (Fig. 9.15c) agreement is found for about 60–80% of the transformation corresponding to a crystallinity level of 20–30%.(19) The polyimide, ODPA 1,[3] (Fig. 9.15d) shows deviation from

[3] The structural formula for this polyimide is given in Table 6.3 (Volume 1).

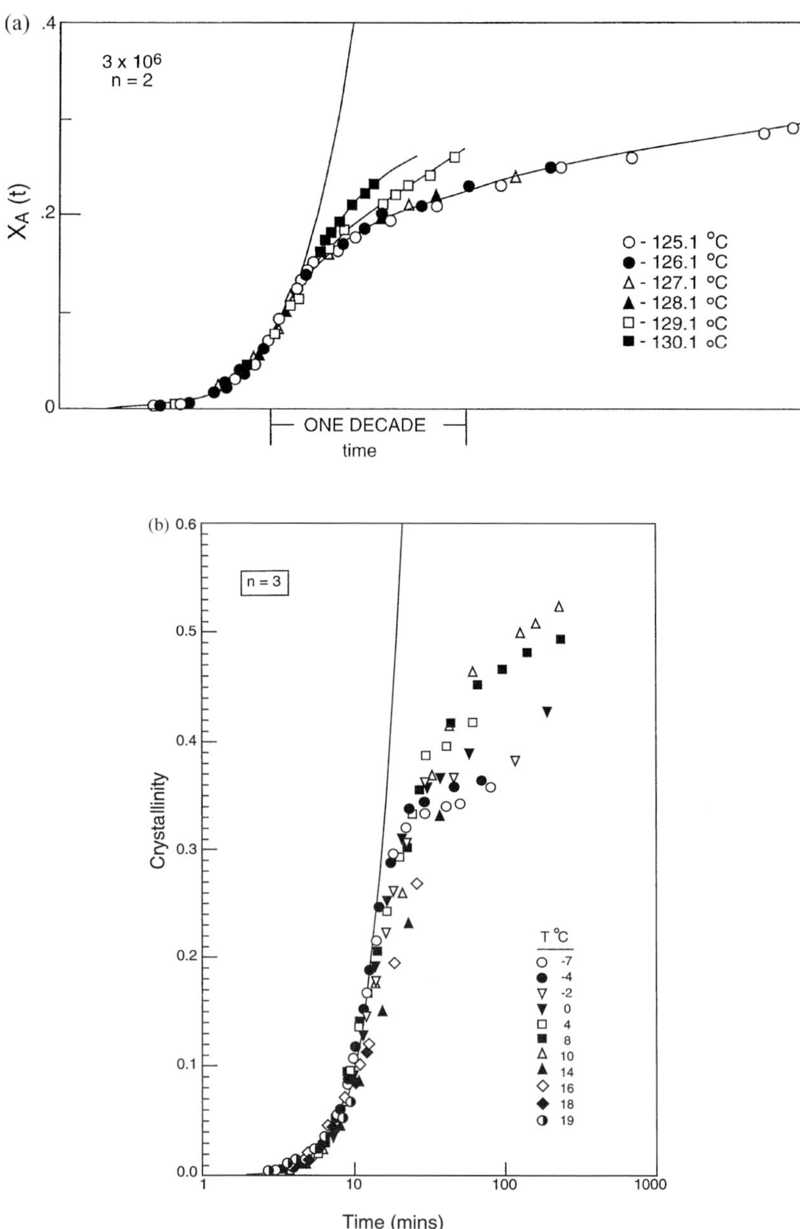

Fig. 9.15 Comparison of derived Avrami equation with experimental results. Plot of fraction crystallinity against log time. (a) Linear polyethylene fraction $M = 8 \times 10^6$ (34); (b) poly(oxetane) (41); (c) natural rubber (19); (d) polyimide ODPA 1 (42); (e) poly(phenylene sulfide) (43).

(c)
n = 3
Relative Crystallinity
Time (mins)
T °C
-5
-11
-22
-27
-33
-38
(d)
n = 2
Crystallinity
Time (mins)
T K
513.2
523.2
533.2
543.2
553.2

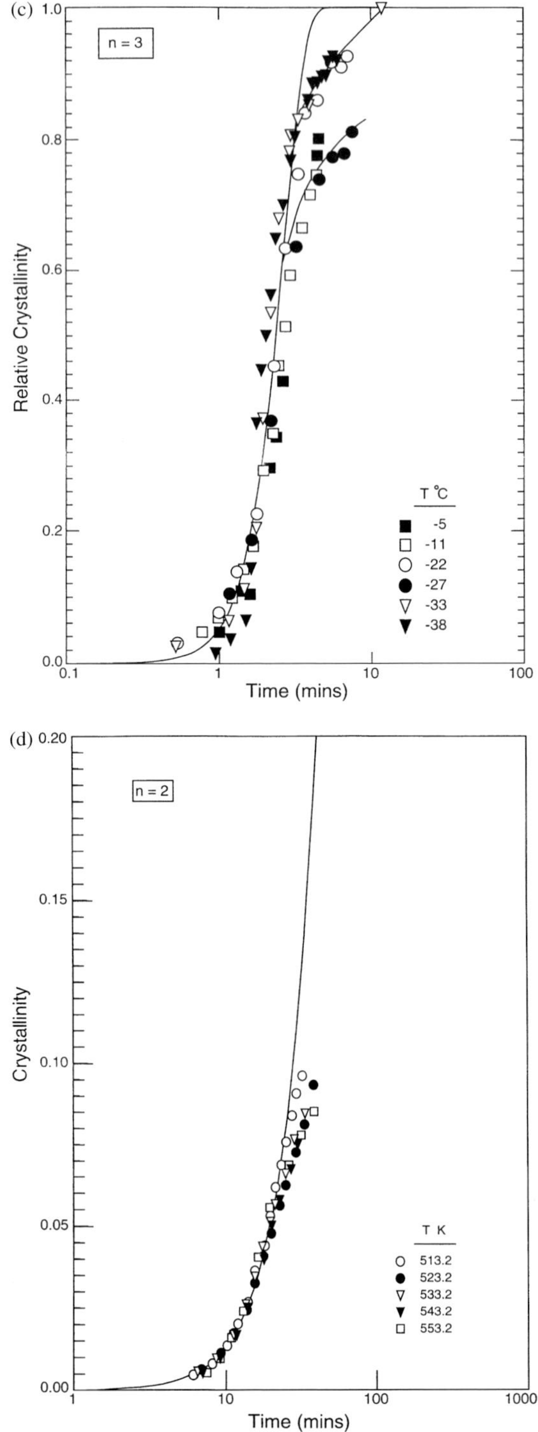

Fig. 9.15 (*cont.*)

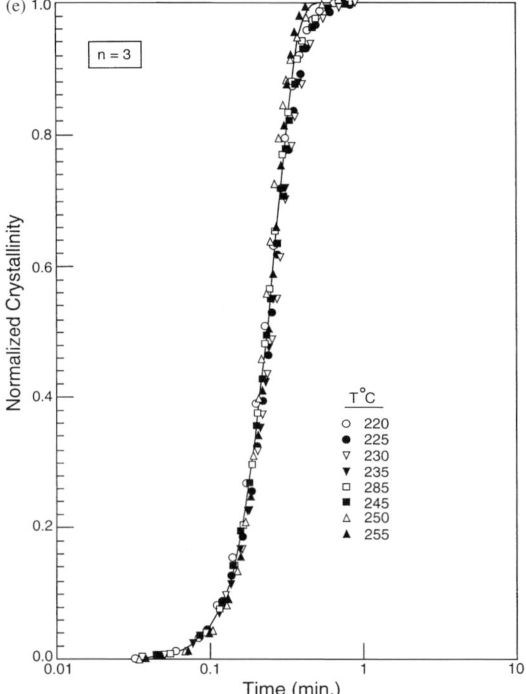

Fig. 9.15 (*cont.*)

theory at crystallinity levels less than 10%.(42) In contrast, poly(phenylene sulfide),(9.15e) when plotted on a normalized basis, adheres to theory, with $n = 3$ over the complete transformation.(43) Other polymers, such as poly(caprolactam),(44) New polyimide (45) and poly(aryl ether ether ketone ketone) (46), among others, also agree with theory when examined on this basis. Different values of n are of course needed for each polymer.

An alternative method to compare theory with experiment, the use of Eq. (9.32), is illustrated by several examples in Fig. 9.16a–d.(23,34,42,43) The results for poly(ethylene adipate)(23) and poly(phenylene sulfide) (43), Figs. 9.16a and b respectively, give the expected straight lines when normalized crystallinity levels are used. However, this is not always the case as shown in Figs. 9.16c and d for a linear polyethylene fraction (34) and the poly(imide) ODPA 1 respectively.(42) In these examples, linearity is not observed over the complete transformation range. However, the fitting procedure indicates that the transformation is continuous. In contrast, as is indicated in Figs. 9.16c and d, deviation from theory in this type of plot results in a sharp break and another straight line with a different slope. The double log type plot exaggerates the final portion of the crystallization and often leads to a misconception in interpreting mechanisms in terms of the kinetics.(42,47) A plot

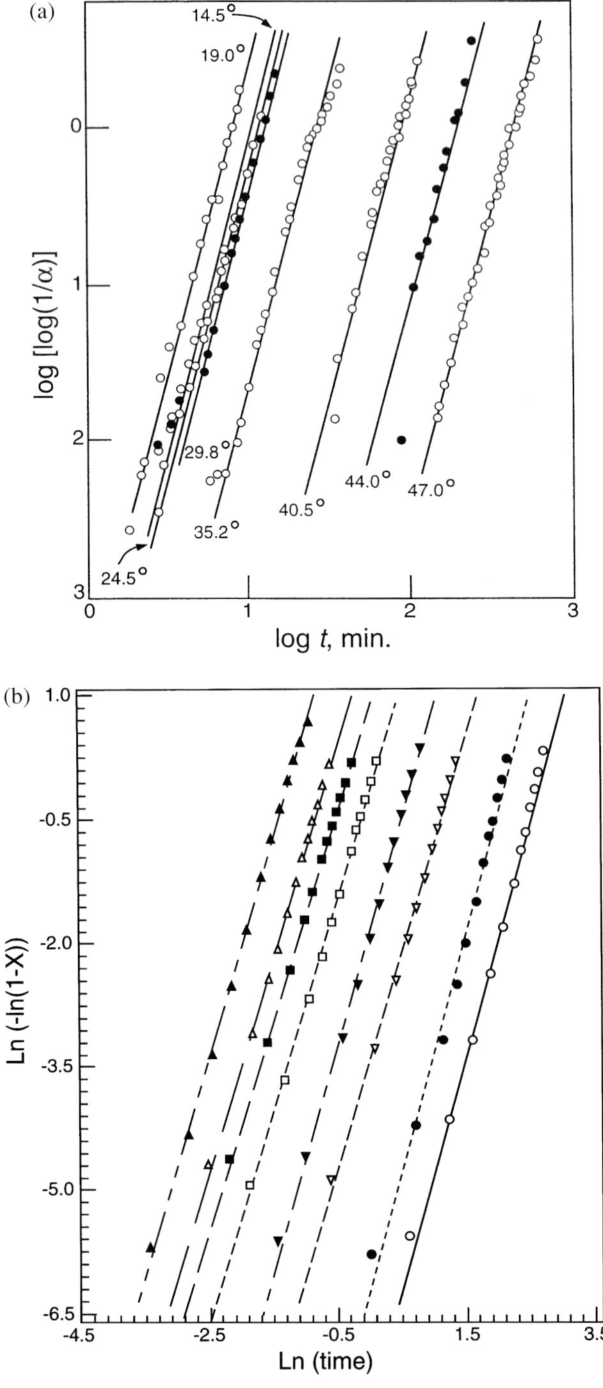

Fig. 9.16 Fit of experimental data to Eq. (9.32). (a) Poly(ethylene adipate) (23); (b) poly(phenylene sulfide) (42); (c) linear polyethylene fraction $M = 47\,000$ (34); (d) poly(imide) ODPA 1 (42).

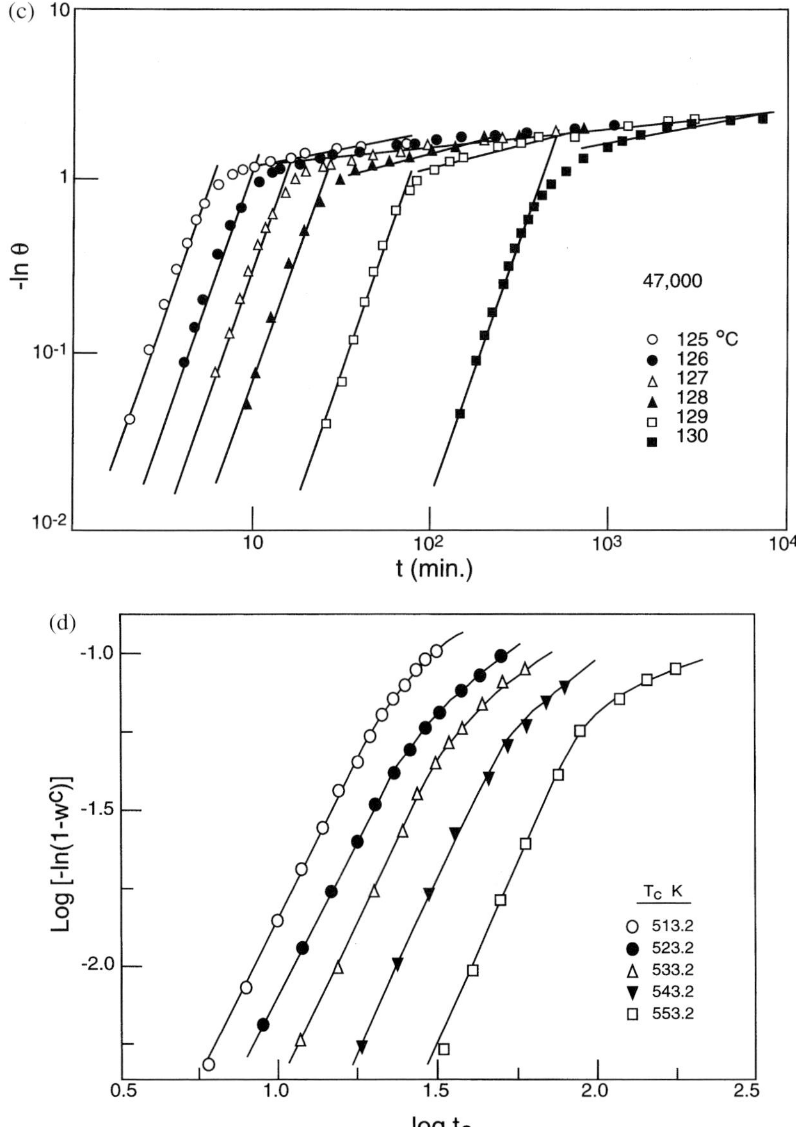

Fig. 9.16 (*cont.*)

according to Eq. (9.32) results in a distorted scale that can allow for a superficially good fit of the experimental data, when in fact the agreement may only hold for the early part of the crystallization.

Another example of this type of analysis is given in Fig. 9.17 for the crystallization kinetics of a mixture of two poly(ethylene oxide) fractions.(47) In Fig. 9.17a the set of superposed isotherms is fitted to the derived Avrami expression, with $n = 3$.

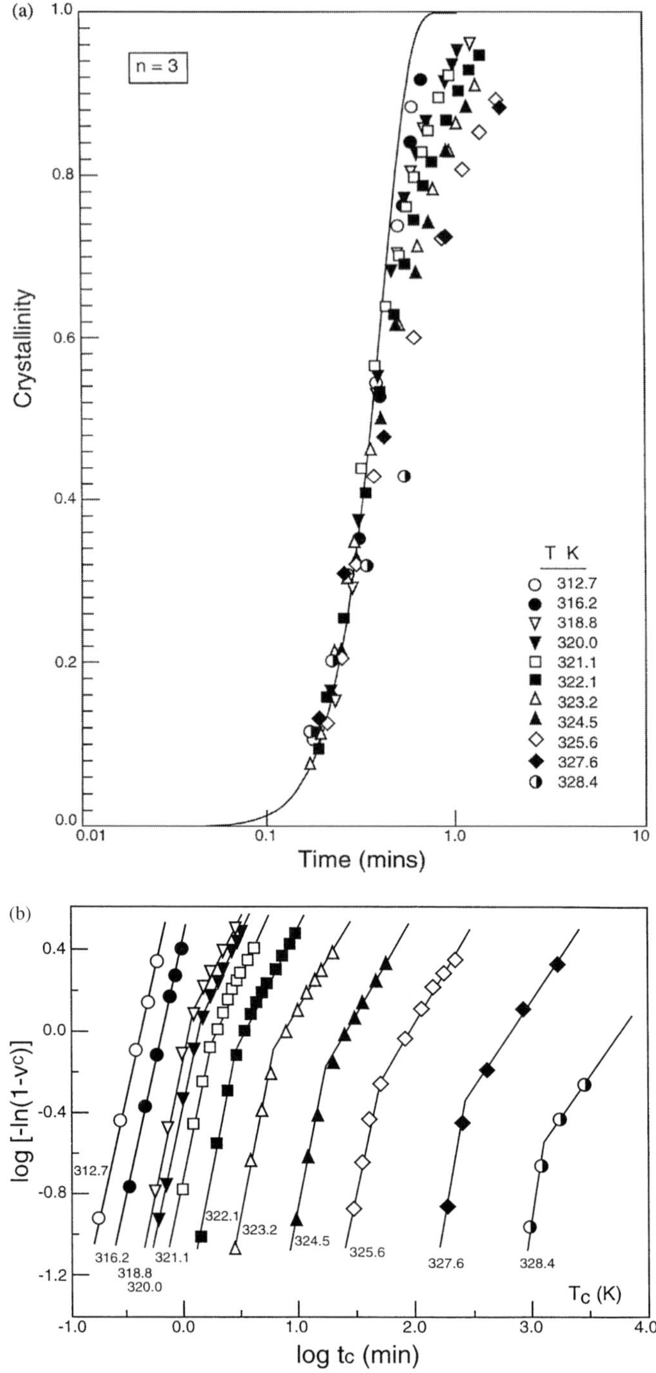

Fig. 9.17 Fit of mixture of two poly(ethylene oxide) fractions to derived Avrami equation. Crystallization temperatures indicated. (a) Solid curve theoretical isotherms with $n = 3$. (b) Use of Eq. (9.32). (Data from Cheng and Wunderlich (47))

Very good agreement is found between theory and experiment up to about 70% of the transformation. The usual deviations are observed at the higher transformation levels as the crystallization rate becomes more protracted than predicted. It is clear that the crystallization is a continuous process. There is no obvious discontinuity when deviation from theory occurs. In contrast, when the same data are analyzed according to Eq. (9.32) two intersecting straight lines result, as is illustrated in Fig. 9.17b. It was concluded from the latter plot that two different mechanisms were involved in the crystallization. These results were then used to support a particular theory.(47) Caution must clearly be exercised in analyzing crystallization kinetic data in this manner.

Another problem encountered in analyzing experimental data according to Eq. (9.32) occurs when attempts are made to obtain the best straight line fit over the complete range. This procedure assumes that the derived Avrami expression is valid over the complete transformation. Usually when this method is adopted, nonintegral values such as 3.8, 3.3, etc. are obtained for the exponent n. Attempts are then usually made to explain the results in terms of growth geometries. As an example, when this method was used in analyzing the results for poly(aryl ether ether ketone ketone) n values in the range 2.3 to 2.5 were obtained.(46) However, when the fitting method was used the superposed isotherm adhered to $n = 4$ over the complete range.

A typical isotherm can be deduced after applying the fitting method to the overall crystallization kinetics of a large number of different homopolymers. A schematic example of such an isotherm is given in Fig. 9.18. The isotherm is drawn so that a pseudo-equilibrium crystallinity level of 0.75 is reached. A striking feature of this isotherm is that despite the distinctly different regions that are observed there are no indications of any discontinuities. In Region I both the Göler–Sachs free growth and the derived Avrami are obeyed. There is really no need to introduce Avrami in this region since nothing is gained. Although the derived Avrami continues along the dashed line, the actual crystallization proceeds at a reduced rate in Region II. Eventually the crystallization slows down and enters Region III, the so-called "tail" region. Here, crystallization proceeds very slowly with time. In this region there is only a small percentage increase in the crystallinity level over many decades of time. There are clearly major impediments present to crystallization in this flat portion of the isotherm. The crystallization is obviously not of the Avrami type. There is, therefore, no point in discussing it in terms of the exponent n. The superposition is not clear in this region but the isotherms appear to join one another after long times. This portion of the isotherm will be discussed in more detail in Volume 3 because of the structural and morphological changes that are involved. It is preferable not to define the complete transformation in terms of primary and secondary crystallization. These expressions have meant different things to different investigators. Sometimes secondary crystallization has meant the deviations

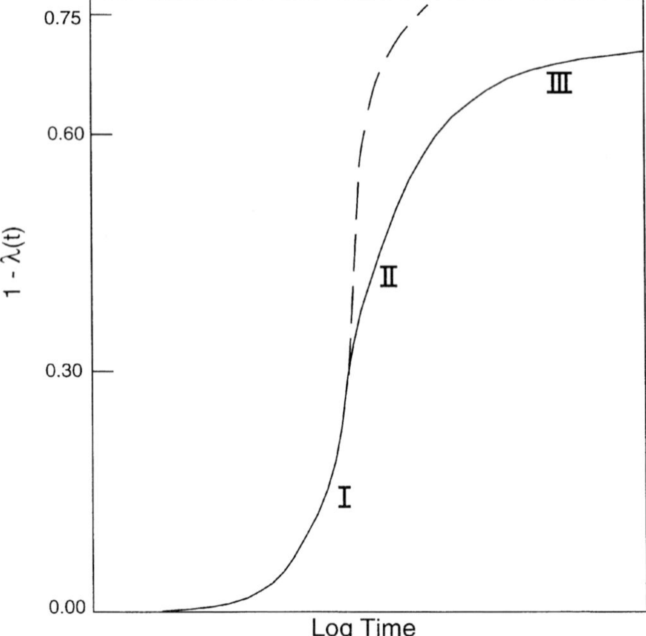

Fig. 9.18 Plot of degree of crystallinity against log time for a typical isotherm.

from a derived Avrami equation. Other times it has meant the "tail" or flat region, where crystallization is very slow. In another definition, secondary crystallization takes place from the noncrystalline regions in the confined environment of already existing crystallites. In this concept primary crystallization is defined as crystallization from the pure melt.(46a,b) The relation between this latter definition and the experimentally observed isotherm is not obvious. A great deal of confusion has resulted from the use of this particular terminology for a continuous process.

With this understanding of the relation between theory and experiment the possible reason(s) for the deviations that are observed between the two can be considered. Also of concern are the advantages, if any, of using the Avrami approach relative to that of free growth. It turns out that important information can be obtained by studying the influence of molecular weight on the kinetics. In this effort a plot is given in Fig. 9.19 of a set of isotherms, superposed to 127 °C, for the indicated molecular weight fractions of linear polyethylene.(34) Here, the absolute level of crystallinity is plotted against log time. The solid curve represents the theoretical derived Avrami for $n = 3$. The crystallinity level at which deviations from the theoretical curve occur decreases as the molecular weight increases. For example, the deviations occur at a crystallinity level of about 0.25 for $M = 1.2 \times 10^6$ and increase to about 0.55 for $M = 1.15 \times 10^4$. Similar results have been found in poly(ethylene oxide)(39) and poly(tetramethyl-*p*-silphenylene siloxane).(48) This

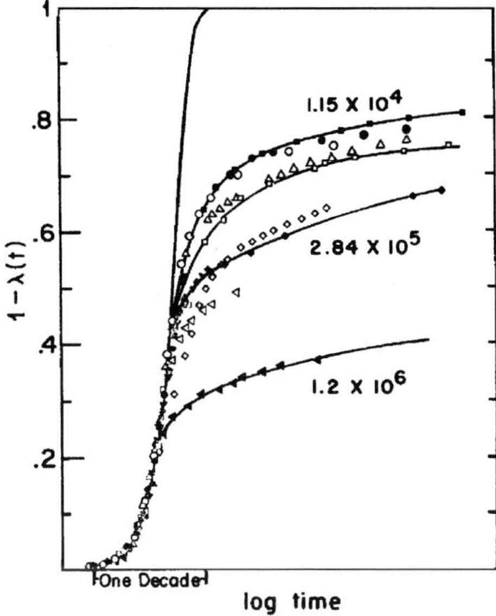

Fig. 9.19 Plot of degree of crystallinity $1 - \lambda(t)$ against log time for indicated molecular weight fractions of linear polyethylene. Isotherms for each molecular weight superposed to 127 °C.(34)

important influence of chain length is given in more detail in Figs. 9.20a and b for linear polyethylene and poly(ethylene oxide) molecular weight fractions respectively.

The crystallinity levels for different situations are plotted against the molecular weight in Fig. 9.20.(34,49) The influence of molecular weight on the crystallinity level that can be attained at an isothermal crystallization temperature is quite evident in this figure. In polyethylene, for example, the crystallinity level that is attained remains constant until $M = 10^5$. There is then a precipitous drop with molecular weight. The deviations from both Avrami and free growth follow a similar pattern. Most important and striking is the fact that, within experimental error, both theories give the same result. Put another way, as far as quantitative agreement between theory and actual experiment is concerned the free-growth approximation does just as well as Avrami. The results for poly(ethylene oxide) follow a similar pattern. Other factors, beside the Avrami termination mechanism, must be involved as the crystallization of long chain molecules progresses.

The profound influence of molecular weight on the crystallization kinetics gives a clue as to a possible reason for the adequacy of the free-growth concept at the early stages and the deviation from the expectations from the Avrami theory. Initially, prior to the onset of crystallization, the melt of a polymer is composed of entangled chains, loops, and knots, as well as other structures that can be considered to be

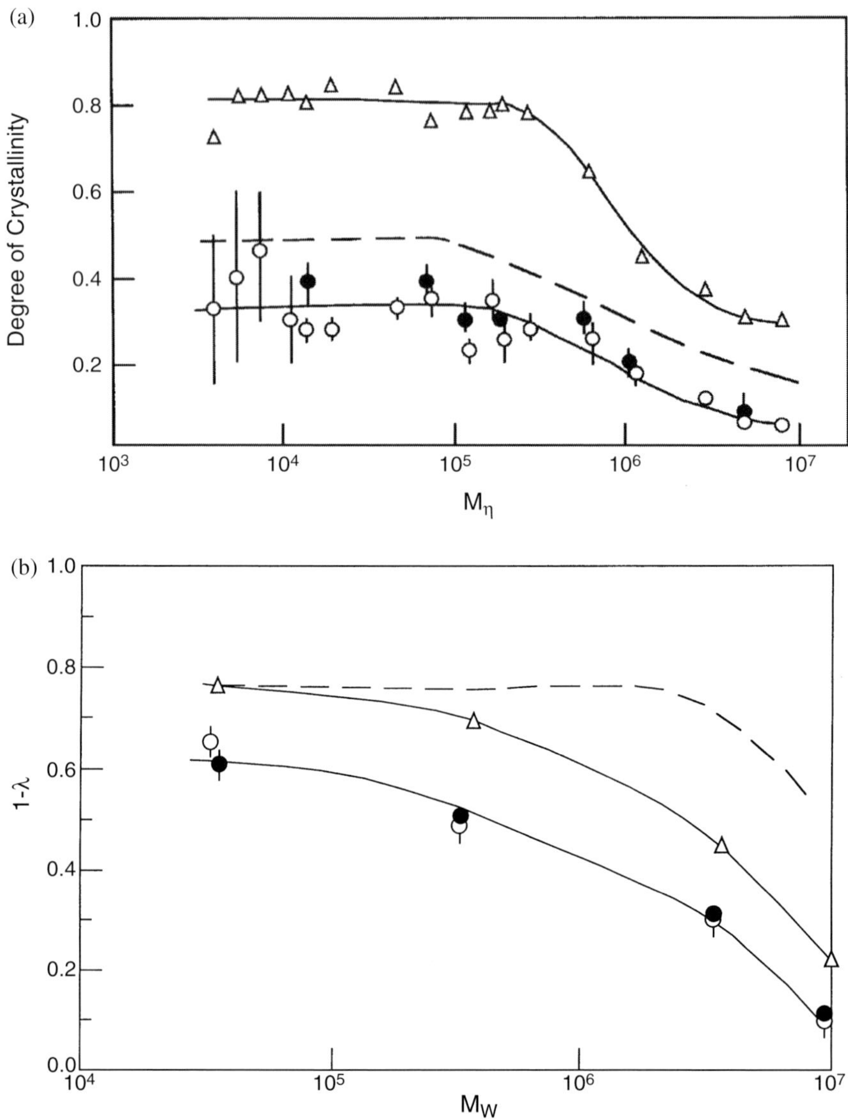

Fig. 9.20 Plots of crystallinity levels as functions of molecular weight. (a) Linear polyethylene fractions (34); (b) poly(ethylene oxide) fractions (49). Pseudo-equilibrium level of crystallinity that is attained △; crystallinity levels at which deviations occur from theory: Göler–Sachs ●; Avrami ○. Dashed curve represents ratio of crystallinity level at which deviation occurs to that actually attained.

topological defects. Although chemically pure, these structures cannot participate in the crystallization. The concentrations of such units are molecular weight dependent and are relegated to the noncrystalline regions. Moreover, there will be regions around such "defects" that are also not transformable as was discussed for cross-linked systems.[4] As the development of crystallinity progresses the availability of transformable units decreases relative to the total number of noncrystalline units. Consequently, the level of crystallinity that can be attained with molecular weight will decrease. Under these circumstances, neither the nucleation nor the growth rates will be invariant with the extent of the transformation. Moreover, small sequence lengths of potentially crystallizable units will be isolated and will not be able to participate in the crystallization.(49a,49b)

The influence of chain entanglement in the melt on the crystallization process has also been described by Galeski and coworkers.(49c) Advantage was taken of the fact that, when linear polyethylene is crystallized under high pressure and temperature, a high level of crystallinity, about 0.90–0.95, is achieved. Upon fusion, the initial entanglement concentration will be low or nonexistent. There is a time interval, of the order of 30 min, before the usual entangled melt is regenerated. During this time the crystallization from the melt is significantly enhanced in comparison to that for a conventional melt. This enhancement has been attributed to the spherulite growth rate. Most of the work was performed with a polydisperse sample, $M_{\mathrm{w}} = 55\,000$, $M_{\mathrm{n}} = 11\,500$, so that the conclusions are tempered by the high concentration of low molecular weight species. Investigations of this type with a series of high molecular weight fractions should be quite illuminating. It has also been claimed that by varying the initial thickness in linear polyethylene the primary nucleation rate increases with the entanglement density.(50a)

Another example that demonstrates the influence of chain entanglements is found in the melt crystallization kinetics of pure poly(dimethyl siloxane) ($M_{\mathrm{n}} = 740\,000$).(50b) It is found that in this case the derived Avrami equation with $n = 3$ can account for 95% of the transformation. This relatively rare event can be explained by the fact the molecular weight between entanglements of this polymer is 12 000 as compared, for example, with 830 for linear polyethylene.(50c) Thus, the number of entanglements per chain is relatively low as compared to other polymers. It will also be shown in Chapter 13 that a derived Avrami expression explains more than 90% of the crystallization of linear polyethylene from dilute solution.

The crystallization of an isotactic poly(styrene) sample that was originally freeze dried from a 0.01 wt percent benzene solution vividly demonstrates the influence of chain entanglements on the kinetics.(50d) Such a sample has a minimal amount of entanglements since it essentially comes from a dilute solution. Consequently it was found that freeze dried samples crystallized, in terms of half-times, approximately nine times faster than the untreated polymer. Both samples crystallized from the pure melt in the conventional manner. This result is consistent with the Flory–Yoon calculations that the chains cannot disentangle each other from the melt during the time scale of the crystallization.(50e,50f) This calculation is consistent with the

[4] This problem, and the influence on the melting temperature, have been discussed in Chapter 7 (Volume 1).

Table 9.2. *Deviation of experiment crystallization kinetic theories for selected polymers*

		$1 - \lambda$ Deviation			
Polymer	$(1 - \lambda)_\infty$	Göler–Sachs[a]	Avrami[b]	Avrami/$(1 - \lambda)_\infty$	Ref.
Poly(ether ether ketone)					
Low temperature	0.18	0.13	0.17	0.94	a
High temperature	0.35	0.23	0.22	0.63	
New poly(imide)					
Low temperature	0.24	0.15	0.23	0.96	b
High temperature	0.25	0.17	0.22	0.88	
Poly(1,3-dioxolane)	0.50	0.30	0.32	0.64	c
Poly(chlorotrifluoro ethylene)	0.60	0.50	0.48	0.72	d
Poly(3,3-dimethyl oxetane)	0.63	0.30	0.48	0.76	e
Poly(oxetane)	0.53	0.28	0.25	0.47	f
Poly(cis-1,4-butadiene)					
Low temperature	0.50	0.50	0.50	1.00	g
High temperature	0.55	0.50	0.45	0.82	

[a] Göler–Sachs from Eqs. (9.12) and (9.13).
[b] Avrami from Eq. (9.31a).

References

a. Cebe, P. and S. D. Hong, *Polymer*, **27**, 1183 (1986).
b. Hsiao, B. S., B. B. Sauer and A. Biswas, *J. Polym. Sci.: Pt. B: Polym. Phys.*, **32**, 737 (1994).
c. Alamo, R., J. G. Fatou and J. Guzman, *Polymer*, **32**, 274 (1982).
d. Hoffman, J. D. and J. J. Weeks, *J. Chem. Phys.*, **37**, 1723 (1962).
e. Perez, E., J. G. Fatou and A. Bello, *Coll. Polym. Sci.*, **262**, 913 (1984).
f. Perez, E., A. Bello and J. G. Fatou, *Coll. Polym. Sci.*, **262**, 605 (1984).
g. Feio, G. and J. P. Cohen-Addad, *J. Polym. Sci.: Pt. B: Polym. Phys.*, **26**, 389 (1988).

influence of chain entanglements on the crystallization process and their effective permanent nature. DiMarzio *et al.* have argued to the contrary.(50g) They calculated that the entanglements present no impediment to the reeling in of molecules to the growing crystal front. It was further concluded that the role of entanglements would be negligible in the crystallization process.(50h) This conclusion is contrary to the abundant experimental results that are now available.

As an aside, it should be noted at this point that chain entanglements have been shown theoretically to also influence many structural properties.(50i) These properties, such as crystallinity levels and lamellar thickness, will be discussed in detail in Volume 3.

The similarity of the two theories in explaining the experimental results, prior to deviation, has been found in many other polymers as is indicated in Table 9.2. Data for a range of molecular weights are not available for these polymers. Tabulated are the final level of crystallinity attained, $(1-\lambda)_\infty$; the crystallinity levels at which the different theories deviate; and the ratio of the $1-\lambda$ Avrami deviation to $(1-\lambda)_\infty$. We find once again that the $1-\lambda$ deviation values for Göler–Sachs and Avrami are close to one another. When the $(1-\lambda)_\infty$ value that is attained is relatively low then the deviation level is close to this value. For example, the New poly(imide) and poly(ether ether ketone) only attain crystallinity levels of the order of 25% or less. The kinetic data for these polymers adhere quite well to the theories over this range. The results for some other polymers such as nylon 11,(50) poly(ϵ-caprolactone),(44) poly(phenylene sulfide) (43) and poly(ether ether ketone ketone) (46) show similar agreement. For polymers that attain high levels of crystallinity agreement between theory and experiment does not encompass the complete transformation range. However, as indicated in the table a significant portion of the transformation can be explained equally well by either the free-growth or Avrami theories.

The introduction of the impingement concept has made substantial improvement in fitting the observed crystallization kinetics of metals, and other monomeric systems, to theory. However, as the analysis of experimental data has indicated, no significant gain over the free-growth approximation is achieved in the crystallization kinetics of polymers. Cessation of crystalline growth because of the impingement of growing centers is thus not the major reason for the observed reduction in the crystallization rate of polymers with the extent of transformation. This is true even when the incomplete transformation is taken into account by the proposed normalization procedure. One must then seek other reasons that are unique to polymers as the source of the deviations. Despite the deviation of experiment from the Göler–Sachs and Avrami theories, the results demonstrate that one is dealing with a classical nucleation and growth process. This general conclusion must be kept in mind as modifications are proposed. In the next section we explore the efforts that have been made to resolve this problem.

9.5 Further theoretical developments: overall crystallization

The realization that the derived Avrami equation does not account for the observed isotherms has led to many proposed modifications. It should be recalled that the basic premise of the Avrami development is that the impingement of two growing centers causes a cessation of their growth. This concept was introduced to allow for the termination of the transformation of monomeric substances. The proposed alterations to Avrami fall into three main categories: re-examination of the impingement and termination steps; allowing for more than one nucleation and growth process to take place; variations in the nucleation and growth rates with the extent of the transformation. We shall consider the major consequences of these endeavors in this

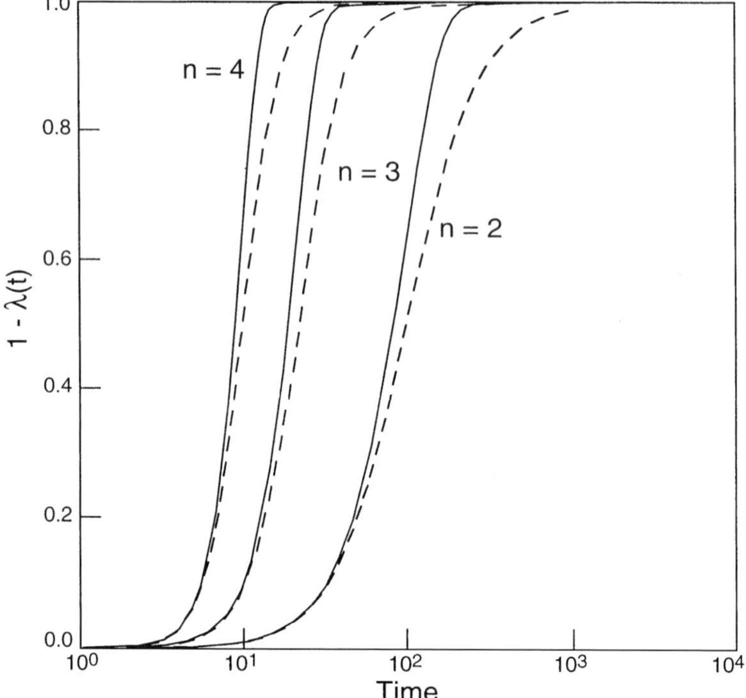

Fig. 9.21 Comparison of Austin–Rickett and derived Avrami equation for different values of exponent n. Plot of degree of crystallinity $1-\lambda(t)$ against log time. Solid curve derived Avrami, Eq. (9.31a). Dashed curve Austin–Rickett, Eq. (9.49).

section. Some of the modifications have their origin in the basic Avrami equation. Others are based on one of the derived expressions.

Closely following the Avrami expression is an empirical relation introduced by Austin and Rickett, based on the experimental results for the decomposition of austenite steel.(51) The relation can be expressed as

$$1-\lambda(t) = 1 - [(kt)^n + 1]^{-1} \tag{9.47}$$

Equation (9.47) is compared with the derived Avrami, Eq. (9.31a), in Fig. 9.21. Here the extent of the transformation is plotted against the log time for integral values of the exponent n. There are only small differences between the two relations, particularly in the usual range of polymer crystallization. Analysis of typical kinetic data indicates that deviations from either theory occur at about the same crystallinity level.

In an alternative approach, Lee and Kim (52) proposed that impingment can be described by a modification of Eq. (9.24). They suggest that

$$\frac{d(1-\lambda(t))}{d(1-\lambda(t))'} = (\lambda(t))^{1-C} \tag{9.48}$$

where C is a constant, termed the impingement parameter. When $C = 0$ the derived Avrami type equation is obtained. The Austin–Rickett equation, (9.47), is obtained when $C = 1$.

Tobin (53) has treated the impingement problem in a somewhat different manner from that of Avrami. The premise that impingement is the cause of the cessation of growth and the termination of the crystallization is still the underlying principle involved. This analysis was initially developed for a constant homogeneous nucleation rate accompanied by either two- or three-dimensional growth. It was subsequently extended to heterogeneous nucleation. For the two-dimensional problem, the transformed area at time t, $A(t)$, can be represented as

$$A(t) = N_0 \int_0^t f(t,\tau)[A_0 - A(\tau)]\,d\tau \tag{9.49}$$

where A_0 is the initial area of the sample. The quantity $f(t, \tau)$ represents the transformed area that develops from each initiating nucleus. The new concept that is introduced resides in the assumption that

$$f(t,\tau) = G^2 \left\{ \frac{A_0 - A(t)}{A_0} \right\} (t - \tau)^2 \tag{9.50}$$

Thus, it is assumed that at time t each growing center has its effective area reduced by the same factor, irrespective of when it was formed. Substituting Eq. (9.50) into Eq. (9.49) and integrating over τ, the nonlinear Volterra integral equation

$$\frac{1 - \lambda(t)}{\lambda(t)} = Kt^3 - 3K \int_0^t (t - \tau)^2\, \lambda(\tau)\, d\tau \tag{9.51}$$

results. Here K is defined as $NG/3$. Equation (9.51) cannot be solved in closed form. The complete solution can only be obtained by numerical methods. However, approximate solutions, adequate for most purposes, are easily obtained.

The zeroth-order solution to this integral equation is

$$\frac{1 - \lambda(t)}{\lambda(t)} = Kt^3 \tag{9.52}$$

or

$$\lambda(t) = Kt^3/(1 + Kt^3) \tag{9.53}$$

A first-order solution can be obtained by setting the zeroth-order solution into Eq. (9.51).

Then

$$\frac{1-\lambda(t)}{\lambda(t)} = 3K \int_0^t (t-\tau)^2 (1+Kt^3)^{-1} d\tau \qquad (9.54)$$

The integral in Eq. (9.54) can be evaluated in closed form yielding the first-order solution. The preceding argument can be extended to three-dimensional growth, with the result that (53)

$$\frac{1-\lambda(t)}{\lambda(t)} = Kt^4 - 4K \int_0^t (t-\tau)^3 \, \lambda(\tau) \, d\tau \qquad (9.55)$$

Corresponding approximate solutions result. In either case the zeroth- and first-order solutions for $1-\lambda(t)$ do not differ from one another by more than 2% over the complete transformation range. Both the first-order and the exact solutions vary as either t^2 or t^3 for large values of t for two- and three-dimensional growth, respectively. It is suggested, therefore, that the first-order solution may be a reasonable approximation to the exact solution. For small values of $1-\lambda(t)$, Eq. (9.54) becomes

$$Kt^3 = \frac{1-\lambda(t)}{\lambda(t)} \simeq \ln \frac{1}{\lambda(t)} \qquad (9.56)$$

and corresponds to the derived Avrami expression for two-dimensional growth. A corresponding result occurs for growth in three dimensions. The equations then reduce to the Göler–Sachs expressions for smaller values of $1-\lambda(t)$.

Theoretical isotherms calculated according to the Tobin zeroth-order solution are given in Fig. 9.11, where a comparison can be made with those of Avrami and Göler–Sachs. The major features of the isotherms are similar to one another. The differences become accentuated as the exponent n is lowered. The Tobin plot reduces to the Avrami formulation at low to modest extents of transformation. At sufficiently low values of $1-\lambda(t)$ they all reduce to the free-growth approximation. However, as the crystallization progresses, small but significant differences develop. The rate calculated by the Tobin equation becomes slower than the Avrami calculation. At still higher levels of transformation the development of crystallinity in the Tobin analysis becomes more protracted and diffuse. There is a clear, analytical termination to the Avrami that is relatively sharp when compared to Tobin. Although there is an apparent termination in the Tobin isotherm in Fig. 9.11, the complete integral equation does not allow for such. The Tobin analysis gives a slight improvement in explaining the experimental results. However, it does not account for the major differences that are observed.

When there is a fixed number, $\bar{N}$, of nuclei, all of which are activated at the same $t = 0$, the Tobin analysis gives (53)

$$\frac{1 - \lambda(t)}{\lambda(t)} = \bar{N} k_2 G^2 t^2 \tag{9.57}$$

For three-dimensional growth under the same conditions

$$\frac{1 - \lambda(t)}{\lambda(t)} = \bar{N} k_3 G^3 t^3 \tag{9.58}$$

When the activation of nuclei follows a first-order rate law

$$\frac{1 - \lambda(t)}{\lambda(t)} = \bar{N} k_2 v \int_0^t (t - \tau)^2 \lambda(t) \exp(-v\tau)\, d\tau \tag{9.59}$$

for two-dimensional growth and the same rate constant that was adopted earlier (Eqs. 9.16 and 9.17). The zeroth-order solution of this integral equation is

$$\frac{1 - \lambda(t)}{\lambda(t)} = \frac{\bar{N} k_2}{v} \int_0^t (t - \tau)^2 \exp(-v\tau)\, d\tau \tag{9.60}$$

The integral in Eq. (9.60) can be evaluated in closed form. In analogy to Eq. (9.17)

$$\frac{1 - \lambda(t)}{\lambda(t)} = A_2 \tag{9.61}$$

The changes that take place in $\frac{1 - \lambda(t)}{\lambda(t)}$ with time follow the same pattern as $\ln(1/\lambda)$ in Eq. (9.38). This analysis can be extended to three-dimensional growth with analogous results.

In quite another approach to the problem it has been postulated that two distinctly different Avrami type crystallizations are operative during the transformation. These processes can occur in either series or parallel with one another. They are based on the derived Avrami expression, Eq. (9.31a). In the series type the first step is termed the primary crystallization, the other the secondary one. However, it is not made clear in many applications where in the model isotherm shown in Fig. 9.18 one process stops and the other begins. The primary process that is initiated at time $t = 0$ is given by (54,55)

$$[1 - \lambda(t)]_1 = [1 - \lambda(\infty)]_1 [1 - \exp(-k_1 t^n)] \tag{9.62}$$

The subscript 1 denotes the primary crystallization process. The crystallinity that develops in the second stage, which is initiated at time $t = \tau$, is expressed in a similar manner.

Thus

$$[1-\lambda(t-\tau)]_2 = [1-\lambda(\infty)]_2[1-\exp(-k_2(t-\tau)^m)] \tag{9.63}$$

In these equations $[1-\lambda(\infty)]_1$ and $[1-\lambda(\infty)]_2$ are the pseudo-equilibrium crystallinity levels that are attained for the respective processes. Each of these has a different value for the Avrami exponent and rate constant. The total transformation at time t, $1-\lambda(t)$, the sum of Eqs. (9.62) and (9.63), is given by

$$[1-\lambda(t)] = [1-\lambda(t)]_1 + \int_t^0 -[\lambda(t)]\left[\frac{\partial(1-\lambda(t-\tau))}{\partial\tau}\right] \tag{9.64}$$

Equation (9.64) can then be rewritten as

$$[1-\lambda(t)] = [1-\lambda(\infty)]_1[1-\exp(-k_1t^n)] + [1-\lambda(\infty)]_2k_2m \times \int_t^0 1-\exp(-k_1\tau^n)^{m-1}[1-\exp(-k_2(t-\tau)^m)]\,d\tau \tag{9.65}$$

Alternatively, it can be expressed as

$$\frac{1-\lambda(t)}{1-\lambda(\infty)} = \frac{[1-\lambda(\infty)]_1}{[1-\lambda(\infty)]}[1-\exp(-k_1t^n)] + \frac{\lambda(\infty)_1}{1-\lambda(\infty)}k_2m \times \int_t^0 [1-\exp(-k_1t^n)](t-\tau)^{m-1}[1-\exp\{-k_2(t-\tau)^m\}]\,d\tau \tag{9.66}$$

Here $1-\lambda(\infty) = 1-\lambda_1(\infty)+1-\lambda_2(\infty)$ and $\frac{1-\lambda(t)}{1-\lambda(\infty)}$ represents the fraction of total crystallinity. The integrals in Eqs. (9.65) and (9.66) cannot be evaluated analytically. However, they can be calculated by approximate numerical and graphical methods.(56,57)

A theoretical isotherm for the two-stage series model, as calculated from Eq. (9.66), is illustrated in Fig. 9.22.(55) The contribution to the total crystallinity of each of the individual steps is also shown. In this example n was taken equal to 1 and 3. For the parameters adopted in this particular example the apparent value of the observed n is about 2. Such two-stage processes can lead to fractional values of the Avrami exponent as well as values that vary with the extent of the transformation. The results depend on the choice of parameters. Considering the number of arbitrary parameters involved, and the selection of the time scale, it is not surprising that agreement can be obtained with different sets of experimental data. The physical significance of this approach is of concern since the model taken is arbitrary. The same kind of result can be obtained for a variety of heterogeneous nucleation mechanisms.

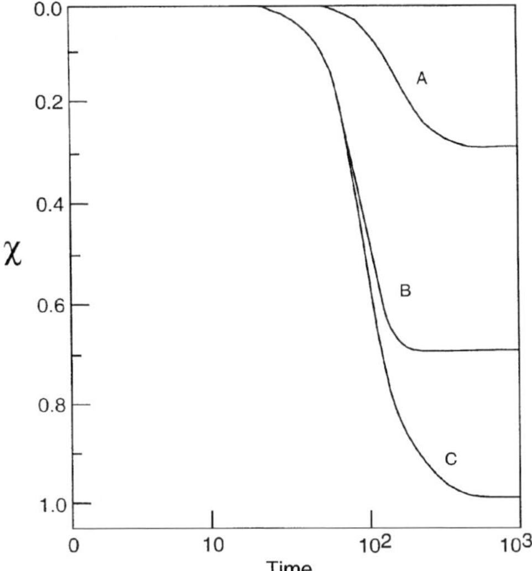

Fig. 9.22 Theoretical isotherms for two-stage series model. Plot of fraction crystallinity χ against log time. Curve A for first-order process, $n = 1$. Curve B primary crystallization, $n = 3$. Curve C total crystallinity. (From Hillier (55))

Pérez-Cardenas *et al.* developed a more realistic multi-stage process by considering an isotherm of the type shown in Fig. 9.18.(58) Region I, where the derived Avrami or free-growth expressions hold, is defined as the primary crystallization. A quality ζ is introduced that represents the weight fraction of polymer that has crystallized at the end of Region II. It is assumed that Region II represents both primary and secondary crystallization. Region III, on the other hand, only represents secondary crystallization. With these assumptions it is found that

$$\frac{1-\lambda(t)}{1-\lambda(\infty)} = 1 - \exp(-k_1 t^n - k_2 t^m)[k_1 n(1-\zeta) \times \int_0^t \exp(k_1\tau^n + k_2\tau^n)\tau^{n-1}\, d\tau + 1] \tag{9.67a}$$

for $\dfrac{1-\lambda(t)}{1-\lambda(\infty)} \leq \zeta$. However, when $\dfrac{1-\lambda(t)}{1-\lambda(\infty)} > \zeta$

$$\frac{1-\lambda(t)}{1-\lambda(\infty)} = 1 - (1-\zeta)\exp(k_2 t^{*m})\exp(-k_2 t^n) \tag{9.67b}$$

Here t^* corresponds to the time where only the secondary crystallization is operative. There are now five arbitary parameters available to fit experimental data. It is not surprising that a good fit can be obtained by this multi-stage process.

The multi-stage processes described above are said to be in series since one starts at $t = 0$ and the other at a later time. If, however, the secondary process is also initiated at $t = 0$, Eq. (9.66) then becomes

$$\frac{1-\lambda(\tau)}{1-\lambda(\infty)} = \frac{[1-\lambda(\infty)]_1}{[1-\lambda(\infty)]}[1-\exp(-k_1 t^n)] + \frac{\lambda(\infty)_1}{1-\lambda(\infty)}[1-\exp(-k_2 t^m)] \quad (9.68)$$

Equation (9.68) represents the two-stage parallel crystallization model, proposed by Velisaris and Seferis.(59) This model has also been found to fit selected experimental data.(60)

Another multi-stage process that has been suggested focuses attention on nucleation rates.[5](61) Two distinctly different steady-state nucleation rates, N_1 and N_2, are assumed. The rates change at time $t = t_c$. Thus at $t < t_c$, $N_2 = 0$, and the derived Avrami formulation will apply. However, at time $t \geq t_c$ the steady-state nucleation rate will be given by

$$N(t) = N_1 t_c + N_2(t - t_c) \quad (9.69)$$

With this nucleation rate the Avrami form becomes

$$1 - \lambda(t) = 1 - \exp\left[\frac{-\pi}{3}\frac{\rho_c}{\rho_l} G^3 (N_1 t_c + N_2 t - N_2 t_c) t^3\right] \quad (9.70)$$

for three-dimensional lineal growth.

The modifications that have been proposed to bring the Avrami formulation into closer agreement with experiment are phenomenological in nature. The main focuses have been on the impingement calculation and the introduction of multi-stage processes. They do not take into account the melt structure and the unique features of polymer chains. Dunning, in 1954, questioned the importance of the basic Avrami premise to polymer crystallization.(62) The issue raised was the importance of the impingement of one crystallite upon another in the termination process when long chain molecules were involved. Other possibilities that are unique to polymers were suggested that could lead to a continuous reduction in the overall crystallization rate with time and provide an effective termination to the transformation.

It was pointed out earlier that the changing constitution of the melt that occurs as the crystallization progresses should have an important influence on the kinetics. This is due to the fact that the topological defects, as well as surrounding regions, cannot be incorporated into the crystal lattice. In effect, there will be liquid-like, or amorphous, regions that cannot be transformed. It can then be assumed that initially, and throughout the course of the transformation, a constant fraction λ^* of the chain units cannot be crystallized. When the fraction transformed plus untransformable

[5] This concept can also be applied to the growth process either by the limiting volume that is available or by nucleation controlled growth.

approaches unity there will be an effective termination to the crystallization. There is therefore a cessation mechanism that is unique to long chain molecules. This effect will be minimal at the early stages of the transformation but will manifest itself more as the crystallization progresses. In contrast to monomeric systems, termination does not require the impingement of crystallites. Other factors can thus intervene before impingement becomes important. It is of interest to apply this concept to both the free-growth and Avrami formulations.

Following the earlier free-growth development, the number of nuclei generated in time $d\tau$ can now be expressed as

$$dn = N(\tau)[\lambda(\tau) - \lambda^*]\,d\tau \tag{9.71}$$

where $N(\tau)$ is the steady-state nucleation rate per transformable volume. The nucleation rate, $N(\tau)$, is still constant. However, the number of nuclei generated in the interval $d\tau$ is now reduced by the factor $\lambda(\tau) - \lambda^*$. The level of crystallinity as a function of time is expressed as

$$1 - \lambda(t) = \int_0^t v(t,\tau)\,N(\tau)(\lambda(\tau) - \lambda^*)\,d\tau \tag{9.72}$$

for three-dimensional linear growth. Here $v(t,\tau)$ is given by Eq. (9.4). A comparison of Eq. (9.72) with Eq. (9.3) indicates that as the transformation progresses the crystallization level will be less than calculated when the melt structure is not taken into account. However, the complete solution of Eq. (9.72) does not yield termination.(62a)

When applying this concept to the Avrami formulation the phantom nuclei have to be taken into account. They will now be located in both the untransformable as well as the transformed regions. The analysis proceeds as previously with either Eq. (9.26) or (9.27). Neither the fraction transformed nor untransformable appear explicity in either of these equations. Hence, in order to introduce the influence of the untransformable fraction on the extent of the transformation a decrease in $N(\tau)$ with the level of crystallinity, or time, needs to be postulated. Several efforts have been made to resolve the problem in this manner.(63–66) However, they all involve postulating an arbitrary retardation in either the nucleation rate, the lineal growth rate, or in both. There is no physical or molecular basis for the functions that have been proposed. The normalization procedure that led to Eq. (9.42) was an effort to account for the fractions of untransformable material in the kinetics. The formalism of nucleation and growth has also been applied to the development of the stable crystalline state from a metastable one, rather than from the pure melt.(66a) Formally, the analysis is qualitatively similar to the two-stage series process that was discussed previously.(54,55)

9.6 Further experimental results: overall crystallization

The Avrami exponent n plays a key role in the analysis of overall crystallization kinetics. Therefore, it is of interest to ascertain the physical significance, if any, of this parameter. The summary given in Table 9.1 shows that an n value does not represent a unique set of nucleation and growth processes. However, there is the possibility that a rationale can be made between the n value, the crystallization mechanisms, and the morphology and structure in the crystalline state. In order to explore this possibility a representative set of n values for different polymers has been compiled in Table 9.3.[6] In a few rare cases the n value changes with the crystallization temperature. This finding reflects a lack of superposition, and that changes occur in either or both the nucleation and growth processes with temperature. Of particular importance is the dependence of n on molecular weight, and the relation, if any, between this exponent and the supermolecular structure that evolves.[7]

Two low molecular weight fractions of linear polyethylene, $M = 4200$ and 5800, have n values of 4 as does the n-alkane $C_{192}H_{386}$.(34,67) Thus, the overall crystallization kinetics of the high molecular weight n-alkanes are similar if not identical to those of low molecular weight polyethylene fractions of comparable molecular weight. Low molecular weight fractions of poly(ethylene terephthalate) can also be fitted with an Avrami exponent of 4.(68) The data for most of the low molecular weight polyether fractions can also be fitted with $n = 4$. Different investigators have reported different n values for low molecular weight poly(ethylene oxides), similar to what has been found for the higher molecular weight homopolymers.(64,69,70) Therefore, a subjective decision with regard to the n value has been made from the data available for this polymer. There appears to be a strong tendency in the results for low molecular weights to adhere to $n = 4$.

The relation between n and the supermolecular structure that is formed is of interest. The low molecular weight polyethylene fractions that have n values of 4 form a unique type of superstructure. They can be represented by either rods or a rod-like assembly of the lamellar crystallites.(5) For molecular weights 7800 and 11 500, crystallized at high temperatures, 129 °C and 130 °C, n is also equal to 4 and similar superstructures are observed. In this molecular weight range the high crystallization temperatures are borderline between the different type superstructures that are formed by linear polyethylene.(5) As the crystallization temperature

[6] The Avrami exponent n is evaluated by the curve fitting procedure that has been described. It only applies to that portion of the isotherm that fits Eq. (9.31a). However, there are situations where a subjective decision has to be made. These are cases where, although a significant portion of the transformation can be fitted by $n = 3$, about half the transformation is also satisfied by $n = 4$. The problem is whether $n = 4$ represents the actual mechanism with deviations from the Avrami equation ensuing.

[7] A detailed discussion of the supermolecular structures, or morphology, and their dependence on molecular weight and crystallization conditions will be given in Volume 3.

Table 9.3. *Values of Avrami exponent n for selected polymers*

Polymer	n	Reference
Ethylene (linear)		
$M = 4200$	4	
$M = 5800$		
		a
$M = 7800\text{–}1.2 \times 10^6$	4	
	3^a	
$M = 3 \times 10^6\text{–}8 \times 10^6$	2	
Ethylene oxide		
$M = 3.5 \times 10^5\text{–}1 \times 10^7$	3	b
$M = 3.5 \times 10^4$	2	b
$M = 4 \times 10^3\text{–}1.64 \times 10^5$	3	b′
Phenylene sulfide		
$M = 24\,000$	2	c
$M = 44\,000$	2	d
$M = 49\,000$	3	c
$M = 64\,000$	3	c
$M = 104\,000$	2	d
1,3-Dioxolane		
$M = 8800\text{–}120\,000$	3	e
Trimethylene oxide		
$M = 8000\text{–}157\,000$	3	f
Hexamethylene oxide		
$M = 2000\text{–}33\,000$	4	g
Octamethylene oxide		
$M = 3400\text{–}90\,000$	4	h
Decamethylene oxide		
$M = 2000\text{–}150\,000$	4	i
3,3-Dimethyl oxetane		
$M = 18\,500\text{–}130\,000$	3	j
Hexamethylene adipamide		
$T_c = 162\,°\text{C}$ to $250\,°\text{C}$	3	k
$T_c = -36\,°\text{C}$ to $-24\,°\text{C}$	2	l
$T_c = 235\,°\text{C}$ to $247\,°\text{C}$	3	l
$T_c = 251\,°\text{C}$ to $252\,°\text{C}$	4	l
Nylon-11	4	m
4,6-Urethane	2.3	n
New polyimide	4	o
Polyimide[b]	~2	p
Polyimide LARC-CPI	2	q
Hexamethylene adipate	3	r
Butylene terephthalate	3	s
Decamethylene terephthalate	3–4	t

(*cont.*)

Table 9.3. (*cont.*)

Polymer	n	Reference
Ethylene terephthalate		
T melt 294 °C[c]		
$T_c = 235$–$250\,^\circ$C	4	u
$T_c = 170$–$220\,^\circ$C	3	
$T_c = 110$–$160\,^\circ$C	2	
$T_c = 108\,^\circ$C	3	
T melt 268 °C		
$T_c = 250\,^\circ$C	4	u
$T_c = 106$–$249\,^\circ$C	3	
T melt 290 °C		
$T_c = 200$–$225\,^\circ$C	3	s
T melt 290 °C		
$T_c = 211$–$223\,^\circ$C	3	v
T glass		
$T_c = 112$–$123\,^\circ$C	3	
ϵ-Caprolactone	4	w
Ether ether ketone	3	x
Ether ether ketone ketone	2	y

[a] $M = 7800$ and 11 500 have values of $n = 4$ for crystallization temperatures of 129.1 °C and 130.1 °C, at lower crystallization temperatures $n = 3$.
[b] Polyimide of 4,4′-oxydiphthalic acid and ethylene glycol.
[c] Initial melt temperature prior to crystallization.

References

a. Ergoz, E., J. G. Fatou and L. Mandelkern, *Macromolecules*, **5**, 147 (1972).
b. Allen, R. C., Masters Thesis, Florida State University, December 1980.
b′. Jadraque, D. and J. M. G. Fatou, *An. Quim.*, **73**, 639 (1977).
c. Lopez, L. C. and G. L. Wilkes, *Polymer*, **29**, 106 (1988).
d. Risch, B. G., S. Srinivas, G. L. Wilkes, J. F. Gerbel, C. Ash, S. White and M. Hicks, *Polymer*, **37**, 3623 (1996).
e. Alamo, R., J. G. Fatou and J. Guzman, *Polymer*, **23**, 374 (1982).
f. Perez, E., A. Bello, J. G. Fatou, *Coll. Polym. Sci.*, **262**, 605 (1984).
g. Marco, C., J. G. Fatou and A. Bello, *Polymer*, **18**, 1100 (1977).
h. Marco, C., J. G. Fatou, A. Bello and A. Bianco, *Polymer*, **20**, 1250 (1979).
i. Tinas, J. and J. G. Fatou, *An. Quim.*, **76**, 3 (1980).
j. Perez, E., J. G. Fatou and A. Bello, *Coll. Polym. Sci.*, **262**, 913 (1984).
k. Magill, J. H., *Polymer*, **12**, 221 (1961).
l. Hartley, P. D., F. W. Lord and L. B. Morgan, *Ricerca Sci. Suppl. Symp. Int. Chim. Macromol. Milan-Turin*, **25**, 577 (1954).
m. Kahle, B., *Z. Electrochem. Phys. Chem.*, **61**, 1318 (1957).
n. Rohleder, J. and H. A. Stuart, *Makromol. Chem.*, **41**, 110 (1960).
o. Hsiao, B. S., B. B. Sauer and A. Biswas, *J. Polym. Sci.: Pt. B: Polym. Phys.*, **32**, 737 (1994).

Table 9.3. (*cont.*)

p. Heberer, D. P., S. Z. D. Cheng, J. S. Barley, S. H-S. Lien, R. G. Bryant and F. W. Harris, *Macromolecules*, **24**, 1890 (1991).
q. Muellerfera, J. T., B. G. Risch, D. Rodriques, G. L. Wilkes and D. M. Jones, *Polymer*, **34**, 789 (1993).
r. Gilbert, M. and F. J. Hybart, *Polymer*, **15**, 407 (1974).
s. Pratt, C. F. and S. Y. Hobbs, *Polymer*, **17**, 12 (1976).
t. Sharples, A. and F. L. Swinton, *Polymer*, **4**, 119 (1963).
u. Hartley, F. D., F. W. Lord and L. B. Morgan, *Phil. Trans.*, **A247**, 23 (1954).
v. Chan, T. W. and A. I. Isayer, *Polym. Eng. Sci.*, **34**, 461 (1994).
w. Chynoweth, K. R. and Z. H. Stakurski, *Polymer*, **27**, 912 (1986).
x. Cebe, P. and S. D. Hong, *Polymer*, **27**, 1183 (1986).
y. Wang, J., J. Cao, Y. Chen, Y. Ke, Z. Wu and Z. Mo, *J. Appl. Polym. Sci.*, **61**, 1999 (1996).

is lowered n becomes 3. Concomitantly, well-developed spherulitic structures are observed for molecular weights up to and including 1.2×10^6 with $n = 3$. The order within the spherulites decreases, however, with increasing molecular weight. For higher molecular weights, $3.8–8 \times 10^6$, $n = 2$ and no defined superstructures are discerned. Consistent with this n value only randomly oriented lamellae that are not correlated with one another are observed.(4,5) For linear polyethylene, there is a very good correlation between the superstructures that are observed and the Avrami exponent n.

Studies with poly(ethylene oxide) give qualitatively similar results.(49,71) At the high molecular weights, 3.8×10^6 and greater, $n = 3$ and spherulites are observed. Presumably, if still higher molecular weights were studied, randomly arranged lamellae, with $n = 2$, would be observed following the pattern established by linear polyethylene. Intermediate structures, between spherulites and hedrites, develop at lower molecular weights. Here, the n values of 3 reflect the domination of the spherulitic type structures. At molecular weights 3.5×10^4 and lower $n = 2$, reflecting the fact that only hedrites form. Presumably, these grow by a two-dimensional rate controlling process. Thus, there is also a correlation between the Avrami exponent n and the superstructure that evolves for this polymer.

For poly(1,3-dioxolane) $n = 3$, independent of the isothermal crystallization temperature and molecular weight, in the range 8800–120 000.(38,72a) A spherulitic type growth is implied by these results. However, light microscopic studies show that there is a superstructure transition, from spherulite to hedrite, at the elevated crystallization temperatures.(38,72) Thus, there must be a change in nucleation type to compensate for the apparent change in growth habit. Alternatively, the growth could still maintain an element of three-dimensional character.

The kinetic data for poly(hexamethylene adipamide) present some interesting correlations.(73) At the highest crystallization temperatures $n = 4$. According to the formal analysis, this value is consistent with sporadic homogeneous nucleation and three-dimensional isotropic spherulitic growth. The spherulite size distribution, as measured by polarized light microscopy, and the fact that the number of spherulites increases with crystallization time confirm this conclusion. At lower crystallization temperatures, 235 °C–247 °C, $n = 3$. In this temperature interval, all the spherulites at any given crystallization time have the same size. This direct observation implies predetermined nucleation initiated at $t = 0$. It is consistent with the measured exponent. At very low crystallization temperatures for this polymer $n = 2$. It is inferred from this result that the crystallization is fibrillar from a sporadic type nucleation.

The studies with poly(ethylene terephthalate) by different investigators demonstrate the influence of the specific sample, the initial state prior to crystallization, and the crystallization temperatures on the kinetics. The results of Hartley *et al.* (74) show that the value of n depends on both the crystallization temperature and the temperature of the melt prior to crystallization. Presumably, at the higher melt temperatures more of the potential heterogeneous nuclei are destroyed. This presumption is borne out by light microscopic observation. For example, for melt temperature of 294 °C $n = 4$ at the highest crystallization temperatures. Consistent with this value, microscopic studies show a distribution of spherulite sizes, the number of which increases with time. At this initial high melt temperature (294 °C) and a crystallization temperature of 220 °C the spherulites are all of the same size and their number does not change with the crystallization time. This microscopic observation is consistent with a kinetic value of $n = 3$. For crystallization at 110 °C $n = 2$. Under these crystallization conditions no discernible or definitive structures are observed under the microscope. This observation reflects the large number of nucleation acts that occur at this low temperature that lead to an undefined fine grained structure. The microscopic observations after crystallization at 108 °C were ambiguous and did not allow for a comparison with the determined value of $n = 2$.(74) In contrast, at a lower melt temperature, 268 °C, and crystallization temperature of 245 °C a value of 3 is observed instead of 4. Concomitantly, a fine grained structure is observed indicating the presence of a large number of pre-determined nuclei. The complementary microscopic and kinetic studies with poly(hexamethylene adipamide) and poly(ethylene terephthalate) demonstrate that in many situations there is a good correlation between the Avrami exponent and the superstructure that evolves.

The overall crystallization kinetics, accompanied by morphological observations, have also been observed for several different polyimides.(42,45,75) For New polyimide $n = 4$, and well-developed spherulites are formed.(45) The polyimide LARC-CPI gives $n = 2$ at all crystallization temperatures, and spherulitic structures are not observed.(75) Rather a hedrite or sheaf-like structure consistent with two-dimensional growth forms. The kinetic and morphological behavior of polyimides based on 4,4′-oxydiphthalic anhydride (ODPA) and ethylene glycol have also been reported.(42) The kinetic data for the three polyimides of this type that were studied could be fitted to an n value of 2. The complete transformation was used to obtain this value. However, for about the first one-third of the total transformation, using the fitting procedures described, the data agree very well with $n = 3$. This n value is consistent with microscopic studies that show the formation of three-dimensional supermolecular structures. The possible reasons for the protracted rate that subsequently develops have already been discussed in detail.

This survey of a variety of polymers demonstrates that when n values are obtained in an appropriate manner there is a self-consistency between the Avrami exponent and the supermolecular structures that are observed.(76) There are, however, cases where such a correlation cannot be made even for the same polymer.(46,76a,76b) This inconsistency can be attributed to different samples of the same polymers as well as differing roles played by heterogeneities.

There have been many reports of nonintegral values for n. In some cases these are natural consequences of heterogeneous nucleation of the kinds that have been described. Very often they are the result of trying to fit the Avrami formulation to the complete transformation. This procedure ignores the basic theoretical assumptions that are involved and the deviations from the Avrami theory that are observed. Values of n less than 1 have also been reported.(77) Although $n = 1$ can be tolerated by the Avrami approach, lower values cannot be explained in any straightforward manner. However, caution must be taken to insure that, in analyzing kinetic data that give these low n values, one is dealing with homopolymers of flexible chains and not a copolymer or a liquid crystal forming polymer. Moreover, the method of measurement must be commensurate with the time scale of the crystallization. Otherwise, it is quite possible that only the final portion of the crystallization is observed. When this occurs an apparent low n value will be deduced.

Studies of the overall crystallization kinetics also give important information as to the role played by molecular weight and polydispersity in the crystallization process. The influence of molecular weight on the crystallization rate, when fractions are used, is illustrated in Fig. 9.23a for linear polyethylene (34) and in Fig. 9.23b for poly(tetramethyl-p-silphenylene siloxane).(78) In these figures the log of the overall crystallization rate, expressed in terms of $1/\tau_{0.01}$ or $1/\tau_{1/2}$, is plotted against the log of the molecular weight for a series of isothermal crystallization temperatures. In both of these examples well-defined maxima are observed. The location of the maximum is dependent on the crystallization temperature, shifting to lower molecular weights as the crystallization temperature is reduced. The depth of the maximum is also a function of the crystallization temperature. Only a very shallow maximum is observed at the lower crystallization temperatures. The

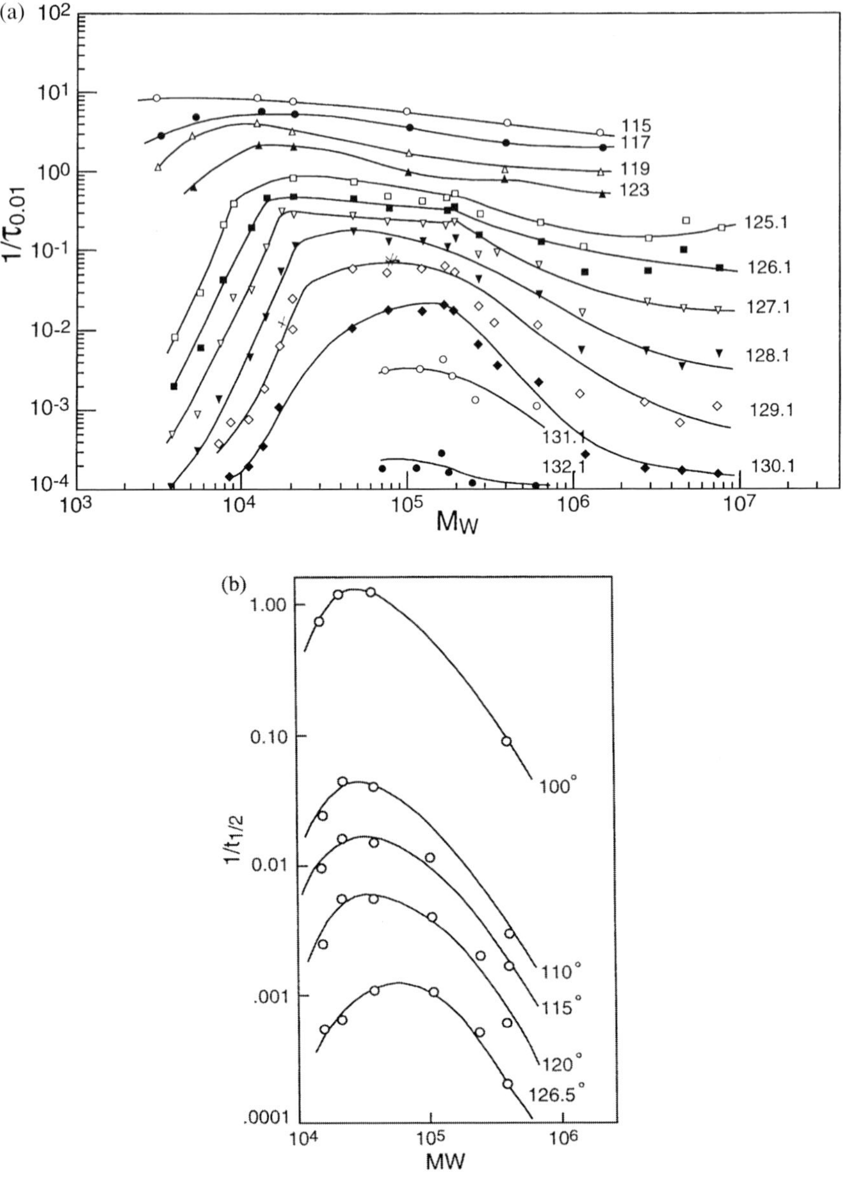

Fig. 9.23 Dependence of overall crystallization rate on molecular weight for a series of isothermal crystallization temperatures. (a) Plot of log $1/\tau_{0.01}$ against log M_{w} for fraction of linear polyethylene at indicated crystallization temperatures.(34) (b) Plot of log $1/t_{1/2}$ against log M_{w} for fraction of poly(tetramethyl-*p*-silphenylene siloxane) at indicated crystallization temperature. (From Magill (78))

results for poly(ethylene oxide) are very similar to those of linear polyethylenes illustrated.(39) Other poly(ethers) behave in a similar manner.(79)

Since the linear polyethylene data are extensive we examine them in some detail. In the lower molecular weight range the crystallization rates increase over several decades with increasing molecular weight until a maximum in the rate is reached. The molecular weight at the maximum value depends on the crystallization temperature. The maximum is in the range of $M = 1\text{–}2 \times 10^5$ for the highest crystallization temperature. The molecular weight at the maximum steadily decreases as the crystallization temperature is lowered. At the lower isothermal crystallization temperatures the maximum in the rate occurs at $M = 1\text{–}2 \times 10^4$. At the left-hand side of the maximum there is an almost linear relation, in the log-log plot, between $1/\tau_{0.01}$ and molecular weight. The slope in this region is essentially independent of the crystallization temperature. In contrast, at the right side of the maximum there is a complex relationship between the crystallization rate and molecular weight. The specifics are dependent on the temperature. At the lower crystallization temperatures in this region, there is only a relatively small change in the time scale with molecular weight. However, at the higher crystallization temperature there is initially a very sharp decrease in $1/\tau_{0.01}$ with molecular weight, followed by a definite leveling off at the very highest molecular weights. The molecular weight at which the leveling off occurs is also dependent on the crystallization temperature. At the higher crystallization temperatures the leveling off begins at about $M \simeq 10^6$, while at the lower ones it begins at lower molecular weights. For example, at a crystallization temperature of 129 °C the time for 1% of crystallinity to develop is about 10 min at the rate maximum. This time increases to about 10^3 min for $M \simeq 10^6$ and levels off at this value up to the highest molecular weight studied, $M = 8 \times 10^6$. In contrast, for crystallization at 125 °C, $\tau_{0,01}$ is only 1 min at the maximum rate. It increases to about 5 min at approximately $M = 7 \times 10^5$ and then levels off. At the very lowest crystallization temperature there is only a slight change in the rate over a thousand-fold variation in chain length.

The dependence of the crystallization rate on molecular weight and crystallization temperature is obviously complex. Depending on the polymer being studied, and the crystallization temperature, the crystallization rate can either increase, decrease, or show very little change with molecular weight. Studies over a limited molecular weight range result in the observation of only one of these possibilities. The maximum in the crystallization rate with molecular weight at constant temperature indicates that at least two competing mechanisms are operative. An analysis of these data requires the results of similar spherulite growth rate studies as a function of molecular weight. In addition, a discussion of nucleation theory appropriate to long chain molecules is needed. Therefore, a detailed analysis of the influence of molecular weight on the crystallization rate will be postponed until these subjects have been discussed.

It has been emphasized that the crystallization process is continuous despite the fact that several distinctly different regions can be recognized. Typical isotherms, as well as simultaneous wide- and small-angle x-ray synchrotron studies with linear polyethylene fractions have demonstrated the continuity. There is no indication of any discontinuity during the course of the crystallization or the abrupt doubling, or quantization of the crystallite thickness.(80,81) The completion of the transformation (Region III) is characterized by a flat or "tail" portion in the isotherm. Here the crystallinity level increases extremely slowly. It can be represented by a linear relation on a log time basis. It would take many decades of time to achieve a 10% increase in the level of crystallinity in this portion of the transformation. Although the contribution of the tail portion to the total crystallinity is small, some important structural changes that affect many physical properties do in fact take place. In addition to the formation of small crystallites, thickening of the already existing ones also takes place.(82) A detailed consideration of these factors, and their influence on properties, will be discussed in conjunction with the morphology and structure of crystalline polymers (Volume 3).

9.7 Nonisothermal crystallization

Attention up to now has been focused on isothermal crystallization. This mode of crystallization is most amenable to theoretical analysis and comparison with experimental results. However, crystallization can also occur by cooling from the melt or heating from the glassy state. Nonisothermal crystallization has many practical applications. It is thus of interest to analyze such processes. There has been a great deal of theoretical and experimental activity in this area, which has been extensively summarized.(66,83,84) Therefore, focus here will be on the basic theoretical principles involved and pertinent experimental results.

Most of the theories are based on a derived Avrami expression, such as Eq. (9.31a), with a variety of modifications.(66,83,84) Cooling from the melt can be thought of as passing through a succession of isotherms each at a given crystallinity level that corresponds to the particular time and temperature. Since the isotherms are superposable, a typical one can be represented by the isotherm illustrated in Fig. 9.18. As long as the nonisothermal crystallization process is restricted to Region I (in Fig. 9.18) at all temperatures, the assumption of a derived Avrami, or the equivalent free-growth expression, is valid. It should be recalled that in this region the free-growth and Avrami expressions fit the data equally well. However, when at a given time and temperature the transformation takes place in Region II or III, the basic underlying assumption, common to most theoretical developments, is no longer valid. The structural factors in the melt that cause deviation from Region I will now become important.

Underlying any theoretical development is the temperature, $T(t)$, at time t. In general

$$T(t) = T_0 - \psi(t) \tag{9.73}$$

Here T_0 is the initial melt temperature and $\psi(t)$ is the time function for either cooling or heating. The simplest case to consider is a constant cooling rate where $\psi(t) = \phi t$, ϕ being a constant. Using the Evans approach to isothermal crystallization kinetics, Ozawa showed that when a sample is cooled from the equilibrium melting temperature to the temperature T, at a constant rate ϕ, the volume fraction of transformed material can be expressed as (85)

$$1 - \lambda(t) = 1 - \exp\left\{-\frac{4\pi}{\phi^3}\int_{T_m^0}^{T} N(\theta)\,[V(T) - V(\theta)]^2\, G(\theta)\, d\theta\right\} \tag{9.74}$$

for three-dimensional growth. Here $N(\theta)$ is the number of nuclei per unit volume that are activated between temperatures T_m^0 and θ, $G(\theta)$ is the radial growth rate and $V(T)$ is given by the expression

$$V(T) = \int_{T_m^0}^{T} G(\theta)\, d\theta \tag{9.75}$$

In this formulation the fraction transformed will depend on the type of nucleation that is taking place. For instantaneous nucleation, the nucleation rate, $N(\theta)$, is independent of time and cooling rate. It only depends on the temperature. Consequently, under these circumstances

$$\ln\{-\ln[1 - \lambda(T)]\} = C_1 - 3\ln|\phi| \tag{9.76}$$

The constant C_1, termed the cooling crystallization function, is given by

$$\ln 4\pi + \ln\left|\int_{T_m^0}^{T} N(\theta)\,[V(T) - V(\theta)]^2\, G(\theta)\, d\theta\right| \tag{9.77}$$

For noninstantaneous nucleation, $N(\theta)$ is a function of temperature and time. The specifics need to be prescribed. For steady-state nucleation, where nuclei are formed at a constant rate per unit volume, $\dot{N}(\theta) = N(\theta)$, it follows that

$$\ln\{-\ln[1 - \lambda(T)]\} = C_2 - 4\ln|\phi| \tag{9.78}$$

where

$$C_2 = \ln 4\pi + \ln \left| \int_{T_m^0}^{T} \left[\int_{T_m^0}^{\theta} \dot{N}(u)\, du \right] [V(T) - V(\theta)]^2\, G(\theta)\, d\theta \right| \tag{9.79}$$

Equations (9.76) and (9.77) can be written in general form as

$$\ln\{-\ln[1 - \lambda(T)]\} = C - n \ln|\phi| \tag{9.80}$$

where n is the Avrami exponent. The mathematics involved in the Ozawa calculation is straightforward. However, the implicit assumption is made that all of the crystallization takes place in Region I, and the particular type, or types, of nucleation has to be assumed. It is also possible that the type of nucleation will change during the time course of the crystallization. If these assumptions are not fulfilled, disagreement with experiment can be anticipated. Moreover, the time lag to reach the desired temperature on cooling (or heating) has not been taken into account. It is implicitly assumed that the temperature is reached instantaneously.

The approach taken by Nakamura *et al.*, in another theoretical development, is also based on the Avrami approach.(86,87) However, in this analysis the general Avrami expression is used. Therefore, the integral in Eq. (9.27) has to be evaluated. The evaluation of this integral has been simplified in the theory by assuming the isokinetic condition. This condition requires that the ratio $G(T)/N(T)$ be a constant. With this requirement, the integral can be evaluated. The result is that the relative extent of the transformation can be expressed as (87,88)

$$\frac{1 - \lambda(T)}{1 - \lambda(\infty)} = 1 - \exp\left[- \int_0^t [K(T)\, d\tau]^n \right] \tag{9.81}$$

where

$$K(T) = k(T)^{1/n} \tag{9.82}$$

the quantity k being defined by Eq. (9.28). Equation (9.81) can be expressed as a function of temperature by introducing the constant cooling rate. Thus, Eq. (9.81) can be written as

$$\frac{1 - \lambda(T)}{1 - \lambda(\infty)} = 1 - \exp\left\{ - \left[\frac{1}{\phi} \int_{T_m^0}^{T} K(T)^{1/n}\, dt \right]^n \right\} \tag{9.83}$$

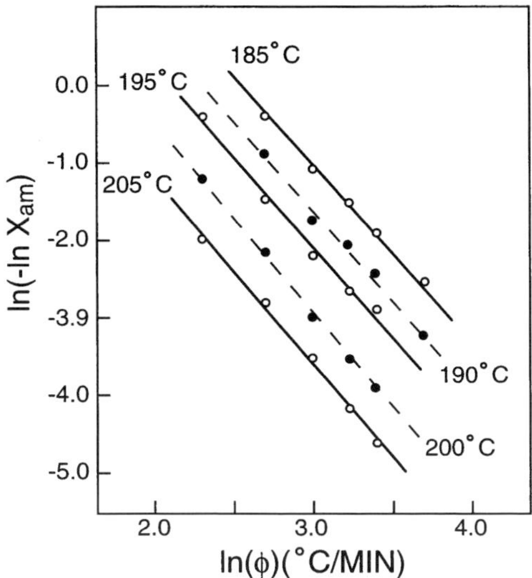

Fig. 9.24 Plot according to Eq. (9.80) of $\ln(-\ln x_{\rm am})$ against $\ln\phi$ for poly(phenylene sulfide), $M = 75\,000$ at indicated temperatures. (From Lopez and Wilkes (43))

The validity of the general Avrami expression over the range of crystallinities studied is assumed, as is the isokinetic process. Hieber has pointed out that the Nakamura equation, Eq. (9.81), contains both the Avrami and Ozawa limits.(88a)

The Ozawa and Nakamura theories are basic to this problem. It is, therefore, appropriate to compare them with experimental results before discussing modifications that have been made. A detailed summary of this comparison has been given.(84) Some typical examples follow. Figure 9.24 is a plot, according to Eq. (9.80), for the nonisothermal crystallization of poly(phenylene sulfide).(43) A parallel set of straight lines results that are completely consistent with the theory. These results indicate that in this case the basic premise of the theory is fulfilled. Furthermore, the Avrami n is in good agreement with the value obtained from isothermal crystallization kinetics. Good agreement with the Ozawa theory is also found with poly(caprolactam) (88), isotactic poly(propylene) (89–93), poly(vinylidene fluoride) (94) and New TPI polyimide,(94a) among others. Reasonable agreement was obtained between the n value and that obtained in isothermal experiments. The results for isotactic poly(propylene) give indication that there is a shift from homogeneous to heterogeneous nucleation as the temperature is increased.(91) The results that agree with the Ozawa theory imply that the crystallization is restricted to Region I. Conflicting results have been reported for poly(ethylene terephthalate). Agreement was found with Ozawa's theory when a narrow and modest cooling rate was used.(85) However, deviations were found at the beginning and end of the crystallization when a wide range in cooling rates was studied.(95)

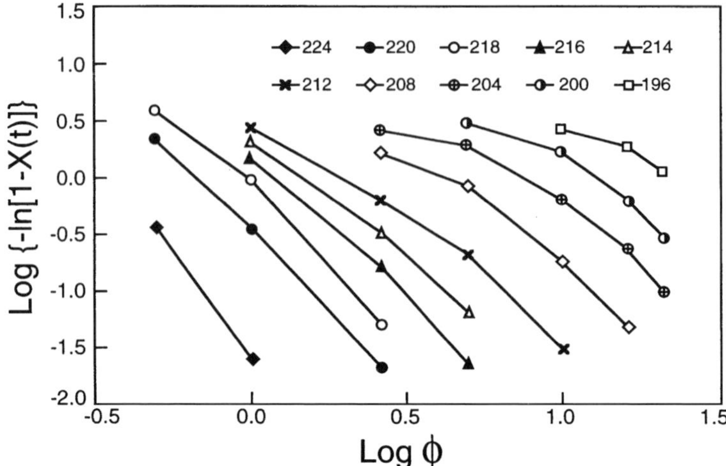

Fig. 9.25 Plot according to Eq. (9.80) of $\log\{-\ln[1-x(t)]\}$ against $\log\phi$ for poly(butylene naphthalene 2,6-dicarboxylate) at indicated temperatures. (From Papageorgiou and Karayannidis (96))

A similar plot for poly(butylene naphthalene 2,6-dicarboxylate) is given in Fig. 9.25.(96) In contrast to the previous figure, these plots are only linear at the lower portion and curvature is observed at the higher levels of crystallinity. These results indicate that in this case the crystallization has not been limited to Region I. Curvature in the Ozawa type plot has also been observed with poly(aryl ether ether ketone) (40), poly(aryl ether ether ketone ketone) (97) and poly(aryl ether ether sulfide).(97a) Curvature and deviation from the theory will be observed, if crystallization occurs beyond Region I, because the derived Avrami equation is no longer valid.

Hieber has shown that a number of studies of nonisothermal crystallization of isotactic poly(propylene) and poly(ethylene terephthalate) can be treated by the Nakamura model.(88a) This model has also been shown to hold for syndiotactic poly(styrene) and poly(caprolactone). (98,98a) The results for linear polyethylene have not been conclusive. All of the studies have involved unfractionated polymers. In one study curvature was observed in the Ozawa type plot.(89) In another study, with a different sample, the Nakamura model was shown to hold.(88)

A variety of theoretical modifications have been made to the two basic theories.(99–100f) These include utilizing the Tobin rather than the Avrami relation, modifying the Avrami equation, introducing a linear combination of homogeneous and heterogeneous nucleation, including an induction or lag time, modifying the isokinetic assumption, and accounting for secondary crystallization among others. Progress in this area requires studies with either molecular weight fractions or well-characterized distributions with the samples being devoid of additives. The wide range in cooling rates needs to be studied. In this way basic experimental facts will be established that are consistent with a given polymer.

9.8 Spherulite initiation and growth: general concepts

Another useful way to study crystallization kinetics is to measure the rate at which supermolecular structures form and grow. Supermolecular structures represent the organization of individual crystallites into well-defined, three-dimensional arrays. Most common among the observed superstructures are spherulites. They represent a spherical organization of the crystallites. Although spherulites are a common manifestation of polymer crystallization they are by no means universal. There are important restraints of molecular weight, distribution, chain structure and crystallization temperature on spherulite formation.(5,6,7) In particular, spherulites are not observed with high molecular weight polymers, which in general do not form superstructures. Low molecular weight species form other kinds of structures. Thus, there is a limitation on the crystallization conditions and molecular constitutions that can be studied by this technique. Since a given spherulite is not completely crystalline the relation between the nucleation and growth of a crystallite and that of a spherulite is not *a priori* obvious.

Carefully conducted experiments have shown that it is possible for spherulites to develop sporadically in both time and space in thin polymer films.(22,23,101,102) Repetitive experiments have indicated that spherulites do not necessarily form in identical positions if complete melting of the sample is ensured. However, a study of poly(decamethylene terephthalate) has shown that although spherulites are formed sporadically in time, they appear in identical positions within the sample.(103) Experimental evidence showed that for this sample the spherulitic centers are initiated from a fixed number of heterogeneities. There is a strong tendency for spherulites to appear in the same position in the field of view after successive crystallizations.(73,74,104–106) In some cases this observation is solely a result of incomplete melting.(33,36,102) In others, it can also be due to the presence of a finite number of nucleation catalysts in the polymer melt.

It is universally observed that in the vicinity of the melting temperature the rate at which spherulites are generated depends very strongly on the crystallization temperature. The rate increases very rapidly as the temperature is lowered. As was pointed out earlier, the rate at which spherulite centers are generated in poly(decamethylene adipate) decreases by a factor of 10^5 as the crystallization temperature is raised from 67 to 72 °C.(22,107) The effect is quite general and is illustrated in Fig. 9.26 for the crystallization of poly(hexamethylene adipamide).(108) Here the number of spherulites that are formed per unit volume is plotted against the time for a series of

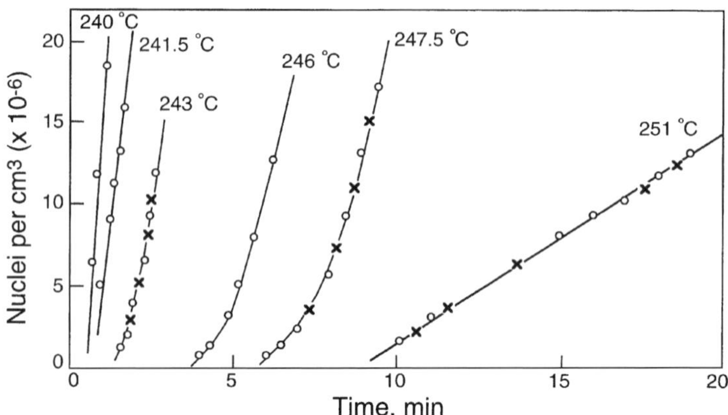

Fig. 9.26 Plot of spherulite nucleation rate in poly(hexamethylene adipamide), $M_n = 14\,600$, at indicated crystallization temperatures. (From McLaren (108))

temperatures. The strong negative temperature coefficient for spherulite initiation in this region is quite apparent.

An impressive body of experimental evidence has demonstrated that the radius of a growing spherulite of a pure homopolymer increases linearly with time at a fixed temperature. A typical example was illustrated in Fig. 9.8 for the spherulitic growth rate of poly(ethylene adipate).(23) The linear growth rates are clearly defined and are typical of other polymers. The linear increase of the spherulite diameter with time implies that the growth is not diffusion controlled. It indicates that the rate of volume change is proportional to the surface area. Growth is thus controlled by processes that occur at the boundary.(109) Figure 9.8 also illustrates that the linear growth rate is very sensitive to the crystallization temperature. In the vicinity of the melting temperature there is a very strong negative temperature coefficient. As the crystallization temperature is lowered for this polymer the growth rate increases. Typically, when isothermal crystallizations are carried out over a wide temperature range a maximum is observed in the spherulite growth rate. This characteristic was illustrated in Fig. 9.9 for poly(tetramethyl-*p*-silphenylene siloxane). Other examples are found with different molecular weight fractions of poly(ethylene terephthalate),(110) isotactic poly(styrene) (111) and many other polymers. With a further decrease in the temperature, as the glass temperature is approached, a marked retardation in the rate of spherulite growth takes place. This type of behavior is independent of molecular weight. Thus, the temperature variation in the spherulitic growth rate is similar to the temperature coefficient of the overall rate of crystallization that was described previously. The temperature at which the maximum in that rate occurs for a given polymer is similar for both types of rate measurements. The plot in Fig. 9.9 indicates that at a fixed crystallization temperature the spherulite growth rate depends on molecular weight. This molecular weight dependence will be discussed subsequently along with that of the overall crystallization.

9.9 Nucleation theory: temperature coefficient in vicinity of T_m^0

9.9.1 Low molecular weight nonchain molecules

The experimental results make evident that the crystallization rate of polymers is strongly influenced by the crystallization temperature. There are several points of interest that are common to all crystallizing polymers. Characteristically, a strong negative temperature coefficient is observed in the vicinity of the equilibrium melting temperature. Because of its low rate in the vicinity of the melting temperature, crystallization can only be carried out at temperatures well below the melting temperature, i.e. at large undercoolings. This is in marked contrast to low molecular weight substances. Another characteristic of the crystallization process is the maximum observed in the rate when the crystallization is carried out over an extended temperature range. A significant retardation in the crystallization rate takes place as the glass temperature is approached.

The analysis of the overall crystallization kinetics led to the conclusion that, in common with low molecular weight species, crystallization of polymers is governed by nucleation and growth processes. This general conclusion establishes the basic framework within which the problem can be analyzed. In principle, two different types of nucleation processes could be operative.(112) One of these is the required initiation of crystallization, termed primary nucleation. There are several different growth mechanisms. These include growth by screw dislocation, growth on an atomically rough surface and growth by nucleation. The latter, termed secondary nucleation, is important to spherulite growth rates, and could differ from the primary one. Material also has to cross the crystal–liquid interface in order for crystal growth to proceed. All of these processes have different temperature coefficients. There is a set of concepts that can be used to explain the temperature behavior of crystallizing substances. These general principles have been established in a straightforward, general manner. However, when applying these principles to a specific class of substances, polymers in our case, one has to be concerned with the details. To rephrase an old saying "the devil is in the details". Because of the general nature of the mechanisms involved one must be careful and exercise caution in reaching conclusions. It can be anticipated that there may be several different mechanisms that can be invoked to explain the same experimental results. Therefore, postulates or assertions that are made need to be carefully examined.

With this introduction, the discussion of the temperature coefficient begins with a detailed analysis of the crystallization kinetics in the vicinity of T_m^0. It will become evident that nucleation is the dominant process in this temperature region. Polymers, and chain molecules in general, bring some unique features to the nucleation problem. These involve the dimensions of a nucleus relative to the molecular length of the chain and the arrangement, or conformation, of the repeating units within the nucleus. To establish the proper background, it is advisable to first examine nucleation processes in low molecular weight, nonchain molecules before discussing polymers. Quantitative descriptions of a variety of nucleation types have been given in detail.(113,114) These apply equally well to low molecular weight substances as well as polymers. We shall select a few, from the many examples that have been analyzed, to illustrate the basic principles involved, and to emphasize the important differences between them.

If two phases A and B of a single component are in equilibrium at T_m and if phase B has the lower free energy at temperatures below T_m, it does not necessarily follow that phase B will spontaneously form when the temperature is lowered. For the macroscopic phase to develop, it must first pass through a stage where it consists of relatively small particles. It is, therefore, possible for small structural entities of phase B to be in equilibrium with phase A at temperatures below T_m. This occurs because the decrease in Gibbs free energy that would normally characterize the development of a large phase can be offset by contributions from the surfaces of the small particles. Hence the relative contributions of the surface area and volume to the Gibbs free energy of the particle determine the stability. Nucleation is the process by which a new phase is initiated within a parent phase, a nucleus being a small structural entity of the new phase. Nuclei can be formed homogeneously in the parent phase by means of statistical fluctuations of molecular clusters. The nucleation process can also be catalyzed by the presence of appropriate heterogeneities. Nuclei can also form preferentially on foreign particles, walls or cavities as well as on the surfaces of already existing crystals. Although heterogeneous nucleation is usually the predominant initiation mechanism for bulk systems, an analysis of homogeneous nucleation in low molecular weight substances provides a deep insight into the principles that are involved.

A simple, but instructive example is found in the formation of a spherical nucleus. The free energy of homogeneously forming such a nucleus from the melt is

$$\Delta G = -(4/3)\pi r^3 \, \Delta G_v + 4\pi r^2 \sigma \tag{9.84}$$

Here ΔG_v is the bulk free energy change per unit volume, σ is the surface free energy per unit area and r is the radius of the sphere. This function is illustrated in Fig. 9.27. As the radius increases, ΔG initially increases until a maximum is reached at $r = r^*$. The free energy then decreases precipitously and becomes

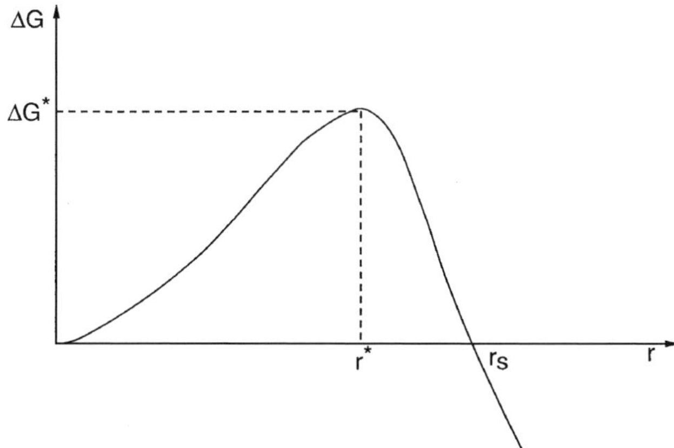

Fig. 9.27 Schematic diagram for the free energy of formation of a spherically shaped nucleus.

negative. At the maximum

$$r = r^* = \frac{2\sigma}{\Delta G_v} \tag{9.85}$$

and

$$\Delta G = \Delta G^* = \frac{16\pi}{3} \frac{\sigma^3}{(\Delta G_v)^2} \tag{9.86}$$

If ΔG_v is expanded in a Taylor series about T_m the first term, $\Delta G_v(T_m) = 0$. The second term is $\Delta G_v(T_m)\, \Delta H_v(T_m - T)/T_m$. Utilizing this expression

$$r^* = \frac{2\sigma T_m}{\Delta H_v\, \Delta T} \tag{9.87}$$

and

$$\Delta G^* = \left(\frac{16\pi}{3}\right) \frac{\sigma^3}{(\Delta H_v)^2} \frac{T_m^2}{(\Delta T)^2} \tag{9.88}$$

Here, $T_m - T$, defined as ΔT, is the undercooling. Thermodynamic stability of the nucleus is attained when $\Delta G = 0$, and

$$r_s = \frac{3\sigma}{\Delta G_v} \tag{9.89}$$

Nuclei smaller than r^* are inherently unstable and disappear from the system since ΔG increases with increasing size. However, nuclei that exceed this critical dimension can easily grow to sizes exceeding the minimum stability requirement. ΔG now decreases very rapidly as the radius increases. ΔG^* represents the free energy barrier that must be surmounted before a stable new phase can develop. We distinguish, therefore, between the dimension of a stable nucleus, $\Delta G = 0$, and the critical dimension at the barrier height, ΔG^*. For spherical geometry the ratio of the respective radii is 3/2. However, the drop in ΔG from r^* to r_s is very steep. The essence of classical nucleation theory, as embodied in Eq. (9.84), is to consider the free energy contribution of a structurally perfect phase of finite size and to add to it the excess free energy due to the presence of a surface.[8] This classical concept, attributed to Gibbs,(115) implies that the nucleus is sufficiently large that its interior is homogeneous and its exterior surface is well defined. When ΔT becomes large, the nuclei sizes become very small and can approach atomic dimensions. The above conditions no longer hold and more sophisticated nucleation theories are necessary.(116,117) However, despite these complications, the classical theory of nucleation has been shown to have widespread applicability, particularly in understanding polymer crystallization.

The simplicity of the foregoing analysis is a consequence of introducing spherical symmetry into the problem. Consequently, only one surface and one surface free energy are involved. There is, however, no real physical requirement that the nucleus be spherical. Asymmetric nuclei, where several different surfaces are involved, can also be treated in the same way. From among the many possibilities, we take as an example the homogeneous formation of a monomeric substance into a cylindrically shaped nucleus. Other three-dimensional geometries can equally well be chosen. These can include cubes and rectangular parallelepipeds. The free energy of forming a cylindrical nucleus that contains ρ molecules in cross-section and ζ molecules in length is given as (12)

$$\Delta G = 2\zeta\sqrt{\pi\rho}\sigma_{\text{un}} + 2\rho\sigma_{\text{en}} - \zeta\rho\Delta G_v \tag{9.90}$$

Here, ΔG_v is the free energy of fusion per molecule, σ_{un} the lateral interfacial free energy per molecule, and σ_{en} the interfacial free energy per molecule at the cylinder ends. The interfacial free energies, σ_{un} and σ_{en}, are those involved in forming a nucleus. They are not the same as the corresponding free energies associated with the mature crystallites of finite size. It is important that they not be identified as such. The contributions of strain and edge free energies are neglected in this example. There is neither a maximum nor minimum in the free energy surface represented by Eq. (9.90). It does, however, contain a saddle point. The coordinates of the saddle point are obtained by setting $(\partial\Delta G/\partial\zeta)_\rho$ and $(\partial\Delta G/\partial\rho)_\zeta$ equal to zero. It then follows that

$$\rho^* = \frac{4\pi\sigma_{\text{un}}^2}{\Delta G_v^2} \tag{9.91}$$

[8] In the sense employed here, the perfect phase possesses the lowest free energy consistent with the constraints imposed on the system. Hence the presence of equilibrium type defects, such as lattice vacancies, is automatically included. The possibility that nonequilibrium type defects may exist in the macroscopic crystal or crystallite that eventually develops is not pertinent to the problem of nucleus formation.

and

$$\zeta^* = \frac{4\sigma_{\mathrm{en}}}{\Delta G_v} \tag{9.92}$$

These dimensions are those of the critical-size nucleus. At the saddle point

$$\Delta G = \Delta G^* = \frac{8\pi\sigma_{\mathrm{un}}^2\sigma_{\mathrm{en}}}{\Delta G_v^2} \tag{9.93}$$

It is convenient to define a set of reduced variables

$$\bar{\rho} = \frac{\rho}{\rho^*} \qquad \bar{\zeta} = \frac{\zeta}{\zeta^*} \qquad \Delta\bar{G} = \frac{\Delta G}{\Delta G^*} \tag{9.94}$$

Equation (9.90) can then be rewritten as

$$\Delta\bar{G} = 2\bar{\zeta}\bar{\rho}^{1/2} + \bar{\rho} - 2\bar{\zeta}\bar{\rho} \tag{9.95}$$

A graphical representation of Eq. (9.95) is given in Fig. 9.28 as a contour map for constant values of $\Delta\bar{G}$. The free energy barrier that must be overcome, $\Delta\bar{G} = 1$, is located at the saddle point, $\bar{\zeta} = \bar{\rho} = 1$. A stable nucleus is achieved when

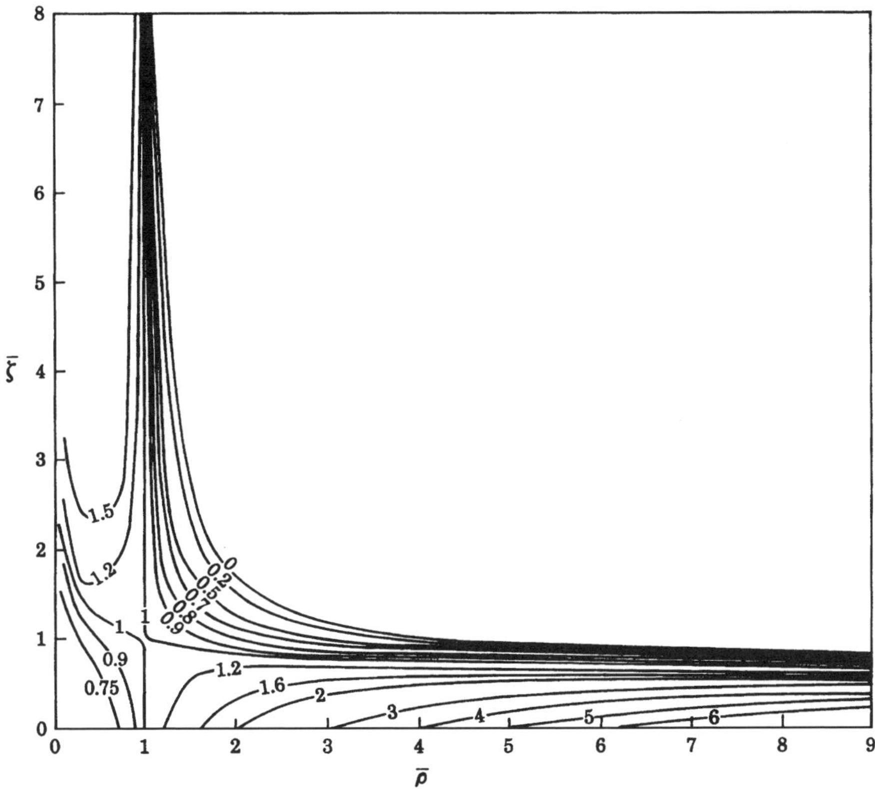

Fig. 9.28 Contour diagram for constant values of $\Delta\bar{G} = \Delta G/\Delta G^*$ according to Eq. (9.95).

the contour line $\Delta\bar{G} = 0$ is crossed. The attainment of stability thus involves an increase in dimensions over that characterizing a critical-size nucleus. There are a large number of paths that originate at the saddle point, and allow for stability to be achieved. However, the pursuit of certain paths is futile. It can be seen from Fig. 9.28 that $\bar{\rho}$ must exceed unity, irrespective of the value of $\bar{\zeta}$, in order for $\Delta\bar{G}$ to become negative. On the other hand, $\bar{\zeta}$ can be fixed at unity and stability will be achieved if $\bar{\rho}$ exceeds 4. Hence growth in the ζ direction does not need to occur beyond the critical size in order for a stable crystallite to develop. This conclusion is an important consequence of this type of nucleation. A firm requirement of Eq. (9.95) is that $\bar{\zeta}$ must exceed $^1/_2$ for a stable nucleus to be formed even if unrestricted lateral growth is allowed. Although forbidden paths can be delineated, it is not *a priori* possible to prescribe a unique path for the growth of an asymmetric nucleus of critical size to a thermodynamically stable crystallite. The conclusions reached for other three-dimensional geometries follow those established for the cylindrical nucleus. Characteristically, for this type of nucleation ΔG^* is always proportional to $1/\Delta G_v^2$. The proportionality factor will depend on the geometry assumed. This type of nucleation has also been termed three-dimensional nucleation because of the geometry involved.

The temperature dependences of ρ^* and ζ^*, as well as ΔG^*, are characteristic of nucleus formation. At the melting temperature T_m, ΔG_v is zero, so that the critical dimensions, and ΔG^*, are infinite. As the melting temperature is lowered, the governing variable is the undercooling that changes dramatically with just small variations in the crystallization temperature T_c. In turn major changes occur in the critical dimensions. Thus, there is a strong temperature dependence of the quantities involved in forming a critical-size nucleus. It can be anticipated that these unique features will have a marked influence on the rate at which stable nuclei are formed.

In homogeneous nucleation, critical-size nuclei are formed by statistical fluctuations in the melt. However, nucleation processes can also be catalyzed by heterogeneities. Heterogeneous nucleation is the most common type that is encountered. There are several different types of heterogeneity that enhance the formation of stable nuclei. Extraneous solids and grain boundaries can catalyze the nucleation process.(118–120) Also, surfaces that are wetted by the nucleus, cavities within which nucleation is favored, and container walls will also enhance the nucleation process. Embryonic nuclei can be retained in cavities above the melting temperature.(121) Nucleation can occur on already formed crystal surfaces. This type of nucleation can take place coherently, incoherently or epitaxially. All of these possibilities have important implications to polymer crystallization.

An example of a foreign surface that is wetted by the nucleating species is illustrated in Fig. 9.29.(122) Here we consider the nucleus of species B to be in

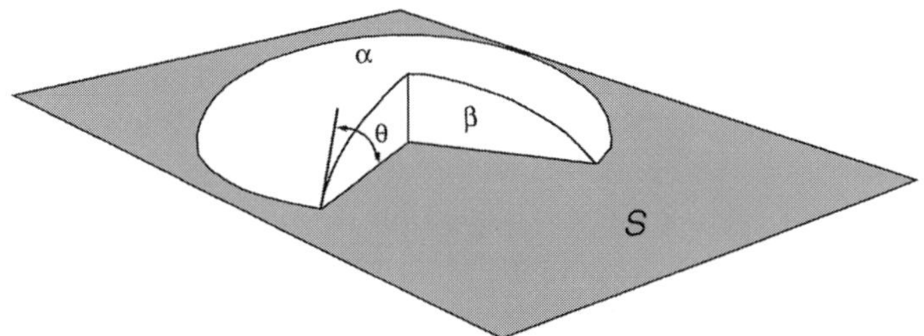

Fig. 9.29 Schematic example of heterogeneous nucleation on a flat substrate. The nucleus makes a contact angle θ with the substrate. (From Uhlmann and Chalmers (122))

the shape of a spherical cap of radius r. It is being formed on the flat nucleating substrate so all are immersed in the liquid phase A. Here, θ is the contact angle for equilibrium with respect to the horizontal force components. The balance of the horizontal forces can be written as

$$-\sigma_{\alpha\beta}\cos\theta = \sigma_{\beta s} - \sigma_{\alpha s} = W_s \tag{9.96}$$

where the σ's are the appropriate surface tensions. Substrates that are characterized by $\theta < 180°$ for a particular nucleant can serve as nucleation catalysts. Ignoring strain, ΔG_{het}, the free energy change in forming such a nucleus, can be written as (122)

$$\Delta G_{het} = \pi r^2(1-\cos^2\theta)W_s + 2\pi r^2(1-\cos\theta)\sigma_{\alpha\beta} - \frac{\pi}{3}r^3(2+\cos\theta)(1-\cos\theta)^2\Delta G_v \tag{9.97}$$

It then follows that

$$r^* = \frac{2\sigma_{\alpha\beta}}{\Delta G_v} \tag{9.98}$$

and

$$\Delta G^*_{het} = \frac{4\pi}{3}\frac{\sigma^2_{\alpha\beta}}{(\Delta G_v)^2}(2+\cos\theta)(1-\cos\theta)^2 \tag{9.99}$$

Thus,

$$\Delta G^*_{het} = \Delta G^* f(\theta) \tag{9.100}$$

where

$$f(\theta) = \frac{(2+\cos\theta)(1-\cos\theta)^2}{4} \tag{9.101}$$

and ΔG^* is the free energy change required to form the nucleus homogeneously.

It is important to note that r^*, given by Eq. (9.98), is identical with that obtained from Eq. (9.85) for homogeneous nucleation of a spherical body. The principal role of a nucleating heterogeneity is to reduce the barrier to nucleation caused by the surface interfacial free energy. Equations (9.99) and (9.100) show that in this situation the free energy barrier to nucleation decreases with decreasing θ, and approaches zero as $\theta \to 0$.

Instead of the spherical cap, a cylindrical nucleus of height h and radius r, forming on the substrate, can also be treated.(123) It follows from similar arguments that

$$r^* = \frac{2\sigma_{\mathrm{un}}}{\Delta G_v} \qquad h^* = \frac{2(\sigma_{\alpha\beta} + \sigma_{\beta\mathrm{s}} - \sigma_{\alpha\mathrm{s}})}{\Delta G_v} \tag{9.102}$$

and

$$\Delta G^*_{\mathrm{het}} = \frac{4\pi\sigma^2_{\mathrm{un}}}{\Delta G^2_v}(\sigma_{\alpha\beta} + \sigma_{\beta\mathrm{s}} - \sigma_{\alpha\mathrm{s}}) \tag{9.103}$$

Here σ_{un} represents surface free energy, and $\sigma_{\alpha\beta}$, $\sigma_{\beta\mathrm{s}}$ and $\sigma_{\alpha\mathrm{s}}$ the respective surface tensions. When $\sigma_{\alpha\mathrm{s}} = \sigma_{\beta\mathrm{s}}$, Eq. (9.103) becomes

$$\Delta G^*_{\mathrm{het}} = 4\pi\sigma^2_{\mathrm{un}}\sigma_{\alpha\beta} \tag{9.104}$$

For isotropy, $\sigma_{\mathrm{un}} = \sigma_{\alpha\beta}$ so that

$$r^* = h^* \tag{9.105}$$

and

$$\Delta G^*_{\mathrm{het}} = \frac{4\pi\sigma^3_{\alpha\beta}}{\Delta G^2_v} \tag{9.106}$$

Nuclei with other geometries that are formed on a wetting substrate can be treated in a similar manner. The critical dimensions are inversely proportional to ΔG_v, for this type of heterogeneity, irrespective of the nucleus geometry. On the other hand, $\Delta G^*_{\mathrm{het}}$ is inversely proportional to the square of ΔG_v. The proportionality constants are, however, dependent on the geometry involved. The expression for $\Delta G^*_{\mathrm{het}}$ is qualitatively similar in form to that of the homogeneous case. The general result is that $\Delta G^*_{\mathrm{het}} = f(\theta)\,\Delta G^*_{\mathrm{hom}}$, with $f(\theta)$ varying between zero and unity. The temperature dependence of the free energy of forming a critical-size nucleus is identical in both cases. However, less free energy is expended in forming the heterogeneous nucleus.

The possibility exists that embryos, or nuclei, can form and be retained within cavities of foreign bodies even though under the same conditions they would be unstable on flat surfaces. The necessary stability conditions have been established for cylindrical and conical shaped cavities.(121) It is found that embryos can be retained in cavities at temperatures above the bulk melting temperature if certain conditions with respect to geometry, dimensions and surface tensions are satisfied.

The retention of embryos within cavities can explain the effect of thermal history on the subsequent nucleation and crystallization, even when the sample is heated above the melting temperature.

Another type of nucleus that is of special interest involves the deposition of a collection of molecules on an already developed crystal face. The nucleus can be either three-dimensional or unimolecular. In the latter case only a monolayer is formed. This type of nucleus was suggested by Gibbs.(123) In the Gibbs type nucleus one dimension is fixed. Therefore, growth from an embryo to a critical-size nucleus is restricted to the two lateral dimensions. We are thus dealing with a two-dimensional nucleation process in contrast to the previous discussion. This type of nucleus can form either coherently, i.e. in register with the already formed crystal face or incoherently, where there is a mismatch between the faces. Since this nucleus is formed on an already existing crystal face it can be thought of as a growth nucleus, so that growth occurs by a secondary nucleation process.(112) Nucleation controlled growth may be involved in some crystallization processes. There are, however, other possible growth mechanisms.(124) An initiating or primary nucleus in one form or another must always be involved in the crystallization process.

In analogy to the model of a spherical nucleus that was used to demonstrate the basic principles of three-dimensional nucleation, a disk-shaped nucleus can be taken to represent the two-dimensional case. This disk has a fixed thickness l and a variable radius r. When deposited coherently as is illustrated in Fig. 9.30 the only new surface formed will be the lateral one. The surface at the top of the disk is compensated by the loss of the original surface of the crystal. Consequently, the free energy change for coherently forming such a disk is given by

$$\Delta G_{\mathrm{D}} = 2\pi r l \sigma_{\mathrm{un}} - \pi r^2 l \, \Delta G_v \tag{9.107}$$

Here σ_{un} is the lateral surface free energy.

The critical values of r^* and the barrier height, ΔG_{D}^*, are given by

$$r^* = \frac{\sigma_{\mathrm{un}}}{\Delta G_v} \qquad \Delta G_{\mathrm{D}}^* = \frac{\pi l \sigma_{\mathrm{un}}^2}{\Delta G_v} \tag{9.108}$$

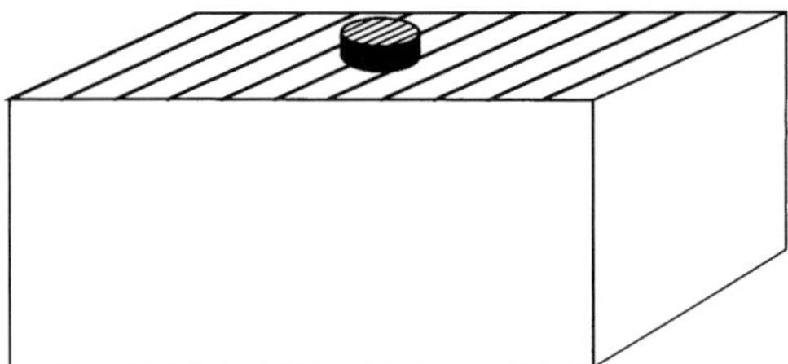

Fig. 9.30 Schematic example of coherent nucleation of a disk on an already formed crystal surface.

It is also possible for the molecules to be deposited incoherently, so that the surfaces of the nucleus and crystal are not in register. In this case the top and bottom surfaces do not completely compensate one another. There is a net interfacial free energy between the pillbox and substrate that we will designate as σ. Therefore, for this type of nucleation (125,126)

$$\Delta G_{\mathrm{D}} = 2\pi r l \sigma_{\mathrm{un}} - \pi r^2 \sigma - \pi r^2 l \, \Delta G_v \tag{9.109}$$

and

$$r^* = \frac{\sigma_{\mathrm{un}}}{\Delta G_v - \dfrac{\sigma}{l}} \tag{9.110}$$

with

$$\Delta G_{\mathrm{D}}^* = \frac{\pi l \sigma_{\mathrm{u}}^2}{\Delta G_v - \dfrac{\sigma}{l}} \tag{9.111}$$

The radius of a critical-size nucleus is greater for a mismatched, incoherent nucleus than one deposited coherently. The free energy barrier required to form such a nucleus is also greater.

With this background the formation of a Gibbs type nucleus, composed of monomers, can be analyzed. The free energy change in forming such a nucleus that is one molecule thick, ρ units broad and ζ units long is given by

$$\Delta G = 2\sigma_{\mathrm{en}}\rho + \zeta\sigma_{\mathrm{un}} - \rho\zeta \, \Delta G_v \tag{9.112}$$

There is no contribution from the interfacial free energy of the upper surface of the nucleus in this case since it is canceled by the matched coverage of the lower surface. This geometric factor is an important characteristic of this type of nucleus. The critical conditions for the formation of a stable nucleus are given by the saddle point of the surface defined by Eq. (9.112). Consequently,

$$\rho^* = \frac{2\sigma_{\mathrm{un}}}{\Delta G_v} \qquad \zeta^* = \frac{2\sigma_{\mathrm{en}}}{\Delta G_v} \tag{9.113}$$

and

$$\Delta G^* = \frac{4\sigma_{\mathrm{en}}\sigma_{\mathrm{un}}}{\Delta G_v} \tag{9.114}$$

For such a coherent two-dimensional nucleus, ΔG^* is inversely proportional to ΔG_v. This is in contrast to a three-dimensional type nucleus, formed either homogeneously or heterogeneously, where ΔG^* is inversely proportional to the square of ΔG_v. Although the proportionality factors are different, the important difference between the two nuclei types is the dependence on ΔG_v. Although the detailed critical dimensions vary according to the model, the strong dependence on the undercooling is the dominant factor. It is inherent to all nucleation theories. An interesting question, and one that will be addressed shortly, is how to distinguish the specific type of nucleus that is involved in a particular crystallization.

An extremely important conclusion reached from Eq. (9.113) is that ζ^* corresponds to the minimum size for the thermodynamic stability of a mature crystallite, even if growth in the lateral dimensions is unrestricted. Thus, if the value of ζ^* given by Eq. (9.113) is maintained, the mature crystallite that evolves will be thermodynamically stable at the crystallization temperature. However, such a crystal will melt at a temperature just infinitesimally above the crystallization temperature, irrespective of the value of ρ. Hence, if this type of surface, or growth, nucleus is involved in a real crystallization process, a mechanism must be provided by which the longitudinal dimension increases beyond critical size in order for a crystallite to be stable above the crystallization temperature. This mandatory requirement is an inherent property of a Gibbs type nucleus. This requirement holds irrespective of the molecular species involved since it is based on a straightforward thermodynamic requirement. It will become evident shortly that this requirement is very important in polymer crystallization. In contrast, thermodynamic stability can be achieved in a three-dimensional nucleus without any need for increasing ζ^*.

Following the general pattern of analysis, the critical parameters for a unimolecular, noncoherent nucleus can be expressed as

$$\rho^* = \frac{2\sigma_{\text{un}}}{\Delta G_v - \sigma/a} \qquad \zeta^* = \frac{2\sigma_{\text{en}}}{\Delta G_v - \sigma/a} \tag{9.115}$$

and

$$\Delta G^* = \frac{4\sigma_{\text{en}}\sigma_{\text{un}}}{\Delta G_v - \sigma/a} \tag{9.116}$$

Here a is the thickness of the monolayer. For this type of nucleation ζ^* is greater than the corresponding value that is obtained for coherent nucleation. Therefore, there is an enhanced stability to the mature crystallite of thicknesses ζ^*.

Coherent nucleation on the face of an already formed crystallite is not restricted to a unimolecular layer. Growth nuclei can also be multilayer, i.e. the thickness is not fixed. In this case a three-dimensional process is involved and ΔG^* is proportional to ΔG_v^{-2}.

The main focus up to now has been on the properties of different types of nuclei. In applying these concepts to crystallization kinetics it is necessary to formulate an expression for the rate at which stable nuclei are formed. Since embryos, or nuclei, will in general grow one molecule at a time as a result of statistical thermal fluctuations, those containing less than the requisite number of molecules will disappear without reaching critical size. Occasionally a series of energy fluctuations will produce a nucleus that exceeds the critical size. Based on a proposal by Becker (127), Turnbull and Fisher (128) treated this rate problem by taking advantage of Eyring's absolute rate theory. They considered the reaction

$$\beta_i + \alpha_l \rightleftharpoons \beta_{i+l} \tag{9.117}$$

at steady state. Here α_l represents the liquid phase and β_i the nucleus of the new phase being formed. It was found that the steady-state rate of forming critical-size nuclei per mole of untransformed material per unit time could be expressed as

$$N = (n_0 kT/h)\exp[-(E_D + \Delta G^*)/kt] \tag{9.118}$$

where n_0 is a constant. Equation (9.118) was derived for a process that does not require the long-range diffusion of molecules. E_D is the activation free energy for the short-range movement of molecules crossing the interfacial boundary in order to join the nucleus. An equation having a similar form is also obtained for heterogeneous nucleation.(118) Equation (9.118) represents steady-state nucleation. There is a transient nucleation process that proceeds reaching steady-state.(17a,17b)

Equation (9.118) accounts for the marked negative temperature coefficient of the crystallization rate that is observed in the vicinity of the melting temperature. Not only is there a very strong temperature coefficient, but it is in the opposite direction to that of conventional chemical reactions. Conventionally, the rate of a process or reaction increases with increasing temperature. On the other hand, nucleation rates increase with decreasing temperature at low to moderate undercoolings. The reason is the dependence of ΔG^* on the inverse of $T_m^0 - T$. At temperatures just below T_m^0, the nucleation rate has a very large negative temperature coefficient primarily as a result of the variation of the term $\exp(-\Delta G^*/RT)$. Small decreases in the crystallization temperature in this region reduce ΔG^* by a relatively large amount. The nucleation rate is then enhanced by the exponential argument. As the temperature is decreased further the nucleation increases at a much slower rate, reaches a maximum value, and then decreases. At temperatures below the maximum, the positive temperature coefficient is a result of the dominance of the transport term $\exp(-E_D/RT)$ in Eq. (9.118). If nucleation is assisted by foreign bodies, or surface heterogeneities, the temperature dependence is the same as in the homogeneous case, but the numerical factors differ. Steady-state nucleation, as embodied in Eq. (9.118), explains in a qualitative way the major aspects of temperature dependence of the crystallization rate in the vicinity of T_m.

The study of nucleation, and consequently crystallization, from the melt can be complicated by heterogeneities. Under these circumstances nucleation will occur at temperatures that are much higher than the homogeneous case. In a classical set of experiments Turnbull showed that small mercury droplets can be supercooled by a much greater amount than the bulk liquid.(121) The principle involved here is that when the sample is divided into small droplets the heterogeneities are isolated. Thus, the vast majority of the droplets will be free of heterogeneities and the transformation of the sample will then occur by homogeneous nucleation. Homogeneous nucleation by this technique has been demonstrated for many metals, inorganic and organic compounds (129–131) including the *n*-alkanes.(132–135b) For example, various metals have been supercooled from about 60 °C to several hundred degrees utilizing this method. Organic and inorganic compounds have been undercooled by as much as 150 °C. Even the *n*-alkanes, which can scarcely be supercooled in the bulk, can be undercooled as much as 20 °C utilizing the droplet technique. The results of such studies with polymers will be discussed shortly.

9.9.2 Long chain molecules

9.9.2.1 Homogeneous nucleation

Nucleation theory involving low molecular weight substances has been introduced to serve as a background for the related problem in polymer crystallization. The basic concepts of classical nucleation theory were, therefore, emphasized. Different nucleation models and nuclei structures were presented to serve as a broad base upon which to discuss nucleation processes involving long chain molecules. This procedure enables both a comprehensive and critical analysis to be made of polymer crystallization. There is no reason to believe that polymers do not follow the same basic principles that govern nucleation theory of monomeric substances. The same laws of thermodynamics and kinetics should apply. However, cognizance must be taken of the molecular character of long chain molecules, particularly the connectivity of the repeating units. One must then ask in what way, if any, does this structural feature modify conventional nucleation theory.

In developing theory pertinent to polymers it is necessary to consider the nature of the ordered portion of a molecule that participates in nucleus formation. It will be asymmetric, i.e. very long in the chain direction and relatively narrow in the lateral dimensions. Therefore, at least two different surfaces will be involved. At a minimum, the chain will be characterized by an interfacial free energy of the surface that is normal to the chain axis, σ_{en}, and an interfacial free energy, σ_{un}, associated with the lateral surface. It is emphasized again that a sharp distinction must be made between the interfacial free energies characteristic of the nucleus, and the corresponding quantities for the mature crystallite. These will be designated as σ_{ec} and σ_{uc}. The surface structure of the mature crystallite does not have to be the same as the nucleus from which it is formed. The interfacial free energies will be altered accordingly. There is no *a priori* reason for the mature crystallite to mimic the nucleus.

Nuclei composed of low molecular weight species contain the requisite number of complete molecules. In contrast, with the exception of extremely low molecular weights, an entire chain molecule will not be located within the nucleus. A possible exception to this generalization is when the complete polymer chain is compacted, or folded up in such a manner that the entire molecule can be placed within the nucleus. This situation will be very rare, if it in fact exists at all. Irrespective of the type of nucleation, be it two- or three-dimensional, or how the chain units are structured within the nucleus, the number of units ζ in the ordered sequence that participates in forming the nucleus must be selected from among all the x repeating units of the chain. This selection requirement introduces an additional entropy term, as compared to monomers, to the free energy for forming a nucleus. Accordingly, the critical dimensions of the stable nucleus will be affected.

This problem can be treated by using the Flory expression for the free energy of fusion, ΔG_{f}, of a crystallite ζ units long and ρ units broad selected from N chains each having x repeating units. As has been shown (1)

$$\frac{\Delta G_{\mathrm{f}}}{xN} = \frac{\zeta\rho}{xN}\Delta G_{\mathrm{u}} + \frac{RT}{x}\ln\left(\frac{1-\zeta\rho}{xN}\right) + \frac{RT\rho}{xN}\left[\ln D + \ln\frac{(x-\zeta+1)}{x}\right] \qquad (9.119)$$

Here ΔG_{u} is the free energy of fusion per repeating unit of the infinitely long chain and $\ln D \equiv \exp[-2\sigma_{\mathrm{en}}/RT]$. The quantity σ_{en} plays the role of an interfacial free energy accounting for the dissipation of order from the crystalline to disordered state, which cannot take place abruptly. Since nucleation only involves low levels of crystallinity $\ln(1 - \frac{\zeta\rho}{xN}) \approx \frac{-\zeta\rho}{xN}$ so that Eq. (9.119) becomes

$$\Delta G_{\mathrm{f}} = \zeta\rho\Delta G_{\mathrm{u}} - RT\frac{\zeta\rho}{x} + RT\rho\ln\left(\frac{x-\zeta+1}{x}\right) - 2\rho\sigma_{\mathrm{en}} \qquad (9.120)$$

The first term in Eq. (9.120) represents the bulk free energy of fusion for the $\zeta\rho$ units that form the nucleus. The next two terms result from the finite length of the chain. The first of these expresses the entropy gain due to the increased volume available to the ends of the molecule after melting. The second results from the fact that only a portion of the repeating units of a given molecule participate in nucleus formation. It represents the entropy gain that arises from the number of different ways a sequence of ζ units can be located in a chain x units long.(1) This latter term is important because it introduces the effect of a finite chain length on the net free energy of fusion and consequently on the free energy of forming a nucleus. The expression for a single chain is given by Eq. (9.120) with $\rho = 1$.

The free energy change that takes place in forming a nucleus from a collection of chain molecules can now be calculated. The selection process that has been outlined must be taken into account irrespective of the nucleation mode and the chain conformation within the nucleus. This selection process is a crucial step in forming a nucleus. Subsequently, additional conditions, or restraints, on the nucleus can be added if deemed necessary. The selection of the sequences by itself leads to a perfectly legitimate nucleus that has been commonly termed a "bundle" nucleus. This kind of nucleus does not place any restraints on the shape or structure of the mature crystallite that develops from it.

After selecting the sequences from ρ chains a nucleus can be formed by merely arranging them in a cylindrical array. This nucleus, the earliest one studied for polymer chains, was selected because it accounts for both the general features and chain asymmetry in a straightforward manner.(12) The number of surfaces involved is minimal. The free energy of forming such a nucleus can be expressed as (136)

$$\Delta G = 2\pi^{1/2}\rho^{1/2}\zeta\sigma_{\mathrm{un}} + 2\rho\sigma_{\mathrm{en}} - \rho\zeta\,\Delta G_{\mathrm{u}} + \frac{RT}{x}\zeta\rho - \rho RT\ln\left(\frac{x-\zeta+1}{x}\right) \tag{9.121}$$

The surface that is described by Eq. (9.121) is virtually identical to that described by Eq. (9.90) and contains a saddle point. The critical dimensions, ζ^* and ρ^*, are defined by the coordinates at the saddle point. Appropriate differentiations yield

$$\rho^{*1/2} = \frac{2\pi^{1/2}\sigma_{\mathrm{un}}}{\Delta G_{\mathrm{u}} - \frac{RT}{x} - \frac{RT}{(x-\zeta^*+1)}} \tag{9.122}$$

and

$$\frac{\zeta^*}{2}\left(\Delta G_{\mathrm{u}} - \frac{RT}{x} - \frac{RT}{x-\zeta^*+1}\right) = 2\sigma_{\mathrm{en}} - RT\ln\left(\frac{(x-\zeta^*-1)}{x}\right) \tag{9.123}$$

It then follows that

$$\Delta G^* = \pi^{1/2}\zeta^*\rho^{*1/2}\sigma_{\mathrm{un}} \tag{9.124}$$

for a chain of length x, when the complete molecule does not participate in the nucleation act. The quantity ΔG_{u} represents the free energy of fusion per repeating unit of the infinite molecular weight chain. On the other hand Eqs. (9.122), (9.123) and (9.124) refer to chains of finite length. Utilizing the first-order approximation, ΔG_{u} can be expressed as

$$\Delta G_{\mathrm{u}} \cong \Delta H_{\mathrm{u}}\frac{(T_m^0 - T)}{T_m^0} \tag{9.125}$$

where T_m^0 is the equilibrium melting temperature of the perfect crystal of infinite molecular weight. Thus the equilibrium melting temperature of the actual chain of length x is not directly involved. The reason for this surprising result is that the $\zeta^*\rho^*$ units that are selected to participate in the nucleus do not recognize the chain

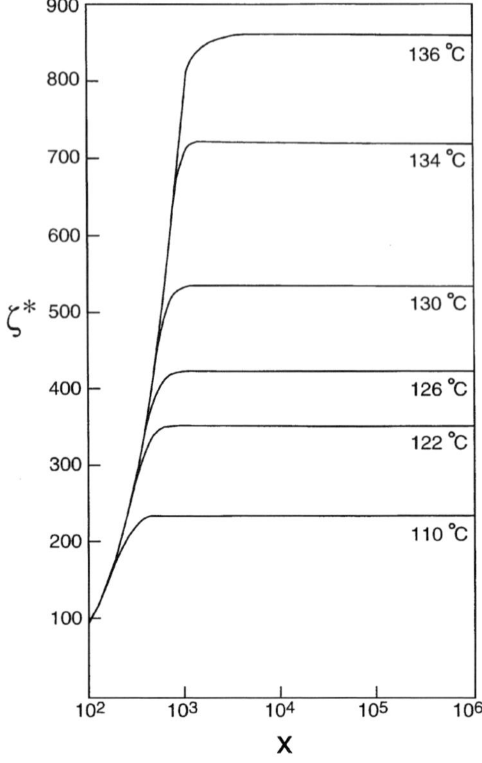

Fig. 9.31 Plot of ζ^* against chain length x for indicated temperatures for linear polyethylene. Parameters used $T_m^0 = 145.5\,^\circ\mathrm{C}$, $\sigma_{\mathrm{un}} = 100\,\mathrm{cal\,mol^{-1}}$, and $\sigma_{\mathrm{en}} = 4600\,\mathrm{mol^{-1}}$.(136)

ends, except for the correction term embodied in Eqs. (9.122)–(9.124). Therefore, in contrast to the conventional nucleation theory appropriate to monomers, the equilibrium melting temperature of the species directly involved is not used with chain molecules. The undercooling is therefore not reckoned in the conventional manner since the same value of ΔG_{u} is used for all chain lengths. The error involved in adopting conventional procedures can be significant for small values of x. It is less important and becomes negligible at the higher molecular weights. In the limit of infinite molecular weight, Eqs. (9.122)–(9.124) reduce to

$$\rho^* = \frac{4\pi\sigma_{\mathrm{un}}^2}{\Delta G_{\mathrm{u}}^2} \tag{9.126}$$

$$\zeta^* = \frac{4\sigma_{\mathrm{en}}}{\Delta G_{\mathrm{u}}^2} \tag{9.127}$$

$$\Delta G^* = \frac{8\pi\sigma_{\mathrm{un}}^2\sigma_{\mathrm{en}}}{\Delta G_{\mathrm{u}}^2} \tag{9.128}$$

These equations reflect the fact that for a chain of infinite molecular weight there is no influence of end-groups in the selection and nucleation processes. Equations (9.126)–(9.128) are identical to those obtained for nonchain low molecular weight species when arranged in the same three-dimensional geometric array.

The influence of chain length on ζ^*, based on Eq. (9.123), is illustrated in Fig. 9.31.(136) Here ζ^* is plotted against the chain length at a series of different temperatures for polyethylene, taken as a model system. The parameters that were used are indicated in the legend. ζ^* is essentially independent of chain length at the higher molecular weights. The range over which ζ^* is constant with chain length increases with decreasing crystallization temperature. Put another way, when compared at the same undercooling, calculated from T_m^0, ζ^* is independent of molecular weight at large undercoolings. For the lower molecular weights a significant increase in ζ^* with chain length is found.

An important characteristic of nucleation, the strong dependence of the critical dimensions on the undercooling in the vicinity of T_m^0, is graphically illustrated in Figs. 9.32 and 9.33. In these figures, ζ^* and ρ^* are plotted against the temperature,

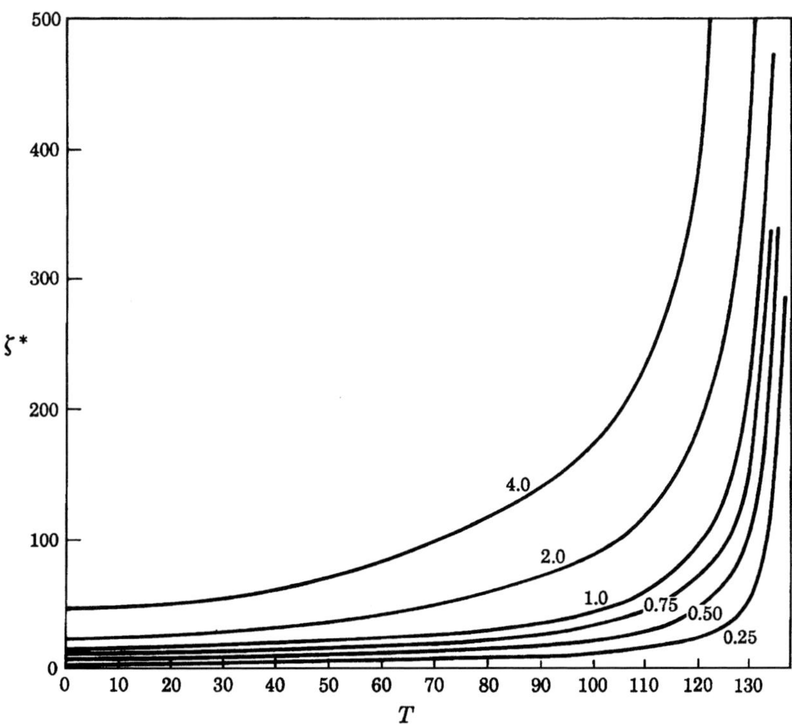

Fig. 9.32 Plot of ζ^* against T according to Eq. (9.126). Values of ratio $\sigma_{en}/\Delta H_u$ are indicated and T_m^0 is taken to be 137.5 °C.

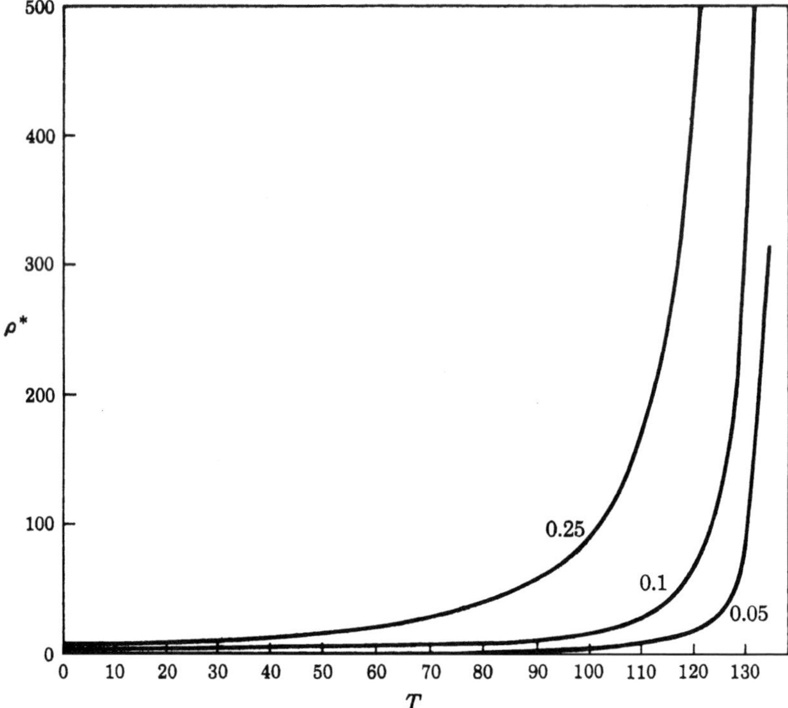

Fig. 9.33 Plot of ρ^* against T according to Eq. (9.127). Values of ratio $\sigma_{\rm un}/\Delta H_{\rm u}$ are indicated and $T_{\rm m}^0$ is taken to be 137.5 °C.

according to Eqs. (9.126) and (9.127), for different values of the ratios $\sigma_{\rm en}/\Delta H_{\rm u}$ and $\sigma_{\rm un}/\Delta H_{\rm u}$. At $T_{\rm m}^0$, taken as 137.5 °C in this example, the critical dimensions are infinite. They decrease sharply, but continuously, as the temperature is lowered. This decrease is very dramatic in the range of low undercoolings. However, for crystallization temperatures well removed from $T_{\rm m}^0$ (ΔT of the order of 40 to 50 °C in this example) the critical nucleus dimensions become small. They are essentially insensitive to further decreases in the crystallization temperature. Thus, with the same values of the parameters over the complete range of crystallization temperatures, the critical dimensions become small and essentially constant at large undercoolings. This result is a natural consequence of classical nucleation theory. It can be seriously questioned whether the classical nucleation concepts are valid at the dimensions of the very small nuclei that form at large undercooling. It is evident that ΔG^* is a continuous function that is strongly dependent on the crystallization temperature. Since Eqs. (9.126) and (9.127) are the same for monomers and high molecular weight polymers, the contour diagram of Fig. 9.28 applies equally well to polymers.

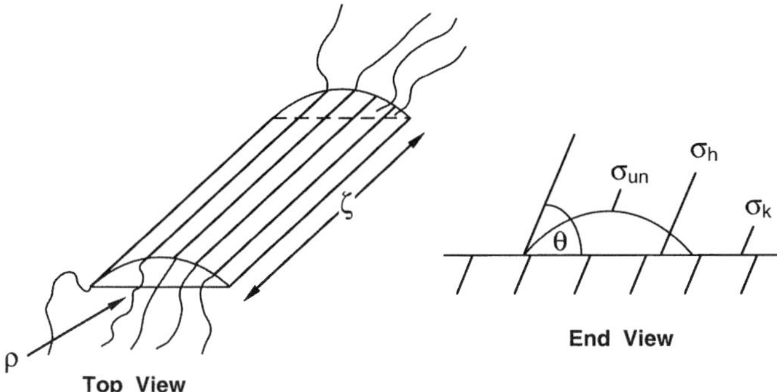

Fig. 9.34 Schematic diagram of the nucleation of a long chain molecule on a flat substrate.(137)

9.9.2.2 Heterogeneous nucleation

Nucleation theory pertinent to chain molecules, including the selection process, can be extended to a variety of heterogeneous processes.(137) The analysis is similar to that used for monomeric substances. Again, each situation has to be treated separately. In one example that has been analyzed the heterogeneity is a flat, incompressible surface. The polymer chains are deposited on this surface and form a nucleus whose shape is defined by the requirement that surface free energy be a minimum. As is shown in Fig. 9.34, such a nucleus contains ρ chains, or ordered sequences, each being ζ repeating units long. For illustrative purposes, the chain length is taken to be sufficiently large that the free energy of fusion is independent of molecular weight.[9] The quantities σ_{un} and σ_{en} are again the interfacial free energies characteristic of the nucleus–liquid interface parallel and perpendicular to the chain direction respectively; σ_{k} and σ_{h} are the corresponding free energies characteristic of the heterogeneity–liquid and heterogeneity–nucleus interfaces. The contact angle is defined by θ. Analysis of the problem leads to Eq. (9.129) for ΔG, the change in free energy required to form such a nucleus(137)

$$\Delta G = 2\rho\sigma_{\mathrm{en}} + K(\theta)2\pi^{1/2}\sigma^{1/2}\zeta\sigma_{\mathrm{un}} - \zeta\rho\Delta G_{\mathrm{u}} \tag{9.129}$$

where

$$K(\theta) = [(1/\pi)(\theta - 1/2\sin 2\theta)]^{1/2} \tag{9.130}$$

[9] The effect of finite chain length can easily be included in the expression for the free energy of fusion. The same physically meaningful results are obtained from the equations that describe the critical nucleus.

The coordinates of the saddle point of the surface described by Eq. (9.129), that correspond to the dimensions of the critical-size nucleus, are

$$\zeta^* = 4\sigma_{en}/\Delta G_u \tag{9.131}$$

$$\rho^* = K^2(\theta)\left(\frac{4\pi\sigma_{un}^2}{\Delta G_u^2}\right) \tag{9.132}$$

It follows that

$$\Delta G^* = \frac{K_f^2(\theta)8\pi\sigma_{un}^2\sigma_e}{\Delta G_u^2} = K_f^2(\theta)\Delta G_H^* \tag{9.133}$$

Here, ΔG_H^* is again defined as the free energy change in a homogeneously formed three-dimensional nucleus of the same geometry. Although ΔG^* and ρ^* are altered by the substrate, ζ^*, the critical nucleus size in the chain direction, is not. Insofar as the nucleation rate is dependent on ΔG^* this type of heterogeneity will have a catalytic effect. However, ζ^*, which is an important quantity in other aspects of morphology and structure, remains unchanged.

The influence of the substrate in lowering the free energy barrier for the formation of stable nuclei is expressed by the ratio $\Delta G^*/\Delta G_H^*$. This ratio depends on $K_f^2(\theta)$ and thus on the contact angle θ. As $\theta \to \pi$, $K_f^2 \to 1$ and the two free energies become equal to one another in this physically tenable situation. However, as $\theta \to 0$, $K_f^2 \to 0$ so that ΔG^* approaches zero. This is a physically untenable situation and results from the breakdown of Eq. (9.130). The change in $\Delta G^*/\Delta G_H^*$ with contact angle is the same as for monomeric species that form a spherical cap nucleus on the substrate.(118) There are similarities between the asymmetric nucleus formed by chain molecules on a flat substrate and monomeric substances that form a spherical cap.

The breakdown in three-dimensional nucleation, which was indicated above, occurs when $\sigma_k - \sigma_h = \sigma_u$. This situation leads to the consideration of the unimolecular deposition of ρ chains on the foreign substrate. In this case

$$\Delta G = 2\zeta\sigma_{un} + 2\rho\sigma_{en} - \rho\zeta\,\Delta G_u + \rho\zeta(\sigma_u + \sigma_h - \sigma_k) \tag{9.134}$$

This mode of nucleation is favored only if $(\sigma_u + \sigma_h - \sigma_k) < 0$. Under these conditions the critical dimensions of the monolayer nucleus are given by

$$\zeta^* = \frac{2\sigma_{en}}{q\,\Delta G_u} \tag{9.135}$$

$$\rho^* = \frac{2\sigma_{un}}{q\,\Delta G_u} \tag{9.136}$$

where

$$q \equiv 1 + \frac{W_s}{\Delta G_u} \tag{9.137}$$

and

$$W_s \equiv -(\sigma_{un} + \sigma_n - \sigma_k) \tag{9.138}$$

It follows that

$$\Delta G^* = 2\sigma_{un}\zeta^* \tag{9.139}$$

When $q > 1$, the critical dimensions and the free energy for nucleation are all reduced relative to the homogeneous process. For the special case $q = 1$, i.e. $\sigma_k - \sigma_h = \sigma_u$, these equations reduce to those for the coherent formation of a monomolecular layer nucleus on the face of an existing crystallite. This important special case will be treated in more detail shortly.

To summarize, for nucleation on a flat surface the mode of nucleation is determined by the relative values of σ_{un}, σ_h and σ_k. When $\sigma_h - \sigma_k \geq \sigma_{un}$ three-dimensional homogenous nucleation is preferred; for $\sigma_{un} > \sigma_n - \sigma_k$ three-dimensional heterogeneous nucleation will occur. When $\sigma_n - \sigma_k \leq -\sigma_u$ two-dimensional nucleation will result. For the unique situation of $\sigma_h - \sigma_k = -\sigma_u$ the nucleation process will be formally identical to the classical Gibbs coherent monolayer nucleus. These are general conclusions applicable alike to both monomers and polymers. A major difference between the two nucleation modes is that for the three-dimensional heterogeneous process ζ^* is not altered relative to homogeneous nucleation. However, it will be reduced relative to the size required for coherent unimolecular nucleation.

The theory for the nucleation of monomeric substances within cavities or crevices that are contained within the mother phase (118,138) can also be extended to long chain molecules.(137) The specific examples that have been treated are illustrated in Fig. 9.35. Each diagram represents a cross-sectional view of the crevice normal to the chain direction. The shaded areas indicate the polymer chain. The free energy of forming a nucleus can again be written in the general form of Eq. (9.129). The constant $K(\theta)$ now depends on the shape assumed by the nucleus in filling the crevice. When $\theta < \alpha + \pi/2$ the crystal–liquid interface can be either convex (case (i)) or concave (case (ii)). It is equal to zero for the special situation of a flat interface where $\theta + \alpha - \pi/2 = 0$.(139) When $\theta > \alpha + \pi/2$, two separate liquid–crystal interfaces must exist as is illustrated by case (iii) in Fig. 9.35. By operating on Eq. (9.129) in the usual manner it is found that the critical conditions are identical to those for nucleation on a flat substrate.

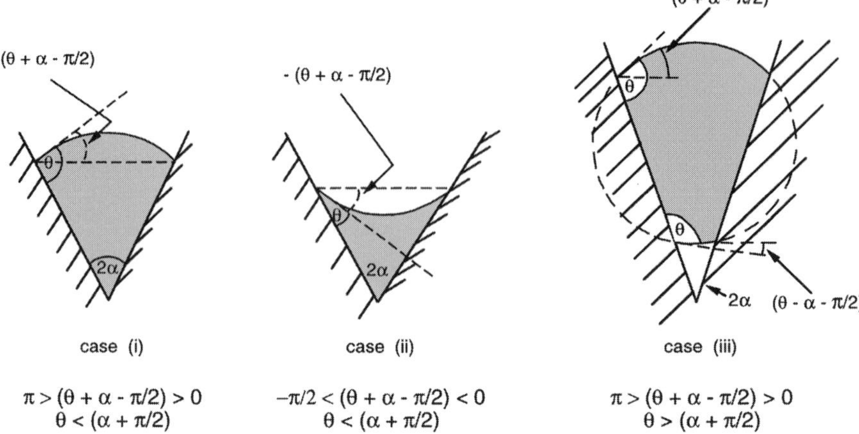

Fig. 9.35 Schematic diagram for nucleation within a conical cavity.(137)

The nuclei geometries appropriate to the heterogeneous nucleation of chain molecules that have been considered are the simplest ones to treat geometrically. They correspond to those treated for monomeric substances. All of the models treated yield the common result that $\zeta^* = \zeta_H^*$. The origin of this conclusion can be shown in the following.(137) Without recourse to a detailed model, one can write

$$\Delta G = 2\rho\sigma_{en} - \zeta\,\Phi(\rho) - \zeta\rho\,\Delta G_u \tag{9.140}$$

where $\phi(\rho)$ is a function of ρ. A detailed analysis shows that in order for $\zeta^* = \zeta_H^*$,

$$\Phi(\rho) = \text{const.}\rho^{1/2} \tag{9.141}$$

Nuclei geometries that satisfy Eq. (9.141) must yield the result $\zeta^* = \zeta_H^*$. The result that $\zeta^* = \zeta_H^*$ is thus not completely general. However, it is satisfied by the models most commonly used to represent nuclei on substrates. Other types of cavities, or crevices, can be proposed that will not satisfy Eq. (9.141).

Under certain circumstances stable embryos of long chain molecules can be retained in cavities above their melting temperature.(137) This conclusion is in accord with results for monomeric substances. A general analysis that encompasses all possible heterogeneous nucleation processes cannot be formulated. The detailed nature of the substrate and the manner in which chain units fill cavities must be specified before the problem can be solved.

9.9.2.3 Gibbs type nucleation

Of special interest in the realm of nucleation theory is the formation of a Gibbs type nucleus by long chain molecules. In analogy to monomers this involves the coherent deposition of a monomolecular layer of chain units on the surface of an already

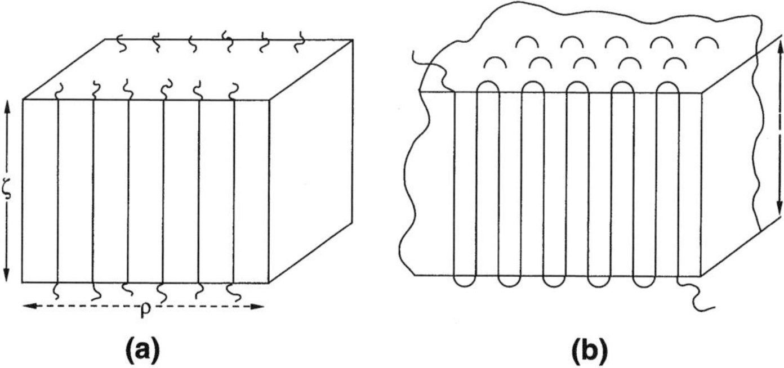

Fig. 9.36 Schematic representation of Gibbs type nucleus of long chain molecules. (a) Bundle nucleus, ζ is thickness of nucleus; (b) regularly folded chain, adjacent re-entry nucleus. l is thickness of nucleus.

existing crystallite. Coherency implies that the sequences being deposited are in register with the crystal surface and the crystallographic features are maintained. Thus, a portion of the crystal surface is replaced by that of the nucleus. This type of two-dimensional nucleus can play the role of a growth or secondary nucleus. The structure of one such nucleus of this type is schematically illustrated in Fig. 9.36a. The nucleus being formed in this example is composed of unfolded chains, which represent a legitimate model for chain nucleation. The free energy of forming such a nucleus from long chain molecules can be expressed as (139)

$$\Delta G = 2\zeta\sigma_{\text{un}} + 2\rho\sigma_{\text{en}} - \zeta\rho\,\Delta G_{\text{u}} + \frac{RT}{x}\zeta\rho - \rho RT\ln\left(\frac{x-\zeta+1}{x}\right) \tag{9.142}$$

The last two terms in Eq. (9.142) have the same significance as was described for three-dimensional nucleation. The dimensions of the critical-size nucleus are

$$\rho^* = \frac{2\sigma_{\text{un}}}{\Delta G_{\text{u}} - \dfrac{RT}{x} - \dfrac{RT}{x-\zeta+1}} \tag{9.143}$$

$$\zeta^* = \left[\frac{2\sigma_{\text{en}} - RT\ln\left(\dfrac{x-\zeta^*+1}{x}\right)}{\Delta G_{\text{u}} - \dfrac{RT}{x}}\right] \tag{9.144}$$

and

$$\Delta G^* = 2\sigma_{\text{un}}\left[\frac{2\sigma_{\text{en}} - RT\ln\left(\dfrac{x-\zeta^*+1}{x}\right)}{\Delta G_{\text{u}} - \dfrac{RT}{x}}\right] = \sigma_{\text{un}}\zeta^* \tag{9.145}$$

As the chain length x gets very large these expressions reduce to the classical results

$$\zeta^* = \frac{2\sigma_{\mathrm{en}}}{\Delta G_{\mathrm{u}}} \qquad \rho^* = \frac{2\sigma_{\mathrm{un}}}{\Delta G_{\mathrm{u}}} \qquad \Delta G^* = \frac{4\sigma_{\mathrm{en}}\sigma_{\mathrm{un}}}{\Delta G_{\mathrm{u}}} \tag{9.146}$$

Equation (9.146) is applicable both to monomers and to polymers of high molecular weight. As with three-dimensional nucleation, the undercooling of finite length chains need to be reckoned from the equilibrium melting temperature of the infinite chain. The contour diagram for the reduced variables of ζ^* and ρ^* for this type of nucleation has the same major features and is qualitatively similar to that in Fig. 9.28 for three-dimensional nucleation. A numerical analysis of Eq. (9.146), indicates that the classical limiting result is obeyed for values of x approximately equal to or greater than 10^4.(139) The conclusions reached for two-dimensional coherent nucleation are valid, irrespective of the molecular chain structure of the mature crystallite that eventually evolves, as long as the complete chain molecule does not participate in the nucleation act.

It was noted previously that although nuclei described by Eqs. (9.143)–(9.146) are stable at the crystallization temperature they are thermodynamically unstable at an infinitessimally higher temperature. In order to be stable above the crystallization temperature, ζ^* must increase, i.e. growth must take place in the chain direction irrespective of the value of ρ^*. The conclusion is an inherent and inviolable characteristic of Gibbs type coherent nucleation. It applies equally well to monomers and polymers, irrespective of the polymer chain conformation within the nucleus. There is no problem in achieving stability with monomers and with the bundle type nucleus that was just considered. The chain conformation within this nucleus places, in principle, no impediment to growth in the chain direction. Hence it is possible for ζ to increase beyond ζ^* isothermally and achieve stability above T_{c}.

In addition to nucleation theory, the steady-state rate at which critical-size nuclei are formed is an important quantity in analyzing crystallization kinetics. The expression for this quantity can be developed in a manner similar to that for low molecular weight systems. Instead of the stepwise addition of atoms or molecules, chain units or sequences of units are added. Consequently, the steady-state rate at which nuclei are formed per unit untransformed volume is again expressed as

$$N = N_0 \exp\left\{\frac{-E_{\mathrm{D}}}{RT} - \frac{\Delta G^*}{RT}\right\} \tag{9.118}$$

The nuclei considered to this point have consisted of unfolded chains arranged either in a three-dimensional bundle or in a monolayer. This kind of nucleus can lead in a natural and straightforward manner to all known mature crystallite structures and related morphologies and superstructures. Other types of nuclei, where the chains are configured and packed in different ways, can also be considered. However, the sequence selection requirement must always be satisfied. Any such modifications will result in an increase in ΔG^*, the barrier height that needs to be overcome to form such a nucleus. If the chains, for example, are assumed to be folded in a regular array, the end surface free energy must be increased because of the additional contributions from the fold and strain free energies.(140–142) The need of gauche rotational states to establish a regularly folded structure in linear polyethylene involves an excessive increase in free energy.

9.9.2.4 Regularly folded chain nucleation

As was described in the introductory chapter, when homopolymers are crystallized either in the bulk or from solution, lamellar-like crystallites are formed. Depending upon the crystallization conditions the thickness of the crystallites can range up to several hundred angstroms. In contrast, the length of the lamellae can be as much as several micrometers, while the width is somewhat reduced from this value. In such crystallites the chain axes are preferentially oriented normal to the basal plane of the lamellae. Studies of the lamellae dimensions, coupled with the observed chain orientation, require that, except for very low molecular weights, some type of chain folding takes place in the mature crystallites. Based on those observations, and the interpretation of transmission electron micrographs, a specific type of chain folding has been postulated.(143–144a) It was assumed that in the crystallite a chain makes a set of sharp, hairpin-like bends, with adjacent re-entry of the ordered sequences. It is thus assumed that a chain in the crystallite resembles a folded fire hose.(145) Detailed studies involving a variety of experimental and theoretical techniques have made quite clear that, despite the lamellar structure and the requirement of some type of folding, regular chain folding with adjacent re-entry is not a common feature of polymer crystallization.[10](141,142,145a)

A nucleation theory involving growth has been proposed.(146–149) It is assumed that the chains within the nucleus are regularly folded with adjacent re-entry. It is based on the premise that the chain conformation and surface structure of the nucleus are the same as those of the mature crystallite that evolves. The assumption that the chains in a lamellar crystallite are regularly folded can be seriously questioned as a general rule in polymer crystallization. Also to be questioned is identifying the nucleus chain structure with that of the crystallite. Crystallite growth is taken to be nucleation controlled consistent with the observed spherulite growth rates.

In this theory a Gibbs type nucleus was selected to represent the growth nucleus. Other types of growth nuclei could just as easily be postulated. Thus, the nucleus

[10] A detailed discussion of the chain conformation within the mature lamellar crystallite, as well as the associated interfacial structure, will be found in Volume 3.

is composed of a unimolecular layer of regularly folded chains that are coherently deposited on the face of an already existing crystallite. A schematic representation of such a regularly folded chain nucleus is depicted in Fig. 9.36b. In this nucleus, the thickness of the chain or monolayer is defined as b_0; l is the fold length; and a is the width of the nucleus. Crystal growth will then be normal to the 100 or 101 planes. There is obviously a fundamental difference between this nucleus and the one composed of unfolded chains. There are some unique features associated with a nucleus of regularly folded chains that warrant a detailed analysis.

The free energy change ΔG_{rf} that is incurred in forming a regularly folded chain nucleus from the melt is given as (146–150)

$$\Delta G_{\mathrm{rf}} = 2b_0 l \sigma_{\mathrm{un}} + 2b_0 a \sigma_{\mathrm{en}} - ab_0 l\, \Delta G_{\mathrm{u}} \tag{9.147}$$

Equation (9.147) can be recognized as also representing the coherent unimolecular deposition of monomers and unfolded polymer chains (see Eqs. 9.112 and 9.142). In fact, the only difference between these equations is that the interfacial free energy, σ_{en}, represents surfaces with different structures. The interfacial free energies σ_{un} and σ_{en}, as well as the free energy of fusion per repeating unit, ΔG_{u}, have their usual meaning. Their values will depend on the model taken. However, most importantly, the equations are formally the same. Specifically, the a dimension for the regularly folded chain can be written as

$$a = \nu a_0 \tag{9.148}$$

where ν can only take on integral values 1, 2, 3, ... and a_0 is the effective width of the chain, i.e. the actual width plus the additional contribution necessitated by the discrete folding. Thus

$$\Delta G_{\mathrm{rf}} = 2b_0 l \sigma_{\mathrm{un}} + 2a_0 b_0 \nu \sigma_{\mathrm{en}} - a_0 b_0 \nu l\, \Delta G_{\mathrm{u}} \tag{9.149}$$

or

$$\Delta G_{\mathrm{rf}} = 2b_0 l \sigma_{\mathrm{un}} + \nu a_0 b_0 [2\sigma_{\mathrm{en}} - l\, \Delta G_{\mathrm{u}}] \tag{9.150}$$

In the high molecular weight approximations of Eqs. (9.149) and (9.150), the terms representing the sequence selection of the requisite number of repeating units to form the nucleus from the x repeating units have been neglected. As was pointed out earlier, the selection step is necessary for any type of nucleus formation. Hidden in σ_{en} is the free energy change required to form the sharp bend. This quantity can be excessive in many polymers and will act to suppress the formation of such a nucleus. Furthermore, since the complete molecule is not contained within the nucleus, the junction points between the ordered and disordered sequences need

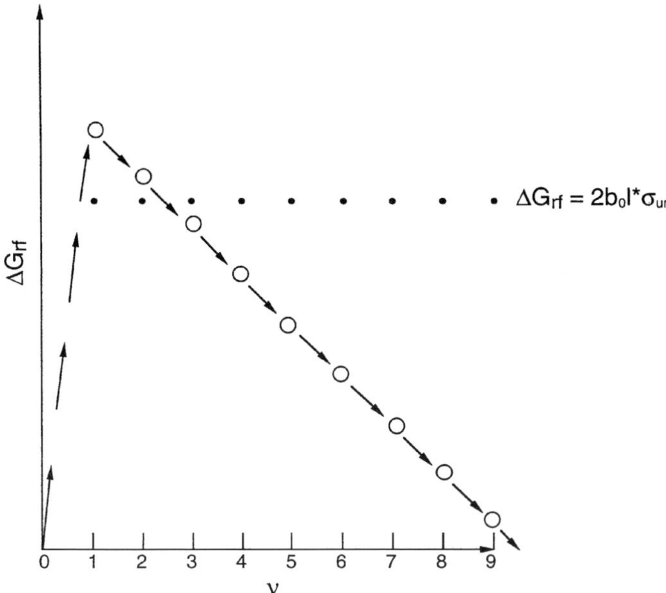

Fig. 9.37 Schematic diagram of the free energy of forming a regularly folded, adjacent re-entry nucleus ΔG_{rf} plotted against the number of sequences ν.

to be taken into account. Neither of these factors have been explicitly taken into account in formulating Eq. (9.150). These factors will cause the total free energy of forming a nucleus of regularly folded chains to be substantially increased.

The surface represented by Eq. (9.150) is not continuous. Therefore, the critical dimensions of the nucleus, and the free energy barrier to be overcome, cannot be obtained in the conventional manner. A different procedure has to be adopted. Consider the quantity $l = l^* = 2\sigma_{en}/\Delta G_u$, which is the critical value for a conventional Gibbs nucleus. For a regularly folded chain nucleus, with $l = l^*$, $\Delta G_{rf} = 2b_0 l^* \sigma_{un}$. For all values of $l \leq l^*$ and of ν, ΔG_{rf} will always be positive. Therefore, a stable nucleus that corresponds to $\Delta G_{rf} = 0$ cannot be formed with l^*. As is indicated in Fig. 9.37, there must be a maximum in ΔG_{rf} in order for a path to exist that allows the nucleus to become stable. The stability requirement can only be satisfied when $l > l^*$. The first step in the nucleation process, $\nu = 1$, is crucial in forming this type of nucleus. It defines the critical value of $l \equiv l_g^* > l^*$. The length l_g^* is the critical dimension in the chain direction of the growth, or secondary, nucleus. It is popularly termed the fold length. According to Eq. (9.150), ΔG_{rf} decreases by the amount $a_0 b_0 [2\sigma_{en} - l_g^* \Delta G_u]$ with each successive step, for all values of $\nu > 1$. Eventually, ΔG_{rf} will become zero and the nucleus will achieve stability. Thus, the value of l^*, the conventional critical dimension, serves as a boundary. Values of $l < l^*$ cannot lead to stability under any circumstances. On the other hand, any value of $l > l^*$ will lead to a stable nucleus and thus to a crystallite. This condition for stability is the same as for all other types of unimolecular coherent nucleation.

The crucial step in this theoretical development is to specify the value of l_g^*. This quantity not only determines the fold length but also the nucleation and growth rates. The schematic diagram in Fig. 9.37 indicates that although larger fold lengths would grow faster, since fewer steps are needed to reach stability, they would be nucleated less frequently because of the high free energy barrier that needs to be overcome. On the other hand, in the extreme where l is just infinitesimally greater than l^*, the barrier is very small so that the initiation of growth is facile. Growth, however, would be very slow under these conditions because of the large number of steps that are needed to achieve stability. A compromise in the value of l between these two extremes needs to be reached. This compromise value, l_g^*, will represent the length of the critical dimension or fold period of the nucleus.

Several different approaches have been made in an effort to resolve this formidable problem. Price (151) has transformed Eq. (9.150) into the form

$$(a - a^*)(l - l^*) = -(\Delta G_{\text{rf}} - \Delta G_{\text{rf}}^*)/\Delta G_{\text{u}} \tag{9.151}$$

A contour diagram can be constructed whose characteristics are similar to those of a three-dimensional nucleus. The objective is to calculate the distribution of l when $l > l^*$. This calculation requires knowing the shape of the steepest profile of the pass on this surface. This profile can be expressed as

$$\Delta G_{\text{rf}} - \Delta G_{\text{rf}}^* = b_0\, \Delta G_{\text{u}}\, (l^*)^2 \left[\frac{l}{l^*} - l\right]^2 \tag{9.152}$$

Under these conditions the ratio of the number of nuclei with free energy ΔG to that having the free energy ΔG^* is

$$P = \exp\{-(\Delta G - \Delta G^*)/kt\} \tag{9.153}$$

Analysis of Eq. (9.153) indicates that the distribution of l/l^* is very narrow. The average value of l/l^* can be expressed as

$$\langle l/l^* \rangle = \frac{\int_1^\infty (l/l^*) P\, d(l/l^*)}{\int_1^\infty P\, d(l/l^*)} \tag{9.154}$$

and leads to the result that

$$\langle l/l^* \rangle = l + \frac{l}{l^*}\left(\frac{kt}{\pi b_0\, \Delta G_{\text{u}}}\right)^{1/2} \tag{9.155}$$

Thus, the average value of l is only slightly greater than l^*. The argument is then made that since the distribution is narrow, $\langle l \rangle$ can be identified with the thickness of the growth nucleus l_g^*. Thus, $l_g^* \equiv \langle l \rangle$, for $l > l^*$, only exceeds l^* by an extremely small amount at low to moderate undercoolings. Eventually l_g^* will be used to calculate ΔG_{rf}^*, and define the steady-state nucleation rate. The upper limit for l/l^* in the integral of Eq. (9.154) has been taken to be infinity. This limit is unrealistic since the largest possible value of l is limited by the chain length. The necessary correction will bring the quantity $\langle l/l^* \rangle$ even closer to unity.

In another approach to the problem, which leads to similar results, the steady-state growth (nucleation) rate is calculated following the principles developed by Turnbull and Fisher for more conventional nucleation processes.(128) The forward and backward rates of depositing a strip are expressed according to standard kinetics.(146) The overall steady-state nucleation (or growth) rates are thus calculated. The free energy diagram given in Fig. 9.37 indicates that, except for the first step, the free energy difference between successive steps is the same. Therefore, the forward and backward steps will have the same rates for each value of ν greater than one. For all steps except the first, the forward rates can be expressed as

$$\alpha_1 = \alpha_2 = \alpha_3 = \ldots \equiv \alpha \tag{9.156}$$

The backward rates can be written as

$$\beta_2 = \beta_3 = \beta_4 = \ldots \equiv \beta \tag{9.157}$$

The forward rate for the first step is defined as α_0 and the backward one as β_1. If N_ν is the occupancy after step ν, the rates of change for the first step can be expressed as

$$\frac{dN_1}{dt} = N_i\alpha_0 - N_1\beta_1 + N_2\beta_2 - N_1\alpha \tag{9.158}$$

and

$$\frac{dN_\nu}{dt} = N_{\nu-1}\alpha + N_{\nu+1}\beta - N_\nu(\beta + \alpha) \tag{9.159}$$

for all values of ν greater than 1. The set of differential equations given by Eqs. (9.158) and (9.159) appear in a variety of other problems.(109) In the steady state, $dN_\nu/dt = 0$ for all values of ν. The net rate of growth or flux, can be expressed as (146,147)

$$S(l) = \frac{N_0\alpha_0(\alpha - \beta)}{\alpha - \beta + \beta_1} \tag{9.160}$$

For many purposes, including the present, β_1 can be set equal to β. Then

$$S(l) = N_0\alpha_0\left[1 - \frac{\beta}{\alpha}\right] \tag{9.161}$$

The individual rates, obtained from transition state theory, can be expressed as

$$\alpha_0 = \frac{kT}{h}\exp\left(\frac{-E_D}{kT}\right)\exp\left(\frac{-\Delta G_1^*}{kT}\right) \tag{9.162}$$

and

$$\frac{\beta}{\alpha} = \exp[-a_0b_0(2\sigma_e - l\,\Delta G_u)] \tag{9.163}$$

It follows that

$$S(l) \equiv N = \frac{N_0kt}{h}\left\{\exp\left[-\frac{(E_D + \Delta G_1^*)}{kT}\right][1 - \exp(-E/kT)]\right\} \tag{9.164}$$

Here ΔG_1^* is the free energy change that occurs with the attachment of the first sequence ($\nu = 1$); i.e. the barrier height. E is defined as

$$E \equiv a_0b_0[2\sigma_{en} - l\,\Delta G_u] \tag{9.165}$$

In Eq. (9.164) the transport term is represented by E_D; ΔG_1^* is the activation free energy required to deposit the first strip. The explicit value of ΔG_1^* is defined by l_g^*. The first exponential term within the bracket is thus the Turnbull–Fisher relation for the steady-state nucleation rate of monomers and nonfolded polymers. This conventional expression for the steady-state nucleation rate is modulated by the second term within the brackets. This term is present because of the basic assumption that the nuclei are composed of regularly folded chains. The extent of the modulation depends on the difference between l_g^* and l^*. When this difference is large there will be a significant effect. However, when $(l_g^* - l^*)$ is small there will scarcely be any influence and the conventional expression will apply. Under these circumstances there is no indication in the equation that the chains within the nucleus are regularly folded.

The relation of l_g^* to l^* is a key factor in this analysis. The value of l_g^* has arbitrarily been defined as $\langle l \rangle$, when all values of $l \geq l^*$ are taken into account. The justification for this identification is the claim that the distribution in l is narrow. The value of l_g^*, so defined, was also assumed to correspond to the thickness of the nucleus that corresponds to the steady-state rate of passage over the free energy barrier ΔG_1^*. It is further stated, but not proved, that l_g^* defined in this manner corresponds to the crystallite thickness that has the maximum growth rate.(152) Based on these assumptions, it follows that

$$l_g^* \equiv \langle l \rangle = \int_1^\infty S(l) l \, dl \Big/ \int_1^\infty S(l) \, dl \tag{9.166}$$

From the assumptions made, and Eq. (9.166), the thickness of the growth nucleus can be expressed as

$$l_g^* = \frac{2\sigma_{en}}{\Delta G_u} + \frac{kT}{2b_0\sigma_{un}} \left[\frac{4\sigma_{un}/a_0 - \Delta G_u}{2\sigma_{un}/a_0 - \Delta G_u} \right] \tag{9.167}$$

Accordingly, l_g^* also defines the thickness of the mature crystallite. In the vicinity of T_m^0, where ΔG_u is small, Eq. (9.167) reduces to (146)

$$l_g^* = \frac{2\sigma_{en}}{\Delta G_u} + \frac{kT}{b_0\sigma_{un}} \tag{9.168}$$

The first term in Eq. (9.168) is identical to the critical thickness of a Gibbs nucleus composed of monomers, or nonfolded polymer chains that are coherently formed as a monolayer on an already existing crystallite. The second term in Eq. (9.168) is a small correction term that allows for thermodynamic stability at temperatures above the crystallization temperature. The nucleus composed of nonfolded chains could be made thermodynamically stable above the crystallization temperature by following a similar procedure. The method used is not unique for regular folded chain nuclei. It follows from Eq. (9.168) that

$$\Delta G^* = \frac{4b_0\sigma_{en}\sigma_{un}}{\Delta G_u} + C \tag{9.169}$$

where C is a very small constant, which can also be introduced into other models. It is trivial at low undercoolings. In essence this calculation tells us that $\langle l \rangle$ represents very thin nuclei and crystallites that are only marginally stable. The identification of $\langle l \rangle$ with l_g^* is in effect an arbitrary decision that is made to remove the instability above the crystallization temperature. This instability is inherent to all coherent unimolecular nuclei models. It is not unique to one composed of regularly folded polymer chains. This value of l_g^* must be greater than l^* and could also be defined as the $\langle l \rangle$. Stability could be achieved by nonfolded chains by a small amount of growth in the chain direction. Experiments show that crystallite thickening actually occurs during isothermal crystallization.[11] Thus, stability can be achieved by thickening, irrespective of the model assumed for the nucleus.

It has been estimated that the second term in Eq. (9.169) is no more than about 10 Å at low to moderate undercooling, and is probably less.(153–156) In the temperature

[11] A discussion of isothermal crystallite thickening will be presented later.

range where nucleation dominates the crystallization process, this term would make only a very small contribution to the crystallite thickness. Since l_g^* is very close to l^* at low undercooling, the steady-state nucleation rate will not deviate in any meaningful manner from the conventional Turnbull–Fisher relation. Formally then, the same results are obtained irrespective of whether the chains within the nucleus are regularly folded or not. It is ironical that a discrimination cannot be made between the different chain conformations within the nucleus from either the steady-state nucleation rate or the critical dimensions. Price has reinforced this conclusion in a more detailed discussion of this point.(113)

The discussion of regularly folded chain nucleation has been limited so far to low undercoolings. At sufficiently large supercoolings l_g^* passes through a minimum and increases with a further decrease in the crystallization temperature. When $\Delta T = (2\sigma_{un}/T_m)(a_0\,\Delta H_u)$, l_g^* approaches infinity. This behavior has been termed the "δ*l* catastrophe". It is solely a consequence of the definition of l_g^* and the subsequent averaging. The predicted increase in l_g^* with decreasing crystallization temperature has never been observed. In an attempt to rectify this anomaly a parameter that varies between zero and one was introduced.(156) Its purpose was to apportion the free energy gain between the forward and backward reaction during the deposition of a sequence. By appropriate choice of this parameter, the subsequent averaging removes the "catastrophe" in the expression for l_g^*. However, this endeavor, as well as others, addresses a moot point. Besides concern for the validity of the regularly folded chain nucleus this "catastrophe" is the result of the calculation leading to the second term in Eq. (9.168) and is only predicted to occur at large undercoolings. In this case the dimensions of a nucleus are extremely small. As was pointed out earlier it is questionable whether the concept of the macroscopic surfaces required for conventional nucleation theory applies for the small nuclei dimensions that are involved.

A variety of modifications has been proposed to the original theory of regularly folded chain nucleation.(109,157). The proposed changes include: fluctuations in the length of the strip deposited;(157,158) deposition by small increments rather than simply laying down the complete strip;(159) computer simulations,(159a,b,c) and the participation of more than one chain in the nucleus; and removal of the condition that the chains within the nucleus must be precisely folded. However, none of the proposed alterations make any fundamental changes in the conclusions. In the general scheme of things, these suggestions can be considered to be relatively minor, since the basic assumption of regularly folded chains, with adjacent re-entry, is still the underlying concept. Rather than making minor corrections to this model major concern should be directed to the question of whether there is any validity to the concept of a regularly folded chain conformation within the nucleus.

The assumption of nuclei being composed of regularly folded chains with adjacent re-entry is intimately related to the concept that mature crystallites of homopolymers have similar chain conformations.(140–149) However, serious concerns have been raised regarding the idea that mature crystallites are composed of regularly folded chains with adjacent re-entry.(140,160–165b) Consequently the reality of nuclei being composed of regular folded chains can be seriously questioned. In addition to the general concerns, specific criticisms have been made of the actual calculation. It has been argued that the basic assumption inherent to this model is irrelevant with respect to crystallization from the pure melt.(159) Binsbergen has pointed out that it is highly improbable that such a nucleus will be formed.(165) It has also been pointed out that the end interfacial free energy σ_{en} will in general be greater for a regularly folded nucleus than the bundle type.

Calvert and Uhlman emphasize the distinction between the interfacial energies of the nucleus and the mature crystallite.(140) This distinction becomes important when the magnitudes of σ_{en} for nuclei of either a bundle type or a regularly folded chain are compared. It has been argued that σ_{en} of a regularly folded chain is substantially less than that for a bundle type nucleus.(149,165c) When the free energy required for adjacent re-entry is properly calculated, the results are dependent on the specific polymer.(141,142) For linear polyethylene a bundle type nucleus can be formed. This conclusion, however, is in contrast to the surface structure of mature crystallites, where the calculation of σ_{ec} is involved.

Besides the high end interfacial free energy, σ_{en}, that is associated with nuclei composed of regularly folded chains, objections have also been made to the restricted form of growth that such nuclei allow.(140) It was pointed out that once a single unit of the chain is attached to the surface, the final position of any other unit in the molecule is effectively predetermined. This is a highly unlikely situation in a high density melt. Competition between different parts of a molecule, or from different chains, can be expected in the natural course of events. A crystal growth theory was developed that starts with a bundle type nucleus and leads in a natural way to lamellar crystallites. It encompasses the major experimental results.

The theory that has been developed for the nucleation of regularly folded chains appears complex when compared with other theories. In fact, despite the apparent complexities, the main conclusions of the theory are relatively simple. They are the same when other chain conformations are assumed for the nucleus. The thickness of the nucleus in the chain direction is the same as that expected for monomers, and for polymers with other chain conformations as long as a Gibbs type nucleus is assumed. In addition, the steady-state nucleation and growth rates have the same functional form and temperature dependence as is found for monomers and polymers with other chain structures within the nucleus.

Sanchez (109) and Price (113) have detailed the fact that the temperature dependence of the growth rate for regularly folded chain nuclei is adequately interpreted by conventional nucleation theory. The basis for this conclusion is found in Table 9.4. In this table the dependence of ΔG^* on ΔG_u, and thus the undercooling,

Table 9.4. *Values of ΔG^* for different type nuclei*

Type Nucleation	Molecular Species	ΔG^*
Coherent monomolecular	Monomer	$\frac{4\sigma_{en}\sigma_{un}}{\Delta G_u}$
	Polymer[a] – bundle	$\frac{4\sigma_{en}\sigma_{un}}{\Delta G_u}$
	Polymer[a] – regularly folded	$\frac{4\sigma_{en}\sigma_{un}}{\Delta G_u}$
Three-dimensional homogenous	Monomer (cylinder)	$\frac{8\pi\sigma_{en}\sigma_{un}^2}{\Delta G_u^2}$
	Polymer[a] – bundle (cylinder)	$\frac{8\pi\sigma_{en}\sigma_{un}^2}{\Delta G_u^2}$
	Polymer[a] – regularly folded (rectangular parallelepiped)	$\frac{32\sigma_{en}\sigma_{un}^2}{\Delta G_u^2}$
Three-dimensional heterogeneous	Monomer – cylinder	$f(\theta)\Delta G_H^*$
	Polymer[a] – cylinder	$k(\theta)\Delta G_H^*$

[a] High molecular weight approximation.

is given for different types of nuclei. Once the type of nucleation is specified (two- or three-dimensional) then the temperature dependence of ΔG^* is the same for monomers and polymers. For polymers it is independent of the chain conformation within the nucleus. The steady-state nucleation or growth rate is obtained by substituting the appropriate value for ΔG^* in the Turnbull–Fisher expression, Eq. (9.118). The only differences in the growth rate expressions are in the actual values assigned to the interfacial free energy σ_{en}. Table 9.4, and related commentary,(113) make abundantly clear that studies of growth rates, and their temperature coefficients, cannot distinguish between nuclei composed of either regular folded or unfolded chains irrespective of whether two- or three-dimensional nucleation is involved. It is ironical that the major conclusions that are reached for regular folded chain nucleation are, for all practical purposes, the same as for the nucleation of nonfolded chains. Therefore, nucleation with regularly folded chains cannot be established solely by analysis of the temperature coefficient of growth rates. Independent support from sources other than crystallization kinetics is needed to justify this assumption. Such support has not as yet been forthcoming. It is inherently difficult to obtain independent values of the interfacial free energies in order to distinguish between the different chain conformations within the nucleus.

An interesting problem to consider is the role of molecular weight in modulating the nucleation of chain molecules. A sequence of an infinite number of repeating units is required to form a critical nucleus at T_m^0. This is true irrespective of whether the nucleus is two or three dimensional, formed homogeneously or heterogeneously, or composed of folded or unfolded chains. As the system is cooled slightly below T_m^0 the dimensional requirements for critical nucleus formation can still exceed the chain length. For a bundle-like nucleus under these conditions, ζ^* would have to exceed the number of repeating units per chain. For a regularly folded chain nucleus the critical chain length must be greater than the product of the number of units in a stem multiplied by the number of stems in the critical-sized nucleus. Hence, depending on the molecular weight, irrespective of the chain configuration within the nucleus, there will already be an undercooling range where nucleation, and thus crystallization, cannot occur. This effect will be more severe the lower the molecular weights. Therefore, unless the highly unlikely matching of chain ends occurs, the chain length limitation is one reason that polymer crystallization only takes place at reasonable rates at larger undercoolings, relative to monomeric substances. Another important reason for the large undercoolings required is the high interfacial free energy that is associated with the end surface.

It has been concluded that the specific chain conformation within the nucleus cannot be determined by analyzing the temperature coefficient of the growth or overall crystallization rates. There are, however, other aspects of the nucleation process that still need to be elucidated. One is the question of whether the nucleation is homogeneous or is assisted by foreign bodies. The other is deciding whether either a two- or three-dimensional nucleation process is involved. An effective method that assesses the role of heterogeneities in nucleation, in any system, is to isolate them in small droplets. If the droplets are made small enough, then the vast majority of them will be free of heterogeneities. Studying crystallization within the droplets will then give a measure of homogeneous nucleation. This approach was very effective in resolving the problem with metals and other low molecular weight substances, including the n-alkanes. A similar strategy has also been applied to polymers. An analysis of the results will be given in Sect. 9.12. In order to address the question of whether a two- or three-dimensional nucleation process is operative it is necessary to analyze the temperature coefficient of crystallization of some typical polymers. This analysis is given in the next section.

9.10 Analysis of experimental data in vicinity of T_m^0

It is quite evident that nucleation plays an important role in polymer crystallization. It is dominant in the vicinity of the equilibrium melting temperature, as is manifested by the marked negative temperature coefficient. It is important, therefore, to define the particulars of the nucleation process that are operative in a given situation. In order to attain this objective, the analysis at this point is limited to isothermal crystallization carried out close enough to the equilibrium melting temperature that the temperature variation of the activation energy in the transport term can be neglected. For three-dimensional nucleation ΔG^* is proportional to $(1/\Delta T)^2$, while for two-dimensional ΔG_u is proportional to $1/\Delta T$. Consequently, the steady-state nucleation rate, N, for a three-dimensional process can be expressed as

$$N = N_0 \exp\left[\frac{-E_D}{RT} - \frac{\kappa_3 T_m^{02}}{T(\Delta T)^2}\right] \tag{9.170}$$

Here the constant κ_3 specifies several quantities. These are: the geometry of the nucleus; whether it is formed homogeneously or heterogeneously; in certain situations whether it is deposited coherently or incoherently; and the enthalpy of fusion per repeating unit. Two-dimensional steady-state nucleation (initiation or growth) can be expressed as

$$N = N_0 \exp\left[\frac{-E_D}{RT} - \frac{\kappa_2 T_m^0}{T\Delta T}\right] \tag{9.171}$$

Here κ_2 plays a defining role for two-dimensional nucleation similar to that of κ_3. Equations (9.170) and (9.171) are general and can be used to investigate spherulite initiation and growth as well as overall crystallization kinetics.

The linear spherulite growth rate can be expressed as

$$G = G_0 \exp\left[\frac{-E_D}{RT} - \frac{g_3 T_m^{02}}{T(\Delta T)^2}\right] \tag{9.172}$$

for three-dimensional nucleation and

$$G = G_0 \exp\left[\frac{-E_D}{RT} - \frac{g_2 T_m^0}{T\Delta T}\right] \tag{9.173}$$

for the two-dimensional case. Here the parameters g_3 and g_2 play comparable roles as κ_3 and κ_2 in Eqs. (9.170) and (9.171). A dependence on $(1/\Delta T)$ usually involves unimolecular deposition. However, this temperature dependence can also result from a three-dimensional nucleus with an extraordinarily large contribution from an edge or line free energy.(146,147,166) The $(1/\Delta T)^2$ dependence of a three-dimensional growth nucleus can also result from coherent deposition in three dimensions.(146,147) The possibility also exists that the presence of a parent crystallite could influence the orientation and organization of the adjacent amorphous region and thus promote nucleation in its vicinity.

The concern at this point is limited to the temperature dependence. Therefore, the specifics of the nucleus structure, which are embodied in the parameters k_2, k_3, g_2 and g_3, do not have to be specified at this time. The second term of the exponential argument should dominate in the present temperature range of interest.

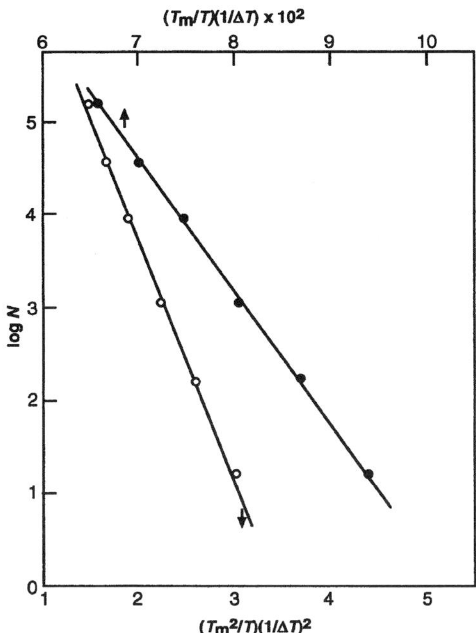

Fig. 9.38 Plot of log N against $(T_m^2/T)\,(1/\Delta T)^2$ ○ and against $(T_m/T)(1/\Delta T)$ ● for poly(decamethylene sebacate). (Data from Flory and McIntyre (22))

Consequently, plots of log N or log G against the appropriate temperature variable should yield straight lines.

The results of Flory and McIntyre (22) for the rate of formation of spherulitic centers in molten poly(decamethylene sebacate) provide a good set of data to examine whether Eq. (9.170) or (9.171) is operative. In these experiments the undercooling ranges from 11 °C to 16 °C. Concomitantly the initiation rate increases by a factor of 10^4. This result is clearly indicative of nucleation control. The log of the rate of spherulite formation, log N, for this polymer is plotted against $(T_m^2/T)(1/\Delta T)^2$ and $(T_m/T)(1/\Delta T)$ in Fig. 9.38. Excellent straight lines are obtained with either plot. Thus, it is not possible to discriminate between the two distinctly different initiation nucleation processes by analysis of the temperature coefficient.

A similar analysis, using Eqs. (9.172) and (9.173), can be made of spherulite growth rates of this polymer. Plots of log G against the two temperature coefficients are given in Fig. 9.39 for the same polymer. Excellent straight lines are again obtained with both of the plots. The spherulite growth rate data for linear polyethylene behave in a similar manner, as is illustrated in Fig. 9.40 for a molecular weight fraction $M = 210\,000$.(167) Similar results are obtained with other molecular weight fractions of linear polyethylene. Analysis of the growth rates of virtually all other polymers gives similar results. Thus, a decision as to whether

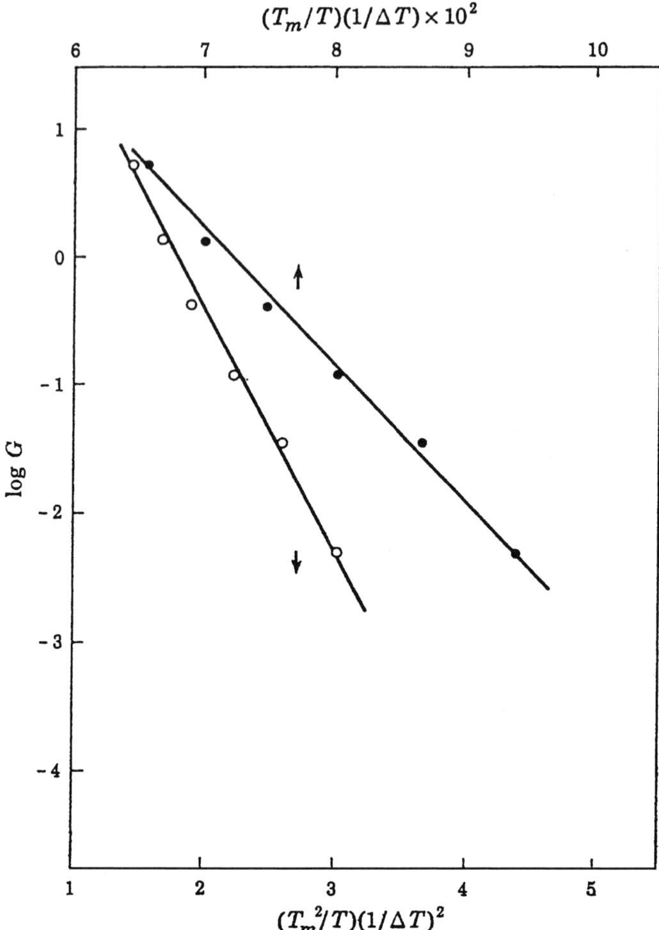

Fig. 9.39 Plot of log G against $(T_{\mathrm{m}}^2/T)(1/\Delta T)^2$ ○ and against $(T_{\mathrm{m}}/T)(1/\Delta T)$ ● for poly(decamethylene sebacate). (Data from Flory and McIntyre (22))

two- or three-dimensional nucleation is operative cannot be made solely on the basis of the temperature coefficient of spherulite growth rates. Analysis of the temperature coefficient of overall crystallization gives a similar conclusion. This conclusion remains unaltered when any reasonable variation is made in the value of T_{m}^0. The analysis can, however, be distorted by the introduction of an arbitrary transport term. Specific but arbitrary transport terms can favor one of the nucleation processes over the other. This procedure is unwarranted because of the small interval in the crystallization temperature that is involved.

In summary, within the limits of the precision of the kinetic data that are available a decision cannot be made as to whether a two- or three-dimensional nucleation process is operative, for either nucleation or growth, irrespective of the experimental

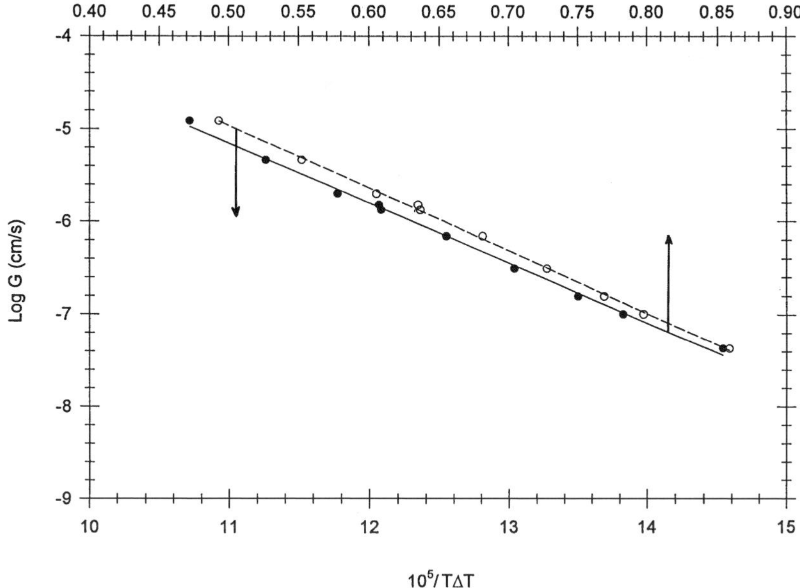

Fig. 9.40 Plot of log G against $(T_m^2/T)(1/\Delta T)^2$ ● and against $(T_m/T)(1/\Delta T)$ ○ for a fraction of linear polyethylene $M = 210\,000$. T_m^0 taken to be 145.5 °C. (Data from Hoffman *et al.* (167))

method. Although a particular choice may be consistent with a specific model it cannot be used as evidence to demonstrate that the specific nucleation process is actually operative. We are thus faced with an unfortunate and frustrating situation since nucleation plays such an extremely important role in controlling kinetics and the resulting structure and morphology. The latter is an important factor in determining the properties of crystalline polymers. Despite the fact that the desirable specifics of nucleation cannot be definitively established, the control role played by nucleation in polymer crystallization is well established. One has to appeal to physical intuition in selecting the type of nucleus. A Gibbs type growth nucleus will be assumed in discussing spherulite growth rates. The temperature coefficients of the nucleation and growth rates, and the value of g_3 are predetermined by the selection of this nucleus type. It must always be recognized that this choice is an assumption.

With the adoption of a Gibbs nucleus, spherulite growth rates in the vicinity of T_m^0 can be analyzed by use of Eq. (9.173). As was shown previously, in the absence of any contribution from the edge interfacial free energy ΔG_{rf}^* is given by Eq. (9.69).(147) ΔG_{rf}^* can be identified with ΔG^*, the classical result, for all meaningful purposes. Therefore, the dependence of the spherulite growth rate, G, on the crystallization temperature will have the same form as that of any coherent unimolecular nucleus, whether it be monomer or polymer, with folded or unfolded chains.

The analysis of the temperature coefficient of the overall crystallization rate is more complex. In this case both the initiation of the crystallite (or spherulite) and its subsequent growth has to be accounted for. Both of these processes are nucleation controlled. However, they are not necessarily of the same type. There are many possibilities that are consistent with a specific set of experimental results. For example, it could be assumed that the initiating or primary nucleus is three-dimensional and formed either homogeneously or heterogeneously. The further assumption can be made that the secondary or growth nucleus also has three-dimensional characteristics. It can be postulated further that the critical free energy required to form a secondary nucleus is less by a factor $\bar{a}$ than that for primary nucleation. With these assumptions

$$\ln k_s = \ln k_0 - \frac{nE_D}{RT} \frac{[1 + (n-1)\bar{a}]\kappa_3 T_m^2}{T(\Delta T)^2} \tag{9.174}$$

It can be postulated equally well that both the primary and secondary nuclei are two dimensional. The possibility that one nucleation is two dimensional and the other three cannot be ruled out. The temperature coefficient of all these possibilities will satisfy the experimental data so that a decision as to which pairs are involved is extremely difficult. This is an unfortunate situation since spherulites do not develop in many high molecular weight homopolymers and in random copolymers as the comonomer content increases.

The kinetic data can now be examined in terms of the concepts that have been developed. Attention is focused on spherulite growth rates since the analysis is simpler. With the understanding of the assumptions involved, ln G is plotted against $T_m^0/T\Delta T$ in Fig. 9.41 for some representative polymers.(72,167,168,169). The temperature range for crystallization has now been expanded slightly but is still in the vicinity of T_m^0. In preparing this figure the accepted values of T_m^0 for each of the polymers were used.(170) Contrary to the theoretical expectations developed so far, none of the plots can be represented by a single straight line. The data for each polymer are, however, well represented by a continuous curve. Similar results are obtained if the overall crystallization rates, in terms of $\ln(1/\tau)$, are analyzed. In some of the plots in Fig. 9.41, as well as those for other polymers, the data can be approximated by two intersecting straight lines. In a few situations the data can be represented by a single straight line.(167) The range of crystallization temperatures that can be studied is important. In the present discussion the polymers do not exhibit a maximum in the crystallization rate. In some cases, the temperature range that can be studied is very restricted. For example, measurements of the spherulite growth rates of linear polyethylene are limited to a 6–8 °C range in crystallization

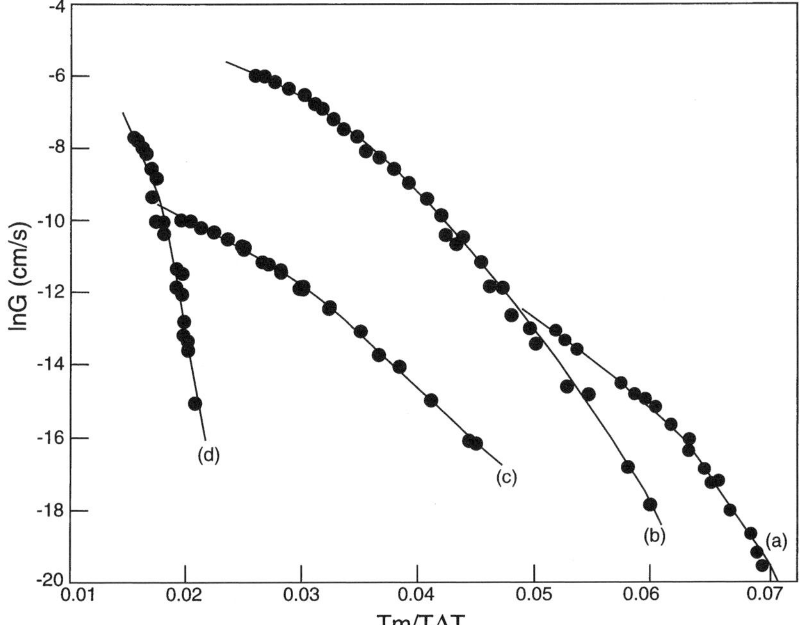

Fig. 9.41 Plot of $\ln G$ against $T_m/T\Delta T$ for some representative polymers. (a) Linear polyethylene fractions $M = 74\,400$. (Data from Hoffman *et al.* (167)); (b) poly(ethylene oxide), $M = 152\,000$ (From Kovacs and Gonthier (168)); (c) poly(chlorotrifluoroethylene) (From Hoffman and Weeks (169)); (d) poly(dioxolane) (From Alamo (72)).

temperatures.(167,171) This is a very small temperature interval and severely limits conclusions that can be derived from the experimental data.[12]

The representative plots in Fig. 9.41 present a dilemma that needs to be resolved before any progress can be made in understanding crystallization kinetics. This concern lies well beyond the question as to how best to represent the data, i.e. by either a continuous curve or two intersecting straight lines. The fundamental issue is to understand why the plots are not linear. There are several possibilities that can be considered in efforts to resolve this problem. One involves including the transport term. In the plots in Fig. 9.41 the activation energy for transport, or the transfer, of a unit from the liquid to crystalline region across the interfacial boundary has been neglected. Moreover, the exact form of E_D, the argument of the exponential term, has to be specified. Another possible reason for the nonlinearity of the plots involves the nucleation term. The argument can be made that spherulite growth involves successive nucleation acts. This would be true irrespective of what

[12] It has also been postulated that nucleation in polyethylene at atmospheric pressure occurs through a hexagonal phase known to be stable only at high temperatures and pressures.

specific chain conformation is involved. This successive nucleation process raises some questions that have not been considered heretofore. The simple nucleation and growth rate expressions, represented by Eq. (9.170) to Eq. (9.173), need to be re-examined. Each of these problems, i.e. nucleation and transport, will be analyzed separately. The nucleation aspects of the problem will be considered first. The role of the transport term will be treated subsequently since it is better examined when crystallization over the complete accessible temperature range is considered. On the other hand nucleation is best analyzed when considering experimental data in the vicinity of T_m^0.

A key factor in considering growth by successive nucleation acts on a crystallite surface is the relation between the rate of nucleation and that of spreading in the direction normal to the chain axes. The spreading rate will be designated as g. This problem was addressed by Nielsen (172), by Hillig (173) and by Calvert and Uhlmann (140) in treating the identical problem in three dimensions for monomeric systems. The magnitudes of the nucleation and spreading rates will be different and each will have different temperature coefficients. The possible temperature coefficients for the different types of nucleation have already been discussed. The general form of the nucleation rate–temperature curve is essentially the same for all polymers. Except for low molecular weights it is independent of chain length. On the other hand, the temperature coefficient of the spreading rate is not as clear. Conflicting reports have been reported with the same set of investigations (173a,b) and between investigators.(173c,174) The issue is whether g depends on the supercooling, or has a temperature coefficient typical of an activated process. There is also the question of whether g depends on molecular weight.(148)

The relationship between the rates as a function of the crystallization temperature, or undercooling, leads to some interesting situations. Schematic representations of three extreme cases are illustrated in Fig. 9.42. No effort has been made in these schematics to represent the interfacial structure. The conclusions reached are not dependent on the structure of the interfacial region. In the example shown in Fig. 9.42a, the spreading rate is much greater than the nucleation rate at all temperatures. Under these conditions a given growth layer will be complete before a new one is initiated. This temperature region corresponds to simple unimolecular nucleation. The growth rate relation previously derived applies. In the literature this region is termed Regime I.(174). In another extreme, limited to large undercoolings, both ΔG^* and the nuclei sizes are extremely small. The nucleation rate is very rapid and effectively constant with crystallization temperature. As is illustrated in Fig. 9.42c a large number of small nuclei will form on the crystallite face. Consequently, there is a limited area into which a nucleus can grow. Spreading is thus effectively retarded in the direction normal to the chain axes. The spreading rate therefore will be essentially zero under these circumstances. This region has

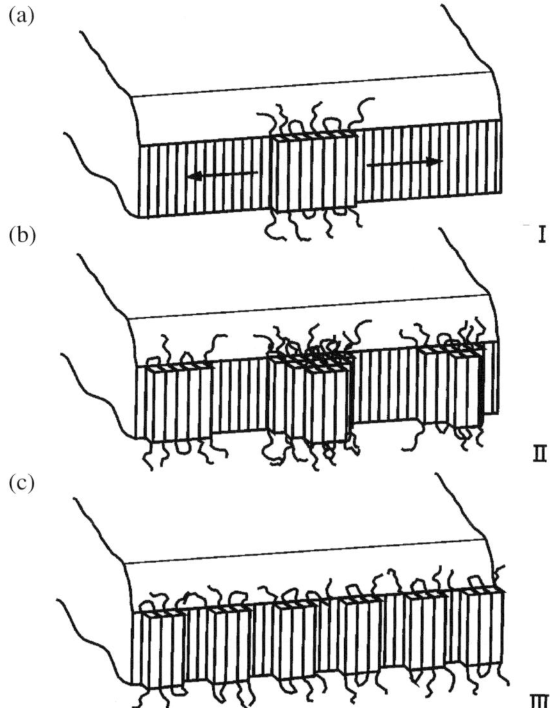

Fig. 9.42 Schematic representation of nucleation spreading rates in three different regimes. (a) Regime I; (b) Regime II; (c) Regime III. Interfacial region is not represented in these schematics.

been termed Regime III. Between the two extreme cases just described, the first of which occurs at low undercoolings, the other at very large undercoolings, there is also another possibility. In this situation the nucleation and spreading rates are comparable to one another. Therefore, several nucleation acts will take place on the same crystallite surface before a given layer is filled and growth can proceed. This case is schematically illustrated in Fig. 9.42b and is termed Regime II. The relation between the nucleation and spreading rates will be reflected in the net growth rate G. In essence, therefore, there has to be concern as to whether single or multinucleation takes place on a given crystal face and the relation between the nucleation and spreading rates. In the realm of small molecule crystallization, where these phenomena are observed, the first case is known as the "small crystal model", the last, the "large crystal model".(140)

There is another case involving polymers that should be considered. This is when g is small, as might be expected for very high molecular weight chains. This regime will be designated as Ia. In this situation nuclei will not spread much beyond the critical dimension in the lateral direction. There will not be any indication of a regime transition if this condition is maintained over the compete range of crystallization temperature.

Despite the different relations that exist between the nucleation and spreading rates, the crystallite and spherulite growth rates are still nucleation controlled processes. However, the actual temperature dependence will reflect the relative positioning of the two rates and the temperature region that is traversed during crystallization. There are, therefore, several possible reasons that the spherulite growth and overall crystallization rates do not in general adhere to the simple formulation of Eq. (9.173). To analyze the kind of experimental result illustrated in Fig. 9.41, a modification of the theory is clearly needed. Before addressing this problem in polymers, it is informative to examine how it has been treated for low molecular weight substances. At this point, the interest is only with growth and crystallization at the right side of the rate maximum. Discussion of the very rapid nucleation region (high undercooling), Regime III, will be discussed shortly.

This problem was initially analyzed by Nielsen (172), Hillig (173) and Calvert and Uhlmann (140) for the crystallization of low molecular weight substances. In the small crystal model, where the rate of spreading is much greater than that of nucleation, the growth step sweeps completely over the crystal substrate of length L and pauses before the next layer is nucleated. Formally, let dn be the number of nuclei that form in the time interval t to $t + dt$ on the surface of the crystal face. Then those that nucleated at time $t = 0$ can be expressed as

$$dn = \pi L^2 N \, dt \tag{9.175}$$

where L is taken as the radius of the substrate. The length L could correspond to one edge of a crystallite. It could just as well be smaller due to some type of obstacle to the spreading, for example imperfections in the lattice. It is difficult to quantitatively define this quantity.[13] If the rate of forming a growth layer is G/b, where b is the thickness of the growth step, then the number of nuclei formed in time b/G must be unity. It then follows that

$$\int_0^{b/G} \pi L^2 N \, dt = 1 \tag{9.176}$$

so that

$$G = \pi b L^2 N \tag{9.177}$$

In this temperature region the growth rate is directly proportional to the steady-state nucleation rate and the area of the substrate. The temperature dependence will then

[13] Other two-dimensional surfaces can be used for the shape of the substrate. Except for the constraints, the results are identical.

be given by the classical nucleation rate Eq. (9.171). It is tacitly assumed that the temperature dependence of L is minimal, and that it remains constant with time.

In the large crystal model new growth steps nucleate before the entire substrate is covered, i.e. multinucleation takes place on a given surface. Under these conditions, the growth rate G depends on both N and g. Therefore, the number of nuclei formed on the crystal surface in the time interval t to $t + dt$, that nucleated at $t = 0$, is given by

$$dn = \pi (gt)^2 N \, dt \tag{9.178}$$

The number of nuclei formed in time b/G can still be approximated by unity, although the initial nucleus has not spread to fill the entire substrate. It follows that

$$\int_0^{b/G} \pi (gt)^2 N \, dt \approx 1 \tag{9.179}$$

and

$$G = (\pi/3)^{1/3} b g^{2/3} N^{1/3} \tag{9.180}$$

Therefore, the temperature coefficients in the two regions will be quite different.

It is a straightforward matter to extend these concepts to polymer crystallization. This adoption can be accomplished without the need to assume a specific chain conformation within the nucleus. The results are quite general in this regard. It is convenient to make the assumption that the spreading in the chain direction is severely retarded relative to that in the lateral direction. This is a reasonable assumption for chain molecules. The more general case, where the spreading rates along and normal to the chain axis are allowed, can also be treated. With this simplifying assumption, Sanchez and DiMarzio (175) showed how to adapt the results for small molecules to polymers. The adaptation involves reducing the growth along the substrate by one dimension so that the thickness of the growth strip, l, or the growth nucleus, is constant, determined only by the crystallization temperature. The growth rate is then controlled by the lateral dimension L of the crystal face as well as the nucleation and spreading rates. Equations (9.177) and (9.180) are then transcribed to polymers. Thus, for unimolecular nucleation the growth rate in Regime I, G(I) is expressed as

$$G(\text{I}) = bLN \tag{9.181}$$

The growth rate in Regime II, G(II) is expressed as

$$G(\text{II}) = b(Ng)^{1/2} \tag{9.182}$$

Equations (9.181) and (9.182) represent the two extreme situations that have been treated.(156,176) They should be considered as asymptotes to the physical situations described by Regimes I and II. The issue of chain folding, and the associated interfacial structure, is not involved in reaching these conclusions.

If it can be assumed that in the vicinity of T_m^0 the temperature coefficient of the nucleation rate will dominate

$$\frac{d(\ln G(\mathrm{II}))}{d(T\,\Delta T)^{-1}} \bigg/ \frac{d(\ln G(\mathrm{I}))}{d(T\,\Delta T)^{-1}} = 1/2 \tag{9.183}$$

and the temperature coefficients of the growth rates will differ by a factor of ½ in these extremes. Equation (9.183) assumes that the temperature dependence of g can be ignored and there is no influence of molecular weight on this quantity. Based on the above analysis it is not surprising that typical growth data, as illustrated in Fig. 9.41, do not adhere to the simple formulation given by Eq. (9.171). The physical situations represented by Eqs. (9.181) and (9.182) are quite reasonable. It should be recognized, however, that they represent extreme or asymptotic situations. The fundamental question arises as to the nature of the transition between the two regimes, i.e. is it sharp or diffuse? The transition from one regime to the other has been tacitly assumed to be sharp.(174) A great deal of experimental data has been analyzed from this viewpoint.(174,177) On the other hand, Point and coworkers have argued that the transition from one to the other is so diffuse that the two regimes may in fact not exist.(173c,178a–d) The argument is based on the continuum theory given by Frank.(179) Although Frank gives an analytical solution to the problem, it was not presented in a tractable form that could be quantitatively applied to experimental data. It is possible, however, to utilize Frank's theory to address the problem in question.(180)

In the Frank theory, growth takes place on a substrate of length L. Each nucleation event creates a pair of steps that travel to the left and to the right with a spreading rate g. There are on the average $l(x)$ of the former and $r(x)$ of the latter. Since no steps enter from outside the limits $x = \pm\frac{1}{2}L$

$$l\left(x - \tfrac{1}{2}L\right) = r\left(x - \tfrac{1}{2}L\right) = 0 \tag{9.184}$$

Between these limits

$$\frac{\delta l}{\delta t} = N + g\left(\frac{\delta l}{\delta x}\right) - 2glr \tag{9.185}$$

$$\frac{\delta r}{\delta t} = N - g\left(\frac{\delta r}{\delta x}\right) - 2glr \tag{9.186}$$

where the three terms on the right represent initiation, drift and annihilation respectively. At steady state Eqs. (9.185) and (9.186) can be equated to zero so that

$$dl/dx = -N/g + 2lr = -dr/dx \tag{9.187}$$

Therefore

$$\frac{d(l+r)}{dx} = 0 \tag{9.188}$$

and

$$l + r = 2c \tag{9.189}$$

where c is a constant. By symmetry, the differential equation to be solved becomes

$$dl/dx = -N/g + 4cl - 2l^2 \tag{9.190}$$

It is convenient to introduce the quantity Q, defined as (179)

$$Q = (P^2 - L^2c^2) \qquad \text{where} \qquad 0 < Q < \frac{\pi}{2} \tag{9.191}$$

and

$$P \equiv L\left(\frac{N}{2g}\right)^{1/2} \tag{9.192}$$

It then follows that

$$Lc = Q \tan Q \tag{9.193}$$

and

$$P = Q \sec Q \tag{9.194}$$

The growth rate G, the quantity that is actually measured, is defined as

$$G = 2bcg \tag{9.195}$$

where b is the width of the chain.

Solutions to this problem were given for values of $Q \to 0$ and $Q \to \pi/2$.(179) The former approximation leads to the growth rate in Regime I, defined by Eq. (9.181). The latter corresponds to growth in Regime II, and to Eq. (9.182). The main interest here, however, is in the nature of the transition between the two regimes. This intermediate region was not explicitly described by Frank. In treating the same problem Lauritzen (176) calculated the upper and lower bounds to Regime II.

The problem is to express G in terms of $(1/T\Delta T)$ by means of the steady-state nucleation rate. To put Frank's solution in a form that can be applied to experimental data, it is noted that (180)

$$GL = 2bgcL = (Q \tan Q)2bg \tag{9.196}$$

Upon rearrangement, Eq. (9.196) gives

$$G^* \equiv G\left(\frac{L}{2bg}\right) = Q \tan Q \tag{9.197}$$

In this equation G^* can be considered to be a reduced value of the growth rate. Furthermore

$$P^2 = L^2\left(\frac{N}{2g}\right) = Q^2 \sec^2 Q = Q^2 + Q^2 \tan^2 Q \tag{9.198}$$

The steady-state nucleation rate can then be expressed as

$$N = P^2\left(\frac{2g}{L^2}\right) = [(Q^2 + Q^2 \tan^2 Q)]\frac{2g}{L^2} \tag{9.199}$$

Upon rearrangement

$$N^* \equiv \frac{NL^2}{2g} = Q^2 + Q^2 \tan^2 Q \tag{9.200}$$

where N^* is now a reduced steady-state nucleation rate.

From the Turnbull–Fisher expression for coherent deposition on an already existing substrate it follows that

$$\ln N^* = -\ln\frac{2g}{L^2 N_0} - \frac{E_D}{RT} - \frac{K_g}{T\Delta T} \tag{9.201}$$

and

$$\ln G^* = -\ln\frac{2bg}{L} + \ln G \tag{9.202}$$

Here $K_g = 4\sigma_{un}\sigma_{en}T_m/\Delta H_u$. If the temperature dependence of the transport term is neglected, for the moment, Eqs. (9.201) and (9.202) can be written as

$$\ln N^* = -K_1 - \frac{K_g T_m}{T\Delta T} \tag{9.203}$$

and

$$\ln G^* = -K_2 + \ln G \tag{9.204}$$

In the last two equations the constants K_1 and K_2 are unknown. They depend on the specific quantities L, b, g, and N_0. The theoretical relation between G^* and

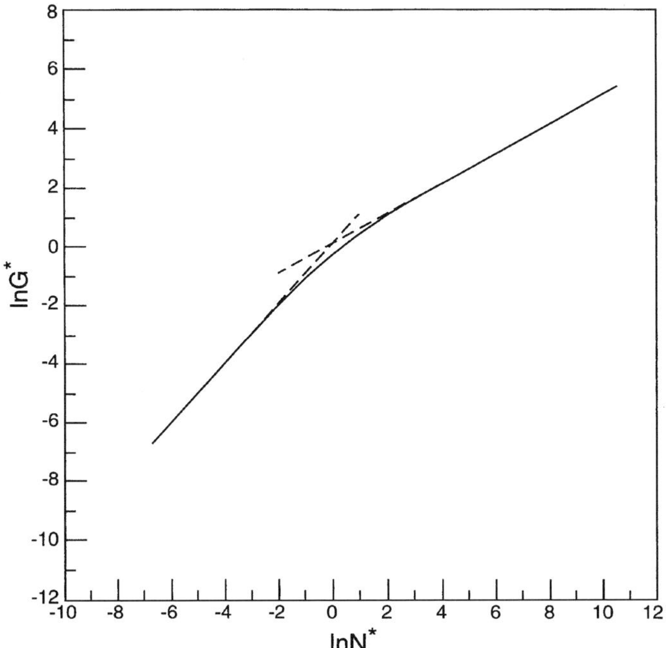

Fig. 9.43 Plot of $\ln G^*$ against $\ln N^*$ according to Eqs. (9.203) and (9.204).

N^* is implicitly given by the quantity Q that can vary from 0 to $\pi/2$. A plot of $\ln G^*$ against $\ln N^*$ is given in Fig. 9.43. This plot in effect represents the complete solution to Frank's equation. It cannot be represented by a single straight line. The two straight lines that are drawn represent Regimes I and II, and their slopes differ by the expected factor of 2. However, the transition from one regime to the other is not sharp. Thus, according to the theory the transition is diffuse. It now remains to ascertain how well the experimental data fit the theory, and how diffuse the transition actually is.

To analyze the experimental growth rates, plots of $\ln G$ against $T_m/T\,\Delta T$ are superimposed on the theoretical plot given in Fig. 9.43. The constants K_g, K_1 and K_2 are chosen so that the experimental data fit the theoretical plot at its asymptotes. Examples, utilizing spherulite growth rate–temperature data for some typical homopolymers, are given in the following.

The spherulite growth rates of linear polyethylene fractions have been extensively used in analyzing regime transitions.(167,171,174,176) However, for reasons that will become apparent shortly, analysis of this data will be postponed until later. The first polymer that will be analyzed by the method outlined above is a high molecular weight poly(ethylene oxide) fraction. Selection of a high molecular weight fraction avoids complications in the nucleation theory and the possibility of a change

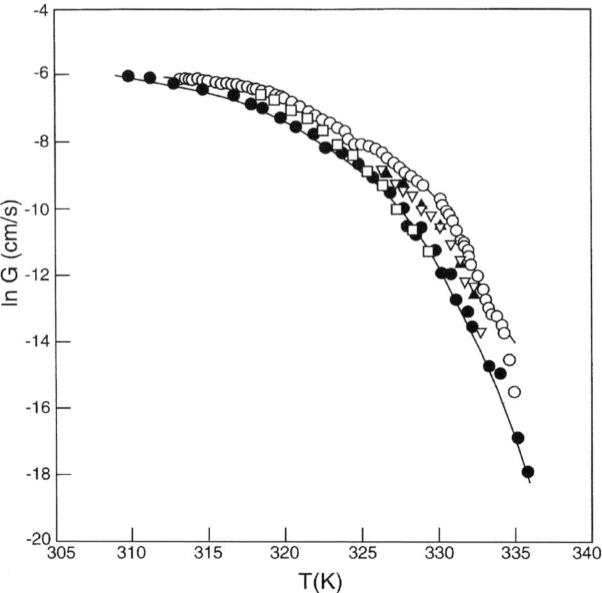

Fig. 9.44 Plot of $\ln G$ against crystallization temperature, T_c, for fraction of poly(ethylene oxide). ● $M = 152\,000$ (From Kovacs and Gonthier (168)); ○ $M = 105\,000$ (From Cheng *et al.* (181)); □ $M = 180\,000$ (From Marentette and Brown (182)); ▲ $M = 130\,000$ (From Booth and Price (183)); ▽ $M = 190\,000$ (From Booth and Price (183)).

from extended to folded chain crystallites with crystallization temperature. Growth rates in the molecular weight range of 10^5 for this polymer are available from several different investigators and are illustrated in Fig. 9.44.(168,181–183) With one exception, all of the $\ln G$–T data delineate smooth curves that are very close to one another. In one data set there are several inflections in the curve that were not observed by the others.(181) For analysis we have selected the data of Kovacs and Gonthier (168) since they cover the largest temperature range, 25 K. The fit to the master plot is shown in Fig. 9.45. The upper x-axis in the figure indicates the actual crystallization temperature. The solid points in the plot represent the experimental data. The high and low temperature data are fitted quite well by straight lines, whose slopes are in the ratio of 1:2. These regions represent Regimes I and II, each of which extends over a significant temperature range. However, the transition from one regime to the other is not sharp. It takes place over a temperature interval of about 4 K. Although this plot establishes that the two regimes exist for this polymer, the transition from one to the other is diffuse.

A similar analysis of the growth rate data for poly(chlorotrifluoroethylene) is shown in Fig. 9.46.(169) Here, the data cover a 35 K range in crystallization temperature. The results for this polymer adhere quite well to the theory. The two

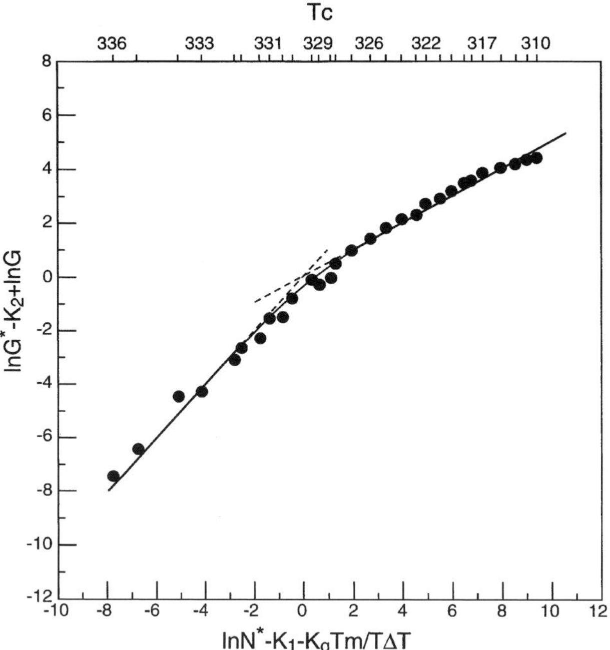

Fig. 9.45 Fit of experimental spherulite growth rate data, (•) to theoretical plot for poly(ethylene oxide), $M = 152\,000$. (Data from Kovacs and Gonthier (168))

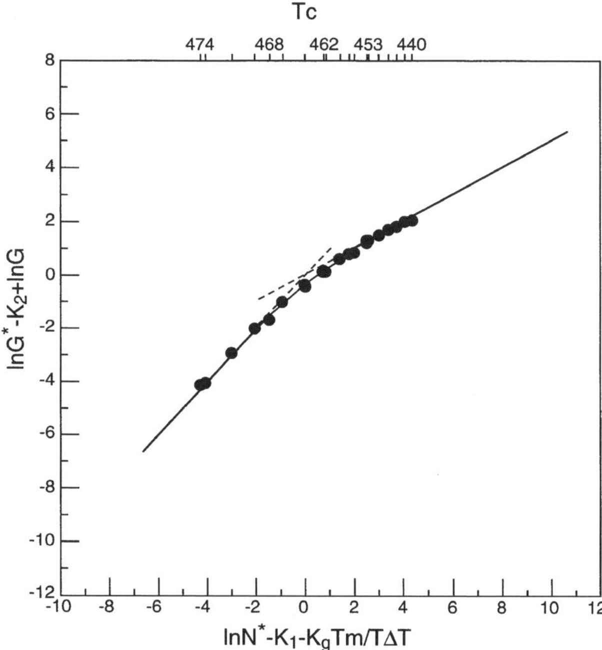

Fig. 9.46 Fit of experimental spherulite growth rate data (•) to theoretical plot for poly(chlorotrifluoroethylene). (Data from Hoffman and Weeks (169))

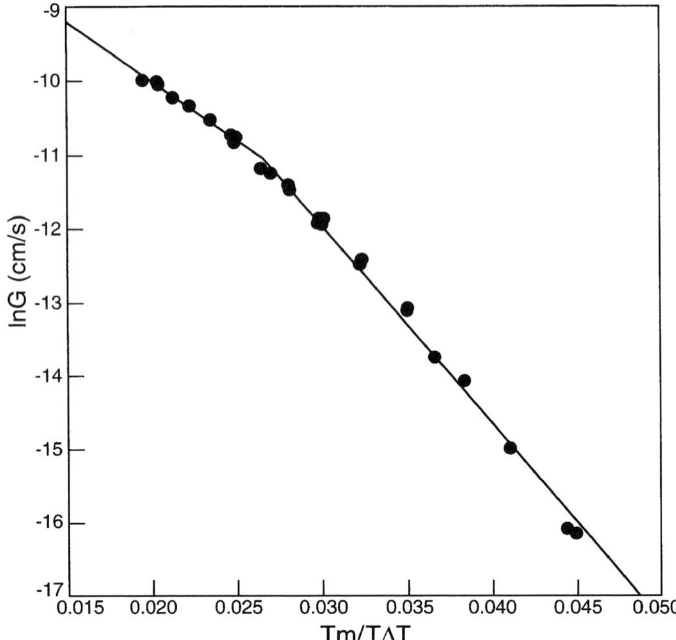

Fig. 9.47 Plot of $\ln G$ against $(T_m/T)(1/\Delta T)$ for poly(chlorotrifluoroethylene). (Data from Hoffman and Weeks (169))

regime regions are again well defined. There is, however, a diffuse transition region of about 5.5 K. The results are thus similar to those found for poly(ethylene oxide). The same data are plotted in the more conventional manner, as $\ln G$ against $T_m/T\ \Delta T$, in Fig. 9.47. As is shown, the data points can be represented by two intersecting straight lines. However, the slope ratio is only 1.7 in this case. This deviation from a ratio of 2 is beyond experimental error and is not acceptable as a demonstration of a I–II transition. However, the analysis given in Fig. 9.46, using Frank's theory, demonstrates that a regime transition does in fact take place. Results similar to those found in Figs. 9.45 and 9.46 are also found in the analysis of growth data for other polymers. These polymers include poly(dioxolane),(38) where there is an 8 K interval between the asymptotes that define Regimes I and II, poly(butene),(184) poly(ϵ-caprolactone),(185) and poly(L-lactic acid) (186) among others. The growth data for many polymers can be represented by two intersecting straight lines when plotted in the conventional manner. In some cases the slope ratios are well removed from 2 and the data cannot be fitted to the Frank theory.(187)

The spherulite growth rates of linear polyethylene fractions have been studied in detail.(167,171) The data obtained have been used extensively in arguments and discussions as to whether or not Regimes I and II exist, and, if they do, whether the transition from one to the other is sharp or diffuse.(178a–e,188) Detailed analysis

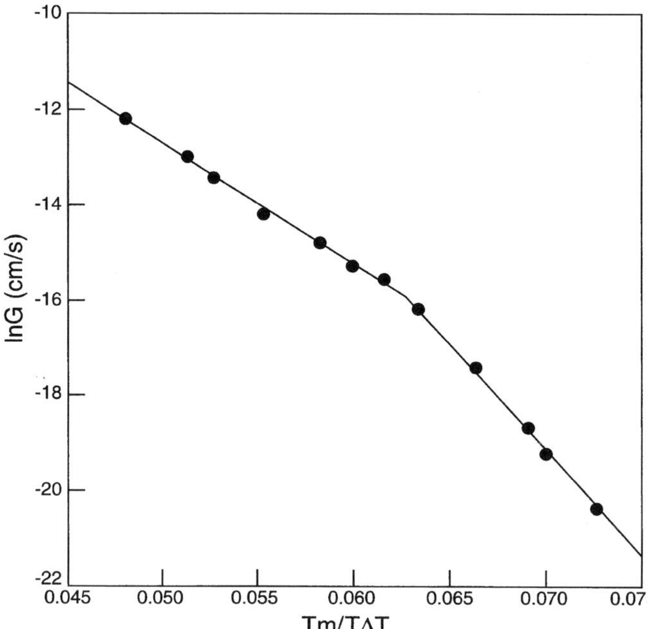

Fig. 9.48 Plot of $\ln G$ against $(T_m/T)(1/\Delta T)$ for a linear polyethylene fraction, $M = 133\,000$. (Data from Labaig (171))

of these results has been postponed because they suffer from a serious limitation. Unfortunately, isothermal spherulite growths of linear polyethylene fractions can only be measured over a small temperature interval. In the cases where the I–II transition in linear polyethylene has been analyzed, the crystallization temperatures have been limited to only a 5–7 K range. Data over such a limited temperature range are not adequate for an analysis of such an important matter. However, because of the widespread use of these data, and the contrary conclusions that have been reached, it is worthwhile to examine the results. Despite the large number of data points that have been obtained with each fraction, over a limited temperature range, the conclusions reached need to be severely tempered. As an illustrative example we take the results for a fraction $M_w = 133\,000$.(171) A conventional plot of the data, as illustrated in Fig. 9.48, can be represented by two intersecting straight lines, apparently indicating a sharp transition. However, the slope ratio of the two straight lines is 1.74. This is not an acceptable ratio for a transition between Regimes I and II, despite the fact that it has been used as such. In contrast, when the same data are analyzed by the Frank theory, as is shown in Fig. 9.49, Regimes I and II are clearly defined. The I–II transition occurs over a 1.2 K interval in this case. Similar results are obtained for the other linear polyethylene fractions that have been studied.(167,171) The conclusions are thus the same as were reached for other polymer types, although the transitions here are much less diffuse.

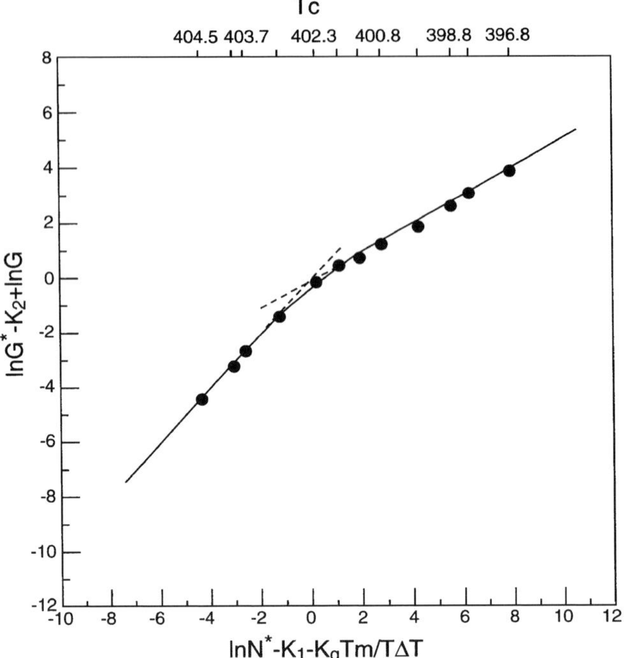

Fig. 9.49 Fit of experimental spherulite growth rate data (●) to theoretical plot of a linear polyethylene fraction $M = 133\,000$. (Data from Labaig (171))

Regimes I–II transitions have also been reported in two lower molecular weight fractions of linear polyethylene, $M \simeq 30\,000$ and $M \simeq 70\,000$. The transition in both cases is diffuse as was illustrated in Fig. 9.49 for a fraction $M = 133\,000$. The transition occurs over a 1–2 K temperature interval for all the linear polyethylene fractions studied. The diffuse nature of the I–II transition, as predicted by the Frank theory, is now well established for all polymers. It is important to recall that the underlying principles governing the I–II Regime transition are not limited to polymers. They are equally applicable to low molecular weight substances. Furthermore, for long chain molecules a regularly folded chain conformation within the nucleus is not required.

In the analysis of the I–II transition, attention so far has been focused solely on the nucleation term in the expression for the growth rate. The transport term has not been considered. Over the limited temperature range that is involved in the I–II transition region, reasonable values of the activation energy have only a slight influence in the analysis. This point is demonstrated in Fig. 9.50 for the poly(ethylene oxide) data shown in Fig. 9.45. Here E_D was taken as 10 000 cal mol^{-1}. The data still fit the Frank theory quite well and the breadth of the transition is still about 4 K. Reasonable values of E_D for this and other polymers do not sensibly change the result.

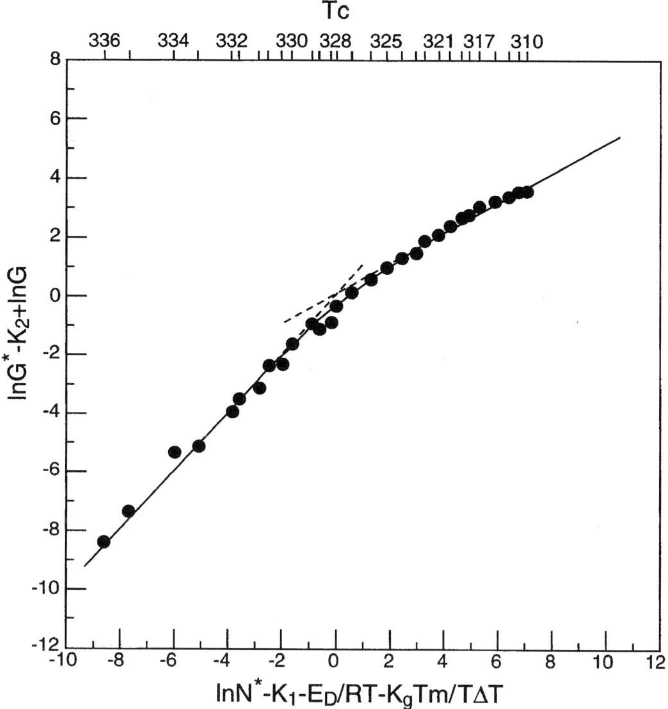

Fig. 9.50 Fit of experimental growth rate data (●) to theoretical plot for poly(ethylene oxide) $M = 152\,000$ with $E_D = 10\,000\,\text{cal mol}^{-1}$. (Data from Kovacs and Gonthier (168))

The analysis given above for some representative homopolymers strongly supports the concept that Regimes I and II exist. These regimes are in fact more than just asymptotes. The different relations between the nucleation and spreading rates manifest themselves in the macroscopic growth rates and explain the observed temperature coefficients. However, the results also demonstrate that the transition between Regimes I and II is not sharp as has been widely thought. The transitions are in fact diffuse in accord with the expectations from the Frank theory.

It is of interest at this point to examine the values of the constants K_1 and K_2. A summary of the values of these constants for some typical homopolymers is given in Table 9.5. It is difficult to ascertain any pattern, or relation, between these values and the chemical nature of the chain repeating unit. Furthermore, it is not possible to obtain values for the specific quantities g, L, and N_0 that specify K_1 and K_2. Lauritzen has shown that when the quantity $Z \equiv NL^2/4g$ is less than 0.01 the system is in Regime I.(176) In order to use the value of Z to determine the regime of a particular data set, three independent quantities, two if the ratio N/g is considered as one, need to be known. Since all of these quantities are independent of one another Z cannot be calculated without making several arbitrary assumptions.(176)

Table 9.5. *Summary of constants K_1 and K_2*

Polymer	K_1	K_2	Reference
Poly(ethylene oxide)	−23	−10	a
Linear polyethylene			
$M_w = 30\,600$	−29	−12	b
$M_w = 35\,000$	−33	−15	c
$M_w = 74\,000$	−34	−16	b
$M_w = 133\,000$	−32	−16	b
Poly(chlorotrifluoro ethylene)	−11	−12	d
Poly(ε-caprolactone)	−9.4	−10	e
Poly(butene)	−11	−11	f
Poly(L-lactic acid)	−23	−15	g
Poly(dioxolane)	−32	−10	h

References

a. Kovacs, A. J. and A. Gonthier, *Koll. Z. Z. fur Polym.*, **250**, 520 (1974).
b. Hoffman, J. D., L. J. Frolen, G. S. Ross and J. I. Lauritzen, Jr., *J. Res. Nat. Bur. Stand*, **70A**, 671 (1978).
c. Labaig, J. J. Ph.D. Thesis, Louis Pasteur University Strasburg (1978).
d. Hoffman, J. D. and J. J. Weeks, *J. Chem. Phys.*, **37**, 1723 (1962).
e. Goulet, L. and R. E. Prud'homme, *J. Polym. Sci.: Pt. B: Polym. Phys.*, **28**, 2329 (1980).
f. Monasse, B. and J. M. Haudin, *Makromol. Chem. Macromol. Symp.*, **20/21**, 295 (1988).
g. Vasanthakumari, R. and A. J. Pennings, *Polymer*, **24**, 175 (1983).
h. Alamo, R., J. G. Fatou and J. Guzman, *Polymer*, **23**, 374 (1982).

It might be expected that the differences in the growth of a crystallite face would alter the morphology of mature crystallites in the two regimes. Such changes have been reported for some polymers (38,167,186,187,189) but not for others.(189a) A more detailed discussion of this point will be found in Volume 3. To summarize, the analysis of the spherulite growth rates in the vicinity of T_m^0 requires the introduction of the concepts characteristic of Regimes I and II and the transitions from one to the other. Similar results are found when overall crystallization rates are analyzed.(189a)

9.11 Kinetics over an extended temperature range

As has been noted previously, when the crystallization is conducted over an extended temperature range most homopolymers display a maximum in both spherulite growth and overall crystallization rates. There are a few polymers that do not show rate maxima under these conditions. In these cases sufficiently low crystallization temperatures cannot be attained in order for a maximum to be observed. The objective in this section is to utilize the general concepts of nucleation and growth to interpret the experimental results. The main points to be addressed are an explanation of the maxima, the reality of a transition from Regime III to II and a consideration of the polymers that do not show the maximum.

To formulate the problem the Turnbull–Fisher approach is adopted once again. The Gibbs type nucleus is assumed to still be operative at the lower temperatures.(129) It is necessary, however, to introduce a transport term that accounts for the fact that as the glass temperature is approached crystallization will become severely retarded. The Arrhenius activation term that has been used up to now fails some 50–80 °C above the glass temperature.(190) To rectify this problem it has been found useful to borrow the Vogel expression.(191) This expression has been useful in explaining the bulk viscosity of glass forming liquids.[14] With this set of assumptions the spherulite growth rate over an extended temperature range can be expressed as (174)

$$G = G_0 \exp\left(\frac{-U^*}{T - T_\infty}\right) \exp\left(\frac{-KT_m^0}{T_c\, \Delta G_u(T)}\right) \tag{9.205}$$

The particular regime involved remains undefined at the moment. Since the interest now is in crystallization over a large temperature range account must be taken of the temperature dependence of the interfacial free energies, embodied in K, and ΔG_u, the free energy of fusion per repeating unit. The former has been tacitly ignored. The latter can be formally expressed by the expansion of $\Delta G_u(T)$ about T_m^0. Up to now, since the analysis was limited to crystallization in the vicinity of T_m^0, only the first term in the expansion has been used. Now, if further terms are needed, the appropriate derivatives of the specific heat can be used. To avoid this formal thermodynamic procedure an empirical relation, based on several assumptions, has been proposed.(192) Utilizing this empiricism Eq. (9.205) can be written as

$$G = G_0 \exp\left(\frac{-U^*}{T - T_\infty}\right) \exp\left(\frac{-KT_m^0}{(T\, \Delta T) f}\right) \tag{9.206}$$

where f is defined as

$$f \equiv \frac{2T_c}{T_m^0 + T_c} \tag{9.207}$$

In the analysis of experimental results that follows the introduction of the parameter f does not sensibly affect the interpretation of the results.

In the above equation T_∞ is the temperature where molecular or segmental motion ceases. It can be defined in terms of the glass temperature, T_g, as

$$T_\infty = T_g - C \tag{9.208}$$

[14] Alternatively, other expressions can be used that serve the same purpose just as well.

Here U^* and C are constants whose values cannot be *a priori* specified.(193) Equation (9.206) can conveniently be rewritten as

$$G = G_0 \exp\left(\frac{-U^*}{T - T_g + C}\right) \exp\left(\frac{-KT_m^0}{T\Delta T}\right) \tag{9.209}$$

There are several points that need to be kept in mind when applying Eq. (9.209). The Vogel equation represents viscous flow and is global in character. On the other hand, in polymer crystallization the transport takes place across a boundary, and is thus localized. The form of the Vogel equation is what is important in the present context. It is not necessary that the constants be the same as those involved in viscous flow of the pure polymer melt. Contrary to what is often stated, Eqs. (9.206) and (9.209) do not represent any basic theory. These equations represent the result of introducing a set of assumptions into the well-founded Turnbull–Fisher theoretical expression for the steady-state nucleation rate. The assumptions involve adoption of the Gibbs type nucleus and the Vogel expression for segmental motion. These equations do not require a specific chain conformation within the nucleus. Despite these restraints, the applicability of Eq. (9.209) to crystallization over an extended temperature range can be investigated.

The spherulite growth rate of isotactic poly(styrene) has been extensively studied over a wide temperature range.(190,194–197) A summary of results from five different investigations is given in Fig. 9.51. There is excellent agreement among these diverse sources. The results of Miyamoto *et al.* (194) have been selected for detailed analysis since they encompass a large temperature range, 13 K above T_g and 22.4 K below T_m^0. The analysis of these data, according to Eq. (9.209), is given in Fig. 9.52. Here the points represent the experimental data and the curve is Eq. (9.209) with $U^* = 1499$ and $C = 39$. These constants are reasonable ones. There is very good agreement between theory and experiment. There is no evidence in this plot of a transition from one regime to another. Another example of this type of analysis is given in Fig. 9.53 for poly(aryl ether ether ketone).(198) Good agreement is again obtained between theory and experiment for this polymer within one regime. The constants in this case are $U^* = 3690$, $C = 73$. As is shown in Fig. 9.54, similar results are obtained with the aliphatic polyester, poly(3-hydroxy butyrate) (199) with $U^* = 4335$ and $C = 71$.

The results shown in Figs. 9.52 to 9.54 indicate that the growth rate data can be represented by Eq. (9.209) without invoking any transition from one regime to another. It can be reasonably assumed that the crystallization is occurring in Regime II. Similar results are obtained for many other polymers, as is summarized in Table 9.6. The U^* and C constants listed in the column marked "No III–II" give excellent smooth curve fits to the experimental data. The values of the constants needed

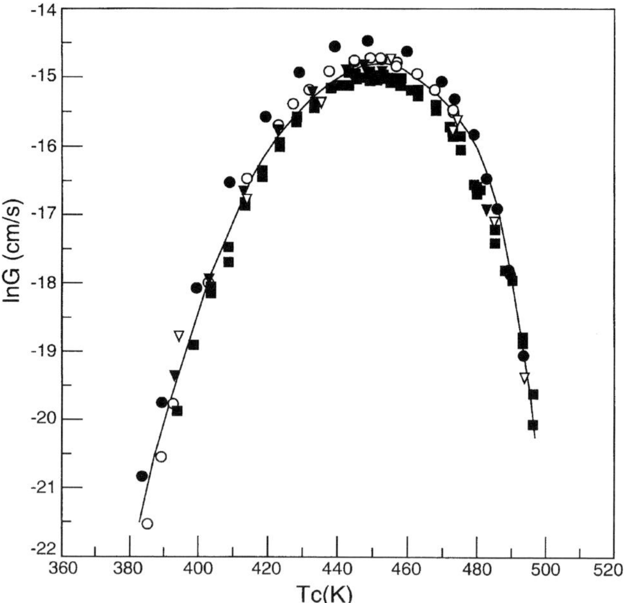

Fig. 9.51 Plot of $\ln G$ against crystallization temperature, T_c, for isotactic poly(styrene), ● Miyamoto *et al.* (194); ○ Suzuki and Kovacs (190); ▼ Boon *et al.* (195); ▽ Edwards and Phillips (196); ■ Iler (197).

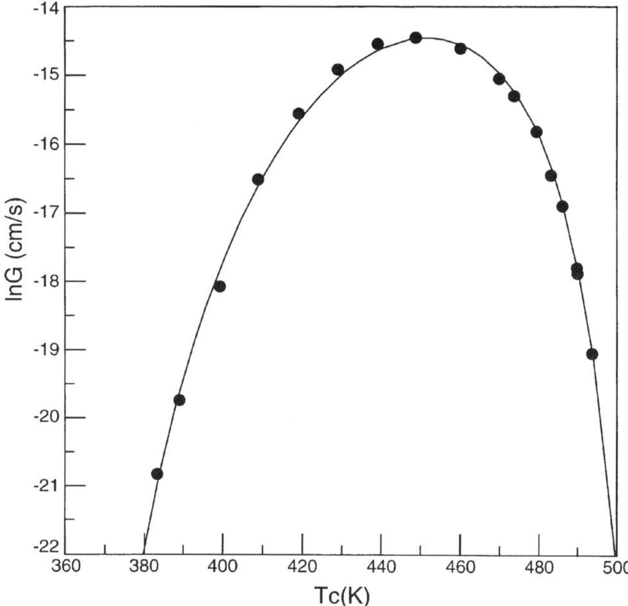

Fig. 9.52 Plot of $\ln G$ against crystallization temperature, T_c, for isotactic poly(styrene). Solid curve according to Eq. (9.209) with $U^* = 1499$, $C = 39$, ● experimental results from Miyamoto *et al.* (194).

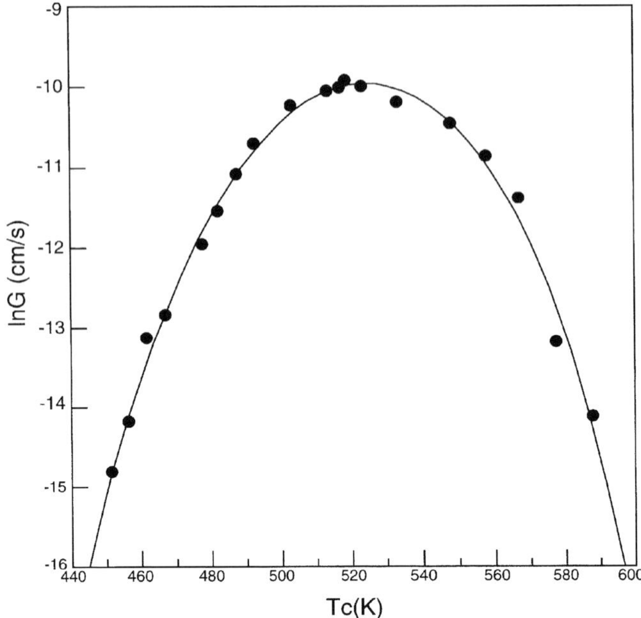

Fig. 9.53 Plot of $\ln G$ against crystallization temperature, T_c, for poly(aryl ether ether ketone). Solid curve according to Eq. (9.209) with $U^* = 3690$, $C = 73$, ● experimental results. (From Medellin-Rodriguez, *et al.* (198))

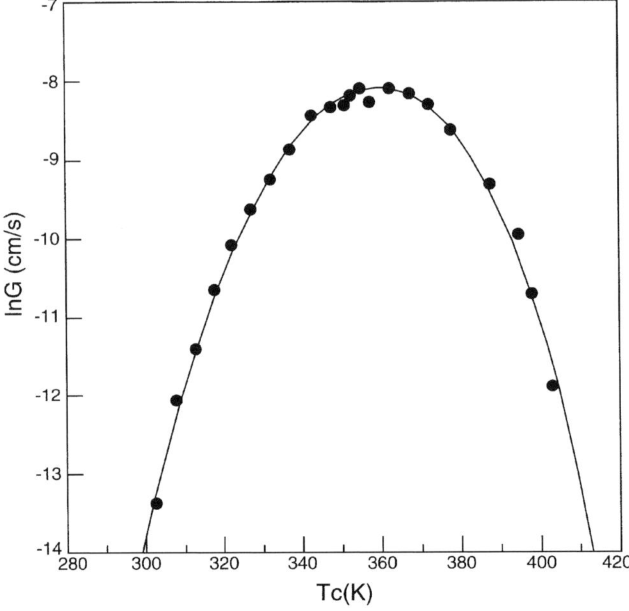

Fig. 9.54 Plot of $\ln G$ against crystallization temperature, T_c, for poly(hydroxy butyrate). Solid curve according to Eq. (9.209) with $U^* = 4335$, $C = 72$, ● experimental results. (From Organ and Barham (199))

Table 9.6. *Analysis of Regime III–II transition*

Polymer	No III–II U/C	Yes III–II U/C	Reference
iso-Poly(styrene)	1499/30	1525/36; 2300/48; 4120/74; 5640/90	a
Poly(tetramethylene-p-silphenylene siloxane) $M = 56\,000$	520/0	1500/21; 4090/68; 5800/90	b
cis-Poly(isoprene) $M = 313\,000$	1936/38	2050/19; 4120/36; 9950/66	c
Poly(caproamide)	672/0	1625/10; 4000/55	d
Poly(caproamide)	2129/60	2000/41; 4000/76; 4940/90	e
Poly(L-lactic acid)	2514/37	2350/0; 4000/17; 17500/90	f
Poly(phenylene sulfide) $M = 51\,000$	1299/27	1590/27; 4000/63; 6390/90	g
Poly(R-epichlorohydrin)	1273/21	1500/7; 4120/49; 8940/90	h
Poly(ethylene terephthalate)	1967/34	2500/29; 4000/46; 9400/90	i
Poly(aryl ether ether ketone)	3690/73	2435/46; 4120/62; 6850/90	j
Poly(ethylene-2,6-naphthalene dicarboxylate	4900/46	6000/44; 8000/57; 10000/36	j
Poly(3-hydroxy butyrate)	4335/72	2550/38; 4050/52; 8500/90	k
Poly(1,3-dioxolane)	7570/54	4120/12; 9800/22	l
TPI (polyimide)	692/0	1490/9; 4120/41; 10850/90	m
Poly(oxymethylene)	1037/0	25500/0	n
cis-Poly(butadiene)	381/0	1480/25; 3950/64; 6070/90	o
Poly(propylene oxide)	461/0	4100/36; 9500/90	p
syn-Poly(propylene)	461/0	1480/11; 4120/58; 6800/90	q
Poly(3-hydroxy valerate)	751/0	1685/0; 4050/31; 9956/67	r

References

a. Miyamoto, Y., Y. Tanzawa, H. Miyaji and H. Kiho, *Polymer*, **33**, 2496 (1992).
b. Magill, J. H., *J. Appl. Phys.*, **35**, 3249 (1964).
c. Phillips, P. J. and N. Vatansever, *Macromolecules*, **20**, 2138 (1987).
d. Burnett, B. B. and W. F. McDevit, *J. Appl. Phys.*, **28**, 1101 (1957).
e. Magill, J. H., *Polymer*, **3**, 655 (1962); *ibid.*, **6**, 367 (1965).
f. Vasanthakumari, R. and A. J. Pennings, *Polymer*, **24**, 175 (1983).
g. Lovinger, A. J., D. D. Davis and F. J. Padden, Jr., *Polymer*, **26**, 1595 (1985).
1. Singfield, K. L. and G. R. Brown, *Macromolecules*, **28**, 1290 (1995).
i. van Antwerpen, F. and D. W. van Krevelen, *J. Polym. Sci.: Polym. Phys. Ed.*, **10**, 2423 (1972).
j. Medillin-Rodriguez, F. J., P. J. Phillips and J. S. Lin, *Macromolecules*, **28**, 7744 (1995).
k. Organ, S. J. and P. J. Barham, *J. Mater. Sci.*, **26**, 1368 (1991).
l. Archambault, P. and R. E. Prud'homme, *J. Polym. Sci.: Polym. Phys. Ed.*, **18**, 35 (1980).
m. Hsiao, B. S., B. B. Sauer and A. Biswas, *J. Polym. Sci.: Pt. B: Polym. Phys.*, **32**, 737 (1994).
n. Inoue, M., *J. Appl. Polym. Sci.*, **8**, 2225 (1964).
o. Cheng, T. L. and A. C. Su, *Polymer*, **36**, 73 (1995).
p. Magill, J. H., *Makromol. Chem.*, **86**, 283 (1965).
q. Supaphol, P. and J. E. Spruiell, *Polymer*, **41**, 1205 (2000).
r. R. Pearce and R. H. Marchessault, *Macromolecule*, **27**, 3869 (1994).

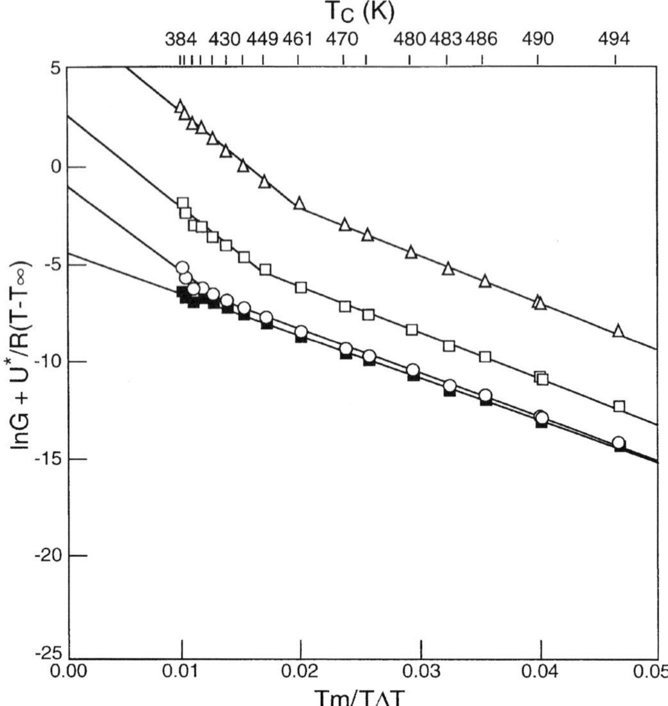

Fig. 9.55 Plot of $\ln G + U^*/R(T - T_\infty)$ against $(T_m/T)(1/\Delta T)$ for isotactic poly(styrene). ■ $U^* = 1499$, $C = 39$; ○ $U^* = 1525$, $C = 36$; □ $U^* = 2300$, $C = 48$; ▼ $U^* = 4120$, $C = 74$. (Data from Miyamoto *et al.* (194))

are reasonable ones. Alternatively, according to Eq. (9.209) a plot of $\ln G + U^*/R(T - T_\infty)$ against $T_m/T\,\Delta T$ gives a straight line when these constants are used. It could be concluded from this analysis that, with the selection of appropriate parameters, the crystallization over the complete temperature range does not require a transition to Regime III.

However, the situation is not as simple as it appears. With just a small change in the value of U^*, for a given polymer, a discontinuity can be observed in the plot of $\ln G + U^*/R(T - T_\infty)$ against $T_m^0/T\,\Delta T$. Such behavior is illustrated in Fig. 9.55 for the same isotactic poly(styrene) data used in Fig. 9.52. The solid squares represent the results obtained using the same constants as in Fig. 9.52. Obviously, a straight line results. However, when U^* is increased from 1499 to 1525 cal mol^{-1}, and C reduced from 39 to 36, a curvature appears in the plot, as indicated by the open circles in the figure. As the constants are varied the data can be represented by two intersecting straight lines. There is a set of constants, $U^* = 4120$ cal mol^{-1} and $C = 74$, that gives a slope ratio of 2.0. This situation corresponds exactly to a III–II Regime transition. Similar results are found for all the polymers listed in Table 9.6.

The column labeled "Yes III–II" lists the sets of U^* and C values that allow for plots of the type illustrated in Fig. 9.55 to be represented by two intersecting straight lines, with slope ratios of 2.0. Each set defines a different transition temperature.

The analysis of the spherulite growth rate data over an extended temperature range thus presents a major dilemma. The reason is that the values of the constants U^* and C are not known *a priori* for a given polymer. It is important to recognize that there is no set of universal constants, as has been proposed,(174, 190,193) although this concept is often invoked. Each polymer will have its own set of unique constants. There are, thus, two conflicting results from the above analysis. In one case for a given set of constants there is no evidence for a regime transition, the crystallization taking place in Regime II. On the other hand, there is another set of constants for a given polymer that allow the data to adhere exactly to a III–II transition. There is no basis at present to discriminate between the two sets of constants.

The physical basis for the existence of Regime III is quite plausible. The nucleation rate continuously increases with decreasing temperature. Thus, at low temperatures, i.e. large undercoolings, the rate is rapid and the nuclei are profuse and small in size. Consequently, there is not too much space into which the nuclei can spread and grow. More quantitatively, the mean separation distance between nucleation sites in Regime II, S_k, can be expressed as (193)

$$S_\text{k} = (2g/N)^{1/2} \tag{9.210}$$

The increase in nucleation rate with ΔT is the dominant term in Eq. (9.210). Consequently, the distance between nucleation sites decreases with the undercooling. Eventually, at sufficiently low crystallization temperatures the site separation becomes comparable to the chain width. There is, thus, a temperature region where the nucleation rate is once again the dominant factor. The growth rate expression then becomes the same as in Regime I. In the type of plots shown in Fig. 9.55 the ratio of the slope in Regime III to that in Regime II should be 2:1. This result is easily obtainable with the data available, as is the contrary result of no regime transition. This enigma has been observed previously where, however, the reality of Regime III has been assumed.(198,199a)

It should be emphasized once again that the issue is not one of physical reality. The physical situation is succinctly expressed by Eq. (9.210). Rather it is a question of whether a definite transition exists between Regimes II and III, or whether the changes are gradual and diffuse. An argument has been made that Regime III does not exist at all.(178a) The contrary argument has been made that there is a sharp transition between Regimes II and III.(193) The analysis of experimental growth rate data is equivocal in the matter. Even with the large amount of suitable data that is available, it is difficult to make an objective choice without knowledge of U^* and C for each polymer. This is a problem that is clearly in need of resolution.

Polymers that crystallize in a temperature interval well removed from T_m, but whose growth rates do not display maxima, also present problems with respect to the existence of the III–II transition. Polymers in this category include poly(butylene terephthalate),(200) poly(trimethylene terephthalate),(201) poly(pivalolactone), (202,203) one study of poly(methylene oxide),(204) linear polyethylene over a slightly more extended temperature range (205, 205a) and a set of reports for isotactic poly(propylene).(206–210)

When plotted according to Eq. (9.209) the data for both poly(trimethylene terephthalate) (201) and poly(pivalolactone) (202,203) yield two intersecting straight lines. The slope ratio in each case was 2. The constants $U^* = 1500$ and $C = 30$ were used for both polymers, along with extrapolated values of T_m^0. The temperature range for crystallization was 35 K. The results were not too sensitive to the chosen values of U^* and C. Poly(pivalolactone) is, however, polymorphic. This is a factor that needs to be investigated further and taken into account if necessary. These results, along with a linear polyethylene fraction, $M = 70\,300$,(205) seem to be exceptions to the general finding. The temperature range studied for the linear polyethylene fraction was extended to 15 K. Analysis of the data for this polymer, using an Arrhenius expression with $E_D = 5736$ and $T_m^0 = 144.7\,^\circ\text{C}$, gave two sharp transitions, with slope ratios that were appropriate to the regimes involved. However, small variations in E_D and T_m^0 yield slightly diffuse transitions, and such a transition has been reported.(205a)

The study of poly(butylene terephthalate) gave quite different results.(200) The data in this case extended over 23 K. Utilizing the extrapolated value of 236 °C for T_m^0 the data could be fitted by two intersecting straight lines. However, in this case the slope ratio was 3.41. This value is not compatible with a III–II transition. However, if T_m^0 is taken to be 260 °C, two intersecting straight lines with a slope ratio of 2 are observed. The analysis of the experimental results for this polymer demonstrates how the same data can yield two intersecting straight lines, but with different slope ratios. The slope ratios can vary over a wide range depending on the choice of T_m^0. The equilibrium melting temperature of poly(butylene terephthalate) has also been reported to be at 247 °C.(211) The importance of T_m^0 in determining whether there is a sharp III–II transition is emphasized by these results. Analysis of the available growth rate data for isotactic poly(propylene), to be discussed later, will further expand on this conclusion.

The study of poly(methylene oxide) by Pelzbauer and Galeski (204) is often quoted as representing a sharp III–II transition.[15](193,212) However, in another

[15] The unusual temperature variables that are used in this report, for both three- and two-dimensional nucleation, should be noted. In the present analysis, the basic growth–temperature data have been extracted from this work.

study of this polymer the growth rate was found to go through a maximum with crystallization temperature.(213) This data set can be treated in the manner previously described (see Table 9.6). The Pelzbauer and Galeski data, encompassing 19 K, can be reasonably represented in the usual type plot by two intersecting straight lines. The slopes depend on the values taken from T_m^0 and the Vogel equation constants. The slope ratio is 2.2 when T_m^0 is taken as 198 °C and the transport term ignored. This ratio can be reduced by increasing T_m^0 and introducing U^* and C. It is then possible to achieve a ratio of 2 for the slopes of the two straight lines.

The growth rate of isotactic poly(propylene) has been studied over the accessible temperature range by many different investigators.(206–210) The fundamental problem here is establishing the value of T_m^0 by conventional extrapolative methods. Isotactic poly(propylene) is unique in displaying melting kinetics.(214) In the fusion range, the observed melting temperature of isotactic poly(propylene) depends on the time the sample is held at a given temperature. It is thus extremely difficult to reliably estimate T_m^0. Estimates of this quantity have ranged from 185 °C to 215 °C. It may in fact be even higher. With this uncertainty in T_m^0, it is not surprising that different conclusions have been reached with regard to the III–II transition in this polymer. Slope ratios, based on two intersecting straight lines, have ranged from 1.4 to 2.6 depending on the value taken for T_m^0. However, by arbitrarily choosing a T_m^0 value each of the poly(propylenes) studied can be represented by two intersecting straight lines, with a slope ratio of 2. Thus, they can be made to represent a sharp III–II transition. The necessary T_m^0 values are 199 °C (210), 189 °C (207), 186 °C (208) and 175 °C.(209)

To further complicate matters the concept of a diffuse III–II transition was introduced by Xu *et al.*(210) Their data are well represented by two intersecting straight lines of slope ratio 2 when T_m^0 is taken to be 199 °C. However, it was concluded for other reasons that T_m^0 for this polymer is properly 212.1 °C. The results of using 212.1 °C for T_m^0 in analyzing this data are shown in Fig. 9.56, utilizing $U^* = 1500$ cal, $C = 30$. Although the major portion of the data points can be represented by two intersecting straight lines of slope ratio 2 there is a set of points that cannot be included. The III–II transition has now become a diffuse one. The transition occurs gradually over a temperature range of approximately 9 K.(210) Thus by selecting appropriate values of T_m^0 the transition can be made either sharp or diffuse. It can be anticipated that for still higher T_m^0 values the transition would become more diffuse. Eventually, a transition, as such, would not be perceptible. The analysis of isotactic poly(propylene) growth data further emphasizes the fact that in order to decide whether the transition is sharp, and with the proper slope ratio, or is diffuse, depends on the confidence in the value of T_m^0. Another problem related to the III–II transition in isotactic poly(propylene) is the fact that vibrational spectroscopic studies have shown that a structural change occurs in the residual melt at about 136 °C.(214a,b) This temperature corresponds to the one usually assigned to

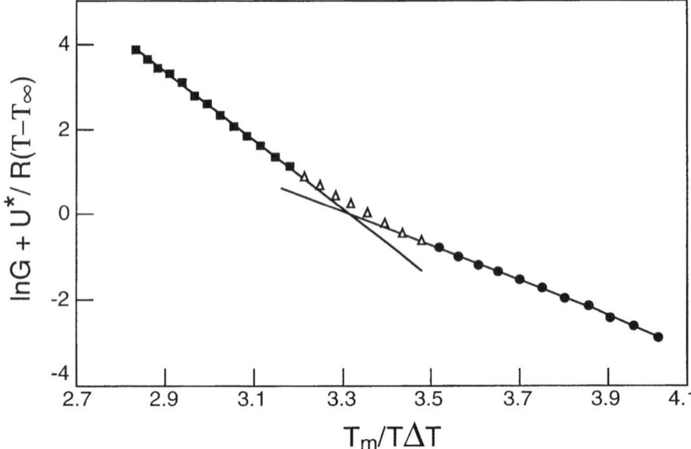

Fig. 9.56 Plot of $\ln G + U^*/R(T - T_\infty)$ against $(T_m/T)(1/\Delta T)$ for isotactic poly(propylene). T_m^0 taken as 212.1 °C. (From Xu *et al.* (210))

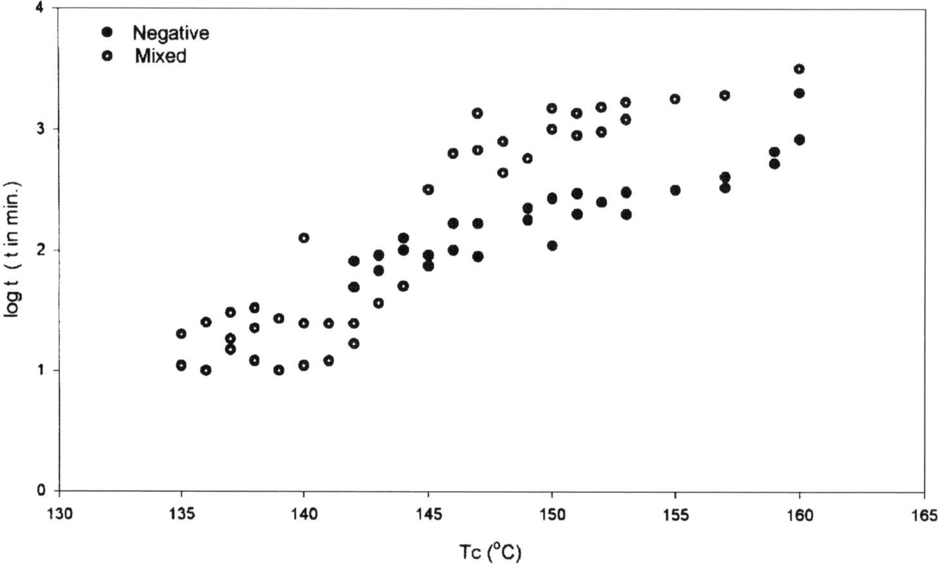

Fig. 9.56a Overall birefringence change as a function of crystallization temperature and time for a metallocene catalyzed isotactic poly(propylene), $M_w = 5.75 \times 10^5$ with 0.3% chain structural defects. (From Alamo and Chi (209))

the regime transition. The structural change would affect ΔG_u and thus ΔG^*. In this connection, Fig. 9.56a is a plot of the time change of the overall birefringence with crystallization temperature of a metallocene catalyzed isotactic poly(propylene), $M_w = 5.75 \times 10^5$, that has 0.3% chain structural irregularities.(209) An obvious discontinuity occurs in the birefringence in the same temperature region where the

II–III Regime transition is deduced from growth rate studies. It is, therefore, very likely from these observations that the discontinuity in the temperature coefficient of the growth rate is a consequence of structural and morphological changes rather than a regime transition.

Regime transitions can also be investigated by analysis of the overall crystallization rate. The advantage in this case is that a much wider molecular weight range can be studied, as compared to spherulite formation and growth. However, the disadvantage is that both the initiation and growth nucleation processes are involved and they cannot be separated from one another. An investigation of the overall crystallization kinetics, utilizing differential scanning calorimetry, has been reported for linear polyethylene over the molecular weight range $M = 3.1 \times 10^3$ to 8.0×10^6.(215,216) Although limited to only a 14 K interval in crystallization temperatures, interesting results were obtained because of the molecular weight range involved.

Figure 9.57 is a plot, on a logarithmic scale, of $1/\tau_{0.25}$ against the nucleation function $T_m^0/T_0\,\Delta T$. Here $\tau_{0.25}$ represents the time for 25% of the transformation to take place. A Gibbs type nucleus is assumed in the analysis. The implied assumption is thus made that the secondary or growth nucleation is controlling. For molecular weight fractions greater than 8.0×10^5 the data are well represented by the same linear relation over the complete range of crystallization temperatures. There is no indication of either a break or discontinuity in the data so that there is no regime change with crystallization temperature for the highest molecular weights studied.

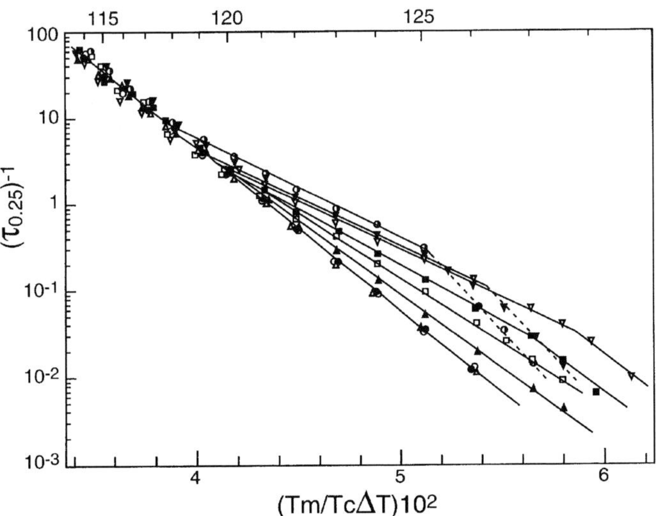

Fig. 9.57 Plot of log $(\tau_{0.25})^{-1}$ against $(T_m^0/T_c)(1/\Delta T)$ for indicated molecular weight. ○ 8×10^6; ● 5×10^6; △ 3×10^6; ▲ 8×10^5; □ 4.25×10^5; ■ 1.07×10^5; ▽ 5.3×10^4; ○ 1.95×10^4. $T_m^0 = 145.5\,°\mathrm{C}$.(215)

It should be noted parenthetically that spherulite growth rates of linear polyethylene cannot be measured in this molecular weight range. In contrast, the data point for molecular weights equal to or less than 8.0×10^5 cannot be represented by a linear relation. The data points for $M = 8 \times 10^5$ and 4.25×10^5 are represented by two intersecting straight lines, suggesting that a I–II Regime transition is involved. Three intersecting straight lines give a good representation of the data for the lower molecular weights, suggesting that Regime III has also been attained. This latter conclusion is consistent with the spherulite growth rate results.(205,205a) In order to decide whether these results represent regime transitions an analysis of the slope ratios needs to be made.

The ratio of the slopes in Regimes II to I and II to III are plotted in Fig. 9.58. It is quite evident in this plot that the two slope ratios depend strongly on the molecular weight. Starting with the highest molecular weights (the straight lines of Fig. 9.57), and an assigned slope ratio of 1, there is a monotonic decrease in both the II/I and II/III ratio with decreasing molecular weight. A more comprehensive plot of the II/I slope ratios is given in Fig. 9.58a as a function of temperature.(216) Included here are the values obtained for low molecular weight linear polyethylene fractions,(216) the higher molecular weights (215) and a high molecular weight n-alkane $C_{192}H_{384}$,(21) all obtained from overall crystallization kinetics as well as the values obtained from spherulite growth rates.(167) Figure 9.58a makes quite clear that the slope ratios of Regime II/I increase linearly with molecular weight.

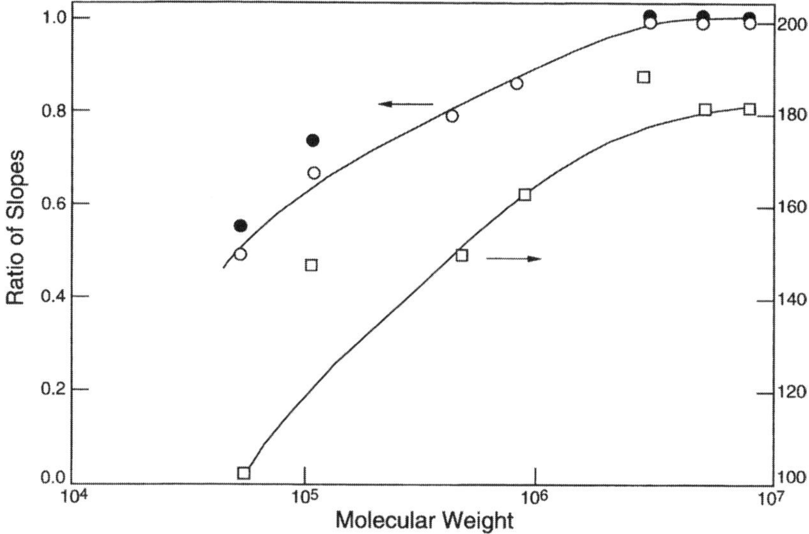

Fig. 9.58 Plot of slope ratios and slope as a function of molecular weight of linear polyethylene fractions. ○ I–II transition; ● II–III transition; □ slope in Regime II. (Data from (215))

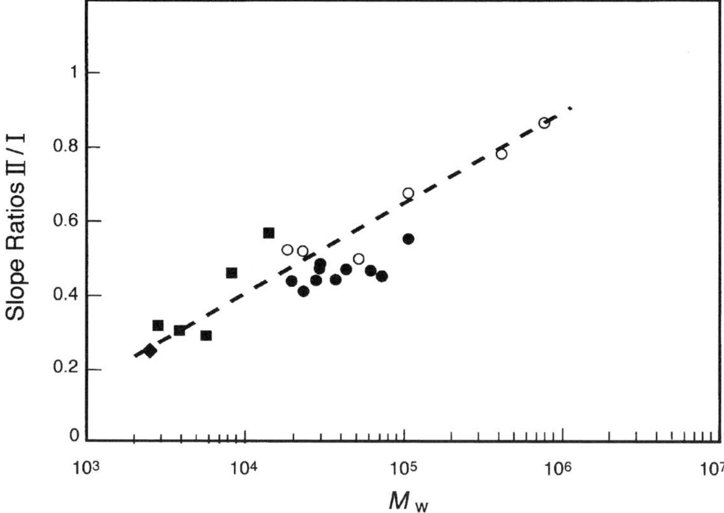

Fig. 9.58a Plot of ratio of II/I slopes for linear polyethylene fractions as a function of molecular weight. From overall crystallization kinetics: low molecular weight ■ (216); high molecular weight ○ (215); *n*-alkane $C_{192}H_{384}$ ♦ (217); Spherulite growth rate ● (167)(216)

The results reported from spherulite growth rate studies, approximately $M \simeq 10^4$ to 10^5, adhere or are very close to the same straight line. A value of unity is approached for very high molecular weights from the straight line in Fig. 9.58a and the ratio of about 0.3 at the lower end. It should also be noted that in the original concept of Regimes I and II that was developed for non-chain-like monomeric systems the II/I ratio was calculated to be 0.33.(173)

The theoretical result, embodied in Eq. (9.183), is based on the assumption that the spreading rate does not depend on molecular weight. The slope ratio depends on the relation between the nucleation rate N and the spreading rate g. The steady-state nucleation rate is not molecular weight dependent in the range of interest. However, the spreading rate could be molecular weight dependent.(148) The slope ratios can be given a qualitative explanation by postulating that g decreases with increasing molecular weight, as a consequence of the changing melt structure. The linear relation that is observed in Fig. 9.57 for the highest molecular weights could be explained by g being so small that crystallization will not proceed much beyond stable nuclei size at all crystallization temperatures. Hence, the growth rate will depend only on the nucleation rate and will be the same as in Regime I and Regime III, but for different physical reasons. As the molecular weight decreases g should increase because of the reduced influence of entanglements on segmental mobility. This argument leads to the expectation that there will be major changes in the slope in Regime II. As is seen in Fig. 9.58 the slope in this regime for linear polyethylene decreases by about 45% as M decreases from 8×10^6 to 5.3×10^4. The variation in g thus makes a major contribution to the molecular weight dependence of the overall crystallization rate of linear polyethylene.

Other polymers, as typified by poly(3,3-dimethyl thietane), show maxima in the overall crystallization rate.(218) The data can be analyzed in a manner comparable to that described for spherulite growth rates and the conclusions are the same. Whether a transition from Regime II to III is discerned depends on the values taken for the Vogel constants.

Another problem concerned with crystallization over an extended temperature range is the temperature maximum, T_{max}, in the crystallization rate. In an early analysis, it was concluded that the ratio T_{max}/T_m should be in the range 0.8–0.9. This conclusion agreed with the experimental data for two polymers that were available.(12) Since then, the results for many more polymers have followed a similar pattern.(219–221a) Plots of T_{max} against T_m are given in Figs. 9.59 and 9.60 for spherulite growth and overall crystallization rates, respectively. In each case the plots represent a set of extensive data for many different polymers. The straight lines drawn obey the relation

$$T_{max} = 0.82 \pm 0.005\, T_m^0 \tag{9.211}$$

in both cases. Thus, based on extensive experimental results, with but a few exceptions, $T_{max}/T_m = 0.82$ for homopolymers. Similar results are obtained for low molecular weight substances.(221a)

The results embodied in Figs. 9.59 and 9.60, and Eq. (9.211) should receive a natural explanation from the growth rate–crystallization temperature expression. For simplicity in analyzing the problem, we initially assume an Arrhenius type activation energy and a Gibbs type nucleation. Equation (9.173) has been shown to be valid for temperatures 50–80 °C above T_g. Utilizing a more complex transport term will not sensibly alter the conclusions.(222) By differentiating Eq. (9.173) and setting the result to zero it is found that (219,220,221)

$$\frac{T_{max}}{T_m} = \frac{(\phi + 1)^{1/2}}{(\phi + 1)^{1/2} + 1} \tag{9.212}$$

Here $\phi = E/K$, where $K = 4b_0\sigma_{en}\sigma_{un}/\Delta H_u$. A plot, according to Eq. (9.212), of T_{max}/T_m against E/K is given in Fig. 9.61.(220) The striking feature of the curve in this figure is the very rapid increase in T_{max}/T_m with E/K. The ratio of T_{max}/T_m levels off at about 0.80–0.90 for E/K values greater than 25. Thus, the functional form of Eq. (9.212) indicates that T_{max}/T_m is relatively insensitive to the ratio E/K and levels off in the experimentally observed range. Therefore, utilizing a simple growth rate expression explains the location of the maximum quite well.

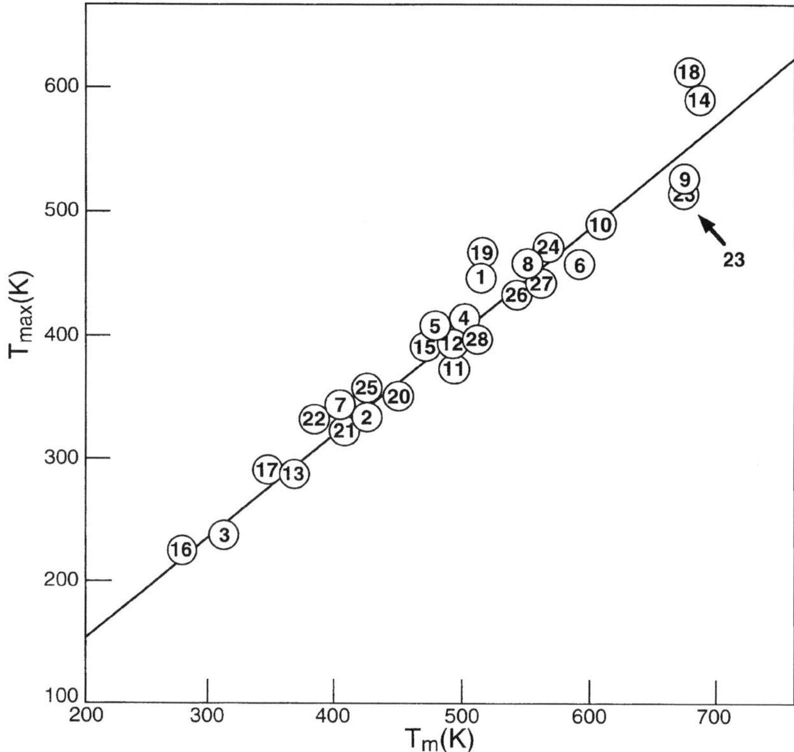

Fig. 9.59 Plot of temperature of maximum spherulite growth rate, T_{max}, against equilibrium melting temperature, T_m, for indicated polymers. (1) isotactic poly(styrene) (a); (2) poly(tetramethyl-p-silphenylene siloxane) (b); (3) poly(cis-isoprene) (c); (4) poly(caproamide) (d,e); (5) poly(L-lactic acid) (f); (6) poly(phenylene sulfide) (g,h); (7) poly(R-epichlorohydrin), poly(S-epichlorohydrin), poly(I-RS-epichlorohydrin) (i); (8) poly(ethylene terephthalate) (j,k,l); (9) poly(aryl ether ether ketone) (m,n); (10) poly(ethylene-2,6-naphthalene dicarboxylate) (n); (11) poly(3-hydroxybutyrate) (o); (12) isotactic poly(methyl methacrylate) (q); (13) poly(dioxolane) (r); (14) New TPI poly(imide) (s); (15) poly(methylene oxide) (t); (16) poly(cis-butadiene) (u); (17) poly(propylene oxide) (v,w); (18) poly(imide) BPDA + 134 APB (x); (19) poly(imide) BPDA + C12 (x); (20) syndiotactic poly(propylene) (y); (21) poly(3-hydroxy valerate) (z); (22) poly(ethylene succinate) (aa); (23) poly(aryl ether ketone ketone) (bb); (24) poly(phenylene ether ether sulfide) (cc); (25) poly(tetramethylene isophthalate) (dd); (26) poly(hexamethylene adipamide) (e,ee); (27) poly(tetrachloro-bisphenol-A adipate) (ff); nylon 6–10 (ee).

References

a. Miyamoto, Y., Y. Tanzawa, H. Miyaji and H. Kiho, *Polymer*, **33**, 2496 (1992).
b. Magill, J. H., *J. Polym. Sci.*, **5**, 89 (1967).
c. Phillips, P. J. and N. Vatansever, *Macromolecules*, **20**, 2138 (1987).
d. Magill, J. H., *Polymer*, **3**, 655 (1962); *ibid.*, **6**, 367 (1965).
e. Burnett, B. B. and W. F. McDevit, *J. Appl. Phys.*, **28**, 1101 (1957).
f. Vasanthakumari, R. and A. J. Pennings, *Polymer*, **24**, 175 (1983).

(*cont.*)

Fig. 9.59 (*cont.*)

g. Lovinger, A. J., D. D. Davis and F. J. Padden, Jr., *Polymer*, **26**, 1595 (1985).
h. Risch, B. G., S. Srinivas, G. L. Wilkes, J. F. Geibel, C. Ash, S. White and M. Hicks, *Polymer*, **37**, 3623 (1996).
i. Singfield, K. L. and G. R. Brown, *Macromolecules*, **28**, 1290 (1995).
j. Hieber, C. A., *Polymer*, **36**, 1455 (1995).
k. Phillips, P. J. and H. T. Tseng, *Macromolecules*, **22**, 1649 (1989).
l. van Antwerpen, F. and D. W. van Krevelen, *J. Polym. Sci.: Polym. Phys. Ed.*, **10**, 2423 (1972).
m. Hsiao, B. S. and B. B. Sauer, *J. Polym. Sci.: Pt. B: Polym. Phys.*, **31**, 901 (1993).
n. Medellin-Rodriguez, F. J., P. J. Phillips, J. S. Phillips and J. S. Lin, *Macromolecules*, **28**, 7744 (1995).
o. Organ, J. J. and P. J. Barham, *J. Mater. Sci.*, **26**, 1368 (1991).
p. Pearce, R., G. R. Grown and R. H. Marchessault, *Polymer*, **35**, 3984 (1994).
q. de Boer, A., G. O. R. Alberda van Ekenstein and G. Challa, *Polymer*, **16**, 930 (1975).
r. Archambault, P. and R. E. Prud'homme, *J. Polym. Sci.: Polym. Phys Ed.*, **18**, 35 (1980).
s. Hsiao, B. S., B. B. Sauer and A. Biswas, *J. Polym. Sci.: Pt. B: Polym. Phys.*, **32**, 737 (1994).
t. Inoue, M., *J. Appl. Polym. Sci.*, **8**, 2225 (1964).
u. Cheng, T. L. and A. C. Su, *Polymer*, **36**, 73 (1995).
v. Magill, J. H., *Makromol. Chem.*, **86**, 2831 (1965).
w. Kennedy, M. A., G. R. Brown and L. E. St-Pierre, *Polym. Composites*, **5**, 307 (1984).
x. Hsaio, B. S., J. A. Kreuz and S. Z. D. Cheng, *Macromolecules*, **29**, 135 (1996).
y. Supaphol, P. and J. E. Spruiell, *Polymer*, **41**, 1205 (2000).
z. Pearce, R. P. and R. H. Marchesault, *Macromolecules*, **27**, 3968 (1994).
aa. Steiner, K.J., J. Lucas and K. Ueberreiter, *Koll. Z.Z. Polym.*, **214**, 23 (1966).
bb. Wang, W., J. M. Schultz and B. S. Hsiao, *J. Polym. Sci.: Pt. B: Polym. Phys.*, **34**, 3095 (1996).
cc. Srinivas, S., J. R. Babu, J. S. Riffle and G. L. Wilkes, *J. Macromol. Sci. Phys.*, **B36**, 458 (1997).
dd. Gilbert, M. and F. J. Hybart, *Polymer*, **13**, 327 (1972).
ee. Lindegren, C. R., *J. Polym. Sci.*, **50**, 181 (1961).
ff. Berghmans, H., E. Lanza and G. Smets, *J. Polym. Sci.: Polym. Phys. Ed.*, **11**, 87 (1973).

Similar results are obtained when a three-dimensional type nucleation is used in conjunction with an Arrhenius transport term.(222)

The analysis is more complex when a Vogel type transport term is used. Now, three parameters are involved, U^*, T_∞ or C, and K. By differentiating Eq. (9.205) and setting the results equal to zero, one obtains (222)

$$\frac{T_{\max}}{T_{\mathrm{m}}} = \frac{2(\psi + 1)}{\psi} - \frac{T_{\mathrm{m}}\psi + T_{\mathrm{m}} + 4T_\infty}{\psi T_{\max}} + \frac{2T_\infty(T_{\mathrm{m}} + T_\infty)}{\psi T_{\max}^2} - \left(\frac{T_{\mathrm{m}} T_\infty^2}{\psi T_{\max}^3}\right) \quad (9.213)$$

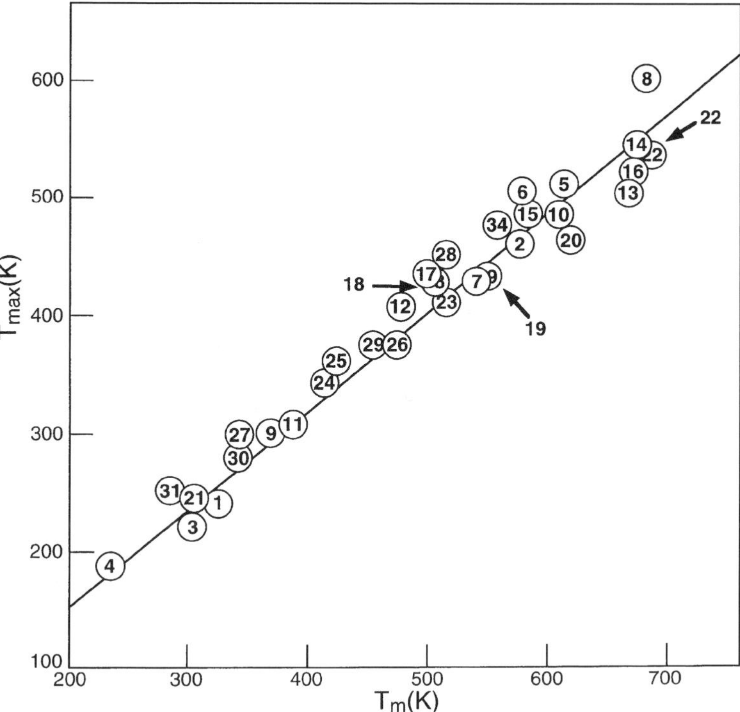

Fig. 9.60 Plot of temperature of maximum, $T_{\max}$, of overall crystallization rate against equilibrium melting temperature T_m. (1) poly(trimethylene oxide) (a,b); (2) poly(2,2′-bis-4,4′-oxyphenyl propane carbonate) (c); (3) poly(cis-1,4-butadiene) (d); (4) poly(dimethyl siloxane) (e,f,g); (5) poly(imide) ODPA $n = 1$ (h); (6) poly(imide) ODPA $n = 2$ (h); (7) poly(imide) ODPA $n = 3$ (h); (8) poly(imide) New TPI (i,j); (9) poly(1-pentene) (k); (10) poly(ethylene-2,6-naphthalene dicarboxylate) (l,m); (11) poly(ethylene succinate) (n); (12) poly(methylene oxide) (o); (13) poly(aryl ether ether ketone) (p,q); (14) poly(aryl ether ketone ketone-terephthalic acid) (r); (15) poly(aryl ether ketone ketone-isophthalic acid) (s); (16) poly(aryl ether ketone ether ketone ketone) (t); (17) isotactic poly(styrene) (u); (18) poly(butylene terephthalate) (v); (19) poly(hexamethylene adipamide) (w); (20) poly(phenylene sulfide) (x,y); (21) poly(cis-isoprene) (z,aa); (22) poly(imide-BPDA + 134APB) (bb,cc); (23) poly(imide - BDPA + C12) (bb,cc); (24) poly(pentamethylene terephthalate) (dd); (25) poly(tetramethylene isoterephthalate) (dd); (26) poly(L-lactic acid) (ee,ff,gg); (27) poly(ethylene adipate) (hh); (28) poly(ethylene terephthalate) (ii,jj); (29) poly(3,3-bis-chloromethyl oxacyclobutane) (kk); (30) poly(β-hydroxy octanoate) (ll); (31) poly(3,3-dimethyl thietane (kk,mm); (32) poly(imide PI-2) (nn); (33) poly(phenylene ether ether sulfide) (oo); (34) syndiotactic poly(styrene) (pp).

References

a. Perez, E., A. Bello and J. G. Fatou, *Coll. Polym. Sci.*, **262**, 605 (1984).
b. Perez, E., J. G. Fatou and A. Bello, *Eur. Polym. J.*, **23**, 469 (1987).
c. Galleg, F., R. Leger and J. P. Mercier, *J. Polym. Sci.: Polym. Phys. Ed.*, **14**, 1367 (1976).

(*cont.*)

Fig. 9.60 (*cont.*)

d. Feio, G. and J. P. Cohen-Addad, *J. Polym. Sci.: Pt. B: Polym. Phys.*, **26**, 389 (1988).
e. Godovskii, Y. K., V. Y. Levin, G. L. Slonimski, A. A. Zhdanov and K. A. Andrianov, *Polym. Sci. USSR*, **11**, 2778 (1969).
f. Andrianov, K. A., G. L. Slonimskii, A. A. Zhdanov, V. Y. Levin, Y. K. Godovskii and V. A. Moskalenko, *J. Polym. Sci., A1*, **10**, 1 (1972).
g. Ebengou, R. H. and J. P. Cohen-Addad, *Polymer*, **35**, 2962 (1994).
h. Heberer, D. P., S. Z. D. Cheng, J. S. Bailey, S. H. S. Lien, R. G. Bryant and F. W. Harris, *Macromolecules*, **24**, 1890 (1991).
i. Huo, P. P., J. B. Friler and P. Cebe, *Polymer*, **34**, 4387 (1993).
j. Hsiao, B. S., B. B. Sauer and A. Biswas, *J. Polym. Sci.: Pt. B: Polym. Phys.*, **32**, 737 (1994).
k. Quinn, F. A., Jr. and J. Powers, *J. Polym. Sci. Polym. Lett.*, **1B**, 341 (1963).
l. Cheng, S. Z. D., *J. Appl. Polym. Sci. Symp.*, **43**, 315 (1989).
m. Buckner, S., D. Wiswe and H. G. Zachmann, *Polymer*, **30**, 480 (1989).
n. Ueberreiter, K., G. Kanig and A. S. Brenner, *J. Polym. Sci.*, **16**, 531 (1955).
o. Inoue, M. and T. Takayanagi, *J. Polym. Sci.*, **47**, 498 (1960).
p. Blundell, D. J. and B. N. Osborn, *Polymer*, **24**, 953 (1983).
q. Day, M., Y. Deslandes, J. Roovers and T. Superunchuk, *Polymer*, **32**, 1258 (1991).
r. Gardner, K. H., B. S. Hsiao, R. R. Matheson, Jr. and B. A.Wood, *Polymer*, **33**, 248 (1992).
s. Hsiao, B. S., K. H. Gardner and S. Z. D. Cheng, *J. Polym. Sci.: Pt. B: Polym. Phys.*, **32**, 2585 (1994).
t. Wang, W., J. M. Schultz and B. S. Hsiao, *J. Polym. Sci.: Pt. B: Polym. Phys.*, **34**, 3095 (1996).
u. Martuscelli, E., G. Demma, E. Drioli, L. Vicolais, S. Spina, H. B. Hopfenberg and V. T. Stannett, *Polymer*, **20**, 571 (1979).
v. Escala, A. and R. S. Stein, in *Multiphase Polymers*, S. L. Cooper and G. M. Estes eds., American Chemical Society (1979).
w. Magill, J. H., *Nature*, **187**, 770 (1960).
x. Risch, B. G., S. Srinivas, G. L. Wilkes, J. F. Gerbel, C. Ash, S. White and M. Hicks, *Polymer*, **37**, 3632 (1996).
y. Lopez, L. C. and G. L. Wilkes, *Polymer*, **29**, 106 (1988).
z. Wood, L. A. and N. Bekkedahl, *J. Appl. Phys.*, **17**, 362 (1946).
aa. DeChirioco, A. P., C. Lanzane, M. Piro and M. Bruzzone, *Chim. Ind. (Milan)*, **54**, 351 (1972).
bb. Hsiao, B. S., J. A. Kreuz and S. Z. D. Cheng, *Macromolecules*, **29**, 135 (1996).
cc. Kreuz, J. A., B. S. Hsiao, C. A. Renner and D. L. Goff, *Macromolecules*, **28**, 6926 (1995).
dd. Gilbert, M. and F. J. Hybart, *Polymer*, **13**, 327 (1972).
ee. Mazzullo, S., G. Paganetto and A. Celli, *Prog. Coll. Polym. Sci.*, **87**, 32 (1992).
ff. Urbanovici, E., H. A. Schneider and H. J. Cantow, *J. Polym. Sci.: Pt. B: Polym. Phys.*, **35**, 359 (1997).
gg. Iannace, S. and L. Nicolais, *J. Appl. Polym. Sci.*, **64**, 911 (1997).
hh. Order, K., R. H. Peters and L. C. Spark, *Polymer*, **18**, 155 (1977).
ii. Hieber, C. A., *Polymer*, **36**, 1455 (1995).

Fig. 9.60 (*cont.*)

jj. Van Antwerpen, F. and D. W. Van Krevelen, *J. Polym. Sci.: Polym. Phys. Ed.*, **10**, 2423 (1972).
kk. Sandiford, D. J. H., *J. Appl. Chem.*, **8**, 188 (1958).
ll. Peres, R. and R. W. Lenz, *Polymer*, **35**, 1059 (1994).
mm. Lazcano, S., J. G. Fatou, C. Marco and A. Bello, *Polymer*, **29**, 2076 (1988).
nn. Cheng, S. Z. D., M. L. Mittleman, J. J. Janimak, D. Shen, T. M. Chalmers, H. S. Lien, C. C. Tso, P. A. Gabori and F. W. Harris, *Polym. Int.*, **29**, 201 (1992).
oo. Srinivas, S., J. R. Babu, J. S. Riffle and G. L. Wilkes, *Polym. Eng. Sci.*, **37**, 497 (1997).
pp. St. Lawrence, S. and D. M. Shinozaki, *Polym. Eng. Sci.*, **37**, 1825 (1997).

where $\psi = U^*/K$. Equation (9.213) does not allow a simple comparison with experimental results. Numerical methods, using fixed constants, need to be used to analyze the results. When the constants U^* and C that give the best fit of the rate data are used, $T_{\text{max}}/T_{\text{m}}$ is found to be in the range of 0.80–0.85, with but a few exceptions. An example is given in Fig. 9.62 for the crystallization of natural rubber and poly(aryl ether ether ketone). These two polymers represent extreme cases of T_{g} and T_{m}^0. $T_{\text{m}}^0 = 308$ K, $T_{\text{g}} = 201$ K for the former polymer and 668 K and 417 K for the latter. Here $T_{\text{max}}/T_{\text{m}}$, calculated according to Eq. (9.213), is plotted against U^*/K for different values of C. All of the curves have similar shapes and are close to one another for the range in the value of C. Even for this more general growth expression there is still only a narrow range of allowable values of $T_{\text{max}}/T_{\text{m}}$.

9.12 Homogeneous nucleation and interfacial free energies

9.12.1 Homogeneous nucleation

It has been pointed out on several occasions that, when a continuous volume of a crystallizable liquid is cooled from the melt, nucleation is most often initiated by heterogeneities. These heterogeneities can be foreign substances, as well as crevices, cracks and surfaces within pure substances. Heterogeneous nucleation is common to all crystallizing substances, low molecular weight species as well as polymers. If, however, the heterogeneities are isolated in the melt, as for example in small droplets, their influence will be minimal. Nucleation and crystallization will then proceed homogeneously. As was alluded to earlier, this technique was pioneered by Turnbull in his study of metals.(120,223) It has since been applied to a variety of organic and inorganic low molecular weight compounds (129,131) including the n-alkanes.(132–135b,224) Before analyzing the application of this technique to polymers it is instructive to examine the results obtained with low molecular weight substances.

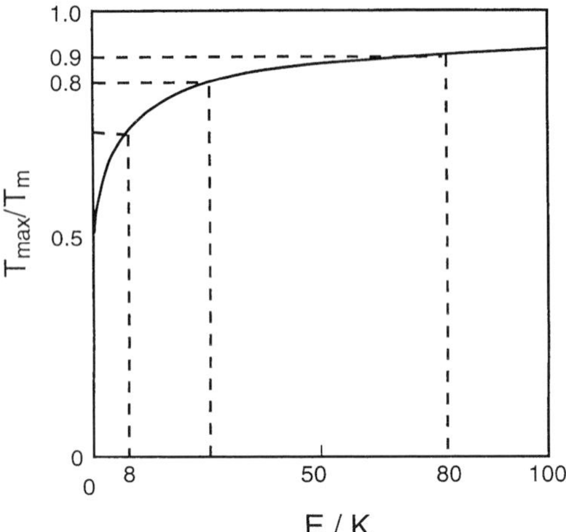

Fig. 9.61 Theoretical plot, from Eq. (9.212), of $T_{\max}/T_{\mathrm{m}}$ against E/K. (Adapted from Okui (220))

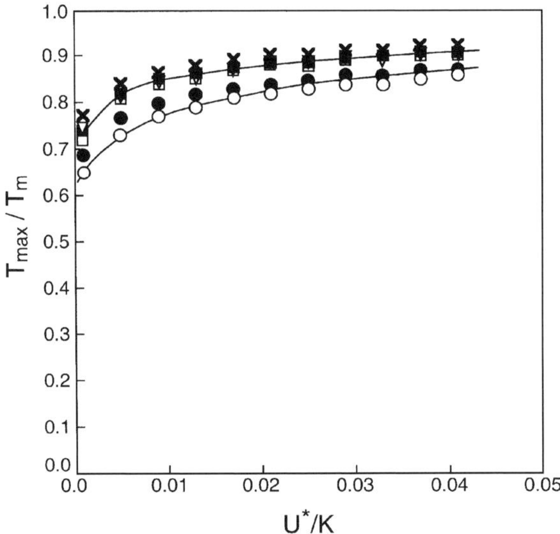

Fig. 9.62 Theoretical plot from Eq. (9.213) of $T_{\max}/T_{\mathrm{m}}$ against U^*/K for poly(aryl ether ether ketone) and natural rubber and different values of the constant C. For natural rubber: ▿ $C = 0$; ● $C = 30$; ○ $C = 51.6$. For poly(aryl ether ether ketone): × $C = 0$; ■ $C = 30$; □ $C = 51.6$.

The analysis of experimental results takes advantage of the fact that the nucleation rate increases rapidly with increasing temperatures. It is then possible to define a homogeneous nucleation temperature, T^*, because the transition from a small fraction of droplets frozen to all of them solidifying occurs over a narrow temperature interval. Experimentally $T^* \approx 0.82T_{\mathrm{m}}^0$ for many substances, such as metals, some organic compounds and alkyl halides. However, this generalization does not include the n-alkanes.(129,225) In analyzing the data from the droplet-type experiment involving low molecular weight substances, including the n-alkanes, it has usually been assumed that the nucleus is spherical. Thus, only one surface is involved. The free energy change in forming such a critical-size nucleus, ΔG^*, is expressed by Eq. (9.88).

$$\Delta G^* = \left(\frac{16\pi}{3}\right)\left(\frac{\sigma^3}{\Delta H_v}\right)^2\left(\frac{T_m}{\Delta T}\right)^2 \tag{9.88}$$

The σ value can be obtained by measuring the nucleation rate as a function of undercooling, and applying the Turnbull–Fisher relation. Alternatively, this value can be obtained by measuring nucleation kinetics under different continuous cooling rates.(134) It is convenient at this point to introduce a reduced temperature T^*/T_m^0 and a reduced undercooling $\Delta T_R^* \equiv (T_m^0 - T^*)/T_m^0$.

Of particular interest in the present context are the results for the n-alkanes, since they represent low molecular weight chain molecules. The nucleation of a series of n-alkanes, from C_5H_{12} to $C_{60}H_{122}$, has been studied by the droplet technique.(131–135b,224) The extension of the n-alkane study to 60 carbon atoms, as well as including low molecular weight polyethylene fractions, has significantly enhanced our understanding of the problem(135–135b) and has lead to important implications with regard to polymer nucleation. The reduced undercoolings, ΔT_R^*, for the n-alkanes are plotted against the carbon number, n, in Fig. 9.63.

The results of the different investigators are in good accord with one another for these direct observations. The plots show three distinct regions. There is a sharp decrease in ΔT_R^* from 0.20 for C_5H_{12} to approximately 0.04 for $C_{18}H_{38}$. This anomalous low value is maintained to about $C_{25}H_{52}$. There is then a gradual increase in ΔT_R^* with carbon number, until a value of 0.086 is reached for $C_{60}H_{122}$, the highest n-alkane studied. There is the suggestion in the data that for the lower carbon number n-alkanes the odd-numbered ones have slightly lower reduced undercoolings than the even-numbered ones. The anomalous behavior found in the lower carbon n-alkanes has been attributed to nucleation of the metastable hexagonal phase and the effect of surface freezing.(135a) Since the principal interest here is in the relation of these results to polymer nucleation, attention is focused on the upsweep that set in at n approximately equal to 25. Also plotted in this figure are the results for what have been termed low molecular weight polyethylenear fractions.[16] The results for the fractions fit in very smoothly with those of the n-alkanes. The plot also indicates that a saturation or leveling off of ΔT_R^* would occur at approximately 0.12. This

[16] These low molecular weight polyethylene fractions are actually hydrogenated poly(butadienes). They are, therefore, random type ethylene-butene copolymers.

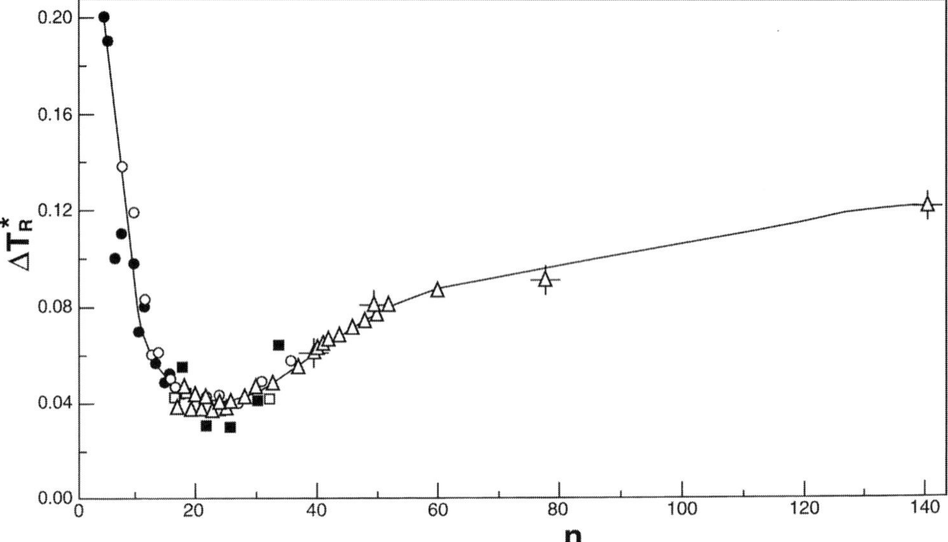

Fig. 9.63 Plot of reduced undercooling, ΔT_R^*, against n, the number of carbons in the n-alkanes and low molecular weight fractions. n-alkanes: ○ Oliver and Calvert (134); ● Uhlmann *et al.* (133); □ Turnbull and Cormia (132); ■ Phipps (224); △ Kraack *et al.* (135,135a); low molecular weight fractions: -△- Kraack *et al.* (135,135b).

value of ΔT_R^* is comparable to that found for fractionated and unfractionated linear polyethylenes (see below).(226,227)

A compilation of the interfacial free energies, σ, assuming spherical nuclei, that were obtained from the droplet experiments by the different investigators is given in Fig. 9.64.[17] The odd–even effect is again observed. Starting at about $n = 25$ there is again a monotonic increase in σ with carbon number. The low molecular weight fractions fit smoothly with the n-alkane data in this plot and a leveling off of σ is observed.

The changes in both the relative undercooling and the interfacial free energy that occur at about $n = 25$ are indicative of a transition between two different nucleation modes. For the shorter n-alkanes the complete molecules can participate in the nucleation in a classical sense. However, it has been proposed that as the molecular length increases the bundle type of nucleation sets in.(135,135a) This proposal is consistent with the low molecular weight fractions following the same pattern as the higher n-alkanes. Small angle x-ray studies have demonstrated that the crystallite thicknesses in this molecular weight range are comparable to the extended chain

[17] For some of the alkanes Oliver and Calvert (134) give results from both isothermal kinetic and cooling rate studies. The data from the isothermal kinetics are used here. Phipps (224) used both the steady-state nucleation rate and one in which a time lag was taken into account. The data from the steady-state nucleation rate were used in Fig. 9.64.

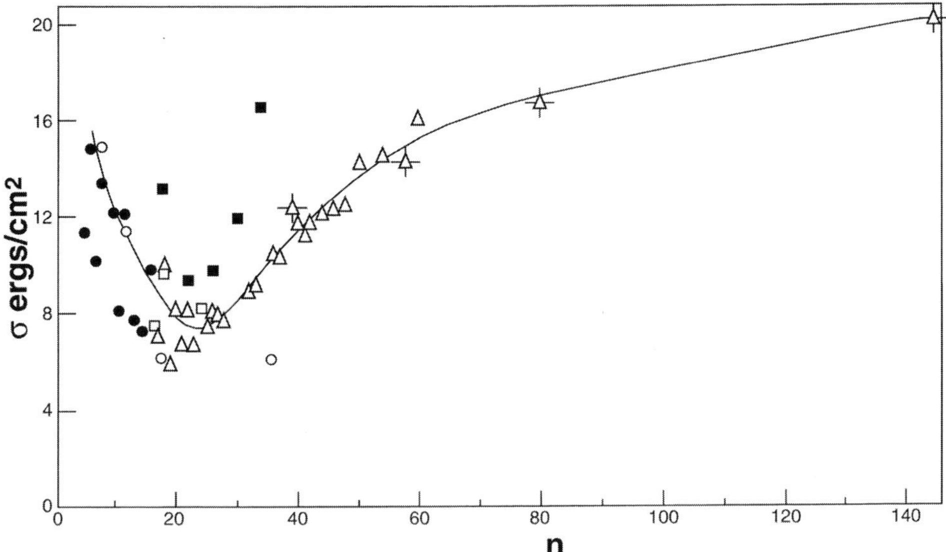

Fig. 9.64 Plot of interfacial free energy, σ, (based on spherical nuclei) against n, the number of carbons in the n-alkanes and low molecular weight fractions. ○, cooling only, Oliver and Calvert (134); ● Uhlmann *et al.* (133); □ Turnbull and Cormia (132); ■ Phipps (224); △ Kraack *et al.* (135,135a); low molecular weight fractions: ⊹△ Kraack *et al.* (135,135b).

length. Folded chain crystallites are only expected to occur at about $n \cong 200$.(225a) Hence, chain folding cannot be involved in nucleus formation in the range of chain length studied and has to be ruled out. The bundle type nucleus has been shown to be acceptable for chain-like molecules. It remains to be seen what information deduced from the results for the n-alkanes, and the low molecular weight fractions, can be applied to high molecular weight polymers.

In order to examine the connection, if any, between the results just described and higher molecular weight polymers, it is necessary to analyze the appropriate droplet experiments. In their pioneering study, Cormia, Price and Turnbull carried out the first successful study of droplet crystallization in polymers.(226) An unfractionated linear polyethylene, with a broad molecular weight distribution, was used in this initial work. An example of the effect of dispersing the sample into droplets on the crystallization of this polymer is illustrated in Fig. 9.65.(226) In this figure the fraction of the droplets solidified is plotted against the undercooling, and temperature, for slow cooling from the liquid state. Less than about 5% of the droplets solidify above 100 °C. This solidification is probably due to heterogeneities that are isolated in these droplets. More than 50% of the droplets freeze in the critical range 85–87 °C and solidification is complete at 84 °C. The solidification process is clearly quite sharp. When crystallized in the bulk, linear polyethylene can only

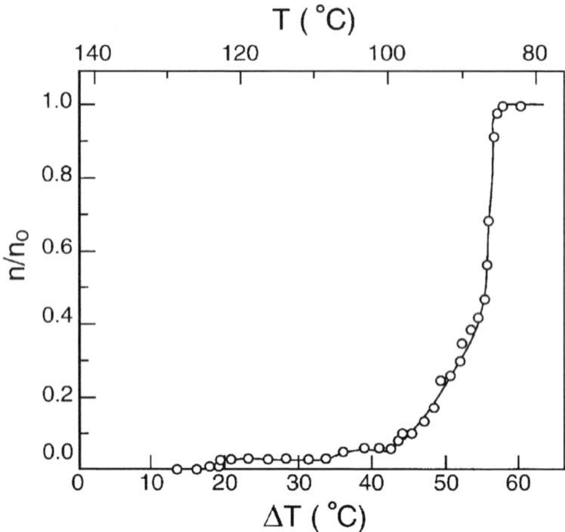

Fig. 9.65 Plot of number fraction solidified, n/n_0, against temperature. Cooling rate 0.58 °C min^{-1} above 100 °C; 0.17 °C min^{-1} below 100 °C. (From Turnbull and Cormia (132))

be supercooled by about 20 °C. However, where the heterogeneities are isolated in the droplets the supercooling that can be attained is increased to 60 °C. For linear polyethylene $T^*/T_m^0 = 0.86$ and the relative undercooling, $\Delta T_R^* = 0.14$. The latter value is comparable to the saturation value observed in Fig. 9.63.

All the polymers that have been studied by this technique show similar levels of supercooling.(227–231) A compilation of the undercooling, ΔT, that can be achieved by the polymers studied, using the droplet method, is given in Table 9.7. The undercoolings obtained for the variety of polymers that are listed range from about 60 °C to 100 °C. In all cases, the supercoolings achieved are much greater than those obtained by the crystallization of continuous bulk samples. The reduced undercoolings for the different polymers range between 0.14 and 0.20. The reduced undercoolings that are characteristic of all polymers are comparable to those for low molecular weight inorganic and organic compounds.(129) However, as has been pointed out, the lower carbon number n-alkanes are an exception to this generalization.

The relationship, if any, between the droplet experiments for the n-alkanes greater than $n = 25$, the low molecular weight fractions, and high molecular weight polyethylenes can now be examined. It has been pointed out that, when plotted against $1/n$, the undercoolings for the n-alkanes with $30 \leq n \leq 60$ extrapolate linearly to $\Delta T = 60$ °C.(135) As is indicated in Table 9.7 this value ($\Delta T = 60$ °C) is the same as found directly in a variety of high molecular weight linear polyethylenes.(135) The reduced undercoolings of these n-alkanes, and the low molecular weight fractions, approach a saturated or asymptotic value of about 0.12

Table 9.7. *Parameters obtained from droplet kinetics of polymers*

Polymer	ΔT °C	ΔT^*	Shape[a]	T_m^0 °C[b]	$\sigma_{un}^2\sigma_{en}$ erg^3 cm^{-6}	Reference
Linear polyethylene	~60	0.14	cylinder	142	15 500	a
Broad molecular weight distribution	~60	0.15	rectangular parallelepiped	141	14 200	b
	~60	0.15	"	142	14 960	b
	~60	0.14	"	143	15 740	b
	~60	0.14	"	144.7	17 200	b
$M_w = 3200$, $M_n = 2140$	~45	0.11	"	131.7	1 180	c
$M_w = 9700$, $M_n = 9150$	~60	0.15	"	141.0	12 400	c
$M_w = 11\,740$, $M_n = 10\,970$	~60	0.15	"	142.0	15 000	c
$M_w = 23\,000$, $M_n = 17\,690$	~60	0.14	"	144.1	18 610	c
$M_w = 30\,600$, $M_n = 24\,710$	~60	0.14	"	144.6	18 600	c
$M_w = 49\,890$, $M_n = 36\,370$	~60	0.14	"	145.2	17 400	c
$M_w = 119\,200$, $M_n = 96\,600$	~60	0.14	"	146.0	20 000	c
$M_w = 249\,000$, $M_n = 179\,00$	~60	0.14	"	146.2	20 500	c
Broad distribution	~55	0.13	—	143	—	d
isotactic Poly(propylene)	~100	0.22	—	186	—	d[c]
isotactic Poly(propylene)	~100	0.22	cylinder	178	25 000	e
Poly(ethylene oxide)	~65	0.19	—	66	—	d
Poly(ethylene oxide)	~65	0.20	—	58	—	f
isotactic Poly(styrene)	~100	0.20	—	240	—	d
Poly(caprolactam)	~90	0.18	—	227	—	d
Poly(oxymethylene)	~84	0.18	—	195	—	d
Poly(3,3-bis chloro-methyl oxycyclobutane)	~90	0.19	—	191	—	d

[a] Geometric shape assumed for nucleus.
[b] Equilibrium melting temperature used in calculation.
[c] Product of interfacial energies not listed because of unusual value used for ΔG^*.

References

a. Cormia, R. L., F. P. Price and D. Turnbull, *J. Chem. Phys.*, **37**, 1333 (1962).
b. Gornick, F., G. S. Ross and L. J. Frolen, *J. Polym. Sci.*, **18C**, 79 (1967).
c. Ross, G. S. and L. J. Frolen, *J. Res. Nat. Bur. Stand.*, **79A**, 701 (1975).
d. Koutsky, J. A., A. G. Walton and E. Baer, *J. Appl. Phys.*, **38**, 1832 (1967).
e. Burns, J. R. and D. Turnbull, *J. Appl. Phys.*, **37**, 4021 (1966).
f. Price, F. P. *Abstract Symposium on Macromolecules*, IUPAC, Wiesbaden, Germany, Sect. 1, paper 1B2 (1959).

or slightly higher. This value is essentially identical to those found with the higher molecular weight polymers (see Table 9.7). The interfacial free energies, σ, of the n-alkanes and the low molecular weight fractions extrapolate smoothly to the value deduced for high molecular weight polymers, about 25–26 erg cm^{-2}.(135,135b,228) This value is based on the assumption of a spherical nucleus in all cases.

The smooth behavior of these quantities, from the higher n-alkanes ($30 \leq n \leq 60$) and low molecular weight polyethylene to high molecular weight polymers, indicates that, irrespective of the nucleation mode of the n-alkanes, it persists with the polymers. It has been pointed out that chain folding cannot be involved in the nucleation of the n-alkanes in the range studied. This conclusion is not restricted by the fact that spherical nuclei have been assumed. A bundle type nucleus appears to be a good candidate for the nucleation mode in long chain molecules. It has been pointed out (Sect. 9.9) that such nuclei can lead to mature, lamellar crystallites in a natural way. It has also been shown that the excessively high values that have been calculated for σ_{en} are not correct.(135b) These results strongly suggest that the bundle, or fringe micelle, concept for critical-size nuclei of long chain molecules needs to be given serious consideration. Although rejected by many, but not all, it should be revisited in terms of theory and experiment. A mature, lamellar crystallite cannot support a fringed micelle structure, because of the need to dissipate the flux of chains emanating from the basal plane.[18] However, this is not a problem with a nucleus because of the small lateral dimensions that are involved.(231a) There is no inconsistency between a bundle type nucleus and a mature lamellar crystallite.

9.12.2 Interfacial free energies

Nucleation in isolated droplets represents the classical homogeneous three-dimensional situation. Only spherical nuclei have been considered up to now in analyzing the results from the droplet experiments. It is more realistic to assume an asymmetric shaped nucleus for the larger alkanes and polymers. For such a nucleus

$$\Delta G^* = \frac{K \sigma_{un}^2 \sigma_{en} T_m^2}{(\Delta H_u \Delta T)^2} \tag{9.214}$$

where the constant K specifies the geometry of the assumed nucleus. The product of the interfacial free energies, $\sigma_{un}^2 \sigma_{en}$, can be obtained from the variation of the nucleation frequency with undercooling. In one type of experiment the isothermal rate of nucleation in the droplets is measured at different temperatures. The product of the interfacial free energies is obtained by applying the Turnbull–Fisher relation.

[18] This problem will be discussed in detail in Volume 3.

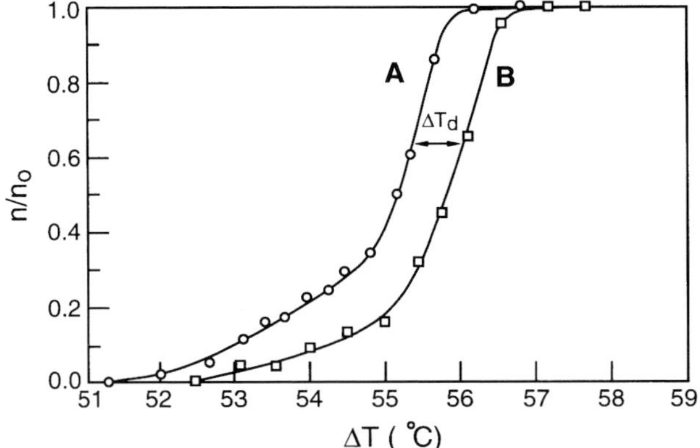

Fig. 9.66 Plot of number fraction crystallized against the supercooling for different cooling rates. Curve A, 0.023 °C min^{-1}. Curve B, 0.130 °C min^{-1}. (From Turnbull and Cormia (132))

Another method that is used to determine the steady-state nucleation rate is to measure the fraction of droplets crystallized at different cooling rates. Plots of the fraction of the droplets frozen against the temperature produces a set of displaced S-shaped curves. An example is illustrated in Fig. 9.66.(132) Cormia, Price and Turnbull showed that for cooling rates r_1 and r_2

$$\ln(r_1/r_2) = \frac{2\Delta G^* \Delta T_{\mathrm{D}}}{kT\,\Delta T_{\mathrm{Av}}} \tag{9.215}$$

where ΔT_{D} is the undercooling and ΔT_{Av} is the displacement of the two curves at their mid-point.

The values of $\sigma_{\mathrm{un}}^2\sigma_{\mathrm{en}}$ that have been obtained from droplet studies are also given in Table 9.7. The values depend directly on the assumed nucleus geometry. For a cylindrical nucleus the factor K in Eq. (9.213) is 8π,(12) it is 32 for a rectangular parallelepiped and is reduced to 30 when it is assumed that the nucleus is built upon the unit cell of polyethylene.(157) Other type geometries, with corresponding values of K, could be used equally well. The actual value obtained for the product of interfacial energies obviously depends on the arbitrary assumption of the nucleus geometry. Except for the lowest molecular weight fractions of polyethylene, the products $\sigma_{\mathrm{un}}^2\sigma_{\mathrm{en}}$ are in the range of 15 000–20 000 erg^3 cm^{-6} for all the polymers studied. The values found for linear polyethylene are in good agreement with one another when cognizance is taken of the different nuclei geometries and different samples that are involved. Not unexpectedly, the interfacial product deduced from the kinetic data is sensitive to the value taken from T_{m}^0. For example, for one set of data $\sigma_{\mathrm{un}}^2\sigma_{\mathrm{en}}$ varies from 14 200 to 17 200 erg^3 cm^{-6} as T_{m}^0 changes from 141 °C to 144.7 °C.(228) It would be expected to be even higher for $T_{\mathrm{m}}^0 = 145.5$ °C.

There is only a slight variation of the listed $\sigma_{\text{un}}^2\sigma_{\text{en}}$ values with molecular weight. In calculating ΔG^* for polyethylene, the equlibrium value, T_{m}^0, should be used for $M \geq 20\,000$. Then, except for the lowest molecular weights, the values of $\sigma_{\text{un}}^2\sigma_{\text{en}}$ determined from the droplet experiments are independent of chain length. The interfacial free energy product obtained for isotactic poly(propylene) is greater than that of linear polyethylene.

The droplet studies demonstrate that homogeneous nucleation in polymers can be observed and the results quantified utilizing basic nucleation theory. Because of the asymmetric nature of chain molecules, more than one surface is involved in nucleus formation. In the simplest case of only two surfaces, the product of interfacial free energies $\sigma_{\text{un}}^2\sigma_{\text{en}}$ is obtained. In order to relate the chemical nature and structure of the chain repeating unit to the free energy of forming a nucleus it is advantageous to know the values of the separate interfacial free energies. The problem of separating this product into its two factors, in a confident manner, is a formidable one. Several methods have been suggested to resolve this problem.

One method that has been used to effect this separation is to invoke the Gibbs–Thomson relation. The interfacial free energy, σ_{ec}, associated with the basal plane of the mature lamellar crystallite is obtained from the relation between the melting temperature and crystallite thickness for bulk crystallized polymers. The inherent assumption is then made that σ_{ec} for the mature crystallite can be identified with σ_{en} of the nucleus. In this manner the separation of the product of interfacial free energy can be achieved. This rather drastic assumption has not been substantiated. There is no requirement that the interfacial structure of a mature crystallite replicates that of the initiating nucleus. In fact this situation is highly unlikely.

In a different approach, the results from droplet experiments that involved both the n-alkanes and linear polyethylene were combined.(226) The σ value of 9.6 erg cm^{-2} obtained for n-octadecane, assuming a spherical nucleus, was identified with σ_{un}, the lateral or side interfacial free energy of polyethylene. A value of σ_{en} of 168 erg cm^{-2} was then obtained by this procedure. However, there is a serious question as to whether the identification between σ and σ_{un} is valid. Moreover, as was pointed out earlier the results for octadecane are anomalous. The value of σ increases monotonically for $n \geq 25$, and extrapolates to $\sigma \approx 20$–23 erg cm^{-2} for high molecular weight polyethylene. There is thus the additional problem of what value to use for σ in following this procedure.

Attempts to obtain σ_{un} for polymers have been made by taking advantage of an empirical relation obtained by Staveley and coworkers for monomeric organic and inorganic compounds (133,232), and by Turnbull for metals.(233) They found, from droplet type experiments, that the empirical relation

$$\alpha = \sigma_{\text{g}}/\Delta H \tag{9.216}$$

satisfies the experimental results. Here ΔH is the molar heat of fusion and σ_g is defined as

$$\sigma_g \equiv A^{1/2} V^{2/3} \sigma \tag{9.217}$$

A is Avogadro's number and V is the molar volume. The interfacial free energy, σ, is the value obtained from the droplet experiment and homogeneous nucleation, assuming spherical nuclei. The quantity σ_g is a measure of the surface free energy of a mole of the substance at the interface. It was found for seventeen compounds that the parameter α varied from 0.23 to 0.37.(133,232) For metals it varied from 0.32 to 0.45.(233) Although α lies in a fairly narrow range, it is far from constant for low molecular weight substances.

In adapting this procedure to polymers, it was proposed that (149,169)

$$\alpha = \frac{\sigma_{un}}{\Delta H_u b} \tag{9.218}$$

where b is the chain thickness. Subsequently, Eq. (9.218) was modified to (156,174)

$$\alpha = \frac{\sigma_{un}}{\Delta H_u (ab)^{1/2}} \tag{9.219}$$

where a is the molecular width, and ab is the cross-sectional area of the chain. It is not clear why in following Eq. (9.216) the molar volume was not used in Eq. (9.219). In order to obtain σ_{un} from Eq. (9.219) the value, or values, of α for different polymers needs to be determined. It is important to ascertain whether α has the same value for all polymers, varies from one type to another or is unique for each polymer. To accomplish this task it is necessary to determine σ_{un} independent of Eq. (9.219).

The method that has been used to determine σ_{un} was to utilize the quantity $\sigma_{un}\sigma_{en}$ determined from spherulite growth rate data.(156,174) The quantity σ_{en} is again identified with σ_{ec}, which is again obtained from the Gibbs–Thomson relation. This procedure was used to analyze the melting temperature–thickness data of linear polyethylenes crystallized from dilute solution (234) and the growth rate data of the same polymer.(167) Following this procedure $\sigma_{ec} = 93 \pm 8$ erg cm^{-2} and $\sigma_{un} = 13.7$ erg cm^{-2}. This value for σ_{un} gives $\alpha = 0.11$ for linear polyethylene.(174) This value is low compared to those obtained for metals and other low molecular weight species. The question is whether this low value is due to either the procedure adopted, and the assumption made to obtain σ_{un}, or is it an inherent property of long chain molecules? In support of this result it was pointed out that, from the homogeneous nucleation of n-octadecane, σ based on spherical nuclei is equal to 9.6 erg cm^{-2}.(135) This leads to $\alpha = 0.12$ for this alkane. However, as shown in Fig. 9.64, for the higher carbon nucleus n-alkanes σ is subsequently greater than the value for n-octadecane. If one takes the leveling off value of about 20 erg cm^{-2} from Fig. 9.64 the value of α is found to be approximately 0.25.

A similar analysis was carried out with poly(pivalolactone).(202) The σ_{en} value was identified with σ_{ec}, the latter again being obtained by the application of the Gibbs–Thomson equation. From the product $\sigma_{un}\sigma_{en}$, obtained from spherulite growth rate studies, σ_{un} is equal to 30 erg cm^{-2}. This deduced value of σ_{un} leads to $\alpha = 0.25$, a much higher value than for linear polyethylene. It was concluded that there are two classes of polymers, one with $\alpha = 0.25$ and the other, typified by linear polyethylene, with $\alpha = 0.11$. Such a generalization cannot be made based solely on the study of two polymers.[19] The results for a variety of polymer types are needed in order to decide whether α is constant, varies from polymer to polymer or falls into a group pattern. In addition, the identification of σ_{en} with σ_{ec} needs to be justified. Caution should be exercised in adopting either $\alpha = 0.11$ or 0.25 for a particular polymer. Even if the value of 0.11 is found to be correct for linear polyethylene it cannot be automatically used with other polymers. The ratio $\sigma_{un}/\Delta H_u$ appears in the growth rate expression for binary mixtures (Chapter 11). However, it is not correct in general to assign it the value of 0.11.(199a,200,233a–c)

A value of σ_{un} can also be determined by comparing the product of interfacial free energies obtained from homogeneous nucleation with that from spherulite growth rates. The factor $\sigma_{un}^2\sigma_{en}$ is obtained from the droplet experiments, while spherulite growth rates give $\sigma_{un}\sigma_{en}$. The ratio of the two gives σ_{un}. The appropriate data are available for linear polyethylene. (167,171,226,228) Analysis of this data indicates that $\sigma_{un}\sigma_{en}$ is independent of molecular weight (see Sect. 9.14). A typical value is 1200 erg^2 cm^{-4}. The values of $\sigma_{un}^2\sigma_{en}$ taken from the droplet experiments depend on the sample studied and the shape assumed for the nuclei. Assuming a cylindrical nucleus, Cormia, Price and Turnbull determined $\sigma_{un}^2\sigma_{en}$ to be 15 500 erg^3 cm^{-6}. For fractions, and a rectangular parallelepiped nucleus, $\sigma_{un}^2\sigma_{en}$ ranged from 18 000 to 20 000 erg^3 cm^{-6} for the higher molecular weights.(228) These data put σ_{un} in the range of 13–17 erg cm^{-2}. Considering all the results for the widely studied linear polethylene there is an uncertainty of at least 30–40% in the value of σ_{un}.

The indirect methods that have been used to determine σ_{un} are inconclusive. A theoretical approach has been proposed that focuses on the chain conformation in the molten state.(235) Attention is given to the entropy change that occurs when a sequence of chain units is transferred from the melt to the nucleus. Specifically, the entropy change in transferring a sequence of n^* units, of length l^*, from the pure melt (state 1) to a surface layer (state 2) is calculated. This surface layer is localized in close proximity to the nucleus (crystal) surface but it is only attached at a few

[19] A similar analysis has also been given for poly(chlorotrifluoroethylene).(174) However, in this analysis the crystallite thickness was identified with the small-angle x-ray scattering long period. Hence there is concern over the use of the Gibbs–Thomson equation with these data.

points. The actual attachment of the sequence to the nucleus in crystallographic register is not considered. It is also assumed that there is no enthalpic contribution to the free energy change involved. Consequently, for the process outlined

$$\Delta G_{1\to 2} = -T\,\Delta S_{1\to 2} \tag{9.220}$$

It is further postulated that

$$\Delta S_{1\to 2} = \frac{-\Delta S_{\mathrm{u}}}{C_\infty} \tag{9.221}$$

where we recall that C_∞ is the characteristic ratio for the given polymer. It is then argued that (235,236)

$$-T\,\Delta S_{1\to 2} = \frac{T\,\Delta S_{\mathrm{u}}}{C_\infty} = T\left[\frac{\Delta H_{\mathrm{u}}}{T_{\mathrm{m}}} a_0 b_0 l_{\mathrm{b}} n^*\right] \Big/ C_\infty \tag{9.222}$$

where l_{b} is the C–C bond length in a carbon backbone chain. When Eq. (9.222) is identified with the expression for adding the first sequence to a Gibbs type nucleus composed of regularly folded chains,

$$\Delta G_{\mathrm{l}}^* = 2b_0\sigma_{\mathrm{un}} l_{\mathrm{u}} n^* \tag{9.223}$$

Then

$$\sigma_{\mathrm{un}} = T\left(\frac{\Delta H_{\mathrm{u}}}{T_{\mathrm{m}}}\right)\left(\frac{a_0}{2}\right)\left(\frac{l_{\mathrm{b}}}{l_{\mathrm{u}}}\right)\frac{1}{C_\infty} \cong \Delta H_{\mathrm{u}}\left(\frac{a_0}{2}\frac{l_{\mathrm{b}}}{l_{\mathrm{u}}}\right)\frac{1}{C_\infty} \tag{9.224}$$

results. Here l_{u} is the projected bond length in the chain direction.

It would be appropriate in deriving Eq. (9.224) to include an additional step, namely the transfer of the sequence from state 2 to the appropriate nucleus surface, state 3. The need for this step is in fact implied by the introduction of the entropy and enthalpy of fusion into the calculation. The quantity σ_{un} only makes physical sense in terms of state 3. The rate determining step is not pertinent to a thermodynamic calculation.

Underlying the above analysis is the assumption that the change in entropy for a sequence of length l^* in going from state 1 to state 2 scales with $1/C_\infty$. It has been pointed out that this assumption implies that the conformational entropy, S_{conf}, should also scale with $1/C_\infty$(237). Calculations using rotational isomeric state theory have shown that for many polymers S_{conf} does not correlate with C_∞. This conclusion is not surprising since the conformational entropy of a statistical chain depends only on the number and relative energies of the bond rotational states. On the other hand C_∞ is a measure of the statistically thermodynamically averaged extension of the chain in space. In addition, C_∞ also depends on the structure and geometry of the chain. This criticism is significant and needs to be given serious consideration.

It should be noted that C_∞ is the characteristic ratio of an infinite or high molecular weight chain. For example the asymptotic or limiting value of the characteristic ratio is not reached for linear polyethylene until the number of methylene units is greater than about 150.(237a) For lower chain lengths the characteristic ratio increases with the number of repeating units. Consequently, if the theory as embodied in Eq. (9.224) is correct, σ_{en} should decrease with n in this molecular weight region, and should be an important factor for the n-alkanes.

As has been indicated there is no direct way of obtaining σ_{un}. It has been proposed that in the present case σ_{un} be obtained from growth rate kinetics and C_∞ calculated from Eq. (9.224). It can then be compared with experiment, or the results from rotational isomeric state theory. Conversely, the known C_∞ can be used to calculate σ_{un} and the results compared with experiment. Either way further assumptions need to be made to establish the value of σ_{un}. The problem is that σ_{un} has to be separated from either the product of $\sigma_{un}\sigma_{en}$ or $\sigma_{un}^2\sigma_{en}$ depending on the type of nucleation that is assumed. Furthermore, since σ_{en} is not known, it is invariably identified with σ_{ec} as obtained from the Gibbs–Thomson equation. This procedure thus involves assumption built upon assumption. It is implicitly assumed that a Gibbs type nucleus is involved and that σ_{ec} can be identified with σ_{en}. With these assumptions one can proceed with the comparison between theory and experiment.(198,235)

Hoffman and coworkers found, using the procedures outlined above, reasonably good agreement between the calculated value of C_∞ and those for polyethylene, isotactic poly-(styrene), poly(l-lactic acid), isotactic poly(propylene), and poly(ϵ-caprolactone).(235) On the other hand the agreement is poor for poly(pivalolactone) (237) as well as for poly(ethylene terephthalate), poly(aryl ether ether ketone) and poly(ethylene naphthalene dicarboxylate). (198) The comparison between theory and experiment, despite the many assumptions made, remains inconclusive.

There is obviously much still to be done in order to establish the value of σ_{un} for a given polymer. The work to date has established the order of magnitude of σ_{un} and indicates that it is much less than σ_{en}. Beyond this general conclusion, the value of σ_{un} remains uncertain for any given polymer.

9.13 Nucleation catalysts

The influence of foreign bodies, surfaces, crevices and similar entities that accelerate the nucleation process has been discussed earlier in this chapter. These heterogeneities are fortuitous since they have not been deliberately introduced into the system. In many instances strong efforts are made to remove them. However, there are also cases where specific species are deliberately introduced in order to accelerate the crystallization and alter properties. Such substances have been termed nucleation catalysts. They are usually low molecular weight organic and inorganic

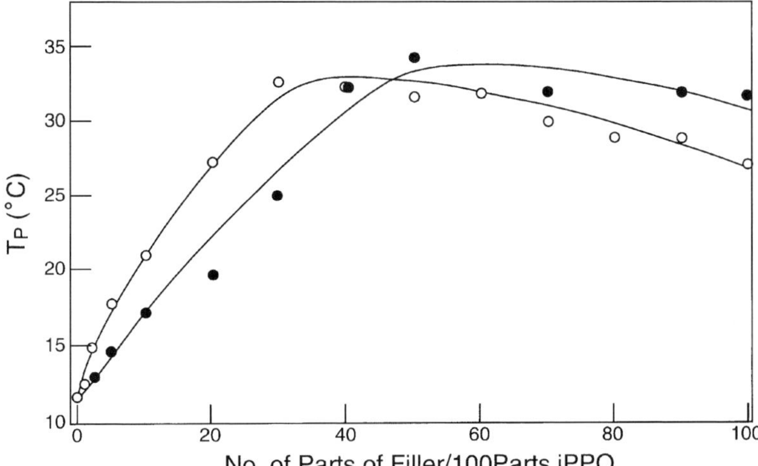

Fig. 9.67 Plots of T_p against filler content for isotactic poly(propylene oxide) with silica additives. ○ polymer plus untreated silica; ● polymer plus Me_4SiCl treated silica. (From Cole and St. Pierre (247))

compounds. However, polymers can also serve in this role.(238–240) Even residual catalysts from the polymerization often serve the same purpose. In special cases nucleation catalysts can direct the development of a given crystalline polymorph, as for example the β polymorph of isotactic poly(propylene).(241) In this section the role of nucleation catalysts in altering polymer crystallization will be examined. Typical examples of the overall crystallization and spherulite growth rates will be analyzed. The main purpose is to determine the kind of structures that are involved and the general principles, if any, that govern the catalysis.

A qualitative assessment of the effectiveness of a particular compound as a nucleating agent can be obtained by cooling the sample in a scanning differential calorimeter, or an equivalent apparatus, at a fixed cooling rate. The temperature, T_p, at which an exotherm is first observed can be taken as a measure of the substance as a nucleating agent. This is in effect a qualitative measure of the crystallization rate. For example, T_p of isotactic poly(propylene) can be raised about 25 °C by appropriate choice of catalyst;(242–244) the T_p of poly(ethylene terephthalate) can be increased by 35 °C;(245) while T_p for poly(dimethyl siloxane) has been increased by 25 °C.(246) As an example, the effectiveness of two silicas as nucleating agents for isotactic poly(propylene oxide) is illustrated in Fig. 9.67.(247) The optimum crystallization temperature steadily increases from 10 °C to 35 °C with increasing concentration of nucleating agent and then essentially levels off. There is also a small but significant difference between the treated and untreated silica.

Cooling rate studies establish that a significant increase in crystallization rates can be achieved by the addition of appropriate compounds. A more quantitative analysis

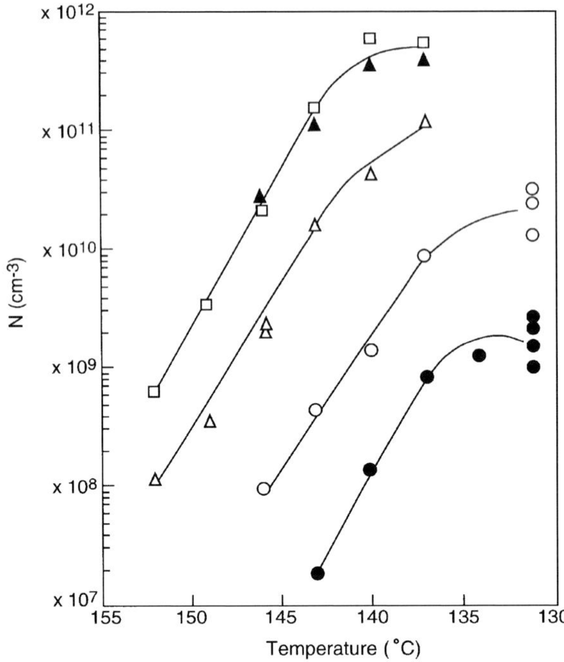

Fig. 9.68 Plot of number of crystallization centers against crystallization temperature for isotactic poly(propylene) containing various amounts of AlOH tertiary butyl benzoate. Weight percent: ● 0.03; ○ 0.01; △ 0.30; ▲ 1.0; □ 3.0. (From Binsbergen and de Lange (248))

of the crystallization kinetics is then important. A first step is to examine the increase in the number of nucleation centers that are formed. A typical example is given in Fig. 9.68. Here, the catalytic effect of finely divided AlOH tertiary butyl benzoate on the number of crystallization centers that are formed in isotactic poly(propylene), at different crystallization temperatures, is shown.(248) At a fixed temperature there is about a four order of magnitude increase in the number of crystallization centers for a 0.03 to 1.0 increase in the weight percent of the catalyst. Other nucleation catalysts for isotactic poly(propylene) initiate centers in a similar manner.(249,250) Nucleation catalysts have a profound effect in isotactic poly(propylene) as well as other polymers.(251)

Nucleation catalysts also have a strong influence on the overall crystallization kinetics, as is demonstrated in the experimental isotherms illustrated in Figs. 9.69 and 9.70 for poly(caprolactam) (238) and poly(ethylene terephthalate) (253) respectively. The isotherms for poly(ethylene terephthalate) are for a fixed crystallization temperature with different types of nucleation catalyst at the indicated weight percent. In all cases the isotherm shapes are similar to one another. There is an enhancement of the crystallization rate that is also found with other catalysts for

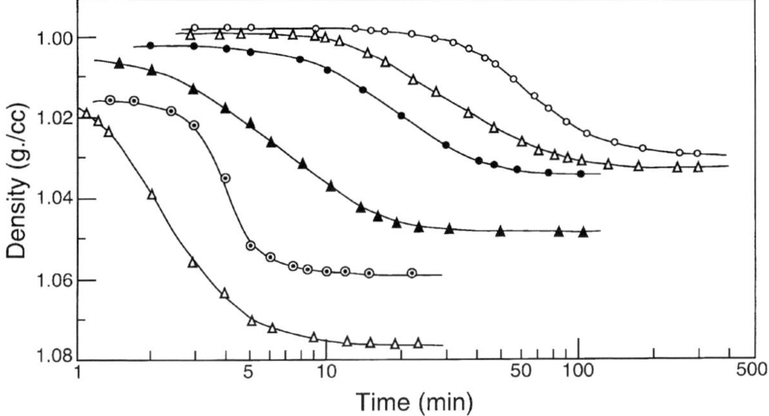

Fig. 9.69 Effect of $Pb_3(PO_4)_2$ as a nucleating agent on the crystallization isotherms of poly(caprolactam). Pure polymer: ○ T_c = 205 °C; ● T_c = 200 °C ; ⊙ T_c = 180 °C. $Pb_3(PO_4)_2$: △ 0.1%, T_c = 205 °C; ▲ 0.1%, T_c = 200 °C; △ 0.1%, T_c = 180 °C. (From Inoue (238))

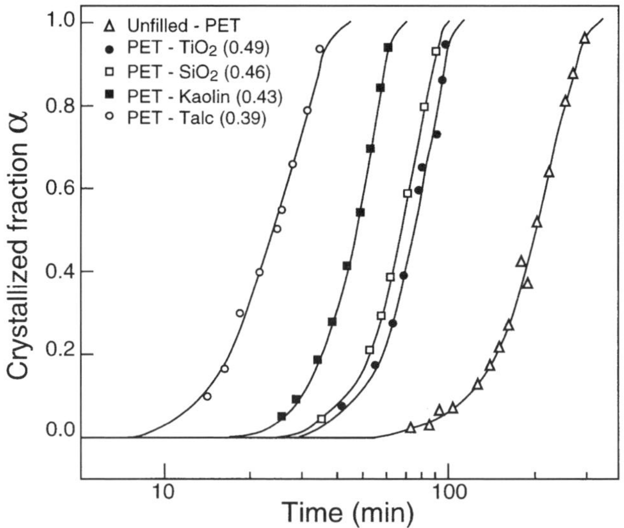

Fig. 9.70 Comparison of crystallization isotherms for pure poly(ethylene terephthalate) and with indicated additives. (From Groeninckx *et al.* (253))

this polymer.(254) Some additives turn out to be more effective than others. The isotherms for poly(caprolactam) are given at three different crystallization temperatures, for the pure polymer and for added 0.1% of $Pb_3(PO_4)_2$. These results are similar and the isotherm shapes are not sensibly affected by the addition of the catalyst. However, the crystallization rates are greatly enhanced at a given crystallization temperature. Similar results are reported for isotactic poly(propylene)

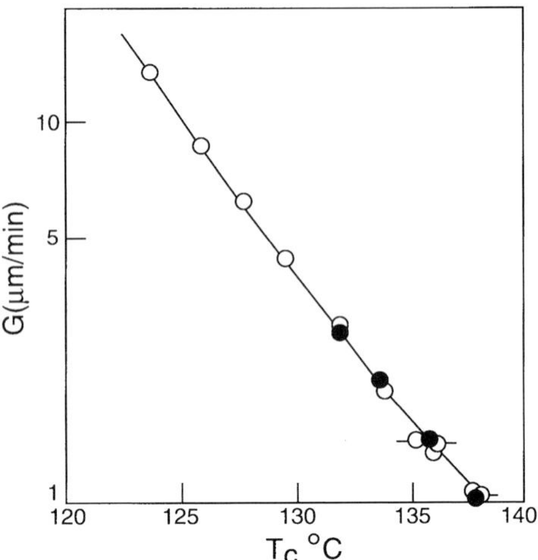

Fig. 9.71 Spherulite growth rates of isotactic poly(propylene) with different sorbitol compounds as nucleation catalysts. ○ pure polymer; ● dibenzylidene sorbitol; ○- (*p*-chloro, *p*-methyl) dibenzylidene sorbitol, -○ bis (*p*-ethylbenzylidene sorbitol. (Data from (249))

(249,250), poly(phenylene sulfide) (252) and isotactic poly(propylene oxide) (247) among others.

In most studies the derived Avrami equation (Eq. 9.31a) has been used to analyze the isotherms. According to Eq. (9.31) a double logarithmic plot is usually made with the data. Most commonly, a nonintegral value of the Avrami exponent was obtained. Despite the inherent shortcoming in analyzing the data by this method, the conclusion can be made that for a given polymer the Avrami n value is independent of the nature and concentration of the catalyst. Thus, the isotherms are superposable with one another and that of the pure polymer.

Spherulite growth rates are an important aspect of polymer crystallization. According to Binsbergen and de Lange (248) the growth rate of nucleated isotactic poly(propylene) is the same as that of the non-nucleated polymer. The constancy of spherulite growth rates is illustrated in Fig. 9.71 for isotactic poly(propylene) that has different sorbital compounds added as nucleation catalysts.(249) There is clearly no change in growth rates between the pure polymer and those containing the nucleation catalysts. Similar results are found when sodium benzoate is added to isotactic poly(propylene) as a nucleation catalyst.(250) With few exceptions, the constancy in spherulite growth rates with added nucleation catalysts appears to be the general rule.

Table 9.8 compares the influence of different nucleating agents on the spherulite growth rate of poly(ethylene terephthalate).(254) The growth rate at 237 °C for the pure polymer is 3.9 ± 0.3 μm min^{-1}. Except for CaO, the spherulite growth rates are

Table 9.8. *Spherulitic growth rates (μm/min) of poly(ethylene terephthalate) samples containing indicated nucleating agents for crystallization at 237 °C (254)*

Concentration of solid in sample, %	CaO	TiO_2	MgO	$BaSO_4$	SiO_2	Al_2O_3
0.2	2.7 ± 0.6	3.4 ± 0.4	2.9 ± 0.5	4.1 ± 0.5	4.3 ± 0.5	4.2 ± 0.4
0.5	2.8 ± 0.6	4.0 ± 0.4	3.6 ± 0.5	3.6 ± 0.5	3.0 ± 0.5	3.7 ± 0.4
0.75	2.6 ± 0.6	3.7 ± 0.4	3.9 ± 0.5	4.0 ± 0.5	3.7 ± 0.5	4.1 ± 0.4
1.0	2.3 ± 0.6	4.2 ± 0.4	3.8 ± 0.5	4.2 ± 0.5	3.7 ± 0.5	4.0 ± 0.4
1.5	2.1 ± 0.6	4.2 ± 0.4	4.4 ± 0.5	4.2 ± 0.5	3.6 ± 0.5	4.2 ± 0.4
2.0	2.1 ± 0.6	4.3 ± 0.4	3.2 ± 0.5	3.5 ± 0.5	3.7 ± 0.5	4.2 ± 0.4
2.5	1.9 ± 0.6	4.1 ± 0.4	4.0 ± 0.5	3.7 ± 0.5	4.1 ± 0.5	3.6 ± 0.4
3.0	2.0 ± 0.6	3.7 ± 0.4	4.1 ± 0.5	3.5 ± 0.5	3.3 ± 0.5	3.8 ± 0.4

independent of the kind and concentration of catalyst and agree quite well with that of the pure polymer. It is interesting to note that although CaO and MgO crystallize in the same system they exert quite a different influence on the growth rates.

The spherulite growth rates of isotactic poly(styrene), nucleated with silica, are unique.(255) At low crystallization temperatures, to the left of the rate maximum, there is no difference in rate between the nucleated and unnucleated polymers. However, at high crystallization temperatures, at the right side of the maximum, the growth rates are reduced significantly by the nucleation catalysts. Moreover, the maximum in the growth rate is lowered by about 8 °C. A qualitative interpretation of these results is based on the concept that the silica particle interacts with the polymer to create entanglements in the melt.(255) The excess entanglements would affect the transport term through the parameter U^*.

In general, the rate at which growing centers are initiated increases with the concentration of nucleation catalyst. In most cases, the spherulite growth rate remains constant. Consequently, the average spherulite size decreases and their distribution becomes more uniform. Effective nucleating agents reduce spherulite diameters by one-fifth to one-tenth that of the non-nucleated polymer. The magnitude and range of spherulite diameters found in poly(caprolactam) with different kinds and concentrations of additives are shown in Fig. 9.72.(238) The size range decreases and narrows dramatically with the initial addition of the catalyst. There is only a small change in diameter with further increase in the additive concentration. Eventually a constant and narrow diameter range is reached. In the example shown there are only small differences between the different catalysts.

The addition of a second component, even in small amounts, does not necessarily result in nucleation catalysis in the sense discussed. Either, or both, of the glass

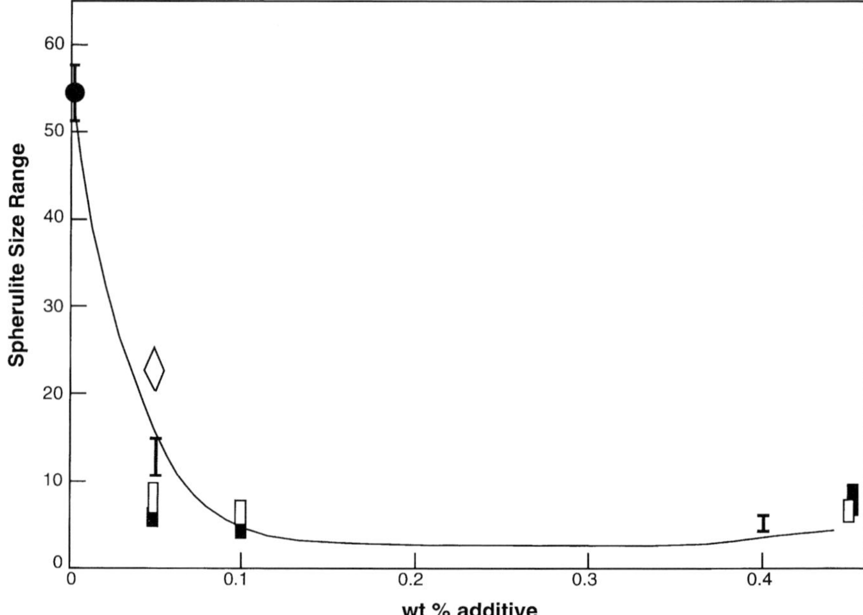

Fig. 9.72 Plot of spherulite size range against wt percent additive for poly(caprolactam) crystallized at 150 °C. **I** pure polymer; ▯ NaH_2PO_4, ▮ $Na_7P_5O_{16}$; ● $Pb_3(PO_4)_2$; ◇ TiO_2. (Data from (238))

and melting temperatures can be altered with concomitant changes in either the overall crystallization or spherical growth rates without affecting the spherulite size. For example, the addition of a small amount of diphenylamine to poly(ethylene terephthalate) alters the growth rate of spherulites but not their size.(256)

To summarize the experimental results the rate of primary crystallization is greatly increased with the addition of a nucleation catalyst while the secondary or growth nucleation remains constant. The increase in the number of growing centers results in a reduction in the spherulite sizes. This results in a reduction in the light scattering and the enhancement in clarity and surface gloss in moldings and film.(254,257,258) There is a direct correlation in the effectiveness of the nucleating agent, as measured by the increase in T_p, and the clarity of the sample. The reduction in particle size reduces the light scattering and hence the turbidity of the sample. Opacity is associated with the large coarse spherulites in non-nucleated polymers, and clarity with small sizes in nucleated polymers. The density also increases with the effectiveness of the nucleation catalyst, indicating an increase in the degree of crystallinity.(254,257)

The question arises as to whether there are any general requirements for an effective nucleating agent. Among the factors that need to be considered are the chemical nature, crystal structure and size of the nucleant and their relation to the corresponding features in the polymer. Several ideas have been proposed with respect to the mode of action of the nucleant. The effective nucleants can be divided into two broad categories. One group is where the interaction leading to nucleation catalysis is physical in character. In this case the nucleant needs to be insoluble in the polymer melt.(259) In the other, the catalysis is the result of chemical interactions. A polymer, such as poly(ethylene terephthalate), can be catalyzed as a result of either chemical reactions (245,260) or by physical interaction.(254) Analysis of the pertinent experimental results should lead to an understanding of the principles that are involved.

Extensive studies of possible nucleating agents for poly(olefins), focusing primarily on isotactic poly(propylene), have allowed for some of the requirements for effective nucleation catalysts to be defined.(248,259,261) Nucleating agents are in the main crystalline and are most efficient in a fine dispersion of small crystals. However, the dispersed nucleating agent needs to be large enough to accommodate nuclei of critical size. The catalysts are usually insoluble in the polymer melt. If not, they crystallize on cooling before the polymer does. The catalyst also has to be of low energy and wet the polymer. Most agents consist of a hydrocarbon group and either a polar group or a condensed aromatic structure. Binsbergen has suggested that most of these compounds consist either of parallel rows or parallel layers of molecules.(259) The layers expose the hydrocarbon groups while the polar groups are confined to the center. Nucleating agents in this category that are effective for isotactic poly(propylene) are also active with other poly(olefins), such as linear polyethylene, isotactic poly(4-methyl-1-pentene) and isotactic poly(styrene).(259,260)

Epitaxial crystallization of a polymer on the surface of a nucleating agent has also been established as a major mechanism for nucleation catalysis.(261a) For epitaxy to occur, there has to be a match between crystallographic planes of the two species. Although this requirement is very specific, certain crystallographic features allow this mechanism to be operative with the poly(olefins), some poly(amides) and poly(esters) as well as certain low molecular weight species. In an interesting observation, finely divided crystalline isotactic poly(propylene) was found to act as a good nucleating agent for linear polyethylene.(258) In a series of papers, Lotz and Wittman pointed out that helical poly(olefins) possess crystallographic faces in which the side-groups form well aligned rows.(262,263) This structural feature can interact with polyethylene, aliphatic poly(esters) and poly(amides). Each of these polymers is characterized by a trans–trans ordered backbone conformation. Therefore, this diverse group of polymers can act as nucleating agents for one another. As a corollary, it is, therefore, not unexpected that among low molecular weight substances the same compound can serve as a nucleating agent for this diverse group of polymers. An epitaxial type crystallization has also been suggested as the catalytic mechanism between poly(caproamide) and its most effective nucleating agent, talc.(264) Talc is also an effective nucleating agent for poly(ethylene terephthalate) by means of epitaxial nucleation.(264a) In this case contact between the (100) plane of poly(ethylene terephthalate) and the (001) talc basal surface has been established as well as an alignment of the polymer chain axis with the [100], [110] and [1$\bar{1}$0] directions in the talc. Although epitaxy is an effective mechanism for nucleation catalysis it is not the only one that can be involved in heterogeneous systems. However, the other cases are not amenable to a general set of rules.(258) There is a set of agents that can act as nucleation catalysts for either the α or β forms of isotactic poly(propylene) and in some cases both.(263a) Detailed crystallographic relations have been established between the polymers and the nucleating species.

The crystallization kinetics for the physical type of nucleation catalysis follows the formulation of heterogeneous nucleation that has been given previously. The results with specific geometric forms have been detailed.(261,265) In contrast to the heterogeneous catalysis just described, the increase in nucleation rate can also be achieved in a homogeneous system. In this case, the additive is soluble in the polymer melt and some type of chemical interaction or reaction takes place between the two. The result is the formation of the nucleating species. This process can be termed chemical nucleation. One group of nucleants in this category is metal salts.(245,260) The chemical reaction usually involves a reduction in molecular weight, through chain scission, and the formation of ionic chain ends that associate into clusters. The nucleation efficiency is closely linked to the presence of ionic clusters. Nucleation that involves this type of catalysis has been demonstrated for poly(ethylene terephthalate) (245,260,266,267), poly(aryl ether ketone) (260) and poly(aryl ether ether ketone).(260) Nucleation catalysis caused by a gelation process has also been suggested in special cases.(268)

A scale has been proposed by which the effectiveness of a nucleation catalyst can be assessed. In addition to comparing the crystallization temperature on cooling the nucleated polymer with that of the pure polymer, the crystallization temperature of the self-seeded polymer is also taken into account.(268,269) The self-seeded polymer (270) is taken to represent the ideal nucleated system. The efficiency of a nucleating agent, E, expressed as a percentage, is given as (268,269)

$$E = 100(T_{\mathrm{p}} - T_{\mathrm{co}})/(T_{\mathrm{cmax}} - T_{\mathrm{co}}) \tag{9.225}$$

Here T_{p} is the crystallization temperature of the polymer with the additive, T_{co} that of the pure polymer, and T_{cmax} that of the self-seeded polymer. A table of E values for additives of the α phase of isotactic poly(propylene) has been given.(268,269) To be meaningful T_{cmax} must be reproducible. If it is then E will scale with T_{p}.

9.14 Influence of molecular weight

In the discussion up to now, the influence of molecular weight on different aspects of the crystallization process has been alluded to. As examples, the shape of the isotherms depends on molecular weight as does the level of crystallinity that can be attained at a given crystallization temperature. The dependence of the overall crystallization rate on molecular weight and crystallization temperature is a complex matter, as was illustrated in Fig. 9.23 for linear polyethylene and poly(tetramethyl-p-silphenylene siloxane). The role of molecular weight will be examined in more detail in this section, within the framework of the nucleation and growth concepts that have been developed. The discussion will be divided into three parts: oligomers represented by high molecular weight *n*-alkanes; low molecular weight fractions of different polymers; and polymers of moderate to high molecular weights.

9.14.1 Crystallization of n-alkanes

The synthesis of high molecular weight *n*-alkanes, containing between 100 and 400 chain carbon atoms, has allowed for a detailed study of the crystallization of these model molecules.(271–273) A high molecular weight *n*-alkane is well suited to serve as a model compound since each chain has exactly the same length and end-group. Either extended chain crystallites or some type of folded one can be formed, depending on the chain length and crystallization conditions. There is a chain length below which extended chain crystals are always formed irrespective of the crystallization temperature. However, the *n*-alkanes that are greater than about 150 carbons can form either extended or some type of folded crystallite, depending on the crystallization temperature.[20](273–275) Thus, it is possible to study the nucleation parameters for the same chain when either folded or extended mature crystallites result. The *n*-alkanes can serve as a bridge to low molecular weight fractions of linear polyethylene, as well as other polymers. In turn the results from studies with low molecular weight chain molecules lead naturally to the larger chain lengths.

The overall crystallization kinetics of the high molecular weight *n*-alkanes yields superposable, classical isotherms of the Avrami type. An example is given in Fig. 9.73, where $\ln(1 - \lambda(t))$ is plotted against $\ln t$ for $C_{192}H_{386}$.(278) The studied temperature interval of 118–124 °C encompasses the time scale of about 0.25–200 min for the detection of crystallinity. Low frequency Raman acoustical mode (LAM) studies indicate that extended chain crystallites are formed at all temperatures in

[20] The nature of the interfacial structure of the folded *n*-alkane crystallites will be discussed in Volume 3. Different possible structures will be presented. At this point it is not necessary to assume either that the crystallite thickness is quantized or that the interface is sharp with adjacent re-entry.(275,276) Other possibilities exist.(277)

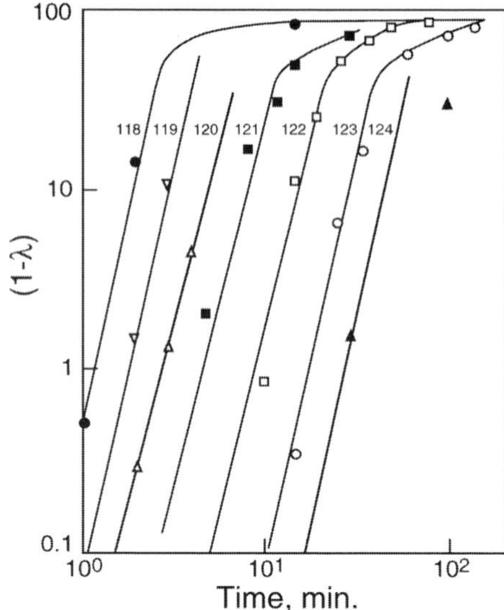

Fig. 9.73 Double logarithmic plot of the degree of crystallinity, $1-\lambda$, against time for melt crystallized $C_{192}H_{386}$. Crystallization temperatures indicated.(278)

the temperature range studied. The Avrami relation is followed for about 60% of the transformation. From the slopes of the parallel straight lines in the figure the Avrami exponent is found to be 4.0 ± 0.2. Thus, the crystallization follows classical concepts since there is no entanglement problem with these chain molecules.

It is also instructive to study the course of the crystallization by following the exotherms that evolve as a function of time. An example is given in Fig. 9.74 for the rapid isothermal crystallization of $C_{192}H_{386}$ at temperatures below those of the previous figure.(278) In Fig. 9.74a, for crystallization at 117 °C, only one exothermic peak is observed. This peak can be attributed to the extended crystallite form. For crystallization at 116 °C, Fig. 9.74b, two exothermic peaks are observed. The peak at about 0.5 min can be assigned to a nonextended form that, with time, develops into the extended form as manifested by the major exotherm observed at 0.89 min. The exotherm for crystallization at 115 °C, Fig. 9.74c, shows a major peak at 0.21 min with the beginning of a new exotherm developing at about 0.75 min. Eventually, after longer crystallization times at this temperature, the extended form develops completely from the folded form. Small-angle x-ray scattering studies with $C_{246}H_{494}$ give qualitatively similar results.(279) The significance of the thickening, as well as thinning, during the crystallization process will be discussed shortly.

The temperature dependence of the overall crystallization rate is of particular interest in terms of the nucleation processes involved. The log of the crystallization rates for the high molecular weight *n*-alkanes $C_{168}H_{338}$, $C_{192}H_{386}$ and $C_{240}H_{482}$ are plotted against the crystallization temperature in Fig. 9.75.(280) In this example,

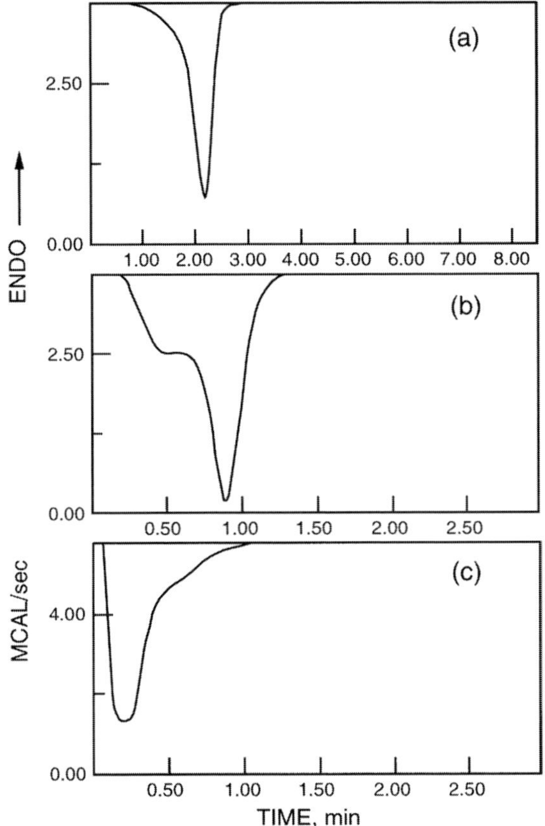

Fig. 9.74 Isothermal crystallization temperatures for melt crystallized $C_{192}H_{386}$ using the exothermic method. (a) At 117 °C; (b) at 116 °C; (c) at 115 °C.(278)

the crystallization rate is taken as the inverse of the time necessary to obtain a crystallization level of 5%. The shapes of the curves that describe each of the alkanes are similar to one another. They are, however, displaced along the time and temperature axes. The crystallization rate decreases with increasing temperature. There are indications of a plateau region in these plots. A discontinuity is also observed in each of the plots.

Results of a similar type of study with $C_{246}H_{494}$ are given in Fig. 9.76.(281) Here the rate is plotted against the crystallization temperature. The rate displays a maximum, followed by a sharp minimum. A rapid increase in the rate then occurs over a very narrow temperature region. Similar results have been reported for normal hydrocarbons ranging from $C_{194}H_{390}$ to $C_{294}H_{590}$.(282) The plots in Fig. 9.75 imply a similar behavior. The observation of a rate maximum in the vicinity of T_m^0 is a unique observation, as is the minimum that follows. Maxima in crystallization rates are quite common and have been described earlier. However, in the previous examples

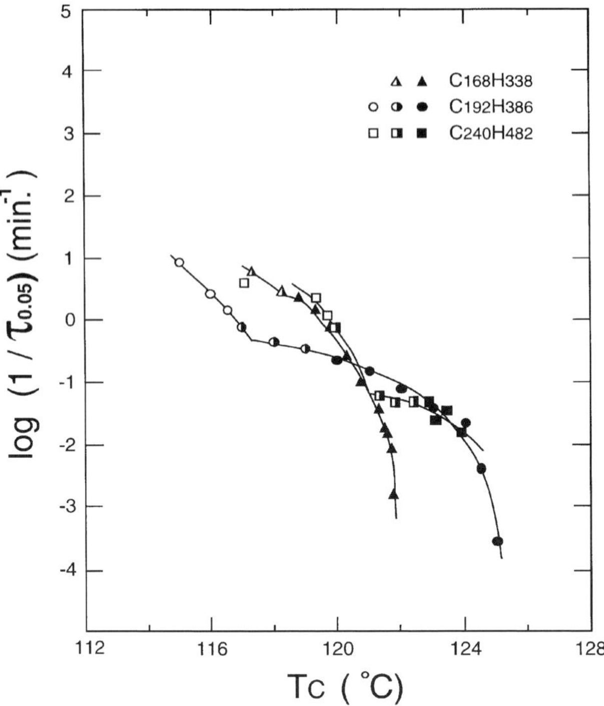

Fig. 9.75 Plot of rate of crystallization as function of temperature for the indicated *n*-alkanes crystallized isothermally from the pure melt. Open symbols represent the formation of once-folded crystals. Closed symbols are data for extended crystals. Half-closed symbols represent the experimental values that may be affected by a rapid transformation from once-folded to extended crystals.(280)

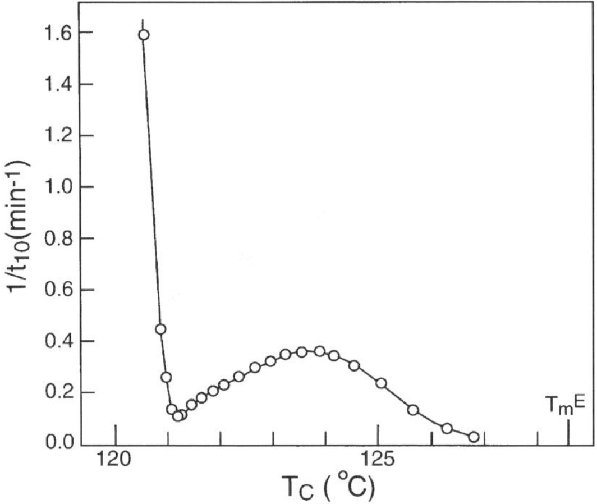

Fig. 9.76 Plot of reciprocal times $1/t_{10}$ against crystallization temperature T_c for $C_{246}H_{494}$ crystallized from the pure melt. t_{10} is the time for 10% of the transformation. T_m^E is the melting point of the extended chain crystal. (From Ungar and Keller (281))

the maximum occurs well below T_m^0 and the rate decreases as the temperature is lowered. Before considering possible reasons for the rate maximum in the n-alkanes, it is helpful to analyze the data presented in Fig. 9.75 in more detail.

In analyzing the data from Fig. 9.75 in terms of nucleation theory it is important to establish the temperature regions where the structure that initially crystallizes from the melt is well defined. Therefore, the temperature regions where initially folded crystals are formed and maintained needs to be specified. Also to be defined are the regions where extended chain crystals are initially formed. It is also necessary to specify the temperature interval where the initially formed folded chain crystallites thicken with time. In order to obtain the necessary information recourse is made to experiments of the type described in Fig. 9.74. The crystallization rates of initially formed folded structures, where the thickening process does not interfere with the rate measurements, are represented by open symbols in Fig. 9.75. The crystallization rates of extended structures that are initially formed from the melt are represented by closed symbols. The temperatures where the rate measurement may be affected by the thickening of folded crystallites to extended ones are represented by the half-closed symbols. There is a temperature interval in each of the curves in Fig. 9.75 where the crystallization rate does not change very much with temperature. In this temperature range most, but not all, of the half-closed symbols are found. This result indicates that the plateaus shown by the log rate–T_c curves (or maxima in the rate–T_c curves) need to be associated, at least in part, with the temperature regions where the transformation of folded to extended crystallites is rapid.

The temperatures over which the isothermal crystallization of the n-alkanes takes place are restricted to a small interval in the vicinity of the equilibrium melting temperature.(283) Therefore, the transport term in the expression for the crystallization rate will remain essentially constant. Thus, the free energy change, ΔG^*, required to form a critical size nucleus controls the nucleation and thus the crystallization rate. The expression for ΔG^* for chains of finite length appropriate to the high molecular weight n-alkanes has been given by Eqs. (9.124) and (9.145). The formation of a coherent Gibbs type nucleus as the controlling nucleation process is taken as an example. Consequently Eq. (9.145) is appropriate. Other types of nucleation models could equally well be used. The appropriate plot is made in Fig. 9.77 where the log of the rate is plotted against the nucleation temperature function for coherent surface nucleation.(280) The same symbols as those in Fig. 9.75 are used. The plots in Fig. 9.77 are well represented by sets of intersecting straight lines. A compilation of the slopes of the straight lines is given in Table 9.9. The high temperature region, where extended crystals are formed, is termed Region I with slope S_I. In analogy, the low temperature region is designated as III and the intermediate region as II. The three regions are clearly demonstrated by the results for $C_{192}H_{386}$. The slopes in the high and low crystallization temperature regions (I and III respectively) are

Table 9.9. *Slopes of straight lines in Fig. 9.77*

n-Alkane	S_I	S_{II}	S_{III}
$C_{168}H_{338}$	−170	−35	—
$C_{192}H_{386}$	−170	−30	−170
$C_{240}H_{482}$	—	−30	−170

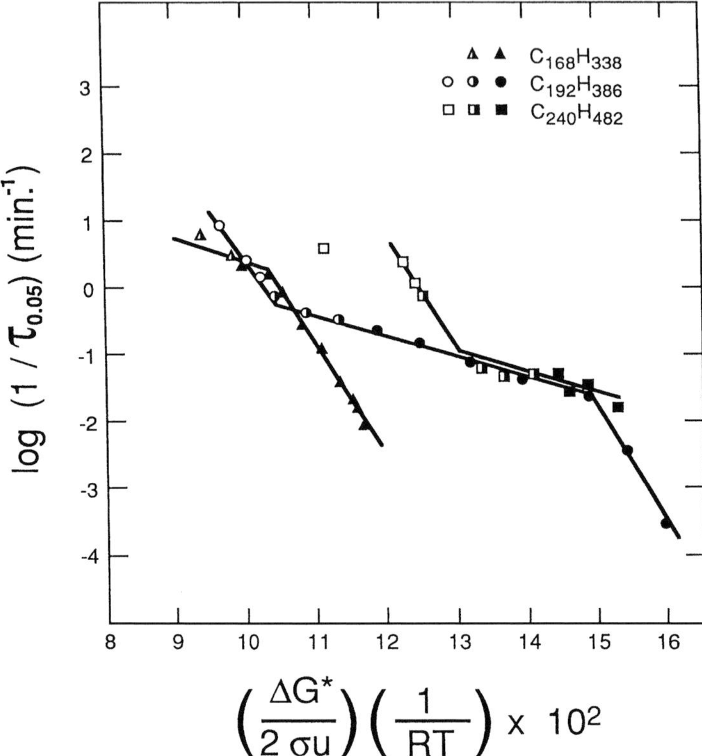

Fig. 9.77 Plot of log crystallization rate against nucleation temperature function for coherent unimolecular nucleation for the n-alkanes indicated. Symbols as in Figure 9.75.(280)

the same.[21] The fact that the slopes are identical indicates that both extended and folded chain crystallites have the same value of σ_{en}. The conclusion is then reached that the same nucleus leads to mature crystallites with different chain structures. Since extended structures are formed it is highly unlikely that the nucleus is of the regularly folded type. These results demonstrate that it is not necessary to assume a

[21] These regions should not to be identified with Regimes I, II and III.

particular chain structure within the nucleus to account for the crystallite structure that is eventually observed.(284,285)

The data for $C_{240}H_{482}$ are well represented by two straight lines, whose slopes are the same as those for $C_{192}H_{386}$ in the low and intermediate crystallization region. It is a reasonable expectation that if data could be obtained at higher crystallization temperatures, the temperature coefficient would be similar to that for $C_{192}H_{386}$. The results for $C_{168}H_{338}$ in the high temperature region support this expectation since the slope is identical to that obtained in the high temperature region for $C_{192}H_{386}$.

There is an intermediate region in the plots in Fig. 9.77 where the slopes, S_{II}, for the three alkanes are the same. However, their values are reduced from those in Regions I and III. As the half-closed symbols indicate, most of the data points in this region represent crystallization temperatures where initially folded chain crystallites form but are then rapidly transformed to extended ones. Consequently, analysis of the kinetics in this region is severely complicated by the isothermal thickening and the structural transformation that takes place during crystallization. This is also the temperature interval where the rate maxima have been reported. Folded chain crystallites are unstable at temperatures in the vicinity of the junction point between Regions II and III and crystallite thickening begins at these temperatures.

In summary, the overall crystallization rate–temperature relations of the high molecular weight n-alkanes can be divided into three regions. In the low temperature region folded structures are formed from the melt. The thickening process is sufficiently slow in this region so as not to affect the rate measurements. In the high temperature region folded structures no longer crystallize. The crystallization rates of well-defined extended crystallites are measured at these temperatures. In the intermediate temperature region that connects the two, the isothermal crystallite thickening and the accompanying structural changes severely complicate the analysis.

The unique maximum in the crystallization rate has already been noted. Several explanations have been put forth to explain this observation. One is the so-called "poisoning effect" proposed by Sadler and collaborators.(286–290) This "self-poisoning" of the crystal growth surface is postulated to be caused by the deposition of almost stable folded crystallites that retard the development of the extended chain crystallites. On this basis this phenomenon would be limited to crystallization temperatures in the vicinity of the junction of Regions II and III. Rate equation models based on the hypothesis, as well as computer simulations, have reproduced the qualitative features of the unique rate–crystallization temperature curves.(284,291)

Another theory that has been proposed to explain this phenomenon is based on a unique application of nucleation theory.(292,293) It is assumed that two types of nuclei are formed. One is an extended chain, nonfolded nucleus, that does not encompass the complete molecule. The other is a once-folded nucleus. The critical size of the chain folded nucleus, and its temperature coefficient follow the theory developed for folded chain nucleation.(146,147) In contrast, it is assumed that the critical-size in the chain direction of the extended type nucleus does not follow nucleation theory. Rather, its rate of formation is governed by an Arrhenius type temperature dependent process. This assumption of two different types of nuclei, each with different temperature coefficients for formation, obviously leads to a maximum in the nucleation rate and concomitantly the crystallization rate. It was further assumed, contrary to experimental results, that at crystallization temperatures below the maximum in $C_{246}H_{484}$ extended chains were nucleated from the original melt.

The proposals made to explain the unique crystallization rate–crystallization temperature relation found in the *n*-alkanes have ignored the role of crystallite thickening. The temperature region where the maximum and inversion in the crystallization rate occurs has been shown to coincide with profound isothermal thickening of a folded structure to an extended one.(280,294) This phenomenon has been shown to quantitatively influence the overall crystallization kinetics. It will also be expected to influence growth rates accordingly. This aspect of the crystallization of the high molecular weight *n*-alkanes has been controversial. However, the crystallization of the *n*-alkanes from solution sheds further light on the problem and will be discussed in Chapter 13.

The growth rates of the supermolecular structures that are formed by the high molecular weight *n*-alkanes have also been studied. They complement the overall crystallization kinetic results.(295–301) The nucleation rate, determined by counting the number of crystals growing within a defined area, has also been reported.(298) Microscopic studies indicate that at low crystallization temperatures the folded chain crystallites develop spherulitic type supermolecular structures. At the high crystallization temperatures, the extended chain crystallites do not form spherulitic structures. Rather, long crystallites are observed that consist of parallel stacks of continuous lamellae. Both types of morphology are observed in the higher carbon number alkanes. Thus, from the growth rate studies with these alkanes, it is possible to determine how the different morphologies and crystallite structures affect the crystallization kinetics and nucleation process.

The *n*-alkanes from $C_{94}H_{190}$ to $C_{246}H_{494}$ can only be crystallized over a narrow range of crystallization temperatures in the vicinity of their respective melting temperatures.(298–301) The available temperature interval varies from 0.25 °C for $C_{94}H_{190}$ to 4 °C for $C_{162}H_{326}$ and $C_{246}H_{494}$. A dramatic increase in the growth rate with decreasing crystallization temperature is observed even over this small temperature interval. For example, the growth rate for $C_{98}H_{198}$ decreases by a factor greater than 30 as the crystallization temperature decreases from 114.9 °C to 113.5 °C.(299)

The growth rates of $C_{122}H_{246}$ to $C_{246}H_{494}$ are reduced by factors of 5 to 7 over the accessible temperature range.(300) However, this is still a substantial decrease. The rapid increase in the growth rate with decreasing crystallization temperature is a quantitative reflection of the well-known observation that n-alkanes in this size range can scarcely be supercooled. The extremely small temperature interval that is available for growth rate studies, a few degrees or less, makes a detailed analysis difficult. However, it has been claimed that the growth rates are directly proportional to the undercooling.(299,300) This conclusion is in accord with theoretical considerations (301) where, however, the implicit assumption is made that the complete molecule participates in the nucleation. This conclusion is not in accord with droplet experiments involving the higher molecular weight n-alkanes.(135a) The growth rate data that were presented adhere quite well to the expectation from classical two-dimensional nucleation theory. The product of interfacial free energies for nucleation, $\sigma_{un}\sigma_{en}$, increases significantly in this range of n-alkanes.

The growth rates of the n-alkanes, $C_{246}H_{494}$ (295,298) and $C_{294}H_{590}$ (295,297), can be studied over a larger temperature range, of 10–11 °C. Folded chain and extended chain crystals are formed with each of these alkanes. The growth rates show both a maximum and a minimum with crystallization temperature, as is illustrated in Fig. 9.78 for $C_{246}H_{494}$.(295) Since both types of crystallites, folded and extended, can be formed with the same alkane it is instructive to analyze the data in terms of appropriate nucleation theory. The growth rate data of $C_{246}H_{494}$ will be studied since they are more extensive.

Accordingly, the growth rate data for $C_{246}H_{494}$ reported by Sutton *et al.* (295) are plotted in Fig. 9.79 according to the analysis of nucleation theory pertinent to low molecular weight chain molecules.(302) For illustrative purposes, a coherent, unimolecular type nucleation is selected. A value of 2000 cal mol^{-1} of sequences is taken for σ_{en}.[22] Only the kinetic data that represent the formation of well-defined structures directly from the pure melt are analyzed. The extended chain crystallites, which satisfy this criterion, develop at the high crystallization temperatures, $T_c \geq$ 125 °C, and are represented by the closed circles in the figure. The folded chains, which give spherulitic structures, are formed at $T_c \geq 120$ °C, and are represented by the open circles. Crystallization at the intermediate temperatures is represented by the half-filled circles. In this latter temperature region the structures that are measured do not represent those that were initially formed from the melt. The situation is analogous to Region II that was defined in the analysis of the overall

[22] In this analysis, as has been pointed out earlier, the value of σ_{en} has to be *a priori* stated. Using the mature crystallite as a reference σ_{en} values of 2000 cal mol^{-1}, 2300 cal mol^{-1} and 2500 cal mol^{-1} were taken for $C_{168}H_{338}$, $C_{192}H_{386}$ and $C_{240}H_{482}$ respectively.(280) The basic analysis is not affected by these choices, since only the relative horizontal location of the plots is influenced. In these calculations T_m^0 was taken as 145.5 °C and $\Delta H_u = 950$ cal mol^{-1}.

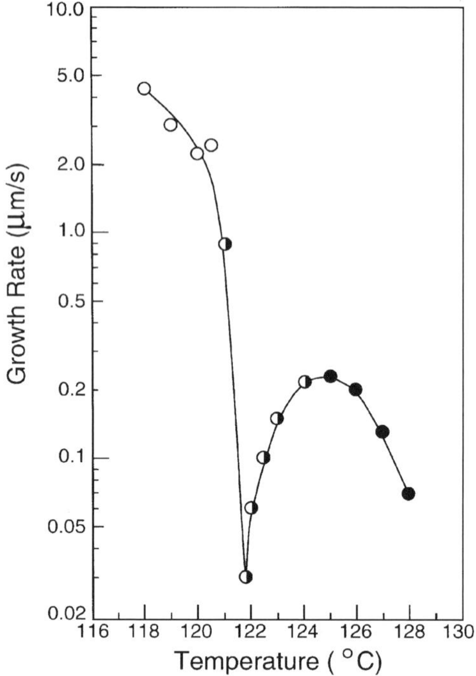

Fig. 9.78 Plot of crystal growth rate as a function of crystallization temperature for $C_{246}H_{494}$ crystallized from pure melt. (Adopted from Sutton *et al.* (295))

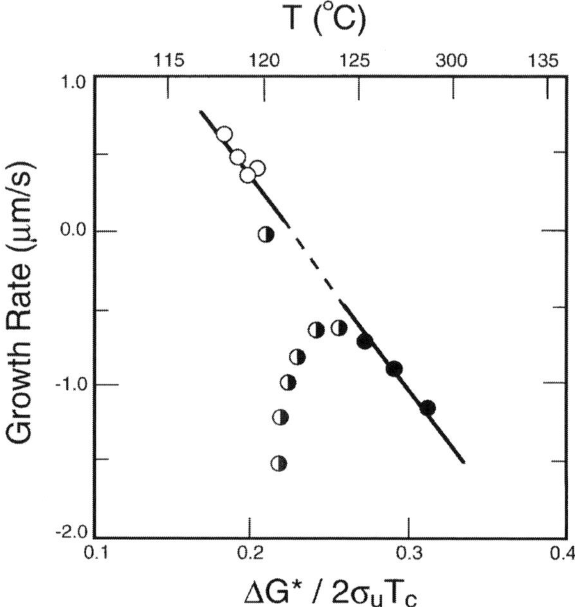

Fig. 9.79 Plot of ln G against $\Delta G^*/2\sigma_{\mathrm{un}}T_{\mathrm{c}}$ for $C_{246}H_{494}$. Upper axis gives corresponding crystallization temperature T_{c}.(302)

crystallization kinetics. In this region, rapid isothermal thickening of the initial folded structure takes place.[23] The measured growth rates will, therefore, be affected by the thickening. Another problem is the possible "poisoning" of the growth surface. Whatever the reason, or reasons, the intermediate temperature region does not represent growth rates of structures that were directly formed from the melt. Consequently, the growth kinetics in this region are not pertinent to our immediate interest. The growth rate data reported by Organ *et al.* (298) for the same n-alkane are displaced slightly from those in Fig. 9.79. However, when analyzed these data give the same temperature dependence of the growth rate.

The important conclusion to be made from Fig. 9.79 is that the growth rate data for both extended and folded chain crystallites, which correspond to nonspherulitic and spherulitic morphologies respectively, adhere to the same straight line. Thus, for a constant value of σ_{un} the interfacial free energies associated with the basal planes of the nuclei, σ_{en}, are the same for the two different chain conformations and the accompanying morphologies that are associated with the mature crystallites. This is the same conclusion that was reached from studies of the overall crystallization kinetics of the n-alkanes $C_{168}H_{338}$, $C_{192}H_{386}$ and $C_{240}H_{482}$. The absolute value of σ_{en} is not of immediate concern here. The fact that the same value of σ_{en} applies in both cases requires that the interfacial structure of the nucleus be the same, irrespective of the chain conformation within the nucleus and of the supermolecular structure of the mature crystallites that evolve. There is no physical law that requires that the structure of the macroscopic crystallite that develops be the same as that of the nucleus from which it is formed. This principle has been discussed in detail with regard to polymer crystallization.(140,274,278)

The ratio ζ^*/x is crucial in determining the thickness of the mature crystallite.(274,278) Since, depending on the crystallization temperature, either extended or folded chain crystallites result, it is reasonable to assume that the nucleus is of the bundle type. For low molecular weights and high crystallization temperatures (low undercoolings) ζ^* will be comparable to x. In this situation there are no major impediments to growth along the chain axis and extended chain crystallites will result. In contrast, for high molecular weights and low crystallization temperatures, ζ^* will always be much less than x. Growth in the chain direction will, therefore, be severely retarded. Under these circumstances the disordered chain units that need to be incorporated into the growing crystal will be highly entangled with one another. The possibility also exists for multiple nucleation acts to occur within the same chain. There are clearly restraints to growth in the chain-axis direction when ζ^* is small. Under these conditions some type of

[23] Crystallite thickening that is observed at this point is distinct from that which occurs during the tail portion of the isotherms for high molecular weight polymers.

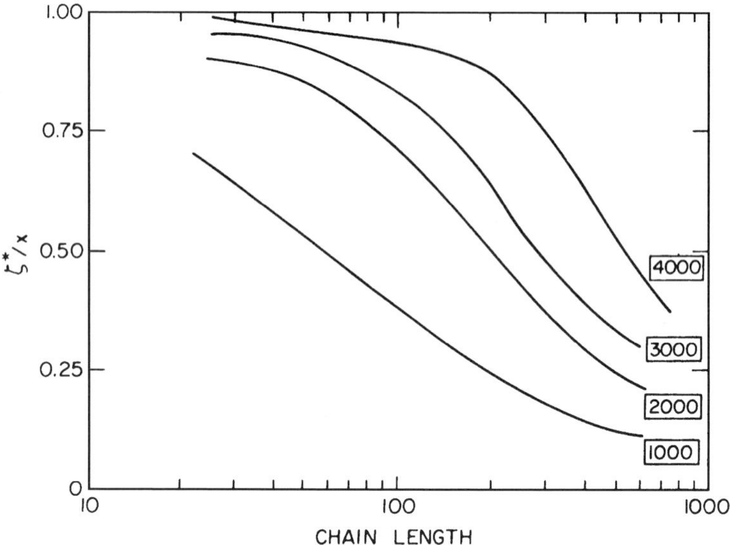

Fig. 9.80 Plot of ζ^*/x as a function of chain length for indicated values of σ_{en} at a constant undercooling, $\Delta T = 20$. Three-dimensional nucleation theory used.(274)

folded chain crystallite that is consistent with a local free energy minimum will result.

Estimates of the ratio ζ^*/x can be obtained from nucleation theory pertinent to chain molecules. These estimates will enable an assessment to be made of the validity of the above explanation. For illustrative purposes the arbitrary assumption is made that when $\zeta^*/x \geq 0.5$ growth along the chain axis is sufficiently rapid to allow extended chain thicknesses to be achieved. The value of 0.5 is only taken as a convenient reference point. Other critical values such as 0.6 and 0.7 could serve equally well to examine the problem. The ζ^* value can be obtained from either Eq. (9.123) or Eq. (9.144) for a given crystallization temperature and chain length. The effect of chain length is illustrated in Fig. 9.80 for three-dimensional nucleation, utilizing Eq. (9.123).(274) Here ζ^*/x is plotted against x, for different values of σ_{en}, in cal mol^{-1} for a fixed undercooling $(T_m^0 - T_c) = 20$. As a point of reference we focus attention on the curve for $\sigma_{en} = 2000$. Even at this fairly large undercooling the ζ^* values for the shorter n-alkanes are very close to the extended chain length and thus extended chain crystallites should result. However, when the chain length is increased, to several hundred units, ζ^* is well below 0.5, and decreases further with increasing chain length. Thus, as the chain length increases, growth in the chain direction will become progressively more difficult and consequently extended chain crystallites will not form easily. However, the chain length at which the nuclei size reaches half the molecular length does change. When σ_{en} is increased to 4000 cal mol^{-1} ζ^*/x does not decrease to 0.5 until a chain of about

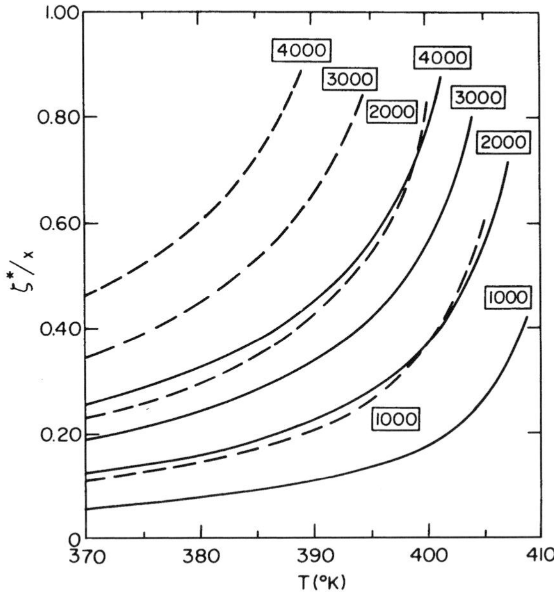

Fig. 9.81 Plot of ζ^*/x against temperature for $x = 168$ and 288 for indicated values of σ_{en}. Two-dimensional nucleation theory used. $x = 168$ (—); $x = 288$ (- - -).(274)

500 units is reached. In contrast, when σ_{en} is reduced to 1000 cal mol^{-1} this condition is achieved at $x = 50$. Similar results are obtained when comparisons are made at $\Delta T = 5\,°\text{C}$. At this low undercooling ζ^*/x values are higher at corresponding chain lengths. This ratio still decreases with x, with the rate of decrease being dependent on the value of σ_{en}.

The conclusions just reached are not limited to three-dimensional nucleation processes. Qualitatively similar results are obtained when a coherent, unimolecular nucleation process (two-dimensional Gibbs type) is considered. An example of the results for this type of nucleation is given in Fig. 9.81 for chain lengths of 168 and 288 units.(274) For this nucleation mode the same trends are observed as found with three-dimensional nucleation. However, smaller nuclei sizes are obtained for the same parameters. Similar to the three-dimensional case nuclei thicknesses greater than half the extended length are obtained at the lower undercoolings and higher interfacial free energies. The transition from extended to smaller size crystallites is again predicted to occur within the range of chain lengths of the high molecular weight alkanes that have been studied.

The nucleation theory that has been used in this analysis is based on the inherent properties of chain molecules. The nuclei sizes are very sensitive to chain length in this molecular weight region when reasonable values are assumed for the nucleation interfacial free energy. The demarcation between extended and folded type crystallites, as it depends on chain length and crystallization temperature, is explicable in a natural way by this analysis. The conclusions are based on the relative nuclei sizes with the implication that they control the thickness of the mature crystallites. Thus, different crystallite structures can evolve from the same type of nucleus.

The *n*-alkanes have other interesting properties that will be discussed subsequently. Their crystallization kinetics from solution will be treated in Chapter 13. The thickening and thinning of the crystallites that occur during isothermal crystallization will be considered in more detail in Volume 3. Also of interest is the structure of the interface of folded type crystallites formed by the *n*-alkanes and the question of whether the crystallite thickness is quantized. These matters will also be treated in Volume 3. Droplet type experiments with these alkanes will obviously be of great interest.

The availability of the high molecular weight *n*-alkanes, with their uniformity of chain length, has produced some new and interesting experimental results. Not unexpectedly several different interpretations have resulted. Most importantly the results serve as a base, or reference, for the crystallization of low molecular weight polymer fractions, the subject of the next section.

9.14.2 Low molecular weight fractions

The crystallization of low molecular weight polymer fractions in themselves is of interest. They also serve as another bridge to the understanding of the crystallization kinetics of the higher molecular weight species. Studies with different types of low molecular weight polymers have been reported and their major features can be compared. In particular the crystallization kinetics of low molecular weight fractions of linear polyethylene can be compared with those of the *n*-alkanes of comparable chain length. It is important to recognize that, when analyzing experimental results for low molecular weight polymers, their chain lengths are not uniform, no matter how well fractionated. Thus, in contrast to the *n*-alkanes end-groups cannot be paired and molecular crystals are not formed. For example, low molecular weight poly(ethylene oxide) fractions have been widely used in studies of crystallization kinetics. The polymer, as prepared, usually has a Poisson distribution (303) with a ratio of weight to number average molecular weight of 1.08. Despite this narrow distribution, only about 40% of the molecules correspond to the most probable or peak molecular weight.(304) Although such polymers are useful by themselves they cannot be treated as possessing uniform chain lengths. Point has demonstrated that in a conventional type poly(ethylene oxide) sample, $M_n = 3490$, fractionation occurs during isothermal crystallization.(305) Thus, low molecular weight poly(ethylene oxides), as well as other polymer fractions, cannot be considered as

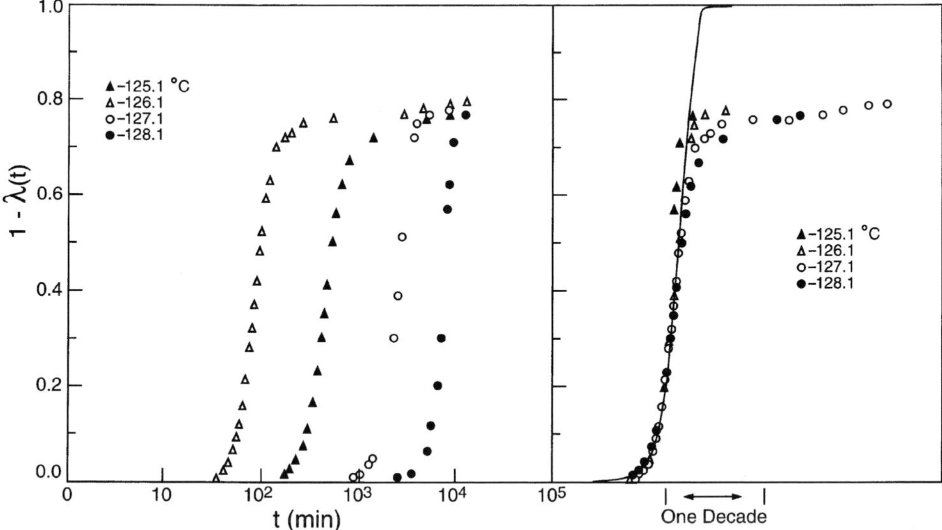

Fig. 9.82 (a) Plot of absolute crystallinity, $1 - \lambda(t)$, against $\log t$, at indicated temperatures for a linear polyethylene fraction, $M = 5800$. (b) Superimposed isotherms. Solid curve theoretical derived Avrami, $n = 4$.(34)

having uniform chain lengths. Caution must then be exercised when interpreting kinetic studies involving such polymers.

Keeping these cautionary concerns in mind an analysis of the overall crystallization kinetics of low molecular weight polymers can be undertaken. A typical set of isotherms is illustrated in Fig. 9.82 for a linear polyethylene fraction having a viscosity average molecular weight of 5800.(34) Here, the absolute level of crystallinity is plotted against the log of time. Typical shaped isotherms are observed that superpose quite nicely, as is indicated by the plot in the right-hand side of the figure. The solid curve represents the derived Avrami expression with $n = 4$. The experimental results fit this Avrami for about 90% of the transformation. The data represent an almost ideal fitting to the simplified Avrami equation. Other polymers of low molecular weight display similar features, i.e. superposition of the isotherms and close adherence to Avrami over almost the complete extent of the transformation with $n = 4$. These polymers include poly(ethylene oxide) (306), other poly(ethers) (306–308) and poly(ethylene terephthalate).(68)

The excellent agreement between the Avrami theory and experiment with $n = 4$, over almost the complete transformation range, can be taken as a reflection of a simple initial melt structure where the chains are essentially disentangled. As has been noted previously, as the molecular weight is increased deviations from the Avrami theory occur at progressively decreasing levels of crystallinity, while n is reduced to 3 and then to 2 at the very high molecular weights with the development of

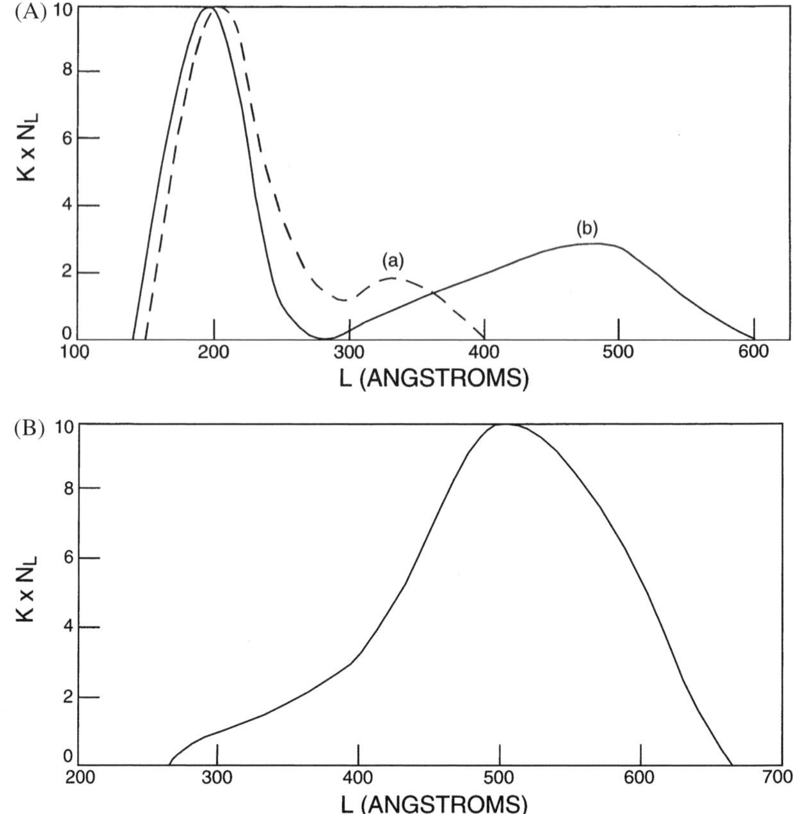

Fig. 9.83 Plot of Raman-derived crystallite size distribution for a linear polyethylene fraction, $M_w = 5800$, $M_n = 5600$ crystallized at 125 °C. (A) (a) after 75 min; (b) after 120 min. (B) after 14 days.(309)

more complex melt structures involving chain entanglements and other topological restraints. The importance of the initial melt structure in the crystallization process is emphasized by these results. There are thus examples with polymer where the derived Avrami is an excellent representation of the experimental results.

As observed with the *n*-alkanes, crystallite thickening in the low molecular weight samples can occur during the entire course of the crystallization. Cognizance of the thickening process must again be taken into account when analyzing kinetic data in this molecular weight range. An example of such thickening is given in Figs. 9.83A and B for a linear polyethylene fraction ($M_w = 5800$, $M_n = 5600$) crystallized at 125 °C.(309) As shown in Fig. 9.83A, at early times the thickness size distribution is centered at about 200 Å. This thickness corresponds to about half the extended chain length. After crystallizing for about 120 minutes, a well-defined bimodal thickness distribution develops. A well-developed new peak at about 500 Å, which corresponds to extended chain crystallites, is now observed. After long time crystallization, the most probable value of the crystallite thickness, shown in Fig. 9.83B, corresponds to the extended chain length. There is clearly a definite thickening kinetics that is part of the overall crystallization. Low molecular weight fractions of poly(ethylene oxide) have also been shown to undergo thickening, as well as thinning, during isothermal crystallization.(310–313) It can be expected that isothermal thickening of all low molecular weight polymers will occur during the entire course of crystallization from the pure melt.

In studying the crystallization kinetics in the low molecular weight range it is important once again to establish whether at a given crystallization temperature the crystallites formed are folded, extended, or if a transformation from one to the other occurs during the time course of the crystallization. This information can be obtained from examining the relationship between the melting and crystallization temperatures.(309) It has been shown that for fractions $M_n = 1586$ and $M_n = 2291$ there is only a slight change in the melting temperature with crystallization temperature.(309) Direct measurements on molecular weight fractions in this range, which have narrow distributions, indicate that the thicknesses of the crystallites are comparable to their extended chain lengths. The result is an invariance in the melting temperature with crystallization temperature. However, when the molecular weights are increased to $M_n = 3769$ and 5600 the melting–crystallization temperature relation changes. The results are shown in Fig. 9.84.(309) A significant increase in the melting temperature is observed with each of these polymers over only a small change in the crystallization temperature. At crystallization temperatures above and below this temperature the melting temperatures do not

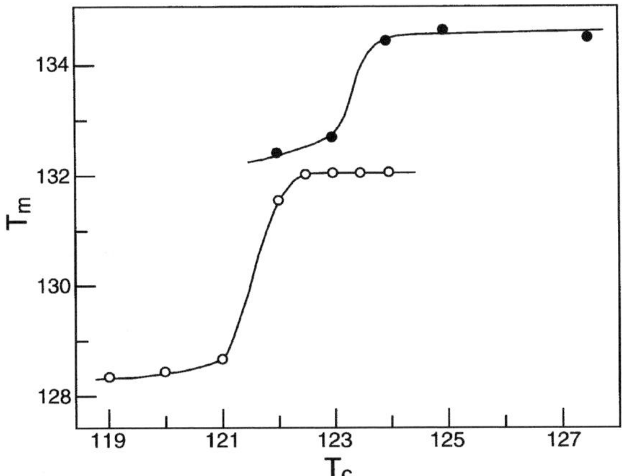

Fig. 9.84 Plot of observed melting temperature, T_m, against crystallization temperature, T_c, for linear polyethylene fractions. ○ fraction $M_w = 4116$, $M_n = 3769$; ● fraction $M_w = 5800$, $M_n = 5600$.(309)

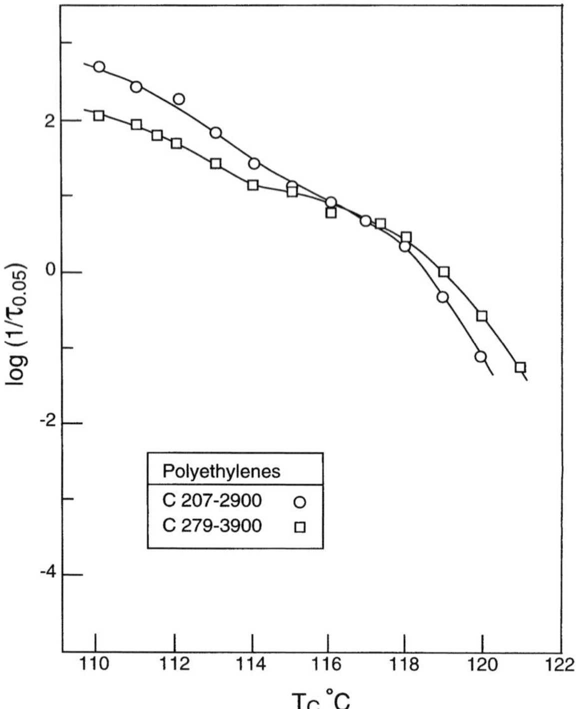

Fig. 9.85 Plot of log overall crystallization rate, $1/\tau_{0.05}$, against crystallization temperature, T_c, for two linear polyethylene fractions. ○ $M_n = 2900$; □ $M_n = 3900$.(316)

change. Extended chain crystallites are formed at the higher crystallization temperatures and some type of folded chain crystallites at the lower ones. These unique melting temperature–crystallization temperature relations have been confirmed for linear polyethylene and the difference in crystallite thicknesses verified.(314) A similar relation has also been reported for poly(ethylene oxide) in the molecular weight range 3000–8000.(315) Thus, for low molecular weight polyethylenes and poly(ethylene oxides) a discrimination between the formation of either folded or extended chain crystallites can be made, based on the crystallization temperature.

Examples of the overall crystallization rates of two linear polyethylene fractions, $M_n = 2900$ and 3900, are plotted as functions of temperature in Fig. 9.85.(316) These fractions correspond to carbon numbers C207 and C279 respectively. The plots for the fractions are similar to those of the n-alkanes that were illustrated in Fig. 9.74. The striking feature in the plots of both the alkanes and fractions is the discontinuity that is observed. A discontinuity in such plots of overall crystallization is typical of the crystallization of low molecular weight species. The data for these fractions are plotted in Fig. 9.86 according to the nucleation theory appropriate to

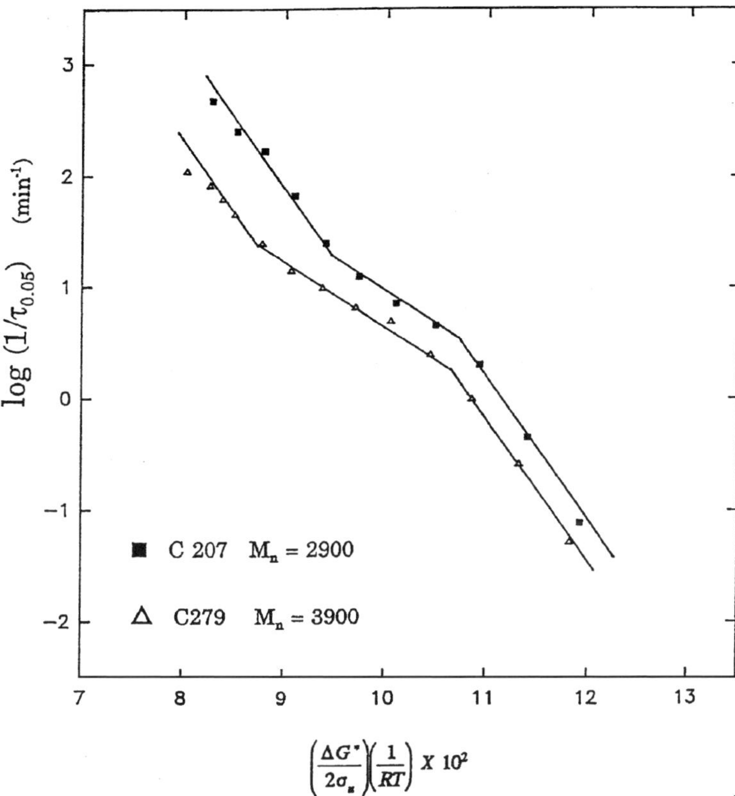

Fig. 9.86 Plot of log overall crystallization rate, $1/\tau_{0.05}$, against nucleation function for two linear polyethylene fractions. ■ $M_n = 2900$; △ $M_n = 3900$.(316)

low molecular weights, assuming a Gibbs type nucleus. The data can be represented by three intersecting straight lines. These are very similar to that found for the n-alkanes, as was illustrated in Fig. 9.77. The slopes in the low and high temperature regions are the same for each polymer. Furthermore, the slopes of each of the two polymers are also the same. These results indicate that the values of σ_{en} are the same for both polymers at all crystallization temperatures. A comparison of the overall crystallization kinetics between the n-alkanes and the polymer fractions of similar carbon numbers is given in Fig. 9.87.(316) The slopes and the crystallization time scales of the polymers and n-alkanes are close to one another. However, the polymers crystallize at slightly slower rates at comparable values of ΔG^*. It can be concluded from the plots in Fig. 9.87 that both the n-alkanes and fractions obey the same nucleation kinetics, for comparable chain lengths.

Discontinuities are also observed in plots of the growth rate against the crystallization temperature for low molecular weight polymers. The polymers studied

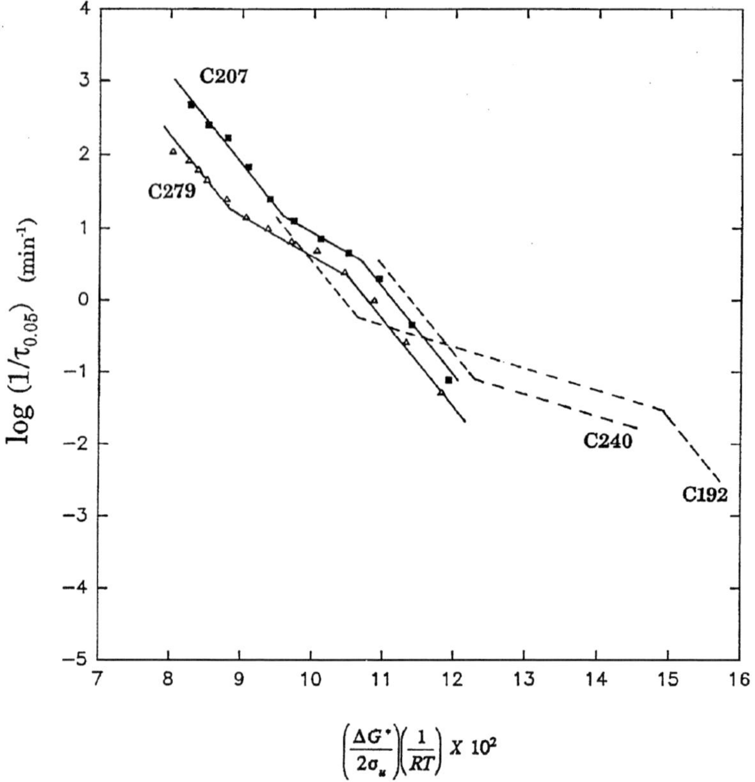

Fig. 9.87 Crystallization rates of n-alkanes and molecular weight fractions of linear polyethylene that have comparable carbon numbers. n-alkanes – – – ; polyethylene —.(316)

include poly(ethylene adipate),(23) poly(ethylene oxide),(317–319a)[24] isotactic poly(propylenes),(320) and linear polyethylene.(314)[25] The growth rates of the low molecular weight poly(ethylene oxides) have received a great deal of attention because of their narrow molecular weight distribution.(317–319a) The extensive investigations by Kovacs and coworkers (318,319) with the low molecular weight poly(ethylene oxides) serve as excellent examples. The log of the growth rates as functions of the crystallization temperatures are plotted in Fig. 9.88 for poly(ethylene oxides) with molecular weights M_n = 1890, 2780, 3900 and 9970.

[24] An inversion in the growth rate of methoxy terminated poly(ethylene oxide), with a molecular weight of 3000 has been reported.(317) The inversion occurred in a 1 °C temperature range where both extended and folded crystallites are observed. The isothermal thickening that takes place here can easily influence the growth rate measurement.

[25] It was pointed out in the discussion of melting temperature depressions that the actual concentration of the chain structural defects needs to be specified. Specification of the pentad fraction is not adequate. Analysis of the crystallization kinetics could be affected in a significant manner, depending on the sequence distribution.

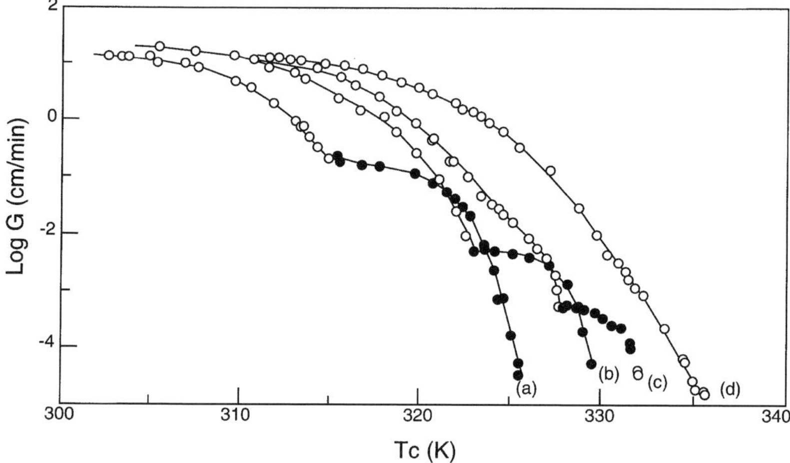

Fig. 9.88 Plots of log growth rate against crystallization temperature for poly(ethylene oxide) molecular weight fractions. (a) $M_n = 1890$; (b) $M_n = 2780$; (c) $M_n = 3900$; (d) $M_n = 9970$. • extended chain crystallites; ○ folded chain crystallites. (Data from Kovacs *et al.* (319))

The plots for the three lowest molecular weights display a well-defined discontinuity, similar to that observed in the overall crystallization kinetics of low molecular weight polyethylene fractions and the n-alkanes. There is a definite flattening of the growth rate at the low crystallization temperatures. The coordinates of the discontinuity move to higher crystallization temperatures as the chain length increases. Following this trend, the discontinuity has disappeared for $M_n = 9970$. As found earlier, the corresponding plot for $M_n = 152\,000$ is also continuous and similar in shape to the one for $M_n = 9970$. The solid symbols in the plots for the three lowest molecular weights represent the fact that extended chain structures are observed eventually. They do not necessarily represent structures that were initially formed. This is particularly true for the lower end of this temperature region. At the lower crystallization temperatures some type of folded structure is formed. For the higher molecular weights, $M \geq 9970$, folded chain crystallites are formed at all crystallization temperatures, resulting in a continuous growth rate–temperature curve.

The data plotted in Fig. 9.88 cover a wide range in undercooling, from about 20 to 50 °C. Consequently, it is necessary to take into account higher order terms in the expansion of ΔG_u. By expanding ΔG_u in a series around T_m^0

$$\Delta G_u(T) = \Delta G_u(T_m^0) + \frac{\Delta H_u}{T_m^0}\Delta T - \frac{\Delta C_p}{2T_m^0}(\Delta T)^2 - \frac{\Delta C_p - T_m^0 \Delta C_p'}{6T_m^{02}}(\Delta T)^3 + \cdots \tag{9.226}$$

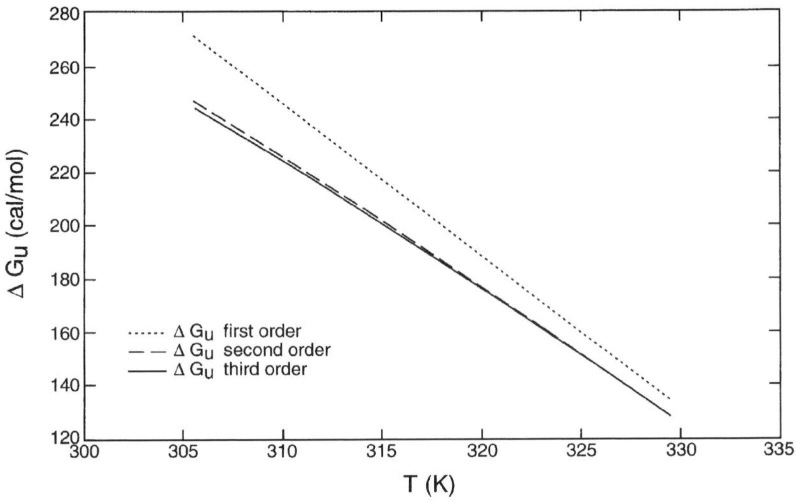

Fig. 9.89 Plot of free energy of fusion of poly(ethylene oxide) to different levels of approximation. - - - first order; – – second order; — third order.(316)

Here ΔC_p is the difference in specific heat between the crystalline and liquid state and $\Delta C_p' = \frac{\partial \Delta C_p}{\partial T}$. The undercooling, ΔT, is properly reckoned from T_m^0. The only restraint in deriving Eq. (9.226) is the number of terms that are retained in the expansion. The series expansion can be carried out to as many terms as is desired. No arbitrary assumptions have to be made with respect to the temperature dependence of either the entropy or enthalpy of fusion. All that is needed to evaluate $\Delta G_u(T)$ are the experimentally available specific heats as a function of temperature, T_m^0 and ΔH_u.

The first term on the right of Eq. (9.226), $\Delta G_u(T_m^0)$, is zero at T_m^0. The second term on the right is the $\Delta G_u(T)$ value that is usually used and is the first-order correction. The next two terms follow in a natural manner and are termed the second- and third-order corrections. Figure 9.89 illustrates how ΔG_u of poly(ethylene oxide) varies with crystallization temperature when the first, second or third orders are considered. The usual values for T_m^0 and ΔH_u are used in this calculation. The specific heat values are those given by Wunderlich.(321) Over the undercooling range of interest there is essentially no difference between the second- and third-order plots. However, ΔG_u will be greater if only the first-order correction is used. As the temperature increases the differences between the orders become progressively smaller. As long as ΔT is not too large the first-order correction, which is commonly used, is adequate for most purposes.

The data in Fig. 9.88 for the four polymers are plotted according to nucleation theory in Fig. 9.90. Here, the temperature dependence of ΔG_u is taken into account by following Eq. (9.226). The values of σ_{en} used are increased slightly with chain

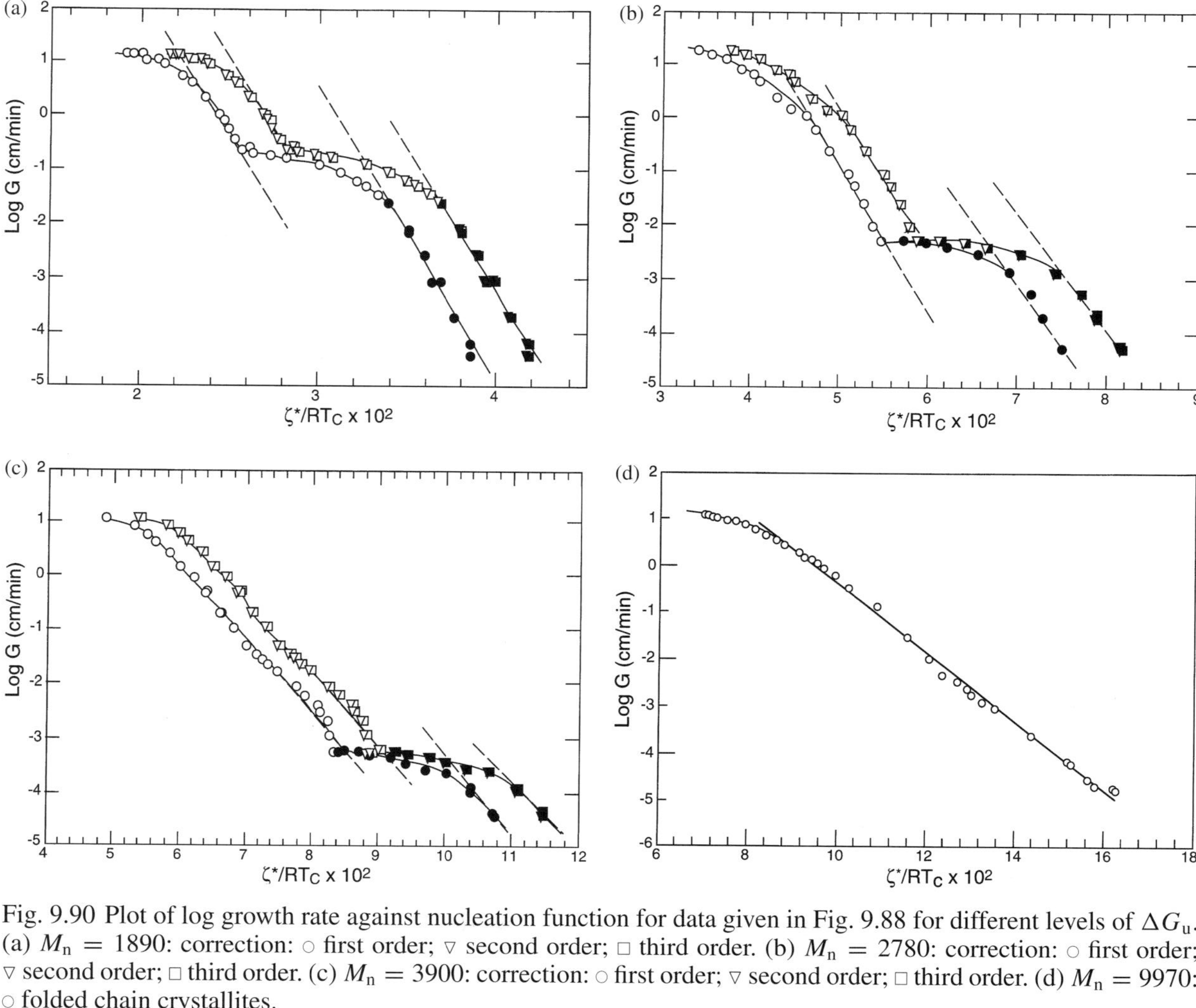

Fig. 9.90 Plot of log growth rate against nucleation function for data given in Fig. 9.88 for different levels of ΔG_u. (a) $M_n = 1890$: correction: ○ first order; ▽ second order; □ third order. (b) $M_n = 2780$: correction: ○ first order; ▽ second order; □ third order. (c) $M_n = 3900$: correction: ○ first order; ▽ second order; □ third order. (d) $M_n = 9970$: ○ folded chain crystallites.

length.(322) The plots for the three lowest molecular weights are similar to one another and also to those of the high molecular weight n-alkanes and low molecular weight polyethylenes. The major feature of each plot is the two parallel straight lines that represent the crystallization temperatures where either extended or folded structures develop. There is also an intermediate temperature region in Fig. 9.90 that connects the two parallel straight lines. Here the data points are represented by a plateau-like curve, as was observed with polyethylene. This temperature region is a reflection of the isothermal thickening, at T_c, of the initially once folded crystallite to an extended one. The shapes of the curves are essentially the same for first-, second- and third-order corrections. Corrections to ΔG_u beyond second order do not have any sensible effect. Deviations from linearity are also observed at the lowest crystallization temperatures. In this undercooling range ΔG_u is large so that the growth rate is relatively insensitive to changes in the crystallization temperature.

The plot for the highest molecular weight, $M_n = 9970$, is different in that there is no discontinuity. For this, and higher molecular weights, only folded chain crystallites form. The plot for this sample is linear, except for the lowest crystallization temperature. This deviation is indicative of a regime transition as was demonstrated for a higher molecular weight poly(ethylene oxide) (see Fig. 9.45). The main conclusion to be reached from the growth rate studies with the low molecular weight poly(ethylene oxides) is that crystallites with different chain conformations have the same σ_{en} value. Thus, they must evolve from the same nucleus. The same conclusion was reached earlier from an analysis of the overall crystallization kinetics of high molecular weight n-alkanes and low molecular weight polyethylene fractions.

The growth rates of two low molecular weight fractions of linear polyethylene have also been reported.(314) Analysis of these data allows for an assessment of the generalization of the results obtained with the poly(ethylene oxides). The log of the growth rate of the two linear polyethylene fractions, $M_w = 5800$, $M_n = 4957$ and $M_w = 3900$, $M_n = 3390$, are plotted against the crystallization temperature in Fig. 9.91.(314) These studies were carried out over a sufficient temperature interval so that the major features can be discerned. Other reports in the literature do not include a large enough temperature range for present purposes.(167,171) The growth rate–temperature patterns are similar for both fractions. The crystallite structures that were formed initially, either folded or extended chains, are represented by the open and closed symbols respectively. The vertical lines in the figures bracket the four to five degree temperature interval where it has been shown that isothermal thickening takes place.(309) The different temperature regions are clear, although the discontinuities are not as marked as were found in the studies of the overall crystallization kinetics. The data from Fig. 9.91 are plotted in Fig. 9.92 according to nucleation theory, again assuming a Gibbs type nucleus. In this figure the plots for the two fractions are similar to one another. The high and low temperature data,

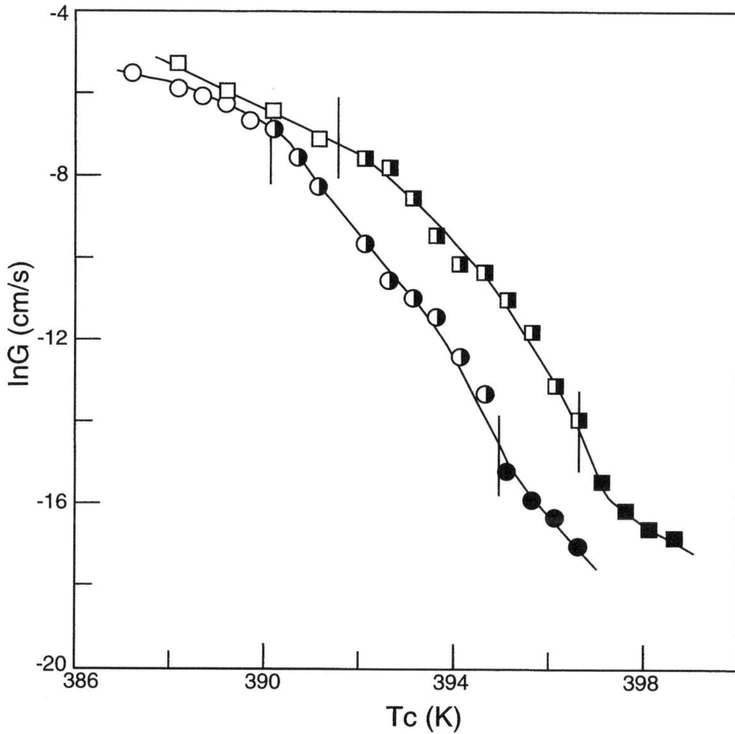

Fig. 9.91 Plot of ln growth rate against crystallization temperature for two linear polyethylene fractions. Circles $M_w = 3900$, $M_n = 3390$; squares $M_w = 5800$, $M_n = 4957$. Open symbols folded chain crystallites; closed symbols extended chain crystallites. (Data from Chiu *et al.* (314))

which represent initially formed extended and folded crystallites respectively, are represented by parallel straight lines, indicative of the fact that the same value of σ_{en} is involved. The slopes of the straight lines in the linear region are the same for both polymers. In the low temperature region they are superposed upon one another. The vertical lines again represent the region of isothermal crystallite thickening. The two linear regions are connected by curves, rather than straight lines. These are consequences of the influence of crystallite thickening on the measured crystallization rate. Thus, the results for the low molecular weight polyethylene fractions follow the same pattern as the growth rates of the poly(ethylene oxides) and the overall crystallization kinetic studies.

The relation between the growth rate and crystallization temperature of low molecular weight fractions of isotactic poly(propylene) follows a pattern similar to poly(ethylene oxide) and polyethylene. Figure 9.93 is a plot of log growth rate against the crystallization temperature for two fractions of isotactic poly(propylene), $M_n = 2000$ and $M_n = 3000$.(319) The curves are similar to those of the other

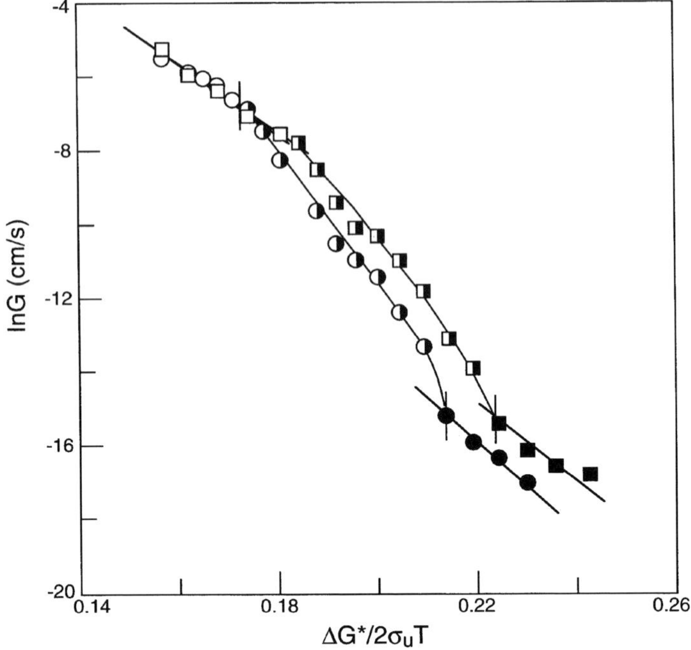

Fig. 9.92 Plot of ln growth rate against nucleation function for data in Fig. 9.91. Symbols the same.

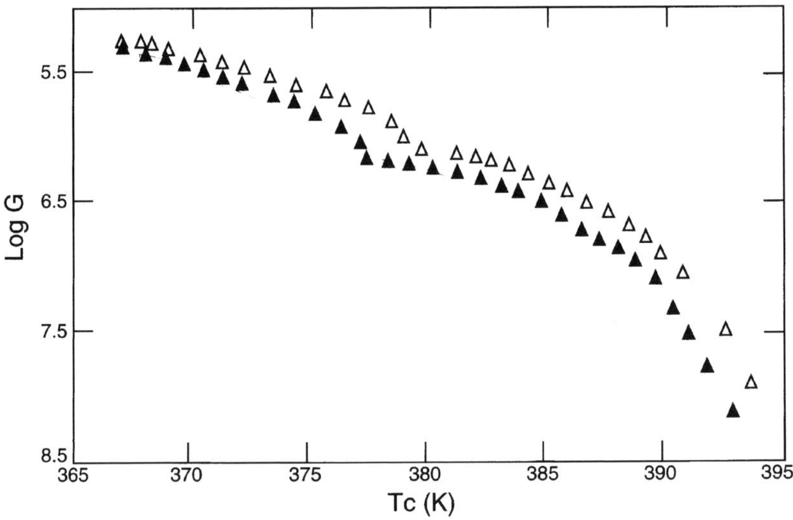

Fig. 9.93 Plot of log growth rate against crystallization temperature for two low molecular weight fractions of isotactic poly(propylene) ▲ $M_n = 2000$; △ $M_n = 3000$. (From Janimak and Cheng (320))

polymers studied. There is the implication that further analysis, according to nucleation theory, would lead to the same conclusions with regard to σ_{en}. However, this analysis is severely hampered in this case by the large uncertainty in T_m^0 for isotactic poly(propylene).(214) It can be shown that by appropriate selection of T_m^0, within the acceptable limits, plots similar to those for polyethylene and poly(ethylene oxide) result and the same conclusions can be reached. This conclusion needs to be tempered in the case of isotactic poly(propylene) because the crystallite thicknesses, and thus the regions of folded and extended chain crystallites, have not been defined for these fractions.

There is substantial evidence that all of the low molecular weight polymers studied follow a similar pattern. However, the complexities introduced by crystallite thickening during the course of the crystallization make it difficult, and somewhat tenuous, to deduce any reliable information about regime transitions in low molecular weight polymers.

9.14.3 High molecular weight

The background that has been developed with respect to the role of nucleation and growth in governing crystallization kinetics, coupled with the behavior of low molecular weights, serves as a basis for analyzing the influence of higher molecular weight on the crystallization process. As will be seen, there is in fact a profound influence of molecular weight on both the overall crystallization and the spherulite growth rates, as well as on the primary nucleation. The experimental results will be presented first. The theoretical explanations and expectations will then be compared with the actual results. Experimental results are available for a variety of polymer types covering a wide range in chain lengths and crystallization temperatures.

Figure 9.23, where the overall crystallization rate of linear polyethylenes is plotted in the form of $1/\tau_{0.01}$ against M_w at different crystallization temperatures, can serve as a convenient point of reference because of the extensive range in molecular weights and crystallization temperatures that were studied.(34) The molecular weights range from less than 10^4 to almost 10^7, while the undercoolings vary from 13 °C to 30 °C. When such a wide range in variables is examined it becomes clear that one is dealing with a complex phenomenon. At low temperatures, large undercoolings, there is only a slight dependence of the crystallization rate on chain length. However, at higher crystallization temperatures the rate actually increases with molecular weight until a broad maximum is reached. The molecular weight at the maximum depends on the crystallization temperature and varies from 10^4 to almost 10^5. Beyond the maximum there is a decrease in rate with molecular weight. This decrease becomes markedly sharper the higher the crystallization temperature.

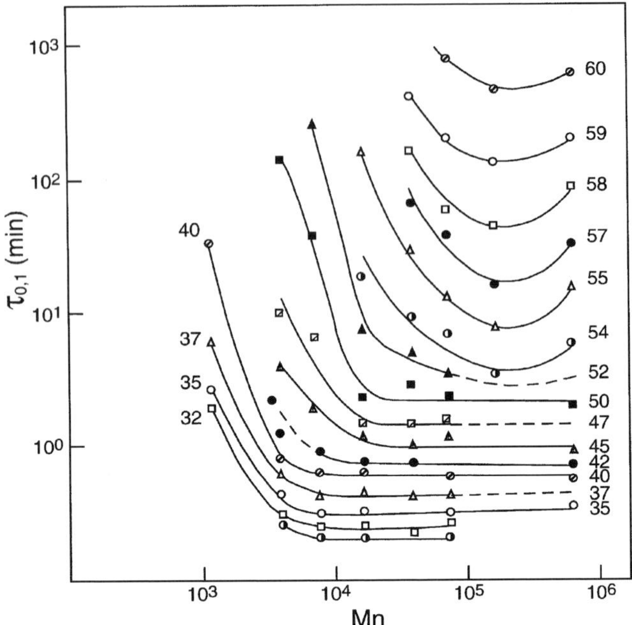

Fig. 9.94 Log-log plot of crystallization rate $\tau_{0.1}$ against number average molecular weight, M_n, for poly(ethylene oxide) fraction at indicated crystallization temperatures. (From Jadraque and Fatou (39))

The overall crystallization rate is essentially constant for molecular weights greater than 2×10^6 at all crystallization temperatures. With this complex behavior, it is important to ascertain whether these results are general for polymers or are limited to linear polyethylene.

Figure 9.94 gives the results of a similar study with poly(ethylene oxide) that covers a molecular weight range from 10^3 to 10^6.(39) The main features found in linear polyethylene are also found in poly(ethylene oxide). There are only minor differences between the two polymers. Other polymers, although not as extensively studied, also show the major features found in Figs. 9.23 and 9.94. For example, maxima with molecular weight in the rates have been observed in poly(ethers) such as poly(oxetane),(41) poly(decamethylene oxide),(306) poly(hexamethylene oxide) (307) as well as in poly(ϵ-caprolactone).(323) A leveling off of the rate at the higher molecular weights was also observed with poly(decamethylene oxide).(306) The crystallization rate does not always decrease with molecular weight. There is a range of molecular weights where the rate actually increases with chain length, and an interval where it is constant. There is also a region where the rate decreases with molecular weight. However, this region is only one part of the total picture.(167,323–325)

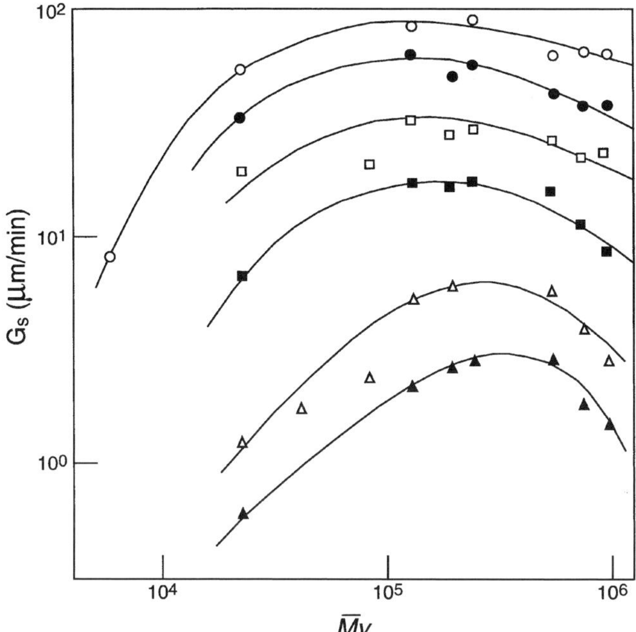

Fig. 9.95 Log-log plot of spherulite growth rate, G_s, against viscosity average molecular weight, M_η, for poly(ethylene oxide) fractions. Crystallization temperatures °C: ○ 53.4; ● 53.5; □ 55.7; ■ 56.9; △ 58.2; ▲ 59.1. (From Maclaine and Booth (326))

The relation between the spherulite growth rate, molecular weight and crystallization temperature follows a similar pattern in general. There are, however, some important differences between specific polymers that depend on the molecular weight range studied. The growth rate results that have been reported fall in several distinct categories that are best described by specific experimental results. A case in point is the results for poly(ethylene oxide) that are given in Fig. 9.95.(326) Several important features are illustrated in this figure. A well-defined maximum is observed at the higher crystallization temperatures. As the crystallization temperature is reduced the maximum is less well defined. The molecular weight corresponding to the rate maximum increases with the crystallization temperature. In the low molecular weight range, below the maximum, there is a sharp increase in the growth rate with chain length. At these chain lengths the rate of change is not very dependent on the crystallization temperature. In contrast, at high molecular weights, at the right-hand side of the maximum, the rate of change is sensitive to the crystallization temperature. At high temperatures, the slopes of the log G– log M plots are about -1. The slope, however, decreases dramatically as the crystallization temperature is lowered. Eventually, at sufficiently low temperatures the growth rate only depends very slightly on molecular weight. Thus, if the growth rate is expressed as

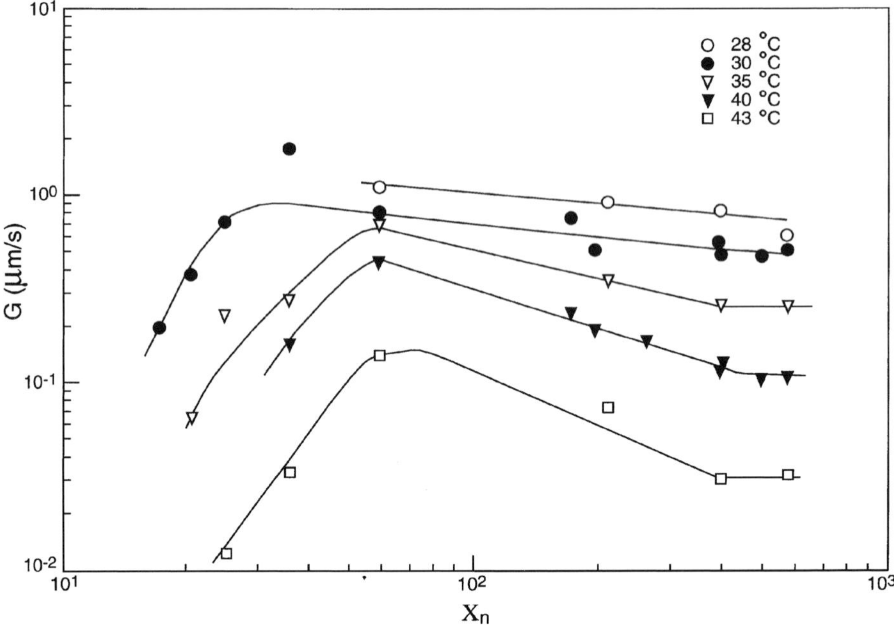

Fig. 9.96 Log-log plot of spherulite growth rate, G, against number average degree of polymerization, x_n for poly(ϵ-caprolactone) fractions. Crystallization temperatures indicated. (From Chen *et al.* (327))

$G = KM^a$, the exponent a varies from -1 to 0 as the crystallization temperature decreases. Except for the invariance in the growth rate at high molecular weights, these results are similar to those found in the studies of overall crystallization rate.

The dependence of the growth rate on molecular weight of poly(ϵ-caprolactone) is similar, as is illustrated in Fig. 9.96.(327) The variation of the maximum with crystallization temperature is the same as that found for poly(ethylene oxide). The growth rate–molecular weight relation is also similar with the slope increasing with the crystallization temperature. The molecular weight is apparently sufficiently high so that the invariance of the growth rate with molecular weight is observed. Similar features are found in the growth rate–molecular weight relation of poly(3,3′-diethyl oxetane).(328)

The growth rates of linear polyethylene, taken from the works of Labaig (171) and Hoffman *et al.* (167), are summarized in Fig. 9.97. With the exception of two inexplicable data points at the highest molecular weights studied, good agreement is found between the two investigations. There is a severe limitation to the temperature interval in which the spherulite growth rate of linear polyethylene can be studied. At the high crystallization temperatures the spherulite growth rate is too slow for most molecular weights, while at lower crystallization temperatures it is much too

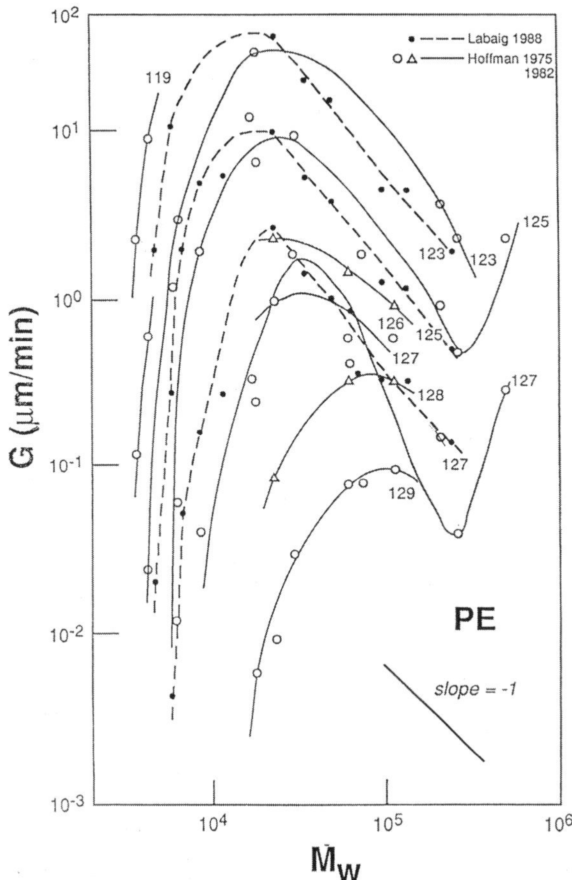

Fig. 9.97 Log-log plot of spherulite growth rate, G, against weight average molecular weight, M_w, for linear polyethylene fractions at indicated crystallization temperatures. ● Data from Labaig (171); ○, △ data from Hoffman *et al.* (167,337).

rapid. A further limitation is that at high molecular weights spherulitic structures are not observed in linear polyethylene at all crystallization temperatures. Thus there is only a very limited data set available for the analysis of linear polyethylene as compared to the growth rates of other polymers. Some of the features found in poly(ethylene oxide), poly(ϵ-caprolactone) and other polymers are also found in linear polyethylene. The maximum in the rate is well defined and its position increases with crystallization temperature. For molecular weights below the maximum the growth rate again increases sharply with chain length and the rate of change is not sensitive to the crystallization temperature. Based on the limited data available it appears that for molecular weights greater than the maximum the slopes only vary slightly with temperature. Estimates of the slopes vary from about -1.5 to -1.7; see below.(324,337) In contrast to the results of overall crystallization studies, an

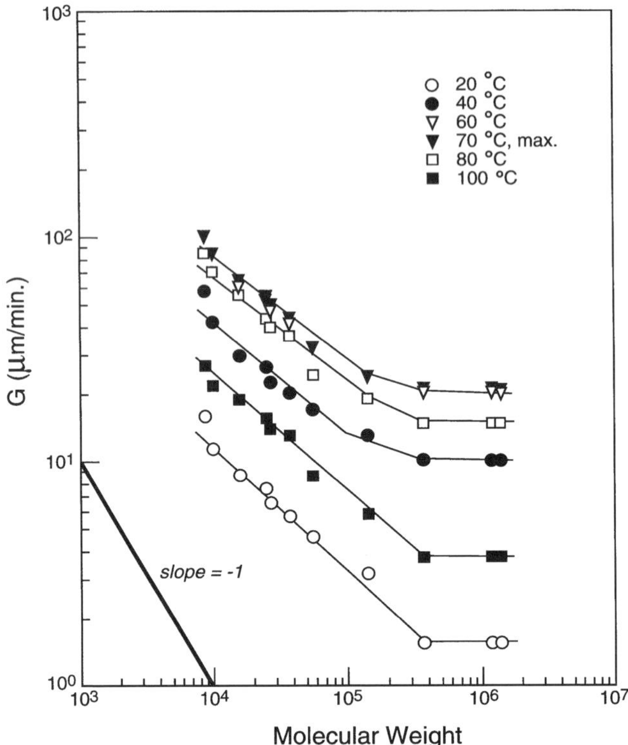

Fig. 9.98 Log-log plot of spherulite growth rate, G, against molecular weight for poly(tetramethyl-*p*-silphenylene siloxane) fractions at indicated crystallization temperatures. (Data from Magill (330))

invariance in the growth rate with chain length is not observed. This is because of the chain length limitation on spherulite formation. Thus, the growth rate of linear polyethylene does not display all of the features observed with other polymers. Hence, because of restrictions of crystallization temperature and molecular weight, the results for linear polyethylene do not give an adequate data base from which to develop a theoretical understanding of the influence of molecular weight on the crystallization kinetics of polymers.

The diversity in the growth rate–molecular weight relations is demonstrated in Fig. 9.98 for poly(tetramethyl-*p*-silphenylene siloxane).(330) Even though molecular weights as low as 10^4 were studied with this polymer no maximum is observed in the growth rate. There is a linear decrease with molecular weight until $M \simeq 2\text{–}3 \times 10^5$. The growth rate becomes invariant at the higher molecular weights, at all crystallization temperatures. The slopes in the linear portion of Fig. 9.98 are approximately -0.55 and are essentially independent of the crystallization temperature. Figure 9.98 is not unique. A very similar dependence of the growth rate

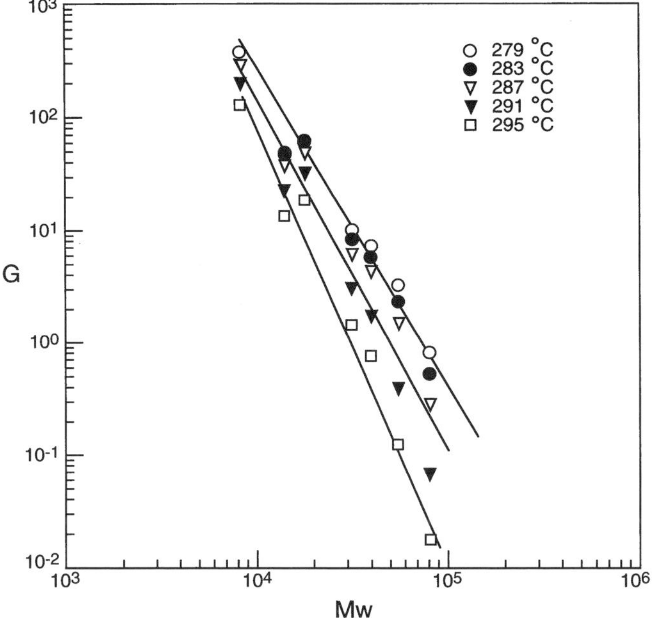

Fig. 9.99 Log-log plot of spherulite growth rate, G, against weight average molecular weight, M_w, for poly(aryl ether ether ketone) at indicated crystallization temperatures. (From Deslandes *et al.* (332))

on molecular weight was also found with the high melting polymorph of trans-poly(1,4-isoprene).(331)

Another example of the growth rate–molecular weight relation is shown in Fig. 9.99 for poly(aryl ether ether ketone).(332) The double log plot of the data for this polymer, over the molecular weight range studied, only shows a linear decrease of growth rate with chain length. Neither the constancy at high molecular weights nor the increase in growth rate at the lower molecular weights are observed. The molecular weight dependence given in Fig. 9.99 can be expressed by the relation

$$\log G(T_c) = \log A(T_c) + B(T_c) \log M_n \tag{9.227}$$

where $A(T_c)$ and $B(T_c)$ are functions of the crystallization temperature. The slopes of the straight lines vary from −2.7 to −3.5 between 279 °C and 295 °C. A similar relation was found with poly(phenylene sulfide), where the slopes varied between −2.1 and −3.1.(43) Qualitatively similar results have also been observed with poly(ethylene terephthalate) (110) and isotactic poly(styrene),(333) with the slopes varying with the crystallization temperature.

Studies of unfractionated isotactic poly(1-butene) over the range M_w= 1.16–3.98 × 10^5, have found that the spherulite growth rate was independent of chain length.(333a) The influence of molecular weight on fraction of isotactic poly(propylene) presents a very interesting situation.(333b) The spherulite growth rates of fractions of the metallocene catalyzed polymer vary from 1.68×10^{-6} to 0.89×10^{-6} cm s^{-1} as the molecular weight increases from 8.6×10^4 to 3.58×10^5. This represents normal behavior in terms of the behavior of the other polymers that have been discussed. In contrast, Ziegler–Natta catalyzed fractions give quite different results. Surprisingly the growth rates of these fractions are independent of molecular weight and the concentration of the chain defects. This result suggests that blocky type copolymer behavior is involved in this case.(333b)

The primary nucleation rates of folded, as well as extended chain crystallites of bulk crystallized linear polyethylene have also been studied.(334,336a) The nucleation rate was obtained by counting the number of crystals that were observed as a function of time within a visual field during isothermal crystallization.(335) The number of crystals observed was assumed to correspond to the number of nuclei that were formed initially. The molecular weight range studied for folded chain crystallite formation was 3.0×10^4 to 9.9×10^4.[26] This molecular weight interval corresponds to the interval where the growth rate decreases with molecular weight at the lower crystallization temperature. At lower undercoolings the growth rate in this molecular weight range increases with molecular weight.(167,171) (see Fig. 9.97) The results, shown in Fig. 9.100, are plotted in the form of log N against $(\Delta T)^{-2}$. This plot is appropriate for primary, initiating nucleation.(12) The major feature in this plot is the set of parallel straight lines. This implies that ΔG^*, the free energy necessary to form a critical size nucleus, is independent of molecular weight. Thus, the product of interfacial free energy $\sigma_{\text{en}}\sigma_{\text{eu}}^2$ is also independent of molecular weight. Furthermore, it is found that for folded chain crystallites the power law

$$N \sim M_{\text{n}}^{-2.3} \tag{9.228}$$

is obeyed. A power law is also observed for the nucleation of extended chain crystallites. However, in this case the exponent is -1.(335)

The steady-state nucleation rate depends on G_0, the transport term, and the free energy of forming a critical-size nucleus. The results shown in Fig. 9.100 indicate that the product of interfacial free energy $\sigma_{\text{en}}\sigma_{\text{un}}^2$ is independent of molecular weight for the range of linear polyethylenes that was studied. The parallel displacement of the straight lines is a reflection of the influence of molecular weight on the transport term. The product $\sigma_{\text{en}}\sigma_{\text{un}}$ can be obtained from the temperature dependence of the spherulite growth rate. The results from two studies of linear polyethylene (167,171) and of poly(ethylene oxide) (326) are given in Fig. 9.101 where a Gibbs type of

[26] The molecular weight range studied has been extended to 1.39×10^5 and the exponent in Eq. (9.228) changed to -2.4.(336a)

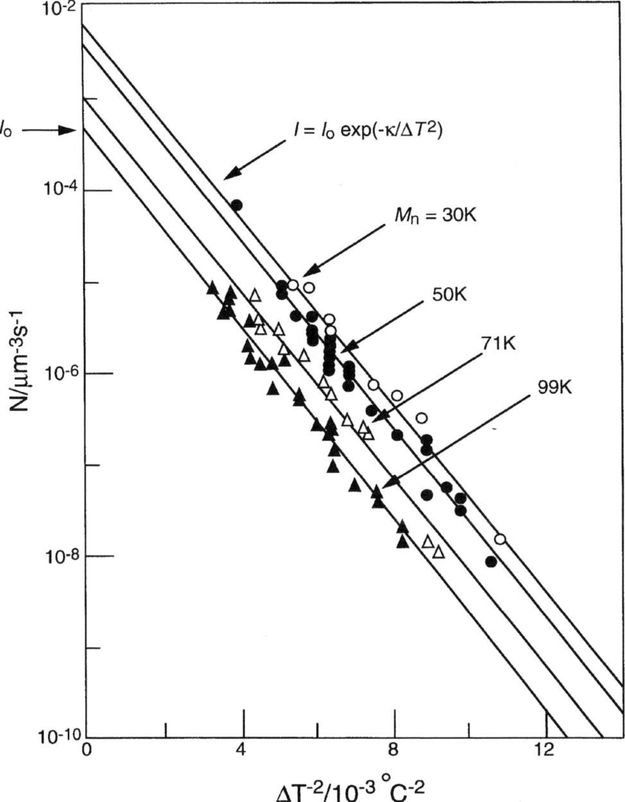

Fig. 9.100 Plot of log of primary nucleation rate, N, against $(\Delta T)^{-2}$ for linear polyethylene fractions of indicated molecular weights. (From Ghosh *et al.* (334))

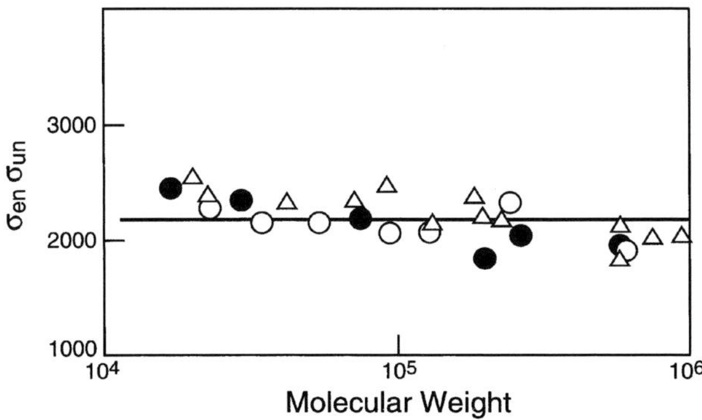

Fig. 9.101 Plot of product $\sigma_{en}\sigma_{un}$ in erg^2 cm^{-4} against log weight average molecular weight, M_w. △ poly(ethylene oxide) data from (326); ● linear polyethylene (167); ○ linear polyethylene (171).

nucleation has again been assumed. Only molecular weights 2×10^4 and greater are considered to avoid the complexities of the finite chain correction term. There is good agreement between the two polyethylene studies and the values are similar to those for poly(ethylene oxide). It is evident in these examples that for moderate and high molecular weights, the product $\sigma_{en}\sigma_{un}$ is independent of molecular weight in agreement with the nucleation studies. Thus, there is no contribution from ΔG^* to the molecular weight dependence of the growth rate. It is important to note that the constancy of $\sigma_{en}\sigma_{un}$ does not imply that the same holds for σ_{ec} or σ_{uc}. The latter are characteristic of the mature crystallite. The basal plane structures of the nucleus and mature crystallite are quite different.

It was pointed out previously that both the overall crystallization rate and the crystallinity level that can be attained depended on molecular weight as a consequence of chain entanglements in the melt. Further investigations have shown that the spherulite growth rates and the primary nucleation rate, both of which involve a transport term, also depend on the entanglement density.(49c,336) The role of entanglements was assessed by taking advantage of the fact that the high pressure-high temperature crystallization of linear polyethylene yields extended chain crystallites with very high levels of crystallinity. Therefore, the chains in the pure melt of such crystallites will be essentially disentangled. Re-entanglement takes place in approximately 5–30 minutes. There is thus a time period wherein the entanglement density increases and appropriate measurements can be made. The spherulite growth rate is increased by 25–45% for a partially disentangled melt relative to a melt with a normal concentration of entanglements.(49c) Similarly, the primary nucleation rate increases with a decrease in the entanglement density. However, ΔG^* remains constant, being independent of the entanglement density.

The molecular weight dependences of many polymers have been presented in order to emphasize the complex behavior and diversity of results. Appropriate theory must explain this behavior. In particular, the increase, decrease and constancy of the overall crystallization kinetics and spherulite growth rates with molecular weight needs to be taken into account. Also to be explained are how the rates are tempered by the crystallization temperatures and the chemical nature of the chain repeating unit. The extensive set of experimental results available allows for a critical and objective examination of this problem and the explanations that have been offered.

An empirical relation has been proposed by Cheng and Wunderlich to explain the dependence of the growth rate on molecular weight.(47) This relation can be expressed as

$$\log G = A \log(\ln M) + B \tag{9.229}$$

The constants A and B depend on the crystallization temperature, or undercooling. In this expression, with A negative, G is a monotonically decreasing function of M. This expression was fitted to the data for several polymers, including poly(ethylene), poly(ethylene oxide), poly(ethylene terephthalate), poly(tetramethyl-p-silphenylene siloxane) and poly(trans-1,4-isoprene).(47) However, the high molecular weight data, where G is independent of M, and the low molecular weight range, where G increases with M, were tacitly ignored. The growth–molecular weight data for poly(aryl ether ether ketone) could not be fitted to Eq. (9.229) since downward curvature results.(332) When straight lines are forced through the data it is difficult to give physical meaning to the resulting slopes. As was pointed out the data for this polymer could be fitted by the empirical relation, Eq. (9.226). The data for poly(phenylene sulfide) behave in a similar manner.(43,332) It can be concluded that, at best, the proposed empirical relation is restricted to the range where the growth rate decreases with molecular weight. Even with this restriction, not all polymers follow Eq. (9.229).

Reptation theory has been applied to the crystallization kinetics of polymers in an attempt to explain the role of molecular weight.(50g,148,150,205a,329,337) The concept of chain reptation was initially introduced by de Gennes to explain the viscoelastic behavior of polymers in the pure melt.(338,339) The problem is to explain the motion of a chain in a medium filled with entangled other chains that form a network-like structure.(340) The chain motion is retarded by the entanglement of the chains with one another as well as by other topological restraints. Thus, no chain can move sideways very far without having to cross the forbidden obstacles. However, the chain can move between these obstacles in a worm- or snake-like motion. This motion has been termed reptation. A detailed molecular theory based on this concept has been developed by Doi and Edwards.(341) In this treatment the motion of a chain is restricted to a hypothetical tube of uncrossable constraints, the diameter of which corresponds to the mesh size of the pseudo-network. The chain is allowed to diffuse through this tube. The theory has successfully explained a variety of viscous and viscoelastic phenomena in polymer melts.(341,342)

The reptation concept has been introduced in an attempt to explain the postulated reeling of chains from the entangled melt. A chain is thought to diffuse through the hypothetical tube to the growing crystal face, where a sequence of length ζ, corresponding to the crystallite thickness, is deposited.(148,337) It is thus assumed that there is a substantial degree of regular chain folding in the mature crystallite. Attention is focused on explaining the spherulite growth rate of linear polyethylene. The available data for this polymer are limited to about only one decade in molecular weight, $\sim 2 \times 10^4$ to $\sim 2 \times 10^5$. The friction coefficient is proportional to the number of chain units, n, in the dangling chain. This number in turn is identified with the complete chain. An important, implicit assumption is made that a chain is involved in only one crystallite. In the initial formulation of the problem the postulate was made that the product of the front factor, G_0, and the transport term in Eq. (9.205) can be expressed as

$$\beta_g = (C_i/n)(kT/h)\exp[-U^*/R(T - T_\infty)] \tag{9.230}$$

The expression for β_g, termed the retardation factor, represents an arbitrary assumption.[27] Equation (9.230) can also be written in the form (337)

$$\beta_g = (C_i/n)(kT/h)\exp[-Q^*_D/RT] \tag{9.231}$$

where C_i is a constant characteristic of the regime and Q^*_D is the activation energy for reptation. The growth rate is then expressed as[28](337)

$$G_i = (C_i/M)\exp(-Q^*_D/RT)\exp\left(\frac{-K_i T^0_m}{T\,\Delta T}\right) \tag{9.232}$$

The dependence of the growth rate on molecular weight is found explicitly in the front factor. This is solely a consequence of the assumption made in proposing Eq. (9.230). There could be perhaps an implicit dependence on chain lengths in the transport term. There is, however, no dependence on the thermodynamic contribution to ΔG^*, which is consistent with the experimental results previously described.

In order to explain the growth rate data for linear polyethylene it was necessary to modify Eq. (9.231).(329) The resulting expression is written as

$$G_i = (C_i/M^{1+\epsilon})(\Delta T)\exp\left(\frac{-Q^*_D}{RT}\right)\exp\left(\frac{-K_i T^0_m}{\Delta T}\right) \tag{9.233}$$

The quantity ϵ is of statistical mechanical origin and arises from the free energy change associated with the first attachment of the chain to the crystallite surface. The undercooling, ΔT, (in the front factor) enters through the retardation term and the force pulling on the molecule. When applying Eq. (9.232), at constant ΔT, to linear polyethylene, it was found that $\epsilon = \pm 0.2$.(329) This result was confirmed by Okada *et al.* for other linear polyethylene data that encompassed a still more limited molecular weight range.[29](324) Additional modifications have been made to further explain the growth rates of linear polyethylene.(150,205a) A fraction $M \simeq 70\,000$ was designated as undergoing perturbed reptation. Based on experience, further modifications to the reptation theory can be expected.

[27] See discussion leading to Eq. (2–8) in Ref. (148).

[28] For the purpose of illustration and simplicity, it is assumed here that we are dealing with a monodisperse system. This simplification avoids having to define the moment of the molecular weight distribution that may be involved. The required moment to be used has been changed several times during the course of the development.

[29] In these analyses ΔT was calculated from the equilibrium melting temperature for the finite molecular weight chain rather than the infinite one.(324)

It has been pointed out that, even limiting the region to where the growth rate decreases with molecular weight, the experimental results for other polymers do not conform to the predictions appropriate to linear polyethylene.(325) Although the form of the decreases is similar for all polymers studied the exponent varies from polymer to polymer. Moreover, contrary to the results for linear polyethylene, the exponent varies significantly with the crystallization temperature for most polymers that have been studied.

The explanation given for the growth rates of linear polyethylene is highly specific. Based on the successive modifications that have been made, there is a serious question as to whether the reptation concept, as it has been applied, has general applicability to the crystallization kinetics of polymers. If the basic ideas of reptation theory are appropriate, major alterations have to be made so that it explains the molecular weight dependence of all polymers.

Studies of the overall crystallization rate of polyethylene, as well as other polymers, indicate that the rates become independent of chain length at high molecular weights. Presumably this is a reflection in part of the contribution of the growth rate. The spherulite growth rates of poly(trans-1,4-isoprene),(331) poly(ϵ-caprolactone),(327) poly(3,3′-diethyl oxetane) (328) and poly(tetramethyl-*p*-silphenylene siloxane) (330) also become invariant at high molecular weights. This aspect of crystallization kinetic cannot be explained by the current adoption of reptation theory. Also of concern, and not to be neglected, is the region where the growth rate increases with molecular weight.

In the molecular weight range where the growth rate decreases with molecular weight reptation theory requires that the chain length dependence resides in the front factor G_0. The form of the dependence can be expressed as $G_0 \sim 1/M^a$. For linear polyethylene $a \simeq 1.2$ at all crystallization temperatures. It has been explicitly shown that $G_0 \simeq 1/M_\mathrm{n}$ for poly(ethylene terephthalate). It is the only polymer where G depends inversely on molecular weight.(110) However, this study was limited to the restricted molecular weight range $M_\mathrm{n} = 1.9 \times 10^4$ to 3.9×10^4. It is thus difficult to generalize the conclusions with regard to G_0 from these results. The G_0 term depends on molecular weight for isotactic poly(styrene) and is proportional to $M_\mathrm{n}^{-0.25}$.(333) Lovering found that up to $M_\mathrm{n} = 150\,000$ there is a pronounced effect of molecular weight on the growth rate of poly(trans-1,4-isoprene). At higher molecular weights the growth rates are essentially independent of chain length.(331) The results in the region where the growth rate decreases with molecular weight could not be explained by either the transport or nucleation terms. Thus, the molecular weight dependence must reside in G_0.

Any influence of molecular weight on the transport term will be through the quantities U^* and T_∞. The latter quantity will depend on the glass temperature. It will only vary in the low molecular weight range. This factor has been shown to have a negligible effect on the growth–molecular weight relation.(329) It is also expected that any U^* dependence on molecular weight will only be in the low molecular weight range.

There are other factors that should be given consideration. The analysis of the overall crystallization rate of linear polyethylene suggested that the spreading rate is influenced by the chain length. This possibility needs to be explored in analyzing the spherulite growth rate. It has also been proposed that sliding diffusion and its molecular weight dependence plays an important role.[30](334,335)

It has been found quite generally that there is a molecular weight range where both the growth and overall crystallization rates increase with chain length. Chain entanglements are not important in this low molecular weight range. Hence, the effect of reptation should be minimal. The results here should be considered in terms of the undercooling, rather than the crystallization temperature. As the molecular weight increases in this range, the finite chain correction to the free energy of forming a nucleus decreases. Therefore, the effective undercooling increases with chain length at a fixed temperature. As a consequence, an increase of the growth rate with molecular weight at a fixed crystallization temperature is to be expected.

The crystallization rates are essentially independent of molecular weight at very low undercooling. At these crystallization temperatures the crystallite thicknesses are small relative to the chain length. Hence a given chain will participate in many crystallites. Therefore, the absolute value of the molecular weight will not be a factor under these circumstances. A comparable argument can be made for the very high molecular weights where the growth and crystallization rates are independent of chain length. Here the ordered sequence to be deposited is also relatively small compared to the chain length. Therefore, the molecule again participates in many crystallites. In this range, the molecular weight is once again not controlling. It has also been noted that in the molecular weight range where $G \simeq M^{-a}$, the exponent a usually varies with the crystallization temperature or undercooling. This result could also be related to the crystallite thickness, which would allow for a chain to participate in more and more crystallites as the crystallization temperature is lowered.

In summary, when a wide range in molecular weights is studied the relation between crystallization rates is obviously quite complex. The rates can increase, decrease or remain constant with chain length, depending on the molecular weight range of interest and the crystallization temperature. The interval where those changes occur is specific to a given polymer. As a result, although a qualitative description can be offered to the experiment results, quantitative explanation is lacking. Although reptation theory, or variants thereof, may play a role in restricted

[30] The concept of sliding diffusion will be discussed in Chapter 13.

molecular weight and crystallization temperature ranges for a given polymer, its general applicability has not as yet been established. Several suggestions have been made that may improve the situation. These remain to be explored.

9.15 Epilogue

A comprehensive discussion of the many different aspects of the crystallization kinetics of homopolymers from the pure melt has been presented in this chapter. A comprehension of crystallization kinetics is central to understanding structure and properties in the crystalline states. A great deal of the observed phenomena can be explained by modifying conventional nucleation and growth processes, characteristic of low molecular weight substances, to the behavior of long chain molecules. Although a well-developed framework has been established, within which to view the crystallization kinetics of polymers, it is quite evident that there are still major problems that remain to be resolved.

A key problem is to theoretically establish the appropriate primary and secondary nucleation processes that are operative. This includes defining the geometry of the rate controlling nucleus, i.e. whether it is two-dimensional or three-dimensional. It should be recalled that the experimental results can fit either model. Of crucial importance is the chain conformation within the nucleus. Conventional type data are not discriminatory. They can be fitted by either a bundle type nucleus, or one in which the chains are regularly folded with adjacent re-entry. At present, either type of nucleus is consistent with the experimental results. It is important that it be recognized that it is not necessary for there to be a one-to-one relation between the nucleus structure and the mature crystallite that evolves. It is not necessary to assume a regularly folded chain nucleus in order for a mature lamellar crystallite to evolve. It is important that the crucial role of nucleation in polymer crystallization be taken from the realm of assumption and placed on a first principled theoretical base. Similar concerns can be expressed with respect to the transport term.

In analyzing temperature coefficient data, when applying nucleation theory, reliable and accurate values of the equilibrium melting temperature need to be known. This quantity is difficult, if not impossible, to determine experimentally. It can be obtained theoretically in special cases. Usually one of several extrapolation methods is used. The extrapolative methods themselves will be discussed in detail in Volume 3. As this point, it should be recognized as a formidable problem that is in need of resolution.

Another key problem is establishing a theoretical isotherm that explains the development of overall crystallinity over the complete extent of the transformation. Since the free growth approximation explains the experimental results as well as the Avrami type theory, the impingement cessation mechanism is not adequate. It is evident that the role of chain entanglement and the changing residual structure of the melt, as crystallinity progresses, needs to be explicitly taken into account in developing a theoretical isotherm from a molecular point of view.

A firm theoretical basis needs to be established to explain the complex molecular weight dependence of the crystallization rates. Theory should encompass and explain the extensive experimental data from the many polymers that are available for analysis.

The analysis and conclusions found in this chapter serve as the basis for the discussion in subsequent chapters of this volume. These include the crystallization of copolymers; polymer–polymer blends; the influence of external force on crystallization kinetics; and the kinetics in polymer–diluent mixtures, including dilute solutions.

References

1. Flory, P. J., *J. Chem. Phys.*, **17**, 223 (1949).
2. Mandelkern, L., A. S. Posner, A. F. Diorio and D. E. Roberts, *J. Appl. Phys.*, **32**, 1500 (1961).
3. Bunn, C. W. *Trans. Faraday Soc.*, **35**, 483 (1939).
4. Voigt-Martin, I. G. and L. Mandelkern, *J. Polym. Sci.: Polym. Phys. Ed.*, **19**, 1769 (1981).
5. Maxfield, J. and L. Mandelkern, *Macromolecules*, **10**, 1141 (1977).
6. Mandelkern, L. and J. Maxfield, *J. Polym. Sci.: Polym. Phys. Ed.*, **17**, 1913 (1979).
7. Mandelkern, L., K. W. McLaughlin and R. G. Alamo, *Macromolecules*, **25**, 1440 (1992).

8a. Cooper, M. and R. St. J. Manley, *Macromolecules*, **8**, 219 (1975).

8b. Cooper, M. and R. St. J. Manley, *Coll. Polym. Sci.*, **254**, 542 (1976).

9. Toda, A., H. Miyaji and H. Kibo, *Polymer*, **27**, 1505 (1986).
10. Bekkedahl, N., *J. Res. Nat. Bur. Stand.*, **13**, 411 (1934).
11. Mandelkern, L., in *Growth and Perfection of Crystals*, R. H. Doremus, B. W. Roberts and D. Turnbull eds., New York, John Wiley & Sons (1958) p. 499.
12. Mandelkern, L., P. A. Quinn, Jr. and P. J. Flory, *J. Appl. Phys.*, **25**, 830 (1954).
13. Wang, Z. G., B. S. Hsiao, E. B. Sirota, P. Agarwal and S. Srinivas, *Macromolecules*, **33**, 978 (2000).
14. Imai, M., K. Kaji and T. Kanaya, *Phys. Rev. Lett.*, **71**, 4162 (1993).
15. Imai, M., K. Kaji, T. Kanaya and Y. Sakai, *Phys. Rev. B*, **52**, 12696 (1995).
16. Wang, Z. G., B. S. Hsiao, E. B. Sirota and S. Srinivas, *Polymer*, **41**, 8825 (2000).
17. Muthukumar, M. and P. Welch, *Polymer*, **41**, 8833 (2000).

17a. Dunning, W. J., in *Nucleation*, A. C. Zettlemoyer ed., Marcel Dekker (1969) p. 1.

17b. Walton, A. G., in *Nucleation*, A. C. Zettlemoyer ed., Marcel Dekker (1969) p. 225.

17c. Pogodina, N. V. and H. H. Winter, *Macromolecules*, **31**, 8164 (1998).

17d. Pogodina, N. V., S. K. Siddiquee, J. W. van Egmond and H. H. Winter, *Macromolecules*, **32**, 1167 (1999).

17e. Kimura, I., H. Ezure, S. Tanaka and E. Ito, *J. Polym. Sci.: Pt. B: Polym. Phys.*, **36**, 1227 (1998).

18. Mandelkern, L., *Chem. Rev.*, **56**, 903 (1956).
19. Wood, L. A. and N. Bekkedahl, *J. Appl. Phys.*, **17**, 362 (1946).

20. Magill, J. H. and D. J. Plazek, *J. Chem. Phys.*, **46**, 3757 (1967).
21. Greet, R. J., *J. Cryst. Growth*, **1**, 195 (1967).
22. Flory, P. J. and A. D. McIntyre, *J. Polym. Sci.*, **18**, 592 (1955).
23. Takayanagi, M., *Mem. Fac. Eng. Kyushu Univ.*, **16** (3), 111 (1956).
24. Magill, J. H., *J. Appl. Phys.*, **35**, 3249 (1964).
25. Von Göler, F. and G. Sachs, *Z. Physik*, **77**, 281 (1932).
26. Johnson, W. A. and R. F. Mehl, *Trans. Am. Inst. Min. Metall. Eng.*, **135**, 416 (1939).
27. Avrami, M., *J. Chem. Phys.*, **7**, 1103 (1939); *ibid.* **8**, 212 (1940); *ibid.* **9**, 177 (1941).
28. Evans, R. U., *Trans. Faraday Soc.*, **41**, 365 (1945).
29. Kolmogorov, A. N., *Bull. Acad. Sci. USSR Phys. Ser.*, **1**, 355 (1937).
29a. Ergoz, E. Ph.D. Dissertation, Florida State University (1970).
29b. Sessa, V., M. Fanfomi and M. Tomelliun, *Phys. Rev. B*, **54**, 836 (1996).
29c. Yu, G. and J. K. Lai, *J. Appl. Phys.*, **79**, 3504 (1996); *ibid.* **78**, 5965 (1995).
30. Lee, Y. and R. S. Porter, *Macromolecules*, **21**, 2770 (1988).
31. Rabesiaka, J. and A. J. Kovacs, *J. Appl. Phys.*, **32**, 2314 (1961).
32. Banks, W., M. Gordon and A. Sharples, *Polymer*, **4**, 289 (1963).
33. Gent, A. N., *Trans. Inst. Rubber Ind.*, **30**, 139 (1954).
34. Ergoz, E., J. G. Fatou and L. Mandelkern, *Macromolecules*, **5**, 147 (1972).
34a. Lehnert, R. J. and H. Hirschmann, *Polymer J.*, **29**, 100 (1997).
34b. Ratta, V., A. Ayamben, R. Young, J. E. McGrath and G. L. Wilkes, *Polymer*, **41**, 8121 (2000).
35. Menczel, J. D. and G. L. Collins, *Polym. Eng. Sci.*, **32**, 1264 (1992).
36. Chew, S., J. R. Griffiths and Z. H. Stachurski, *Polymer*, **30**, 874 (1989).
37. Risch, B. G., S. Srinivas, G. L. Wilkes, J. F. Geibel, C. Ash and M. Hicks, *Polymer,* **37**, 3623 (1996).
38. Alamo, R., J. G. Fatou and J. Guzman, *Polymer*, **23**, 374 (1982).
39. Jadraque, D. and J. G. Fatou, *Anales de Quimica*, **73**, 639 (1977).
40. Cebe, P. and S. D. Hong, *Polymer*, **27**, 1183 (1986).
41. Perez, E., A. Bello and J. G. Fatou, *Coll. Polym. Sci.*, **262**, 605 (1984).
42. Heberer, D. P., S. Z. D. Cheng, J. S. Barley, S. H. S. Lien, R. G. Bryant and F. W. Harris, *Macromolecules*, **24**, 1890 (1991).
43. Lopez, L. C. and G. L. Wilkes, *Polymer*, **29**, 106 (1988).
44. Turska, E. and S. Gogolewski, *Polymer,* **12**, 629 (1971).
45. Hsiao, B. S., B. B. Sauer and A. Biswas, *J. Polym. Sci.: Pt. B: Polym. Phys.*, **32**, 737 (1994).
46. Könnecke, K., *J. Macromol. Sci. Phys.*, **B33**, 37 (1994).
46a. Hsiao, B. S., B. B. Sauer, H. G. Zachmann, S. Seifel, B. Chu and P. Harvey, *Macromolecules*, **28**, 6931 (1995).
46b. Verma, R. K., H. Marand and B. S. Hsiao, *Macromolecules*, **29**, 7767 (1996).
47. Cheng, S. Z. D. and B. Wunderlich, *J. Polym. Sci.: Pt. B: Polym. Phys.*, **24**, 595 (1986).
48. Magill, J. H., *Macromol. Chem. Phys.*, **199**, 2365 (1998).
49. Allen, R. C., Masters Thesis, Florida State University, December (1980).
49a. Gordon, M. and I. Hillier, *J. Chem. Phys.*, **38**, 1376 (1963).
49b. Gornick, F. and J. L. Jackson, *J. Chem. Phys.*, **38**, 1150 (1963).
49c. Psarski, M., F. Piorkowska and A. Galeski, *Macromolecules*, **33**, 916 (2000).
50a. Yamazaki, S., M. Hikosaka, A. Toda, I. Wataoka and F. Gu, *Polymer*, **43**, 6585 (2002).
50b. Feio, G., G. Buntinx and J. P. Cohen-Addad, *J. Polym. Sci.: Pt. B: Polym. Phys.*, **27**, 1 (1989).

50c. Fetters, L. J., D. L. Lohse and R. H. Colby, in *Physical Properties of Polymers Handbook*, J. E. Mark ed., American Institute of Physics (1966) Chapter 24, p. 335.
50d. Bu, H., F. Gu, L. Bao and M. Chen, *Macromolecules*, **31**, 7108 (1998).
50e. Flory, P. J. and D. Y. Yoon, *Nature*, **272**, 226 (1978).
50f. Flory, P. J. and D. Y. Yoon, *Faraday Discuss. Chem. Soc.*, **68**, 288, 109, 389, 402 (1979).
50g. DiMarzio, E. A., C. M. Guttman and J. D. Hoffman, *Faraday Discuss. Chem. Soc.*, **68**, 210 (1979).
50h. Hoffman, J. D. and R. L. Miller, *Polymer*, **38**, 3151 (1997).
50i. Iwata, K., *Polymer*, **43**, 6609 (2002).
51. Austin, J. D. and R. L. Rickett, *Trans. Am. Inst. Min. Eng.*, **135**, 396 (1939).
52. Lee, E. S. and Y. G. Kim, *Acta Metall. Mater.*, **38**, 1669 (1990).
53. Tobin, M. C., *J. Polym. Sci.: Polym. Phys. Ed.*, **12**, 399 (1974); *ibid.*, **14**, 2253 (1976); *ibid.*, **15**, 2269 (1977).
54. Gordon, M. and I. H. Hillier, *Phil. Mag.*, **11**, 31 (1965).
55. Hillier, I. H., *J. Polym. Sci., Pt. A*, **3**, 3067 (1965).
56. Price, F. P., *J. Polym. Sci., Pt. A*, **3**, 3079 (1965).
57. Hsiao, B., *J. Polym. Sci.: Pt. B: Polym. Phys.*, **31**, 237 (1993).
58. Pérez-Cardenas, H., L. F. Castillo and R. Vera-Graziano, *J. Appl. Polym. Sci.*, **43**, 779 (1991).
59. Velisaris, C. N. and J. C. Seferis, *Polym. Eng. Sci.*, **26**, 1574 (1986).
60. Cebe, P., *Polym. Eng. Sci.*, **28**, 1192 (1988).
61. Chynoweth, K. R. and Z. H. Stakurski, *Polymer*, **27**, 912 (1986).
62. Dunning, W. J., *Trans. Faraday Soc.*, **50**, 1115 (1954).
62a. Howard, L., private communication.
63. Kim, S. P. and S. C. Kim, *Polym. Eng. Sci.*, **31**, 110 (1991).
64. Cheng, S. Z. D. and B. Wunderlich, *Macromolecules*, **21**, 3327 (1988).
65. Malkin, A. Y., V. P. Beghisher and I. A. Keapin, *Polymer*, **24**, 81 (1983).
66. Ding, Z. and J. E. Spruiell, *J. Polym. Sci.: Pt. B: Polym. Phys.*, **35**, 1077 (1997).
66a. Kashshiev, D. and K. Sato, *J. Chem. Phys.*, **109**, 8530 (1998).
67. Stack, G. M., L. Mandelkern, C. Kröhnke and G. Wegner, *Macromolecules*, **22**, 4351 (1989).
68. Vilanova, P. C., S. M. Riba and G. M. Guzman, *Polymer*, **26**, 423 (1985).
69. Price, F. P., *J. Appl. Phys.*, **36**, 3014 (1965).
70. Godovsky, Y. K. and G. L. Slonimsky, *J. Polym. Sci.: Polym. Phys. Ed.*, **12**, 1058 (1974).
71. Allen, R. C. and L. Mandelkern, *J. Polym. Sci.: Polym. Phys. Ed.*, **20**, 1465 (1982).
72. Alamo, R., Ph.D. Dissertation, University of Madrid (1981).
72a. Prud'homme, R. E., *J. Polym. Sci.: Polym. Phys. Ed.*, **15**, 1619 (1977).
73. Hartley, F. D., F. W. Lord and L. B. Morgan, *Ric. Sci. Suppl. Symp. Intl. Chim. Macromol. Milan-Turin*, **25**, 577 (1954).
74. Hartley, F. D., F. W. Lord and L. B. Morgan, *Philos. Trans.*, **A247**, 23 (1954).
75. Muellerfera, J. T., B. G. Risch, D. Rodriques, G. L. Wilkes and D. M. Jones, *Polymer*, **34**, 789 (1993).
76. Cheng, S. Z. D., *J. Appl. Polym. Sci. Appl. Polym. Symp.*, **43**, 315 (1989).
76a. Gilbert, M. and F. J. Hybart, *Polymer*, **13**, 327 (1972).
79b. Lu, X. F. and J. N. Hay, *Polymer*, **42**, 9423 (2001).
77. Grebowicz, J., S. Z. D. Cheng and B. Wunderlich, *J. Polym. Sci.: Pt. B: Polym. Phys.*, **24**, 675 (1986).
78. Magill, J. H., *J. Polym. Sci.: Polym. Lett.*, **6B**, 853 (1968).

79. Marco, C., J. G. Fatou and A. Bello, *Polymer*, **18**, 1100 (1977).
80. Bark, M., H. G. Zachmann, R. Alamo and L. Mandelkern, *Makromol. Chem.*, **193**, 2363 (1992).
81. Barham, P. J. and A. Keller, *J. Polym. Sci: Polym. Phys. Ed.*, **27**, 1029 (1989).
82. Stack, G. M., L. Mandelkern and I. G. Voigt-Martin, *Polym. Bull.*, **8**, 421 (1982).
83. Wasiak, A., *Chemtracts-Macromolecular Chemistry*, **2**, 211 (1991).
84. DiLorenzo, M. L. and C. Silvestre, *Prog. Polym. Sci.*, **24**, 917 (1999).
85. Ozawa, T., *Polymer*, **12**, 150 (1971).
86. Nakamura, K., T. Watanabe, K. Katayama and T. Amano, *J. Appl. Polym. Sci.*, **16**, 1077 (1972).
87. Nakamura, K., K. Katayama and T. Amano, *J. Appl. Polym. Sci.*, **17**, 1031 (1973).
88. Kozlowski, W. J., *J. Polym. Sci.*, **38C**, 47 (1982).
88a. Hieber, C. A., *Polymer*, **36**, 1455 (1995).
89. Eder, M. and A. Wlochowicz, *Polymer*, **24**, 1593 (1983).
90. Monasse, B. and J. M. Haudin, *Coll. Polym. Sci.*, **264**, 117 (1986).
91. Lim, G. B. A. and D. R. Lloyd, *Polym. Eng. Sci.*, **33**, 523 (1993).
92. Lim, G. B. A., K. S. McGuire and D. R. Lloyd, *Polym. Eng. Sci.*, **33**, 537 (1993).
93. Hammahi, A., J. E. Spruiell and A. K. Mehrotra, *Polym. Eng. Sci.*, **35**, 797 (1995).
94. Liu, Z., P. Maŕechal and R. Jérome, *Polymer*, **38**, 5149 (1997).
94a. Friler, J. B. and P. Cebe, *Polym. Eng. Sci.*, **33**, 587 (1993).
95. Sajkiewicz, P., L. Carpaneto and A. Wasiak, *Polymer*, **42**, 5365 (2001).
96. Papageorgiou, G. Z. and G. P. Karayannidis, *Polymer*, **42**, 2637 (2001).
97. Liu, T., Z. Mo, S. Wang and H. Zhang, *Polym. Eng. Sci.*, **37**, 568 (1997).
97a. Srivinas, S., J. R. Babu, J. S. Riffle and G. L. Wilkes, *Polym. Eng. Sci.*, **37**, 497 (1997).
98. Wesson, R. A., *Polym. Eng. Sci.*, **34**, 1157 (1994).
98a. Patel, R. M. and J. E. Spruiell, *Polym. Eng. Sci.*, **31**, 730 (1991).
99. Ziabicki, A., *Appl. Polym. Symp.*, **6**, 1 (1961).
100. Dietz, W., *Coll. Polym. Sci.*, **259**, 413 (1981).
100a. Choe, C. R. and K. H. Lee, *Polym. Eng. Sci.*, **29**, 801 (1989).
100b. Hammami, R. and A. K. Mehrotra, *Polym. Eng. Sci.*, **35**, 170 (1995).
100c. Phillips, R. and J. E. Manson, *J. Polym. Sci.: Pt. B: Polym. Phys.*, **35**, 875 (1997).
100d. Mubareb, Y., E. M. A. Harkin Jones, P. J. Martin and M. Ahmad, *Polymer*, **42**, 3177 (2001).
100e. Kamal, M. R. and E. Chu, *Polym. Eng. Sci.*, **23**, 27 (1983).
100f. Wasiak, A., P. Sajkiewicz and A. Wozniak, *J. Polym. Sci.: Pt. B: Polym. Phys.*, **37**, 2821 (1999).
101. Takayanagi, M. and T. Yamashito, *J. Polym. Sci.*, **22**, 552 (1956).
102. Richards, R. B. and S. W. Hawkins, *J. Polym. Sci.*, **4**, 515 (1949).
103. Sharples, A., *Polymer*, **3**, 250 (1962).
104. Allen, P. W., *Trans. Faraday Soc.*, **48**, 1 (1952).
105. Price, F. P., *J. Am. Chem. Soc.*, **74**, 311 (1952).
106. Burnett, B. B. and W. F. McDevit, *J. Appl. Phys.*, **28**, 1101 (1957).
107. McIntyre, A. D., Ph.D. Thesis, Cornell University, Ithaca, N.Y. (1956).
108. McLaren, J. V., *Polymer*, **4**, 175 (1963).
109. Sanchez, I. C., *J. Polym. Sci.*, **59C**, 109 (1977).
110. van Antwerpen, F. and D. W. van Krevelen, *J. Polym. Sci.: Polym. Phys. Ed.*, **10**, 2423 (1972).
111. Boon, J., G. Challa and D. W. van Krevelen, *J. Polym. Sci. A2*, **6**, 1791 (1968).
112. Nielsen, A. E., in *Kinetics of Precipitation*, Pergamon Press (1964) pp. 40ff.

113. Price, F. P., in *Nucleation*, A. C. Zettlemoyer ed., Marcel Dekker (1969) pp. 405ff.
114. Binsbergen, F. L., in *Progress in Solid State Chemistry*, vol. 8, J. O. McCaldin and G. Somarjai eds., Pergamon Press (1974) p. 189.
115. Gibbs, J. W., in *Collected Works*, vol. 1, Longmans, Green & Co., Inc. (1931) pp. 55–371.
116. Tolman, R. C., *J. Chem. Phys.*, **16**, 758 (1948); *ibid.* **17**, 118, 333 (1949).
117. Buff, F. P. and J. G. Kirkwood, *J. Chem. Phys.*, **18**, 991 (1950).
118. Turnbull, D., *J. Chem. Phys.*, **18**, 198 (1950).
119. Hollomon, J. H., in *Thermodynamics in Metallurgy*, American Society for Metals, Cleveland (1950) pp. 161–177.
120. Turnbull, D., in *Solid State Physics*, vol. 3, Academic Press Inc. (1956).
121. Turnbull, D., *J. Chem. Phys.*, **20**, 411 (1952).
122. Uhlmann, D. R. and B. Chalmers, in *Nucleation Phenomena*, A. S. Michaels ed., American Chemical Society (1966) p. 1.
123. Reference 115, p. 325.
124. Woodruff, D. P., in *The Solid–Liquid Interface*, Cambridge University Press (1973) p. 151.
125. Sears, J. W., *J. Chem. Phys.*, **31**, 157 (1959).
126. Lacmann, R., *Z. Kristallogr.*, **116**, 13 (1961).
127. Becker, R., *Ann. Physik.*, **32**, 118, (1935).
128. Turnbull, D. and J. C. Fisher, *J. Chem. Phys.*, **17**, 7 (1949).
129. Jackson, K. A., in *Nucleation Phenomena*, A. S. Michaels ed., American Chemical Society (1966) p. 37.
130. Thomas, D. C. and A. K. Staveley, *J. Chem. Soc.*, 4569 (1952).
131. Reference 124, p. 28.
132. Turnbull, D. and R. L. Cormia, *J. Chem. Phys.*, **34**, 820 (1961).
133. Uhlmann, D. R., G. Kritchevsky, R. Straff and G. Scherer, *J. Chem. Phys.*, **62**, 4896 (1975).
134. Oliver, M. J. and P. D. Calvert, *J. Cryst. Growth*, **30**, 343 (1975).
135. Kraack, H., E. B. Sirota and M. Deutsch, *Polymer*, **42**, 8225 (2001).
135a. Kraack, H., E. B. Sirota and M. Deutsch, *J. Chem. Phys.*, **112**, 6873 (2000).
135b. Kraack, H., M. Deutsch and E. F. Sirota, *Macromolecules*, **33**, 6174 (2000).
136. Mandelkern, L., J. G. Fatou and C. Howard, *J. Phys. Chem*, **68**, 3386 (1964).
137. Devoy, C. and L. Mandelkern, *J. Chem. Phys.*, **52**, 3827 (1970).
138. Volmer, M., *Z. Electrochem.*, **35**, 555 (1929).
139. Mandelkern, L., J. G. Fatou and C. Howard, *J. Phys. Chem.*, **69**, 956 (1965).
140. Calvert, P. D. and D. R. Uhlmann, *J. Appl. Phys.*, **43**, 944 (1972).
141. Flory, P. J., D. Y. Yoon and K. A. Dill, *Macromolecules*, **17**, 862 (1984).
142. Yoon, D. Y. and P. J. Flory, *Macromolecules*, **17**, 868 (1984).
143. Keller, A., in *MTP International Review of Science*, vol. 8, *Macromolecular Science*, C. E. H. Bawn ed., Butterworth (1972) p. 105.
144. Keller, A., *J. Polym. Sci.*, **C51**, 7 (1975).
144a. Keller, A. and G. Goldbeck-Wood, in *Comprehensive Polymer Science*, Second Suppl., S. L. Aggarwal and S. Ruso eds., Elsevier (1996) Chapter 7, p. 241.
145. Frank, F. C., in *Growth and Perfection of Crystals*, R. H. Doremus, B. W. Roberts and D. Turnbull eds., John Wiley (1958) p. 1.
145a. Mandelkern, L., *Chemtracts-Macromolecular Chemistry*, **3**, 347 (1992).
146. Lauritzen, J. I., Jr. and J. D. Hoffman, *J. Res. Nat. Bur. Stand.*, **64A**, 73 (1960).
147. Hoffman, J. D. and J. I. Lauritzen, Jr., *J. Res. Nat. Bur. Stand.*, **65A**, 297 (1961).

148. Hoffman, J. D., C. M. Guttman and E. A. DiMarzio, in *Faraday Discuss. Chem. Soc.*, 68, 177 (1979).
149. Hoffman, J. D., *SPE Trans.*, 375 (1964).
150. Hoffman, J. D. and R. L. Miller, *Polymer*, **38**, 3151 (1997).
151. Price, F. P., *J. Polym. Sci.*, **42**, 49 (1960).
152. Gornick, F. and J. D. Hoffman, *Ind. Eng. Chem.*, **58**, 36 (1966).
153. Hoffman, J. D. and J. J. Weeks, *J. Res. Nat. Bur. Stand.*, **66A**, 13 (1962).
154. Mandelkern, L., *Faraday Discuss. Chem. Soc.*, **68**, 310 (1979).
155. Alamo, R. G., B. D. Viers and L. Mandelkern, *Macromolecules*, **28**, 3205 (1995).
156. Lauritzen, J. I., Jr. and J. D. Hoffman, *J. Appl. Phys.*, **44**, 4340 (1973).
157. Lauritzen, J. I., Jr. and E. Passaglia, *J. Res. Nat. Bur. Stand.*, **A64**, 73 (1960).
158. Frank, F. C. and M. Tosi, *Proc. Roy. Soc. (London)*, **A263**, 323 (1960).
159. Point, J. J., *Faraday Discuss. Chem. Soc.*, **68**, 167 (1979).
159a. Yamamoto, T., *J. Chem. Phys.*, **107**, 2653 (1997).
159b. Chen, C. M. and P. G. Higgs, *J. Chem. Phys.*, **108**, 4305 (1998).
159c. Doye, J. P. K. and D. Frenkel, *J. Chem. Phys.*, **110**, 2692 (1999).
160. Flory, P. J., *Faraday Discuss. Chem. Soc.*, **68**, 488 (1979).
161. Strobl, G., in *The Physics of Polymers*, Springer (1996) p. 188.
162. Magill, J. H., *Macromol. Chem. Phys.*, **199**, 2365 (1998).
163. Mandelkern L., in *Comprehensive Polymer Science*, vol. 2, C. Booth and C. Price eds., Pergamon Press (1989) Chapter 11.
164. Mandelkern, L., *Acc. Chem. Res.*, **23**, 380 (1990).
165. Binsbergen, F. L., *J. Cryst. Growth*, **13/14**, 44 (1972).
165a. Gedde, U. W., *Polymer Physics*, Chapman and Hall (1995) p. 185.
165b. Welch, P. and M. Muthukumar, *Phy. Rev. Lett.*, **87**, 218302 (2001).
165c. Zachman, H. G., *Koll. Z. Z. Polym.*, **216-217**, 180 (1967).
166. Bradley, R. S., *Quart. Rev. (London)*, **5**, 315 (1951).
167. Hoffman, J. D., L. J. Frolen, G. S. Ross and J. I. Lauritzen, Jr., *J. Res. Nat. Bur. Stand.*, **79A**, 671 (1975).
168. Kovacs, A. J. and J. Gonthier, *Koll. Z. Z. Polym.*, **250**, 530 (1974).
169. Hoffman, J. D. and J. J. Weeks, *J. Chem. Phys.*, **37**, 1723 (1962).
170. Mandelkern, L. and R. G. Alamo, in *American Institute of Physics Handbook of Polymer Properties*, J. E. Mark ed., American Institute of Physics Press (1996).
171. Labaig, J. J., Ph.D. Thesis, Louis Pasteur University, Strasbourg (1978).
172. Ref. 112, p. 46.
173. Hillig, W. B., *Acta Metall.*, **14**, 1868 (1966).
173a. Toda, A., *J. Polym. Sci.: Polym. Phys. Ed.*, **27**, 1721 (1984).
173b. Toda, A., *Polymer*, **28**, 1645 (1987).
173c. Point, J. J. and J. J. Janimak, in *Crystallization of Polymers*, M. Dosiéré ed., Kluwer Academic (1993) p. 119.
174. Hoffman, J. D., G. T. Davis and J. I. Lauritzen, in *Treatise on Solid State Chemistry*, vol. 3, N. B. Hannay ed., Plenum (1970) Ch. 7, p. 497.
175. Sanchez, I. C. and E. A. DiMarzio, *J. Res. Nat. Bur. Stand.*, **76A**, 213 (1972).
176. Lauritzen, J. I., Jr., *J. Appl. Phys.*, **44**, 4653 (1973).
177. Phillips, P. J., *Rep. Prog. Phys.*, **53**, 549 (1990).
178a. Point, J. J. and M. Dosiéré, *Polymer*, **30**, 2292 (1989).
178b. Point, J. J. and M. Dosiéré, *Macromolecules*, **22**, 3501 (1989).
178c. Point, J. J. and J. J. Janimak, *J. Cryst. Growth*, **131**, 501 (1993).
178d. Point, J. J. and J. J. Janimak, *Polymer*, **39**, 7123 (1998).
179. Frank, F. C., *J. Cryst. Growth*, **22**, 233 (1974).

180. Haigh, J. A., L. Mandelkern and L. Howard (unpublished results).
181. Cheng, S. Z. D., J. Chen and J. J. Janimak, *Polymer*, **31**, 1018 (1990).
182. Marentette, J. M. and G. R. Brown, *Polymer*, **39**, 1405 (1998).
183. Booth, C. and C. Price, *Polymer*, **7**, 85 (1966).
184. Monasse, B. and J. M. Haudin, *Makromol. Chem. Macromol. Symp.*, **20/21**, 295 (1988).
185. Goulet, L. and R. E. Prud'homme, *J. Polym. Sci.: Pt. B: Polym. Phys.*, **28**, 2329 (1980).
186. Vasanthakumari, R. and A. J. Pennings, *Polymer*, **24**, 175 (1983).
187. Phillips, P. J. and N. Vatansever, *Macromolecules*, **20**, 2138 (1987).
188. Hoffman, J. D. and R. L. Miller, *Macromolecules*, **22**, 3502 (1989).
189. Toda, A., *Coll. Polym. Sci.*, **270**, 667 (1992).
189a. Allen, R. C. and L. Mandelkern, *Polym. Bull.*, **17**, 473 (1987).
190. Suzuki, T. and A. Kovacs, *Polymer J.*, **1**, 82 (1970).
191. Vogel, H., *Physik. Z.*, **22**, 645 (1921).
192. Hoffman, J. D., *J. Chem. Phys.*, **29**, 1192 (1958).
193. Hoffman, J. D., *Polymer*, **24**, 3 (1983).
194. Miyamoto, J., Y. Tanzawa, H. Miyaji and H. Kiho, *Polymer*, **33**, 2496 (1992).
195. Boon, J., G. Challa and D. W. vanKrevelen, *J. Polym. Sci. A2*, **6**, 1791 (1968).
196. Edwards, B. L. and P. J. Phillips, *Polymer*, **15**, 351 (1974).
197. Iler, D. H, Ph.D. Dissertation, Virginia Tech. (1995).
198. Medellin-Rodriguez, F. J., P. J. Phillips and J. S. Lin, *Macromolecules*, **28**, 7744 (1995).
199. Organ, S. J. and P. J. Barham, *J. Mater. Sci.*, **26**, 1368 (1991).
199a. Lovinger, A. J., D. D. Davis and F. J. Padden, Jr., *Polymer*, **26** 1595 (1985).
200. Runt, J., D. M. Miley, X. Zhang, K. P. Gallagher, K. McFeaters and J. Fishburg, *Macromolecules*, **25**, 1929 (1992).
201. Huang, J. M. and I. C. Chang, *J. Polym. Sci.: Pt. B: Polym. Phys.*, **38**, 934 (2000).
202. Roitman, D. B., H. Marand, R. L. Miller and J. D. Hoffman, *J. Phys. Chem.*, **93**, 6919 (1989).
203. Huang, J., A. Prasad and H. Marand, *Polymer*, **35**, 1986 (1994).
204. Pelzbauer, Z. and A. Galeski, *J. Polym. Sci.*, **38C**, 23 (1972).
205. Hoffman, J. D. and J. P. Armistead, *Preprint Mater. Sci. Eng.*, **81**, 230 (1999).
205a. Armistead, J. P. and J. D. Hoffman, *Macromolecules*, **35**, 3895 (2002).
206. Clark, E. J. and J. D. Hoffman, *Macromolecules*, **17**, 878 (1980).
207. Cheng, S. Z. D., J. J. Janimak, A. Zhang and H. N. Cheng, *Macromolecules*, **23**, 298 (1990).
208. Monasse, B. and J. M. Haudin, *Coll. Polym. Sci.*, **263**, 822 (1985).
209. Alamo, R. G. and C. Chi, in *Molecular Interactions and Time-Space Organization in Macromolecules*, Y. Morishima, T. Norisaye and K. Tashiro eds., Springer (1999) p. 29.
210. Xu, J., S. Srinivas, H. Marand and P. Agarwal, *Macromolecules*, **31**, 8230 (1998).
211. Yasuniwa, M., T. Murakami and M. Ushio, *J. Polym. Sci.: Pt. B: Polym. Phys.* , **37**, 2420 (1999).
212. Hoffman, J. D. and R. L. Miller, *Macromolecules*, **22**, 3502 (1989).
213. Inoue, M., *J. Polym. Sci.*, **47**, 498 (1960).
214. Huang, T. W., R. G. Alamo and L. Mandelkern, *Macromolecules*, **32**, 6374 (1999).
214a. Hanna, L. A., P. J. Hendra, W. Maddams, H. A. Willis, V. Ichy and M. E. A. Cudby, *Polymer*, **29**, 1843 (1988).
214b. Zhu, X, Y. Fang and D. Yan, *Polymer*, **42**, 8595 (2001).

215. Fatou, J. G., C. Marco and L. Mandelkern, *Polymer*, **31**, 1685 (1990).
216. Fatou, J. G., C. Marco and L. Mandelkern, *Polymer*, **31**, 890 (1990).
217. Stack, G. M., L. Mandelkern, C. Kröhnke and G. Wegner, *Macromolecules*, **22**, 4351 (1989).
218. Lazcano, S., J. G. Fatou, C. Marco and A. Bello, *Polymer*, **29**, 2076 (1988).
219. Okui, N., *Polym. J.*, **19**, 1309 (1987).
220. Okui, N., *Polymer*, **31**, 92 (1990).
221. Okui, N., in *Crystallization of Polymers*, M. Dosiéré ed., Kluwer Academic (1993) p. 593.
221a. Privalko, V. P. and Y. S. Lipatov, *Russ. J. Phys. Chem.*, **46**, 8 (1972).
222. Haigh, J. A. and L. Mandelkern (unpublished calculations).
223. Turnbull, D. and R. E. Cech, *J. Appl. Phys.*, **21**, 804 (1950).
224. Phipps, L. W., *Trans. Faraday Soc.*, **60**, 1873 (1964).
225. Buckle, E. R., *Nature*, **186**, 875 (1960).
225a. Ungar, G., J. Stejny, A Keller, I. Bidd and M. C. Whiting, *Science*, **229**, 386 (1985).
226. Cormia, R. L., F. P. Price and D. Turnbull, *J. Chem. Phys.*, **37**, 1333 (1962).
227. Ross, G. S. and L. J. Frolen, *J. Res. Nat. Bur. Stand.*, **79**, 701 (1975).
228. Gornick, F., G. S. Ross and L. J. Frolen, *J. Polym. Sci.*, **18C**, 79 (1967).
229. Koutsky, J. A., A. G. Walton and E. Baer, *J. Appl. Phys.*, **38**, 1932 (1967).
230. Burns, J. R. and D. Turnbull, *J. Appl. Phys.*, **37**, 4021 (1966).
231. Price, F. P., *Abstract Symposium on Macromolecules*, IUPAC, Wiesbaden, Germany (1959) Sect. 1, paper 1B2.
231a. Flory, P. J., *J. Am. Chem. Soc.*, **84**, 2857 (1962).
232. de Nordwall, H. J. and L. A. K. Staveley, *J. Chem. Soc.*, 224 (1954).
233. Turnbull, D., *J. Appl. Phys.*, **21**, 1022 (1950).
233a. Phillips, P. J., G. J. Rensch and K. D. Taylor, *J. Polym. Sci.: Pt. B: Polym. Phys.*, **25**, 1783 (1987).
233b. Xing, P., L. Dong, Y. An, Z. Feng, M. Avella and E. Martuscelli, *Macromolecules*, **30**, 2726 (1997).
233c. Silvestre, C., E. D. Pace, R. Napolitano, B. Pirrizi and G. Cesario, *J. Polym. Sci.: Pt. B: Polym. Phys.*, **39**, 415 (2001).
234. Huseby, T. W. and H. E. Bair, *J. Appl. Phys.*, **39**, 4969 (1968).
235. Hoffman, J. D., R. L. Miller, H. Marand and D. B. Roitman, *Macromolecules*, **25**, 2221 (1992).
236. Hoffman, J. D., *Polymer*, **33**, 2643 (1992).
237. Tonelli, A. E., *Macromolecules*, **25**, 7199 (1992).
237a. Flory, P. J., *Statistical Mechanics of Chain Molecules*, Wiley (1969) p. 149.
238. Inoue, M., *J. Polym. Sci.: Part A*, **1**, 2013 (1963).
239. Chattergee, A. M., F. P. Price and S. Newman, *J. Polym. Sci.: Polym. Phys. Ed.*, **13**, 2369 (1975); *ibid.*, **13**, 2385 (1975); *ibid.*, **13**, 2390 (1975).
240. Lotz, B. and J. C. Wittman, *Makromol. Chem.*, **185**, 2043 (1984).
241. Brückner, S., S. V. Meille, V. Petraccone and B. Pirozzi, *Prog. Polym. Sci.*, **16**, 361 (1991).
242. Beck, H. N., *J. Appl. Polym. Sci.*, **11**, 673 (1967).
243. Al-Ghazawi, M. and R. P. Sheldon, *J. Polym. Sci.: Polym. Lett. Ed.*, **21B**, 347 (1983).
244. Kuhre, C. J., M. Wales and M. E. Doyle, *Soc. Plast. Eng. J.*, October, 1113 (1964).
245. Garcia, D., *J. Polym. Sci.: Polym. Phys. Ed.*, **22**, 2063 (1984).
246. Yim, A. and L. E. St.-Pierre, *J. Polym. Sci.: Polym. Lett. Ed.*, **8B**, 241 (1970).

247. Cole, J. H. and L. E. St. Pierre, in *Polyethers*, E. J. Vandenberg ed., American Chemical Society (1975) p. 58.
248. Binsbergen, F. L. and B. G. M. de Lange, *Polymer*, **11**, 309 (1970).
249. Kim, C. Y., Y. C. Kim and S. C. Kim, *Polym. Eng. Sci.*, **33**, 1445 (1993).
250. Jang, G. S., W. J. Cho and C. S. Ha, *J. Polym. Sci.: Pt. B: Polym. Phys.*, **39**, 1001 (2001).
251. Song, S. S., J. L. White and M. Cakmoli, *Polym. Eng. Sci.*, **30**, 994 (1990).
252. Cheng, S. and R. A. Shanks, *J. Appl. Polym. Sci.*, **47**, 2149 (1993).
253. Groeninckx, G., H. Berghmans, N. Overbergh and G. Smets, *J. Polym. Sci.: Polym. Phys. Ed.*, **12**, 308 (1974).
254. Przygocki, W. and A. Wlochowicz, *J. Appl. Polym. Sci.*, **19**, 2683 (1975).
255. Kennedy, M. A., G. Turturro, G. R. Brown and L. E. St.-Pierre, *J. Polym. Sci.: Polym. Phys. Ed.*, **21**, 1403 (1983).
256. Van Antwerpen, F. and D. W. Van Krevelen, *J. Polym. Sci. Phys. Ed.*, **10**, 2423 (1972).
257. Beck, H. N. and H. D. Ledbetter, *J. Appl. Polym. Sci.*, **9**, 2131 (1965).
258. Last, A. G. M., *J. Polym. Sci.*, **39**, 543 (1959).
259. Binsbergen, F. L., *Polymer*, **11**, 253 (1970).
260. Mercier, J. P., *Polym. Eng. Sci.*, **30**, 270 (1990).
261. Binsbergen, F. L., *J. Polym. Sci.*, **59C**, 11 (1977).
261a. Wittman, J. C. and B. Lotz, *Prog. Polym. Sci.*, **15**, 909 (1990).
262. Wittman, J. C. and B. Lotz, *J. Polym. Sci.: Polym. Phys. Ed.*, **19**, 1837 (1981); *ibid.*, **19**, 1853 (1981); *ibid.*, **21**, 2495 (1983); *ibid.*, **25**, 1079 (1987).
263. Lotz, B. and J. C. Wittman, *Makromol. Chem.*, **185**, 2043 (1984).
263a. Mathies, C., A. Thierry, J. W. Wittman and B. Lotz, *J. Polym. Sci: Pt. B: Polym. Phys.*, **40**, 2504 (2002).
264. Murthy, N. S., J. P. Sibilia and A. M. Kothiar, *J. Appl. Polym. Sci.*, **31**, 2569 (1986).
264a. Haubruge, H. G., A. M. Jonas, R. Legras, J. C. Wittman and B. Lotz, *Macromolecules*, **36**, 4452 (2003).
265. Binsbergen, F. L., *J. Polym. Sci.: Polym. Phys. Ed.*, **11**, 117 (1973).
266. Legras, R., J. P. Mercier and E. Nield, *Nature*, **304**, 432 (1983).
267. Legras, R., B. Bailly, M. Daumerie, J. M. Dekominck, J. P. Mercier, V. Zichy and E. Nield, *Polymer*, **25**, 835 (1984).
268. Thierry, A., B. Fillon, C. Straupe, B. Lotz and J. C. Wittman, *Prog. Coll. Polym. Sci.* **87**, 28 (1992).
269. Fillon, B., B. Lotz, A. Thierry and J. C. Wittman, *J. Polym. Sci.: Pt. B: Polym. Phys. Ed.*, **31**, 1395 (1993).
270. Blundell, D. J., A. Keller and A. Kovacs, *J. Polym. Sci.: Polym. Lett.*, **4B**, 481 (1966).
271. Bidd, J. and M. C. Whiting, *J. Chem. Soc. Chem. Commun.*, 543 (1985).
272. Bidd, I., D W. Holdup and M. C. Whiting, *J. Chem. Soc. Perkin Trans.*, **1**, 2455 (1987).
273. Lee, K. S. and G. Wegner, *Makromol. Chem. Rapid Commun.*, **6**, 203 (1985).
274. Stack, G. M. and L. Mandelkern, *Macromolecules*, **21**, 510 (1988).
275. Keller, A., S. J. Organ and G. Ungar, *Makromol. Chem. Macromol. Symp.*, **48/49**, 93 (1991).
276. Ungar, G., J. Stejny, A. Keller, I. Bidd and M. C. Whiting, *Science*, **229**, 386 (1985).
277. Mandelkern, L. and R. G. Alamo, in *Structure–Properties Relations in Polymers: Spectroscopy and Performance*, M. W. Urban and C. D. Carrier eds., American Chemical Society Advances in Chemistry Series (1993) pp. 236, 157.

278. Stack, G. M., L. Mandelkern, C. Kröhnke and G. Wegner, *Macromolecules*, **22**, 4351 (1989).
279. Ungar, G. and A. Keller, *Polymer*, **27**, 1835 (1986).
280. Alamo, R. G., L. Mandelkern, G. M. Stack, C. Kröhnke and G. Wegner, *Macromolecules*, **27**, 147 (1994).
281. Ungar, G. and A. Keller, *Polymer*, **28**, 1899 (1987).
282. Boda, E., G. Ungar, G. M. Brooks, S. Burnett, S. Mohammed, D. Procter and M. C. Whiting, *Macromolecules*, **30**, 4674 (1997).
283. Mandelkern, L., A. Prasad, R. G. Alamo and G. M. Stack, *Macromolecules*, **23**, 3696 (1990).
284. Hoffman, J. D., *Polym. Commun.*, **27**, 39 (1986).
285. Hoffman, J. D., *Macromolecules*, **19**, 1124 (1986).
286. Sadler, D. M. and G. H. Gilmer, *Polymer*, **25**, 1446 (1984).
287. Sadler, D. M. and G. H. Gilmer, *Polym. Commun.*, **28**, 243 (1987).
288. Sadler, D. M., *J. Chem. Phys.*, **89**, 1771 (1987).
289. Sadler, D. M., *Polymer*, **28**, 1440 (1987).
290. Sadler, D. M., *Nature*, **326**, 174 (1987).
291. Higgs, P. G. and G. Ungar, *J. Chem. Phys.*, **100**, 640 (1994).
292. Hoffman, J. D., *Polymer*, **32**, 2828 (1991).
293. Miller, R. L., *Polymer*, **33**, 1783 (1992).
294. Alamo, R. G., in *Crystallization of Polymers*, M. Dosiére ed., Kluwer Academic Publishers (1993) p. 73.
295. Sutton, S. J., A. S. Vaughan and D. C. Bassett, *Polymer*, **37**, 5735 (1996).
296. Bassett, D. C., R. H. Olley, S. J. Sutton and A. S. Vaughan, *Macromolecules*, **29**, 1852 (1996).
297. Bassett, D. C., R. H. Olley, S. J. Sutton and A. S. Vaughan, *Polymer*, **22**, 4993 (1996).
298. Organ, S. J., A. Keller, M. Hikosaka and G. Ungar, *Polymer*, **37**, 2517 (1996).
299. Teckoe, J. and D. C. Bassett, *Polymer*, **41**, 1953 (2000).
300. Hosier, I. L. and D. C. Bassett, *Polymer*, **41**, 8801 (2000).
301. Hoffman, J. D., *Macromolecules*, **18**, 772 (1985).
302. Mandelkern, L., R. G. Alamo and J. A. Haigh, *Macromolecules*, **31**, 762 (1998).
303. Flory, P. J., *Principles of Polymer Chemistry*, Cornell University Press (1953) p. 336.
304. Mandelkern, L., in *Crystallization of Polymers*, M. Dosiére ed., Kluwer Academic Publishers (1993) p. 494.
305. Point, J. J., *Faraday Discuss. Chem. Soc.*, **91**, 2565 (1995).
306. Tinas, J. and J. G. Fatou, *Anales de Quimica*, **76**, 3 (1980).
307. Marco, C., J. G. Fatou and A. Bello, *Polymer*, **18**, 110 (1977).
308. Marco, C., J. G. Fatou, A. Bello and A. Blanco, *Polymer*, **20**, 1250 (1979).
309. Stack, G. M., L. Mandelkern and I. G. Voigt-Martin, *Macromolecules*, **17**, 321 (1984).
310. Kovacs, A. J. and C. Straupe, *Faraday Discuss. Chem. Soc.*, **68**, 225 (1979).
311. Cheng, S. Z. D., A. Zhang, J. Chen and D. P. Heberer, *J. Polym. Sci.: Pt. B: Polym. Phys.*, **29**, 287 (1991).
312. Cheng, S. Z. D., A. Zhang, J. S. Barley, J. Chen, A. Habenschuss and P. R. Zschack, *Macromolecules*, **24**, 3397 (1991).
313. Cheng, S. Z. D., J. Chen, J. S. Barley, A. Zhang, A. Habenschuss and P. R. Zschack, *Macromolecules*, **25**, 1453 (1992).

314. Chiu, F. C., Q. Fu, M. Leland, S. Z. D. Cheng, E. T. Hsieh, C. C. Tso and B. S. Hsiao, *J. Macromol. Sci. Phys.*, **B36**, 553 (1997).
315. Buckley, C. P. and A. J. Kovacs, *Coll. Polym. Sci.*, **254**, 695 (1976).
316. Peña, B., R. G. Alamo and L. Mandelkern (unpublished work).
317. Cheng, S. Z. D. and J. Chen, *J. Polym. Sci.: Polym. Phys. Ed.*, **29**, 311 (1991).
318. Kovacs, A. J. and A. Gonthier, *Koll. Z. Z. Polym.*, **250**, 530 (1972).
319. Kovacs, A. J., A. Gonthier and C. Straupe, *J. Polym. Sci.*, **50C**, 283 (1975).
319a. Fraser, M. J., A. Marshall and C. Booth, *Polymer*, **18**, 93 (1977).
320. Janimak, J. J. and S. Z. D. Cheng, *Polym. Bull.*, **22**, 95 (1989).
321. Gaur, V. and B. Wunderlich, *J. Phys. Chem. Ref. Data*, **10**, 119 (1981); *ibid.*, **10**, 1001 (1981).
322. Mandelkern, L. and G. M. Stack, *Macromolecules*, **17**, 871 (1984).
323. Ou-Yang, W. C., L. J. Li, H. L. Chen and J. C. Hwang, *Polym. J.*, **29**, 889 (1997).
324. Okada, M., M. Nishi, M. Tabahashi, M. Matsuda, A. Toda and M. Hikosaka, *Polymer*, **39**, 4535 (1998).
325. Chen, T. L. and A. C. Su, *Polymer*, **36**, 73 (1995).
326. Maclaine, J. Q. G. and C. Booth, *Polymer*, **16**, 191 (1975).
327. Chen, H. L., L. J. Li, W. C. O. Yang, J. C. Hwang and W. Y. Wong, *Macromolecules*, **30**, 1718 (1997).
328. Gomez, M. A., J. G. Fatou and A. Bello, *Eur. Polym. J.*, **22**, 661 (1986).
329. Hoffman, J. D. and R. L. Miller, *Macromolecules*, **21**, 3039 (1988).
330. Magill, J. H., *J. Polym. Sci. Pt. A-2*, **5**, 89 (1967).
331. Lovering, E. C., *J. Polym. Sci.*, **30C**, 329 (1970).
332. Deslandes, Y., F. N. Sabir and J. Roover, *Polymer*, **32**, 1267 (1991).
333. Lemstra, P. J., J. Postma and G. Challa, *Polymer*, **15**, 757 (1974).
333a. Acierno, S., N. Grizzuti and H. H. Winter, *Macromolecules*, **35**, 5043 (2002).
333b. Alamo, R. G., J. A. Blanco, P. K. Agarwal and J. C. Randall, *Macromolecules*, **36**, 1559 (2003).
334. Ghosh, S. K., M. Hikosaka and A. Toda, *Coll. Polym. Sci.*, **279**, 382 (2001).
335. Nishi, M., M. Hikosaka, A. Toda and K. Yamada, *Polym. J.*, **31**, 749 (1999).
336. Yamazaki, S., M. Hikosaka, F. Gu, J. K. Ghosh, M. Arakaki and A. Toda, *Polym. J.*, **33**, 906 (2001).
336a. Ghosh, S. K., M. Hikosaka, A. Toda, S. Yamazaki and K. Yamada, *Macromolecules*, **35**, 6985 (2002).
337. Hoffman, J. D., *Polymer*, **23**, 656 (1982).
338. de Gennes, P. G., *J. Chem. Phys.*, **55**, 672 (1971).
339. de Gennes, P. G., *Scaling Concepts in Polymer Physics*, Cornell University Press (1978) p. 223f.
340. Graessley, W. W., in *Physical Properties of Polymers*, Second Edition., J. E. Mark ed., American Chemical Society (1993) p. 97.
341. Doi, M. and S. F. Edwards, *Theory of Polymer Dynamics*, Clarendon Press, Oxford (1980).
342. Graessley, W. W., *Adv. Polym. Sci.*, **47**, 68 (1980).

10

Crystallization kinetics of copolymers

10.1 Introduction

It was noted in Chapter 5 (Volume 1) that both the course of fusion and the equilibrium melting temperature depend on the sequence propagation probability parameter, p, and whether the crystalline phase remains pure. It is known that in simple liquids the introduction of a second component significantly alters the nucleation rate and consequently the overall rate of crystallization from the melt. The phase diagrams for such two-component systems govern to a large extent the crystallization process. In copolymers the equivalent of a second component is built into the chain. Hence, changes in the crystallization process of copolymers, relative to homopolymers, can be expected, and are in fact supported by kinetic studies. With copolymers, however, it is the sequence propagation probability rather than the composition that is important.

In this chapter the overall crystallization kinetics and spherulite growth rates of the major copolymer types will be presented and analyzed. It should be recalled from the point of view of polymer crystallization that chains containing structurally different repeating units, although chemically identical, are also treated as copolymers. These include branch points and regio defects as well as stereo and geometric isomers. As a matter of convenience, the influence of cross-links will also be included in this chapter.

10.2 Random type copolymers

10.2.1 Overall crystallization

The chain microstructure of random copolymers can be represented by $p = X_{\mathrm{A}}$, the mole fraction of the major crystallizing repeating unit. Studies of the overall crystallization kinetics of random copolymers are of particular importance since as the comonomer content increases the spherulitic structure becomes poorly developed

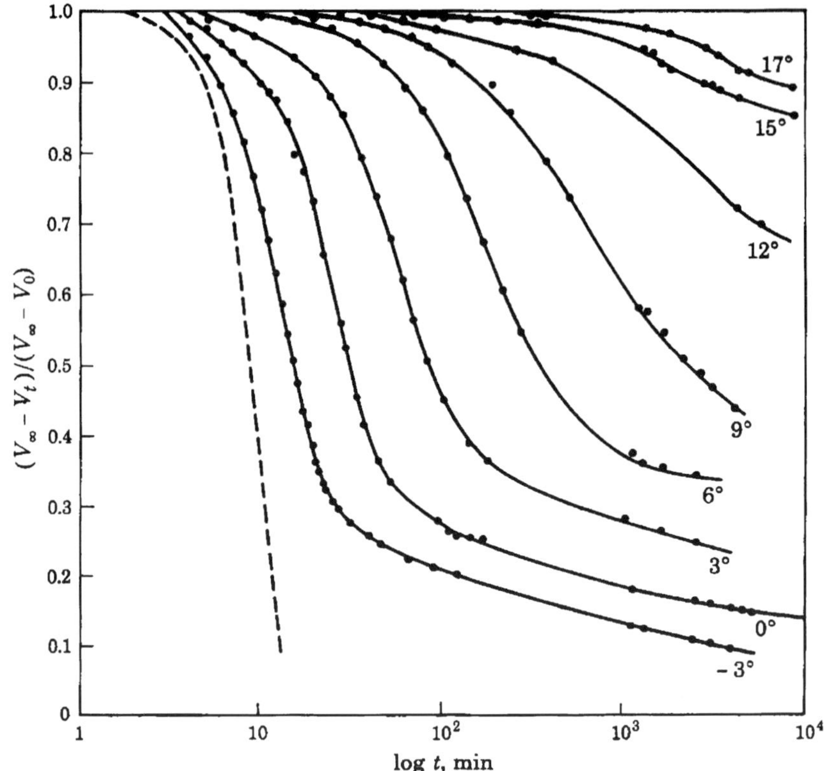

Fig. 10.1 Plot of quantity $(V_\infty - Vt)/(V_\infty - V_0)$ against log t for the crystallization of a poly(butadiene) containing 80% 1,4-trans unit. Temperature of crystallization is indicated for each isotherm.(1)

and eventually disappears. Thus, although the measurements of spherulite growth rates are quite informative, they can only be studied over a limited composition range. Distinction also has to be made as to whether or not the comonomer enters the lattice. It can be assumed in the discussion in this section that, unless otherwise noted, the crystalline phase remains pure.

Typical examples of the overall crystallization kinetics of random type copolymers will be given in the following. Included will be the characteristics of the isotherms, and the influence of both composition and molecular weight. Figure 10.1 gives a set of isotherms for an emulsion polymerized poly(butadiene).(1) This polymer contains 80% of crystallizable 1,4-trans units. The remaining units, that are not crystallizable, are distributed among 1,4-cis and 1,2 addition. The observed melting temperature of this polymer is 37 ± 1 °C. It is evident that in this case the isotherms are quite different from those characteristic of a pure homopolymer. The isotherms are no longer superposable, their shapes being dependent on the crystallization temperatures. However, when the crystallization is conducted at the

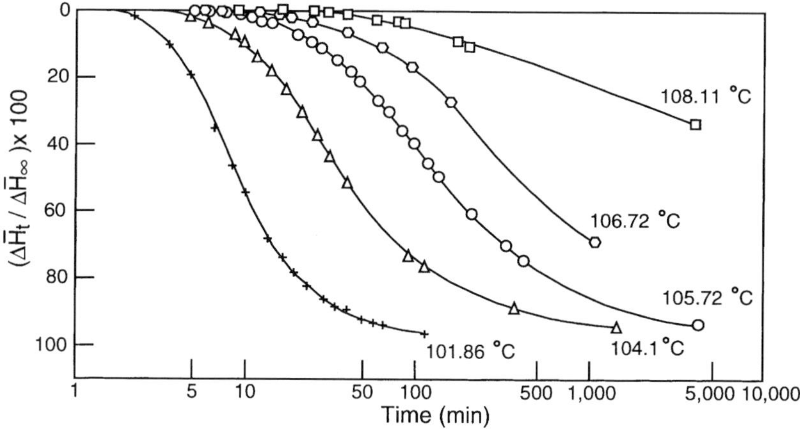

Fig. 10.2 Plot of percentage change in crystallization level against log t for the crystallization of a long chain branched polyethylene. Temperature of crystallization is indicated for each isotherm. (From Buchdahl *et al.* (2))

larger undercoolings the isotherm shapes begin to resemble those of a homopolymer. This effect is illustrated by the dashed curve in Fig. 10.1, which was calculated from Eq. (9.31a) with $n = 3$. However, as the crystallization temperature is raised, the transformation proceeds more slowly and the isotherms spread out in fan-like fashion along the time axis.

The overall crystallization kinetics of polyethylenes that contain long chain branches follow a similar pattern, as is illustrated in Fig. 10.2.(2) In this case, the methine carbons, to
which the long chain branches are attached, serve as the structural irregularity. The lack of superposability of the isotherms is again apparent. Their characteristics are similar to those of poly(butadiene). A comparison of the isotherm shapes in Fig. 10.2 with those for linear polyethylene, Fig. 9.4 or 9.5, demonstrates the strong influence small amounts of structural irregularities, usually about 1% of branch points, have on the overall crystallization kinetics.

The same type isotherms are found when comonomers are introduced into the chain. An example is given in Fig. 10.3. Here the extent of the transformation, $1 - \lambda$, is plotted against log time for three different hydrogenated poly(butadienes), which are random ethylene–butene copolymers, that have narrow molecular weight and composition distributions. In this example, the branch content is fixed but the molecular weight is varied. Again, the isotherms for a given copolymer do not superpose. In this example, for a given molecular weight the isotherms at different crystallization temperatures do not merge at long times, as they do in linear polyethylene fractions. The crystallization rate and the temperature interval over which isothermal crystallization can take place in a practical time scale are

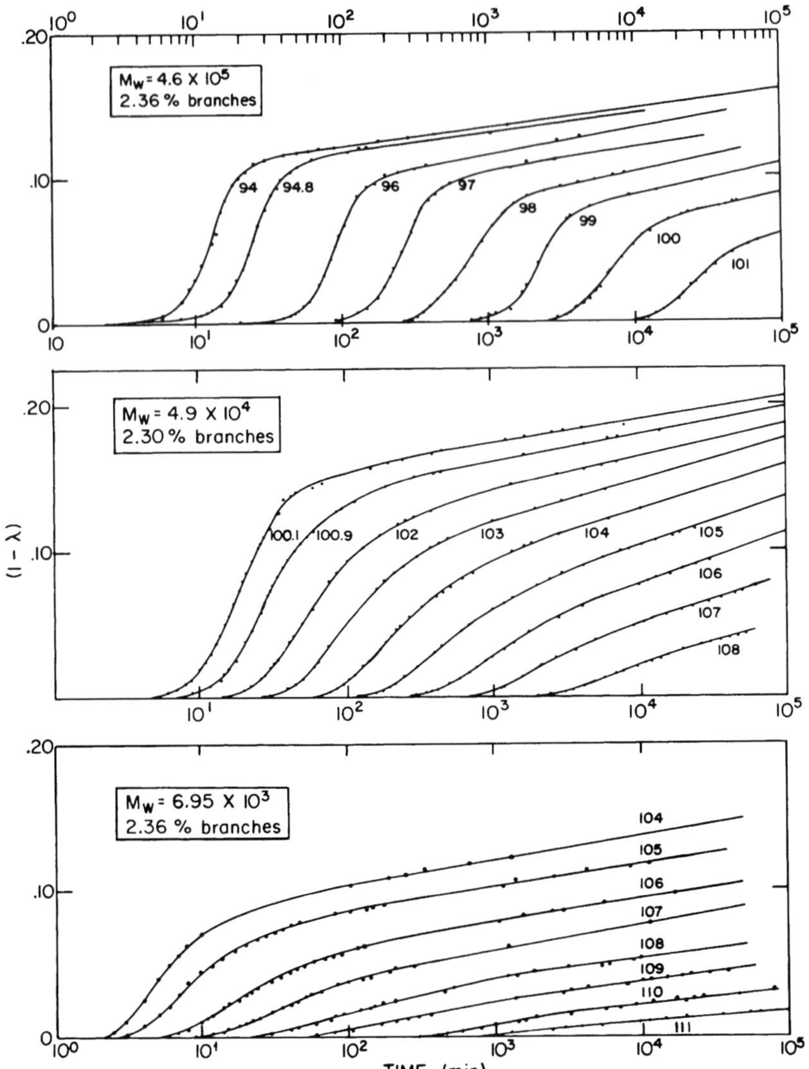

Fig. 10.3 Plot of the extent of transformation, $(1 - \lambda)$, against log time for the crystallization of hydrogenated poly(butadienes) at the indicated temperatures. The weight-average molecular weights and mol percent branch points are indicated.(3)

dependent on molecular weight. For the highest molecular weight illustrated, the crystallization interval is 94–101 °C; it shifts to 100–108 °C for $M = 4.9 \times 10^4$ and to 104–111 °C for $M = 6.95 \times 10^3$.

Analysis of experimental data for some copolyesters (4,5,6) and syndiotactic poly(propylene)(6a) indicates a superposition of isotherms. However, in these examples the crystallization was conducted at relatively large undercoolings. Superposition is to be expected under these circumstances.

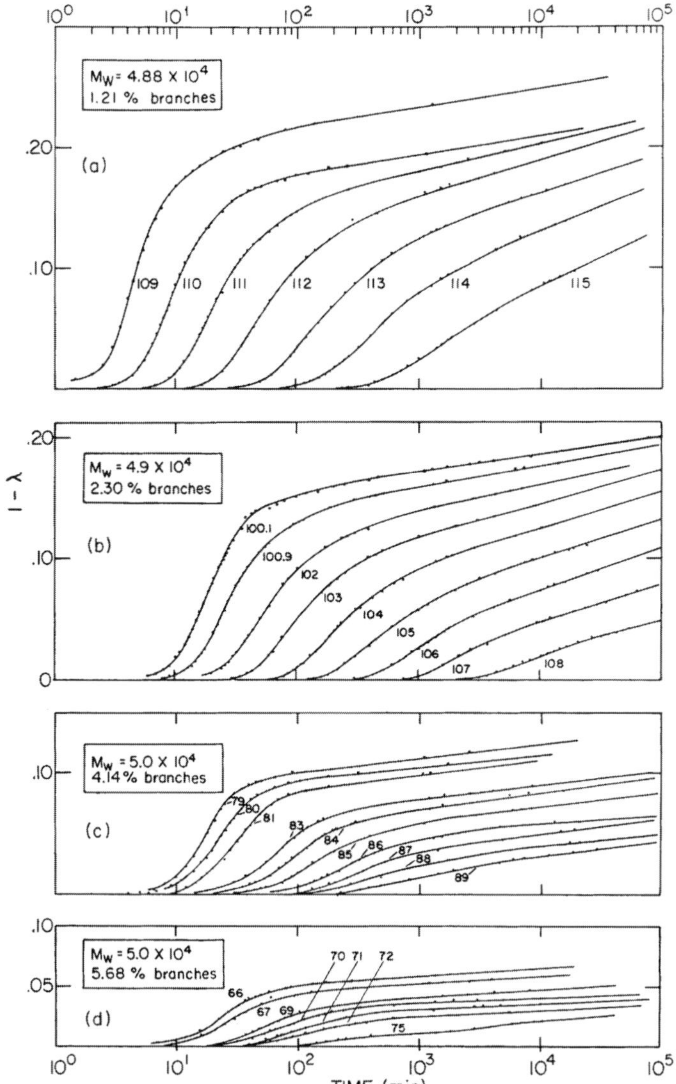

Fig. 10.4 Plot of extent of transformation, $(1 - \lambda)$, against log time for the crystallization of hydrogenated poly(butadienes) and one ethylene-1-hexene copolymer (1.21 mol % branch points) at the indicated temperatures. The weight-average molecular weight and mol percent branch points are indicated.(3)

The crystallization rate is also dependent on the co-unit content at a fixed molecular weight. This point is illustrated in Fig. 10.4 for random ethylene copolymers, where the co-unit concentration varies, but the molecular weight is held fixed at about $M = 5 \times 10^4$.(3) The major features of copolymer crystallization are observed again. However, these features are accentuated with increasing co-unit content. For example, although the temperature interval where isothermal crystallization can be carried out in a reasonable time is the same for all the copolymers, the location depends on copolymer composition. There is a change of about 40 °C in the accessible crystallization range as the mole fraction of branch points varies from 1.21 to 5.68 mol percent. Only a small part of this change can be attributed to the decrease in the equilibrium melting temperature and thus to the decrease in undercooling. The major reason for this change is the concentration of sequences available for crystallization at a given temperature as a consequence of the chain microstructure. The isotherm shapes change drastically with increasing co-unit content, reflecting the fact that crystallization is becoming more protracted. Studies of the overall crystallization kinetics of a set of random ethylene–octene copolymers give similar results.(7,8) In different molecular weight ranges, at a fixed undercooling, the crystallization rate decreases as the comonomer content increases. When the comonomer concentration is fixed the rate decreases with increasing molecular weight. These kinetic studies make quite clear that the molecular weight and copolymer compositions are independent variables that govern the crystallization. This in turn will be reflected in both microscopic and macroscopic properties. Moreover, it is important when studying structure and properties of random type copolymers that the complete range of isothermal crystallization needs to be considered.

The examples given up to now have been limited to crystallization temperatures in the vicinity of T_m. In this temperature region a strong negative temperature coefficient, similar to that of homopolymers, is observed for the crystallization process. As an example, for the copolymers illustrated in Fig. 10.3 there is about a four orders of magnitude time change over only a 6–7 °C temperature interval. Other copolymers behave in a similar manner. Copolymers show features that are similar to those of homopolymers when crystallized over the complete temperature range. In particular, a maximum in the crystallization rate is still observed. For some copolymers one or the other species crystallizes, depending on the composition, and a maximum in the overall crystallization rate can occur.(9a) An example is given in Fig. 10.5 for the crystallization of the random copolyester, ethylene terephthalate–azelate.(9) Here, the half-time of the crystallization, $t_{1/2}$, is plotted against the crystallization temperature for different compositions of this copolymer. At high temperatures the very strong negative temperature coefficient indicates nucleation control. The rate in this region is significantly reduced at a fixed crystallization temperature as the comonomer content increases. As the crystallization temperature is lowered a maximum in the rate is observed (a minimum in $t_{1/2}$). The locations of the rate maxima depend on the copolymer composition. As the temperature is lowered below the maxima the crystallization rate decreases once again. However, an interesting phenomenon occurs, in that an inversion in the rate with temperature takes place. The crystallization rate now increases with comonomer

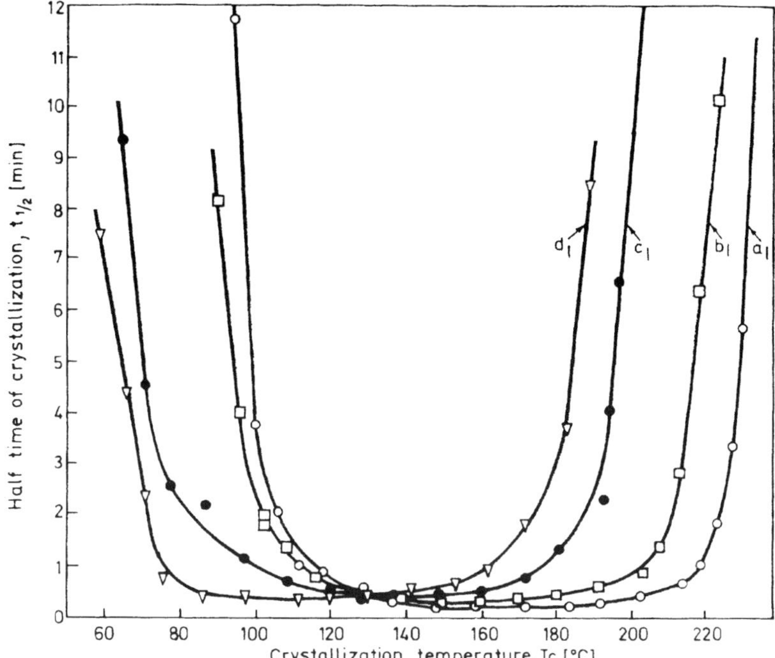

Fig. 10.5 Plot of crystallization half-time, $t_{1/2}$, against crystallization temperatures, T_c, for random copolyesters of ethylene terephthalate–azelate at indicated mol percent azelate. Composition: ○ 6 mol%; □ 11 mol%; ● 25 mol%; ▽ 31 mol%.

content. This inversion in rate with composition can be attributed to the decrease in the glass temperature with increasing comonomer content. For this system the glass temperature decreases from 70 °C for the homopolymer to 14 °C for the copolymer with the highest comonomer content studied. The crystallization behavior of this copolyester demonstrates dramatically the important role of the equilibrium melting temperature and nucleation at the high crystallization temperatures, and that of the glass temperature, and thus the transport term, at the low temperatures.

In order to understand the nature of the copolymer isotherms, the basis for superposability needs to be examined. The Avrami type isotherm, and the deduction of superposition, is based on the implicit assumption that the composition and structure of the melt do not change during the course of the transformation. Consequently, in the simplest case the nucleation and growth rates were taken to be independent of the extent of transformation. This underlying assumption has already been questioned for homopolymer crystallization where entanglements and other topological defects are not crystallizable. This underlying assumption of the constancy of nucleation and growth rates clearly will not apply to random copolymers. In this case both the composition and sequence distribution of the residual melt change continuously during the course of the crystallization. Consequently, the increasing accumulation of the noncrystallizable species in the melt results in a continuous depression of the melting point. Thus, the effective undercooling decreases, even though the crystallization temperature itself is held constant. The question as to whether this expectation leads to isotherm shapes that have been observed in copolymers needs to be examined.

Based on these considerations the kinetic equations appropriate to the crystallization of a copolymer composed of A and B units arranged in random sequence distribution has been investigated.(10) In this analysis only the A units are allowed to enter the crystal lattice. It is convenient to characterize the extent of the transformation at time t by a parameter $\theta \equiv [1 - \lambda(t)]/[1 - \lambda(\infty)]$, where $\lambda(\infty)$ is the fraction of the noncrystalline material at the completion of the transformation and $\lambda(t)$ is the corresponding quantity at time t. With these conventions, the steady-state nucleation rate per untransformed unit volume, at the conversion θ, can be expressed as

$$N(\theta) = N_0 X_{\mathrm{A}}^{\theta} \exp\left(-\frac{E_D}{RT} - \frac{\Delta G_{\theta}^*}{RT}\right) \tag{10.1}$$

Here X_{A}^{θ} is the mole fraction of A units in the melt and ΔG_{θ}^* is the free energy of forming a nucleus of critical size at θ. In terms of the composition of the completely molten phase,

$$N(\theta) = N(0)\frac{X_{\mathrm{A}}^{\theta}}{X_{\mathrm{A}}^{0}} \exp\left(-\frac{\Delta G^* - \Delta G_{\theta}^*}{RT}\right) \tag{10.2}$$

$N(0)$ is the steady-state nucleation rate at $t = 0$, corresponding to the initial melt composition X_{A}^{0}, and ΔG_{θ}^* is the corresponding critical free energy. For a homogeneously formed cylindrically shaped nucleus

$$\Delta G_{\theta}^* = \frac{8\pi \sigma_{\mathrm{un}}^2 \sigma_{\mathrm{en}}}{\Delta G_{\mathrm{u}}(\theta)^2} \tag{10.3}$$

where $\Delta G_{\mathrm{u}} \cong \Delta H_{\mathrm{u}}(T_{\theta} - T/T_{\theta})$, with T_{θ} being defined as the equilibrium melting point of the system at conversion θ. From the ideal expression for the melting point depression of random copolymers, it follows that

$$\frac{1}{T_{\theta}} - \frac{1}{T_{\mathrm{m}}} = -\frac{R}{\Delta H_{\mathrm{u}}} \ln \frac{X_{\mathrm{A}}^{\theta}}{X_{\mathrm{A}}^{0}} \tag{10.4}$$

The ratio of $X_{\mathrm{A}}^{\theta}/X_{\mathrm{A}}^{0}$ can be expressed as

$$\frac{X_{\mathrm{A}}^{\theta}}{X_{\mathrm{A}}^{0}} = \frac{1 + (w_{\mathrm{B}}/w_{\mathrm{A}})(M_{\mathrm{A}}/M_{\mathrm{B}})}{1 + (w_{\mathrm{B}}/w_{\mathrm{A}})(M_{\mathrm{A}}/M_{\mathrm{B}})\{1 - \theta[1 - \lambda(\infty)]/w_{\mathrm{A}}\}} \tag{10.5}$$

where w_{A} and w_{B} are the initial weight fractions of the A and B units respectively, and $(M_{\mathrm{A}}/M_{\mathrm{B}})$ is the ratio of molecular weights of the units. The attenuation of the nucleation rate is then given by

$$\frac{N(\theta)}{N(0)} = \left(\frac{X_A^\theta}{X_A^0}\right) \exp\left\{\frac{C}{T}\left[\left(\frac{T_m}{T_m - T}\right)^2 - \left(\frac{T_\theta}{T_\theta - T}\right)^2\right]\right\} \tag{10.6}$$

where $C = 8\pi\sigma_{un}^2\sigma_{en}/\Delta H_u^2$. The use of Eq. (10.6) involves the assumption that T_θ depends solely on the ratio of X_A^θ/X_A^0, i.e. on the chemical composition of the melt. However, as is known from theory, and experiment, a further reduction in T_θ will occur in copolymers at finite levels of crystallinity because of the change in the sequence distribution of the crystallizable units (see Chapter 5). In the present context, therefore, the value of T_θ for random copolymers is underestimated. In this example a homogeneous, three-dimensional nucleus is assumed for illustrative purposes. However, qualitatively similar results are obtained when other type nuclei are assumed.

The reduction in the nucleation rate can be readily calculated, provided the appropriate parameters are known for the parent homopolymer. The results of such a model calculation for random ethylene copolymers are given in Fig. 10.6 taking

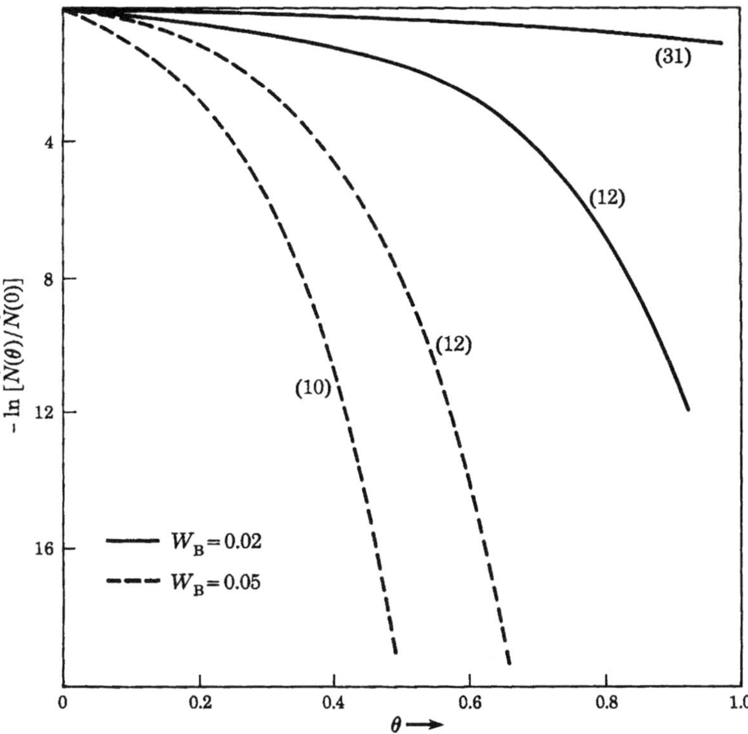

Fig. 10.6 Theoretical plot of $-\ln[N(\theta)/N(0)]$ against θ for two random copolymers of indicated composition. The initial undercoolings are indicated by the numbers in parentheses.(10)

$M_A/M_B = 0.44$ and $1 - \lambda(\infty) = 0.75$.[1](10) Two compositions, and two undercoolings for each, are taken as examples. When small amounts of co-ingredients are added, a slight but perceptible attenuation of the nucleation rate is predicted at large undercoolings. This manifests itself primarily at the latter portions of the transformation. However, if the undercooling is decreased, or if the concentration of B units is slightly increased, the attenuation in the nucleation rate becomes significant. Changes are now expected for even small amounts of crystallization. These effects result from the decrease in the true undercooling $(T_\theta - T)$ with the extent of the transformation. For large nominal undercoolings, the relative change with the transformation is small, so that the nucleation rate is only slightly reduced. However, at small nominal undercoolings, the change in T_θ with conversion is such that an appreciable reduction in the nucleation rate is predicted. This effect becomes more significant as the initial concentration of the B units is increased.

To proceed further in the analysis, it is necessary to develop the general mathematical formulation for the kinetics of phase changes, allowing for the exclusion of noncrystallizing units from the lattice. In essence, this involves evaluating the integral appearing in Eq. (9.16) while accounting for the changes in the nucleation and growth rates with the extent of the transformation. To accomplish this, the primary nucleation rate is expressed as

$$\frac{N(\tau)}{N(0)} = h(\tau) \tag{10.7}$$

The linear growth rate $G(z)$ can be written as $G(0)\,g(z)$, where $G(0)$ is the initial rate at $t = 0$. Then

$$v(t, \tau') = K\,G(0)^{n-1}[\Phi(t, \tau')]^{n-1} \tag{10.8}$$

where

$$\Phi(t, \tau) = \int_{\tau'}^{t} g(z)\,dz \tag{10.9}$$

The more general expression for the kinetics of phase change then becomes (10)

$$\frac{1 - \lambda(t)}{1 - \lambda(\infty)} = 1 - \exp\left\{\frac{nk}{1 - \lambda(\infty)}\int_0^t h(\tau)[\Phi(t, \tau)]^{n-1}\,d\tau\right\} \tag{10.10}$$

where $k = (1/n)(\rho_c/\rho_l)K\,N(0)\,G(0)^{-1}$. Since the growth rate is expected to be nucleation controlled, both $h(\tau)$ and $g(\tau)$ are, in general, functions of $1 - \lambda(t)$. When $g(\tau)$ and $h(\tau)$ are equal to unity, Eq. (9.42) is regenerated.

[1] The parameters used are given in Table 1 of Ref. (10).

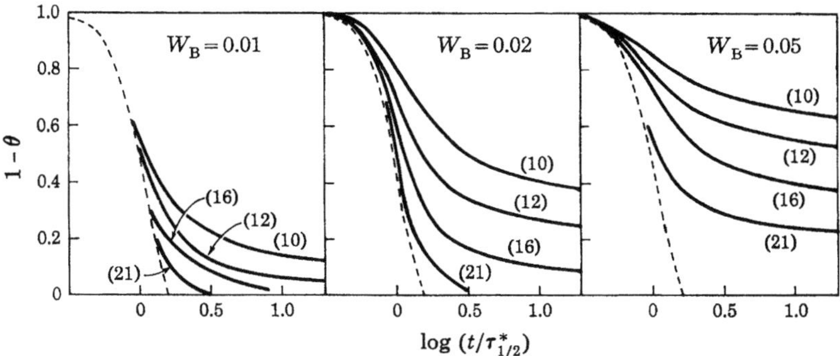

Fig. 10.7 Theoretical plot of $1 - \theta$ against $\log(t/\tau^*_{1/2})$ for three different copolymers. Initial undercoolings are indicated in the parentheses. Dotted curves calculated from Eq. (10.10).(10)

Equation (10.10) is not susceptible to any simple analytical solution. It must be evaluated by numerical methods. For purposes of calculation, it has been assumed that growth proceeds by means of a three-dimensional nucleation process. Qualitatively similar results are obtained if coherent unimolecular growth nucleation is assumed. The attenuation of the nucleation rate previously described has been utilized. The results are summarized in Fig. 10.7 where $1 - \theta$ is plotted against the reduced time rate variable $(t/\tau^*_{1/2})$, $\tau^*_{1/2}$ being the hypothetical half-time that would be observed if Eq. (9.31a) were obeyed.

It is evident from Fig. 10.7 that noncrystallizing units incorporated into the chain cause major changes in the isotherm shape as compared to those of homopolymers. Deviations from superposable behavior are predicted. They become more pronounced with increasing concentration of the noncrystallizable comonomer and decreasing undercooling. When compared with homopolymers, the calculated isotherms show a retardation in the crystallization process. A characteristic fanning out along the time axis is invariably observed. The dashed curves in Fig. 10.7 represent isotherms calculated from Eq. (9.31a), with $n = 4$ where the nucleation and growth rates are assumed to be invariant with the extent of the transformation. It can be seen that at large values of undercooling copolymer isotherms typical of homopolymer are approached. The isotherms also tend toward superposability in the limit of low levels of crystallinity. The calculations are thus in qualitative accord with the experimental observations for the overall crystallization of random type copolymers.

The extreme sensitivity of the crystallization isotherms to small amounts of irregular structure is demonstrated in Fig. 10.7 for $W_B = 0.01$. Although deviations from Eq. (9.31a) are not observed until $1 - \theta \simeq 0.6$, a small but significant departure from the homopolymer theory develops as the level of crystallinity increases

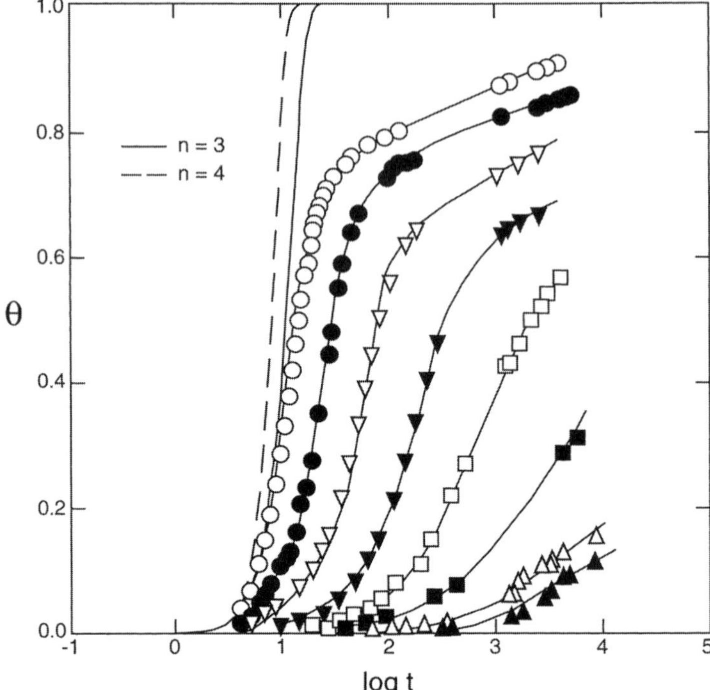

Fig. 10.8 Plot of the relative amount transformed of poly(1,4-butadiene) against log time for indicated crystallization temperature. Data from Fig. 10.1. Crystallization temperatures: ▲ 17 °C; △ 15 °C, ■ 12 °C, □ 9 °C, ▼ 6 °C, ▽ 3 °C, ● 0 °C, ○ −3 °C. Derived Avrami equation: — $n = 3$; − − $n = 4$.

further. Crystallization kinetics thus presents a method by which small amounts of noncrystallizing chain units can be detected.

With this theoretical background it is of interest to analyze, by means of the Avrami formulation, the overall crystallization kinetics of some typical random type copolymers. The results are illustrated in Figs. 10.8, 10.9 and 10.10 for poly(trans-1,4-butadiene) (1), long chain branched polyethylene (2) and an ethylene–octene copolymer (11), respectively. The data for the first two polymers cover the indicated range of crystallization temperatures. Only one crystallization temperature, 115 °C, was studied with the ethylene–octene copolymer. In all three examples illustrated, the early portions of the isotherms fit derived Avrami, with an integral value of n. Quite often when Eq. (9.31a) is used to analyze the data nonintegral values of the exponent n are obtained.(12,13,14) This type of analysis leads to an erroneous interpretation of the basic mechanisms involved. It is often concluded that the nucleation and growth rates change with the extent of the transformation. However, the results can be explained by the fact that the composition of the melt, and the sequence distribution, is continuously changing. The analysis of the ethylene–octene

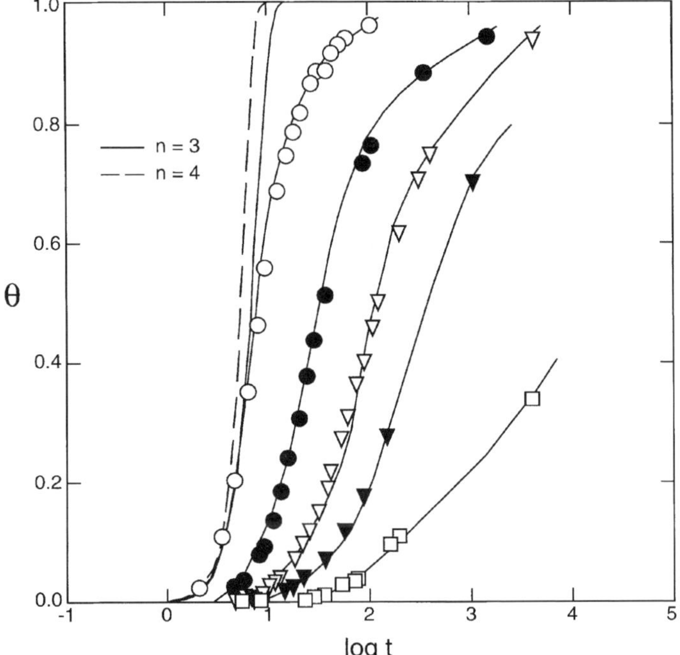

Fig. 10.9 Plot of the relative amount transformed of long chain branched polyethylene against log time for indicated crystallization temperatures. Crystallization temperatures: □ 108.11 °C; ▼ 106.72 °C; ▽ 105.72 °C; ● 104.1 °C; ○ 101.88 °C. Data from Fig. 10.2. Derived Avrami equation: — $n = 3$; – – $n = 4$.

copolymer, given in Fig. 10.10, indicates that the derived Avrami relation, with $n = 3$, holds for better than half of the transformation.

Earlier in this section it was shown that the overall crystallization rates of random copolymers are dependent on molecular weight. This behavior is now examined in more detail by the plots given in Fig. 10.11.(3) The data are for a set of hydrogenated poly(butadienes) with approximately 2.3 mol percent branch points. The time required to develop 10% of the absolute amount of crystallinity, $\tau_{0.10}$, is plotted against the molecular weight. There are several distinguishing features in this figure. Except for the lowest temperatures, the crystallization rate decreases ($\tau_{0.10}$ increases) with increasing molecular weight. This dependence is strong at the highest crystallization temperatures. At the lower ones there are only modest, or negligible, rate changes with molecular weight. These results are qualitatively similar to those observed for linear polyethylene, as well as other homopolymers, as was described in Chapter 9. Linear polyethylene, however, yields a definite rate maximum at the higher crystallization temperatures. At low crystallization temperatures, the maximum is no longer observed in linear polyethylenes and the rate is essentially independent of chain length. The crystallization temperatures at which

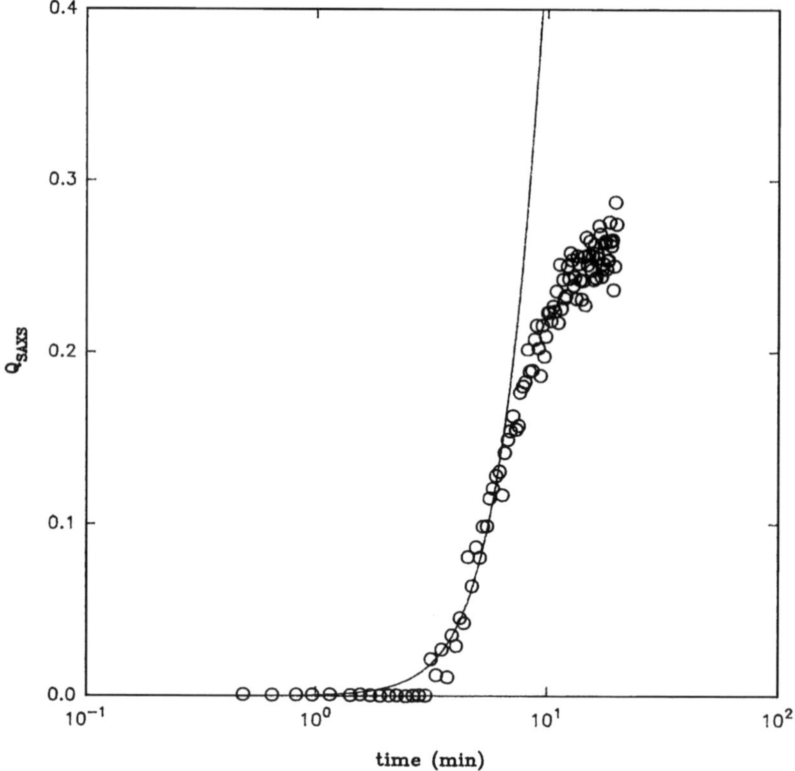

Fig. 10.10 Plot of relative amount transformed for an ethylene–octene random type copolymer, (0.21 mol percent octene) as a function of log time crystallized at 115 °C. (Data from Akpalu *et al.* (11))

the maxima with molecular weight are observed in linear polyethylene correspond to undercoolings of about 20 °C or less. The crystallization of the hydrogenated polybutadienes with about 2.3 mol percent branch points are conducted at undercoolings in the range of 26–43 °C. Maxima would not be observed in linear polyethylene for crystallization carried out at these high undercoolings. The large difference in undercooling appears to be the reason that maxima are not observed for this set of copolymers. In analogy with homopolymer, it can be concluded that the overall crystallization rate of random copolymers depends on molecular weight in a complex manner.

The comonomer concentration also has a profound influence on the crystallization rate. It governs the temperature interval where the observation of crystallinity is practical. Typical behavior is illustrated in Figs. 10.12a and b for hydrogenated poly(butadiene) and hexamethylene adipamide-co-hexamethylene terephthalate respectively.(3,15) Here, the reciprocal of the crystallization rate is plotted as a

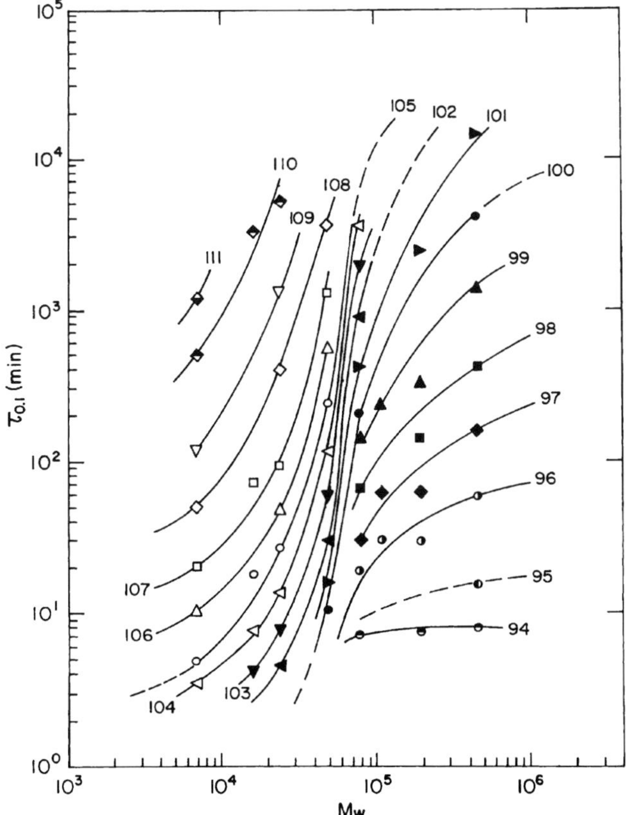

Fig. 10.11 Plot of log crystallization rate ($\tau_{0.1}$) against log M_w for hydrogenated poly(butadienes) with ~2.3 mol percent branch points. Isothermal crystallization temperatures are indicated.(3)

function of the undercooling. The plots illustrate the major changes that take place in the rate, at a fixed undercooling and molecular weight, due only to copolymer composition. The shapes of the curves for either polymer are similar to one another. The curves are displaced along the ΔT axis according to composition. In order to maintain a given crystallization rate a substantial increase in the undercooling is required as the co-unit content increases. For example, in order to maintain $\tau_{0.10} = 10^2$ min for the hydrogenated poly(butadienes) requires a ΔT of 28 °C for the 1.21 mol percent copolymer and a ΔT of 52 °C for the 5.68 mol percent copolymer. Conversely, at a given undercooling the crystallization rate decreases with increasing co-unit content. Examination of Fig. 10.12b indicates that the crystallization rates of the copolyamides behave in a similar manner. It can be expected that other random type copolymers will show similar behavior.

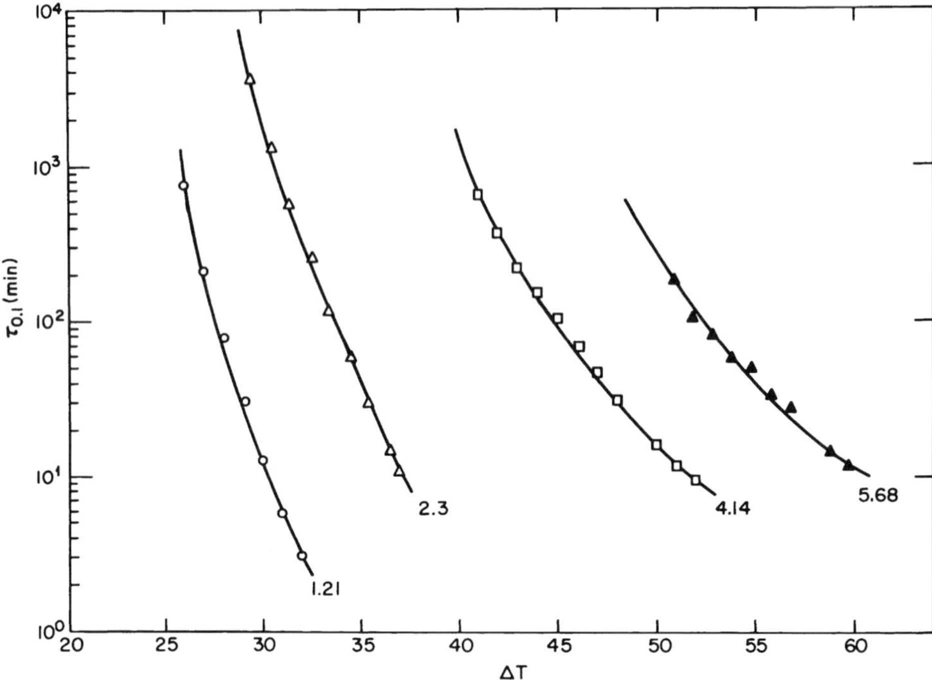

Fig. 10.12a Plot of crystallization rate, $\tau_{0.1}$, as a function of undercooling, ΔT, for copolymers with $M_w \simeq 5 \times 10^4$: ○ ethylene–hexene; △, □, ▲ hydrogenated poly(butadiene). Mol percent branch points indicated.(3)

The plots in Fig. 10.12 emphasize the strong negative temperature dependence in the crystallization rate in the vicinity of the melting temperatures. These results are indicative of nucleation controlled crystallization in this temperature region. Accordingly, we analyze the kinetics in terms of steady-state nucleation theory and the Turnbull–Fisher relation

$$N = N_0 \exp[-E_D/RT - \Delta G^*/RT] \tag{9.118}$$

It is then necessary to assume a model for the nucleus and to calculate ΔG^* for a random copolymer. We take a cylinder as a model for a three-dimensional nucleus. Other geometries could be selected equally well. Except for a constant factor, the basic conclusions that are reached are independent of the geometry that is assumed for the nucleus. In the model selected, the nucleus is composed of ρ polymer chains each of ξ units that are aligned parallel to the length of the cylinder. As in the case of homopolymers, the free energy, ΔG_d, for forming a small crystallite or nucleus is obtained from the expression given by Flory.(16) For a random copolymer comprising A and B units, with only the A units participating in the crystallization,

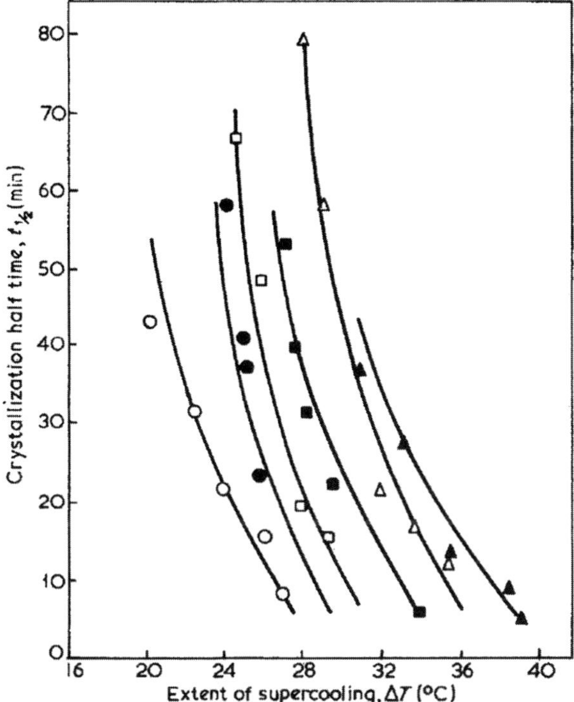

Fig. 10.12b Plot of crystallization rate of poly(hexamethylene adipamide) and its copolymer with hexamethylene terephthalate against undercooling, ΔT. Pure homopolymer ○. Comonomer composition: ● 2 mol%; □ 5 mol%; ■ 10 mol%; △ 15 mol%; ▲ 20 mol%. (From Harvey and Hybart (15))

the free energy, ΔG_d, for forming a three-dimensional nucleus homogeneously is given by

$$\Delta G_\mathrm{d} = 2\xi\sigma_\mathrm{un}\pi^{1/2}\rho^{1/2} - \xi\rho\,\Delta G_\mathrm{u} + \frac{RT}{x}\frac{Z_\mathrm{A}}{\bar{Z}}\xi\rho + 2\rho\sigma_\mathrm{en} - RT\rho\ln\frac{(x-\xi+1)}{x} - RT\rho\xi\ln X_\mathrm{A} \tag{10.11}$$

Here, ΔG_u is the free energy of fusion per repeating unit of an infinite molecular weight homopolymer of A units. In this equation, X_A is the mole fraction of the A structural units, Z_A is the number of segments in an A unit, Z_B is the number of segments in a B unit and $\bar{Z} \equiv Z_\mathrm{A} + (1 - X_\mathrm{A})$ is the average number of segments per unit. The total number of units of both types (A and B) per polymer molecule is given by x; σ_un is the nucleation interfacial free energy of the lateral surface and σ_en is the corresponding free energy for the end surface.

The dimensions of the critical size nucleus, ξ^* and ρ^*, are determined by the saddle point of the surface given by Eq. (10.11), and are expressed as

$$\rho^{*1/2} = \frac{2\sigma_{\mathrm{un}}\pi^{1/2}}{\Delta G_{\mathrm{u}} - RT[(1/x) + 1/(x - \xi^* + 1) - \ln X_{\mathrm{A}}]} \tag{10.12}$$

$$\frac{\xi^*}{2}\left[\Delta G_u - \frac{RT}{x} + \frac{RT}{(x - \xi^* + 1)} + RT \ln X_{\mathrm{A}}\right] = 2\sigma_{\mathrm{en}} - RT \ln\left(\frac{x - \xi^* + 1}{x}\right) \tag{10.13}$$

The free energy change at the saddle point, ΔG^*, is then

$$\Delta G^* = \pi^{1/2}\xi^*\sigma_{\mathrm{un}}\rho^{*1/2} \tag{10.14}$$

Following the same procedure the properties of a Gibbs type two-dimensional, unimolecular, coherent nucleus can also be calculated. For this type of nucleus

$$\Delta G_{\mathrm{d}} = 2\xi\sigma_{\mathrm{un}} - \xi\rho\,\Delta G_{\mathrm{u}} + \frac{RT}{x}\frac{Z_{\mathrm{A}}}{\bar{Z}}\xi\rho + 2\rho\sigma_{\mathrm{en}} - RT\rho \ln\left(\frac{x - \xi + 1}{x}\right) - RT\rho\xi \ln X_{\mathrm{A}} \tag{10.15}$$

and the dimensions of the critical size nucleus become

$$\rho^* = \frac{2\sigma_{\mathrm{un}}}{\Delta G_{\mathrm{u}} - RT[(1/x) + 1/(x - \xi^* + 1) - \ln X_{\mathrm{A}}]} \tag{10.16}$$

and

$$\xi^* = \frac{2\sigma_{\mathrm{e}} - RT \ln[(x - \xi^* + 1)/x]}{\Delta G_{\mathrm{u}} - (RT/x) + RT \ln X_{\mathrm{A}}} \tag{10.17}$$

so that

$$\Delta G^* = 2\xi^*\sigma_{\mathrm{un}} \tag{10.18}$$

In the limit of infinite molecular weight, $x \rightarrow \infty$, Eqs. (10.14) and (10.18) reduce to

$$\Delta G^* = 8\pi\sigma_{\mathrm{en}}\sigma_{\mathrm{un}}^2/(\Delta G_{\mathrm{u}} + RT \ln X_{\mathrm{A}})^2 \tag{10.19}$$

$$\Delta G^* = 4\sigma_{\mathrm{en}}\sigma_{\mathrm{un}}/(\Delta G_{\mathrm{u}} + RT \ln X_{\mathrm{A}}) \tag{10.20}$$

for the two types of nucleation being considered. If $X_{\mathrm{A}} = 1$, the equations reduce to those obtained for homopolymers. The analysis represented by Eqs. (10.11) to (10.18) can again be considered to represent a selection process. It represents the initial step in selecting the minimum length and number of crystallizable sequences that are required to form a nucleus of critical size. If necessary, additional steps can be added to the crystallization process.

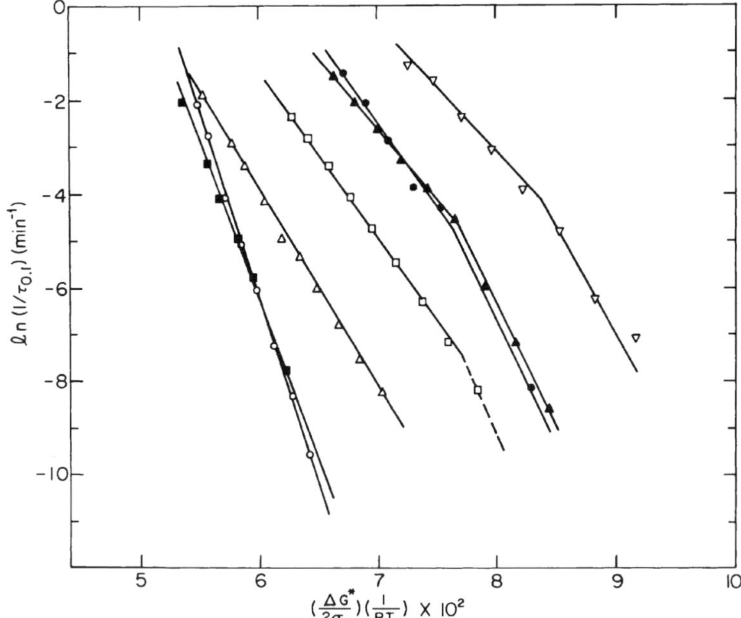

Fig. 10.13 Plot of ln crystallization rate, $1/\tau_{0.1}$, against nucleation temperature function for coherent, unimolecular surface nucleation for hydrogenated poly(butadiene) having ~2.3 mol percent branch points for different molecular weights. Molecular weights: ▽ 6.95×10^3; ▲ 2.4×10^4; ● 1.6×10^4; □ 4.9×10^4; △ 7.9×10^4; ■ 1.94×10^3; ○ 4.6×10^5.(3)

The experimental results for a set of hydrogenated poly(butadienes) serve as good examples with which to test the applicability of the nucleation theory that has been developed. These copolymers have approximately 2.3 mol percent branch points with weight average molecular weights that vary from 6.95×10^3 to 4.60×10^5. A two-dimensional coherent, unimolecular nucleation process has been selected for illustrative purposes. The value of σ_{en} used in the following example is 2000 cal mol^{-1}. Varying σ_{en} over the range of 1000–4000 cal mol^{-1} does not sensibly alter the results. Similar conclusions are reached if a three-dimensional nucleation process is used in the analysis. In analyzing the experimental results $\tau_{0.1}$ is taken as a measure of the crystallization rate. This quantity is plotted against the appropriate temperature function in Fig. 10.13. The latter is taken from Eq. 10.18. T_m^0 is taken as 145.5 °C and $\Delta H_u = 950$ cal mol^{-1}. There are some important features to this figure. The data for the three lowest molecular weights can be represented by two intersecting straight lines. The distinct possibility exists that the fraction $M = 49\,000$ can also be represented in a similar manner. On the other hand, the data for the three highest molecular weights define a single straight line. These results are qualitatively similar to those obtained for molecular weight fractions of linear polyethylene analyzed in a similar manner.(17) The only difference in the results between the two polymer types is the molecular weight range where the change from linearity occurs. For linear polyethylene a single straight line represents the data for fractions $M_w \geq 8 \times 10^5$. The data can be represented by two intersecting straight lines for lower molecular weights. From the point of view of crystallization kinetics the temperature dependence between the two polymer classes is similar to one another. However, the random copolymers appear to behave as much higher molecular weights when compared to the homopolymers. The change in slope between the two straight lines is reminiscent of a Regime I–II transition that was discussed in detail with respect to homopolymer crystallization. The regimes represent asymptotic conditions in the analysis of homopolymer crystallization. The transition from one to the other is not as sharp as appears in Fig. 10.13.

It is also of interest to examine the temperature dependence of this set of copolymers when the co-unit content is varied at a fixed molecular weight. An appropriate set of data are plotted in Fig. 10.14 using coherent, unimolecular nucleation theory in the infinite molecular weight approximation. For comparative purposes data for a linear polyethylene fraction of similar molecular weight are also given. The data for all the copolymers, whose compositions range from 1.21 to 5.68 mol percent

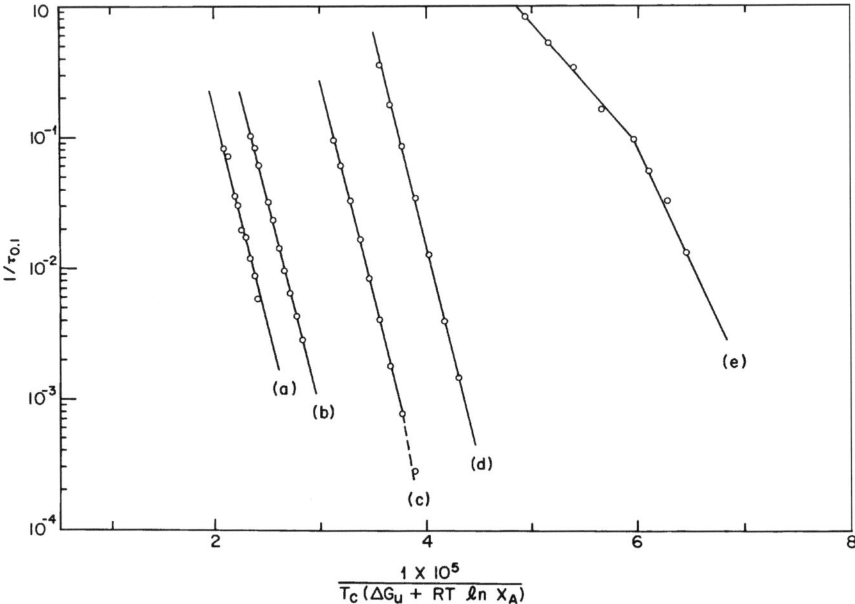

Fig. 10.14 Plot of log crystallization rate, $1/\tau_{0.1}$, against nucleation temperature function for coherent, unimolecular surface nucleation for copolymers with $M_w \simeq 5 \times 10^4$. Hydrogenated poly(butadiene): (a) 5.68 mol% branch points; (b) 4.14 mol% branch points; (c) 2.30 mol% branch points. Ethylene–hexene, 1.21 mol% branch points (d). Linear polyethylene (e).(3)

Table 10.1. *Comparison of slopes from Fig. 10.14*[a]

Mol % branch points	10^{-5} slope, cal mol^{-1}
0.0	4.2 (I)[b]
	2.1 (II)
1.21	7.6
2.30	7.5
4.14	7.6
5.68	8.3

[a] From Ref. (3).
[b] I and II refer to Regimes I and II respectively.

branch points, are well represented by a set of parallel straight lines.[2] The slopes are listed in Table 10.1, along with those for the linear polyethylene. The straight lines are displaced according to composition, reflecting the changing time scale. The slopes of the straight lines are constant within experimental error. Thus, it can be concluded that the product $\sigma_{en}\sigma_{un}$ is independent of copolymer composition. Experiment only yields the product $\sigma_{en}\sigma_{un}$ so that the value of σ_{en} can only be obtained by invoking arbitrary assumptions. If the assumption is made that σ_{un} is essentially independent of copolymer composition then it can be concluded that σ_{en} is also constant independent of composition.

The conclusion that σ_{en} is independent of copolymer composition has very important ramifications. Morphological studies indicate that as the co-unit content increases the lamellar crystallites degrade and eventually become micellar in character. Concomitantly, spherulites become more poorly developed. Eventually, at sufficiently high comonomer content, they do not form at all.[3](18,19) Despite the loss of the lamellar-like crystallite structure the value of σ_{en}, or more properly the product of $\sigma_{en}\sigma_{un}$, remains constant. These results demonstrate that it is not required, or necessary, to relate the chain conformation within the nucleus to that within the mature crystallite. The nucleus is an extremely small entity as compared to the crystallite. The value of σ_{en} only reflects the contribution between the junction of the ordered and disordered sequences, and perhaps a few units beyond the interfacial region of the nucleus. It is not dependent on the nominal copolymer composition. However, the structure of the interfacial region of a mature crystallite can be expected to be different since a relatively large surface area is involved. In this case

[2] Included in this data set are the results for an ethylene–hexene copolymer of similar molecular weight with 1.21 mol percent branch points.
[3] A more detailed discussion of the morphology and structure of random copolymers will be given in Volume 3.

there will be contributions from many sequences, some from the same chain and some from others. Hence, the value of σ_{ec} could very well vary with copolymer composition. There is no requirement that the value of σ_{en}, or the product of $\sigma_{en}\sigma_{un}$, that is deduced from kinetics be the same as that of a mature crystallite.

The conclusion that the value of σ_{en} is independent of whether or not lamellar crystallites are formed is similar to the conclusion reached in analyzing both the growth and overall crystallization rates of high molecular weight n-alkanes (see Sects. 9.14.1 and 9.14.2). In these instances, as well as with low molecular weight fractions of linear polyethylene, the same interfacial free energy for nucleation is involved, irrespective of whether extended or folded chain crystallites are formed. It becomes clear that it is not necessary to postulate that a nucleus is composed of regular folded chains in order to form lamellar-like crystallites.

The analysis of the temperature coefficient data emphasizes the importance of nucleation in the crystallization of copolymers. Equations (10.13) or (10.17) indicate that not all the potentially crystallizable sequences in the untransformed melt can participate in the nucleation act. The thickness, ξ^*, of a critical size nucleus is determined by the copolymer composition and crystallization temperature. Only sequences containing ξ^*, or a larger number of units, can be involved in forming a critical size nucleus. A significant number of chain units, therefore, cannot participate. This limitation on the sequences, and thus the crystallizable units, that can participate in nucleation has important implications for many aspects of the crystallization process. The extent of this limitation is illustrated in Fig. 10.15. Here ξ^*, as calculated from Eq. (10.17), is plotted against the mol percent of branch points,

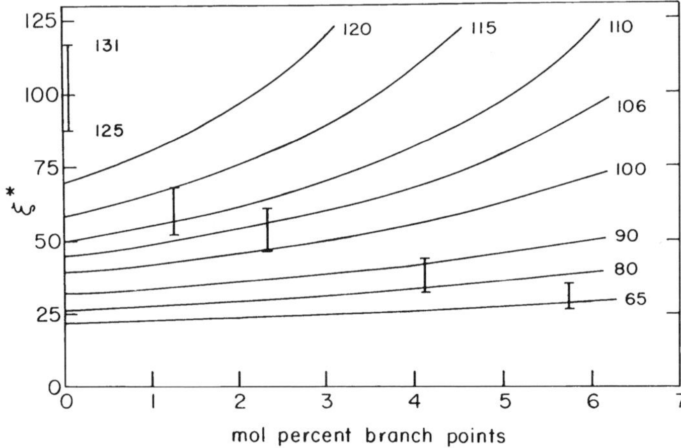

Fig. 10.15 Plot of critical sequence length, ζ^*, for coherent unimolecular surface nucleation against mol percent branch points of hydrogenated poly(butadiene), or similar random type copolymers. Vertical bars represent temperature intervals for isothermal crystallization of each copolymer.(3)

for $M \cong 5 \times 10^4$, at different crystallization temperatures. The parameters used are appropriate to the hydrogenated poly(butadiene).(3) The bars in the figure represent the experimental temperature range for isothermal crystallization of each of the copolymer fractions studied, as well as for linear polyethylene. For illustrative purposes in this calculation we have used a constant value of σ_{en}. It can be deduced from Fig. 10.15 that at low crystallization temperatures, i.e. high undercoolings, the value of ξ^* is small and not sensitive to copolymer composition. Thus at large undercoolings ΔG^* is also small so that the nucleation rate is rapid and independent of copolymer composition. It is comparable to that of the homopolymer under similar conditions. In contrast, at the high crystallization temperatures much longer sequence lengths are required for nucleation and their size increases significantly with co-unit content. Since the effective undercooling also decreases during the isothermal crystallization, this effect will become more profound with the extent of the transformation. Thus, in addition to the equilibrium requirements even more stringent restraints are placed on the sequence length that can actually participate in the nucleation (crystallization) process. Qualitatively similar results are obtained for other modes of nucleation and values of σ_{en}. The Gibbs type nucleus that is used here for illustrative purposes does not form stable crystallites at temperatures infinitesimally above the crystallization temperatures unless either thickening occurs or there is a reduction in the interfacial free energy as the mature crystallite develops. It is difficult to resolve this dilemma for random copolymers.

The value of ξ^* sets a limit on the fraction of the sequences that can participate in the nucleation. This fraction can be calculated irrespective of how they are distributed among the crystallites.(3) The number of sequences, ν, of CH_2 groups, each of which contains a number of units equal to or greater than ξ^*, can be expressed as

$$\nu = \nu_a(1 - X_A)X_A^{\xi^*-1} \tag{10.21}$$

Here ν_a is the number of crystallizable units and X_A is their mole fraction. The fraction of crystallizable sequences, f_c, is found by dividing Eq. (10.21) by the total number of CH_2 sequences. Thus,

$$f_c = \nu/(\text{no. of branches} + 1) \tag{10.22}$$

The results of this calculation are shown in Fig. 10.16, where f_c is plotted against the mol percent branch points for a family of crystallization temperatures. As is illustrated in the figure, the fraction of sequences that can participate in nucleation decreases rapidly with co-unit content at a fixed molecular weight. This factor will significantly retard the crystallization rate and will require a reduction in temperature in order for the transformation to proceed. The experimental temperature range

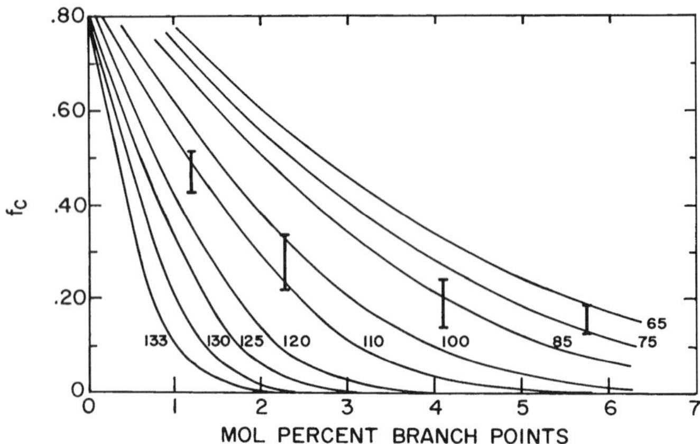

Fig. 10.16 Plot of fraction of sequences, f_c, that can participate in nucleation against mol percent branch points at indicated crystallization temperature in °C.(3)

of the crystallization is also indicated in the figure. For linear polyethylene the corresponding temperature interval is about 120–130 °C for f_c equal to unity. The value of f_c rapidly drops to the 0.2–0.3 range for copolymers with 2.3 mol percent branches and greater. For these copolymer fractions, the crystallization temperature is adjusted accordingly to allow for the fractions of sequences required. The 1.21 mol percent copolymer requires a higher value of f_c. Thus, in effect, f_c establishes the temperature range over which isothermal crystallization can be conducted in a reasonable time frame, i.e. from minutes to days.

The same factors of the melt structure, such as entanglements and other topological structures, that are important in homopolymer crystallization will also affect copolymers. However, for random copolymers the ξ^* requirement, and the concentration of eligible sequences, dominates the crystallization kinetics.

The isotherms of the copolymers with the lower co-unit content initially follow, with but minor exceptions, the Avrami or Goler–Sachs formulation with $n = 3$. In this respect they are similar to homopolymers of modest molecular weight 10^4–10^6. However, for the higher co-unit copolymers n changes from 3 to 2. Electron microscopy studies have shown that in this range of copolymer composition lamellar-like crystallites are no longer formed.(19,20) This is another example of the value of the exponent n reflecting the crystallite growth.

The role of sequence selection during the isothermal crystallization of random type copolymers can be monitored by means of differential scanning calorimetry. (21–24) The evolution of the melting endotherms with the time of crystallization shows some unique features. An example of this development is shown in Fig. 10.17 for a hydrogenated poly(butadiene) with 2.3 mol percent branch points crystallized at 89.8 °C.(24) In this experiment the fusion process was initiated from the

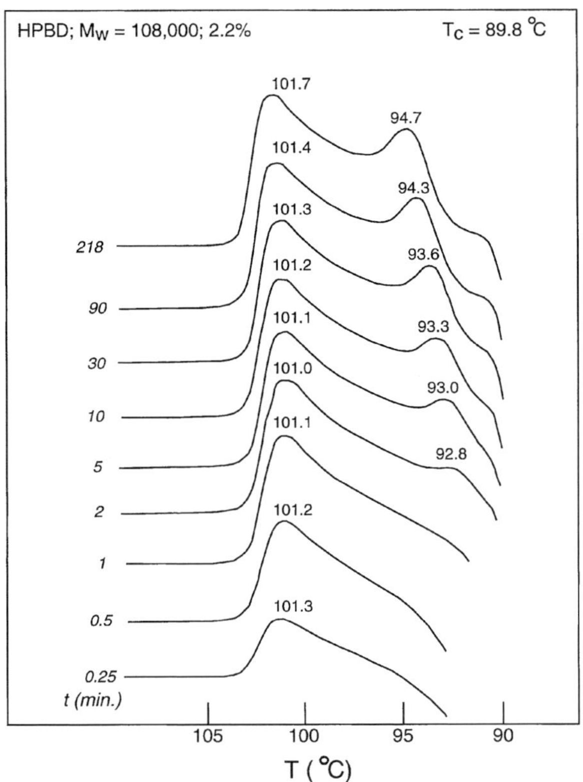

Fig. 10.17 Thermograms from differential scanning calorimetry for a hydrogenated poly(butadiene), 2.2 mol percent branch points; $M_w = 108\,000$, crystallized at 89.8 °C. Times of crystallization are indicated, as are the peak temperatures of the two endotherms.(24)

isothermal crystallization temperature, without any cooling. Initially, at this crystallization temperature, only a single broad endothermic peak, centered at 101 °C is observed on fusion. However, with time, as the crystallization progresses another peak develops, which is centered at about 92.8 °C. The intensity of this peak increases with the crystallization time and is shifted to higher temperatures. The temperature increase is 2 °C in this example. In contrast, the peak of the higher melting endotherms only increases about 0.4 °C.

The results for a similar type of experiment, where the crystallization temperature is increased to 100.8 °C are shown in Fig. 10.18. In this case only a single, broad endotherm is observed, even after crystallization for more than 4000 minutes. The endothermic peak increases by only 0.8 °C over this time span. Figure 10.19 illustrates the increase of the respective endothermic peaks for the copolymer crystallized at different temperatures. The curve describing the results for crystallization at the highest temperature only shows a slight increase in the endothermic

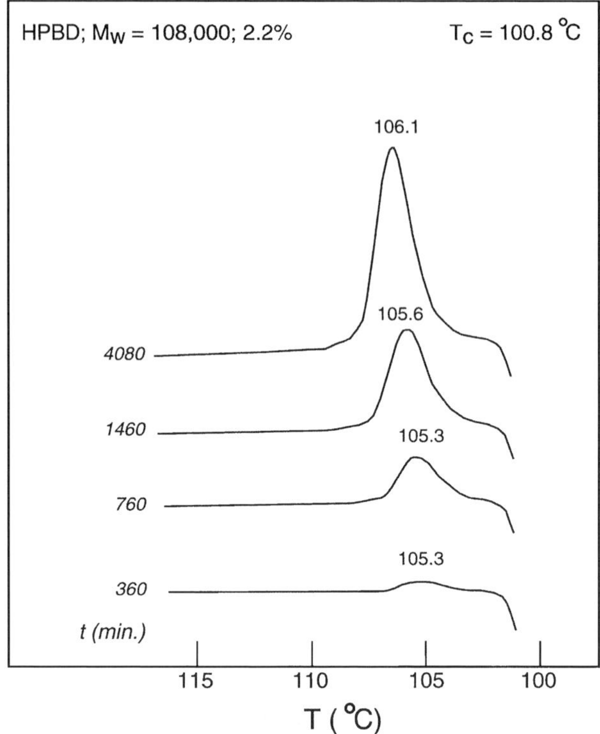

Fig. 10.18 Thermograms from differential scanning calorimetry for a hydrogenated poly(butadiene), 2.3 mol percent branch points; $M_w = 108\,000$, crystallized at 100.8 °C. Times of crystallization are indicated, as are the peak temperatures of the endotherms.(24)

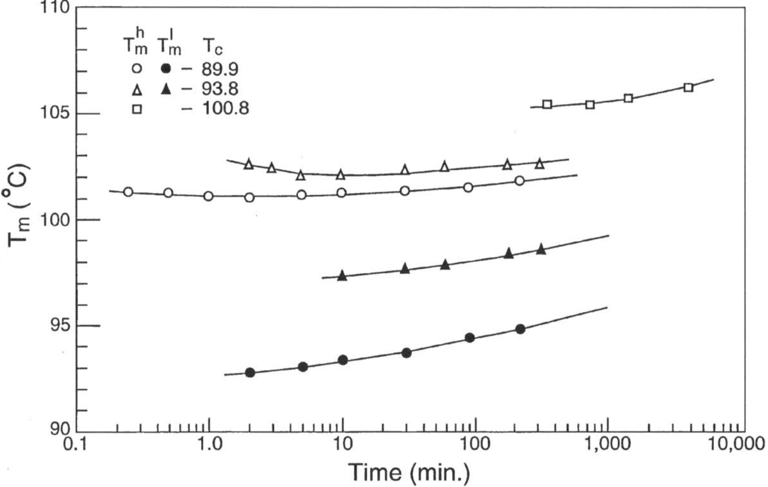

Fig. 10.19 Plot of melting temperature against time after isothermal crystallization, at indicated temperatures, of a hydrogenated poly(butadiene), 2.3 mol percent branch points, $M_w = 108\,000$.(24)

peak in the time interval studied. However, it does portend further increases with longer crystallization time. The results for the lowest crystallization temperature show a substantial increase in the position of the low melting endotherm, but only a slight change in the higher one. It has been demonstrated that the two peaks are a consequence of isothermal crystallization. They are not due to melting or partial melting and subsequent recrystallization.(22,23)

The development of multi-peaks after isothermal crystallization has also been observed in ethylene–1-alkanes and other ethylene copolymers.(24–28) This phenomenon can be expected in all types of random copolymers. The development of the low temperature endotherm has been termed secondary crystallization.(28) It obviously bears no relation to the secondary crystallization processes that have been attributed to the crystallization of pure homopolymers. The development of the low temperature peak does not manifest itself directly in other overall crystallization studies.(3) The isothermal development of two endothermic peaks has also been observed during the crystallization of certain high melting polymers.(29–32) The low temperature peak found in this group of polymers has been attributed to the crystallization of constrained sequences. There is the suggestion, based on the results for the random copolymers, that structural irregularities, such as long branches among others, may be incorporated in these chains.

The development of two endothermic peaks during the isothermal crystallization of random type ethylene copolymers raises the question as to why at least two broad crystallite populations form. The answer lies in the recognition of the sequence distribution of the initial melt inherent to this class of copolymers. Depending on the copolymer composition and crystallization temperature only sequences equal to or greater than a critical length can participate in the nucleation and thus the crystallization. The crystallization rates within the allowable set of sequences will depend on the temperature. At the high crystallization temperatures, only the longer sequences will have an undercooling sufficiently high that they can crystallize at a reasonable rate. However, the undercooling for the shorter sequences will be low, retarding their crystallization over a reasonable time scale. Thus, at the higher crystallization temperature only a single endotherm is observed. As the crystallization temperature is lowered, the effective undercooling is increased thus allowing for the crystallization of the short sequences, albeit at a more protracted rate. The nature of the endotherms and their dependence on the crystallization temperature can be explained in a qualitative manner by focusing attention on the initial sequence distribution. The increase in the melting peaks of both the high and low temperature endotherms can be attributed to the changing sequence distribution in the melt; in particular the removal of the shorter sequences to the crystalline state. It does not necessarily follow that because the melting temperature of a random copolymer increases that the crystallites are thickening.(26) The chemical potential of the melt is of equal importance with that of the crystallite state in determining the melting temperature. Changes in the sequence distribution in the residual melt will alter its chemical potential.

In summary, certain features of the overall crystallization kinetics of copolymers are similar to those of homopolymers, while in other respects there are some fundamental differences. A major controlling factor of both the equilibrium and kinetics of copolymer crystallization is the chain microstructure. What is important is the sequence distribution of the crystallizable units. The importance of the distribution of sequence lengths of the crystallizing species in all aspects of copolymer crystallization cannot be overemphasized. Even on an equilibrium basis, not all the crystallizable sequences can participate in the crystallization process. Equilibrium theory makes clear that only sequences that exceed a certain critical length (not to be confused with the critical sequence length for nucleation) can participate in the crystallization at a given temperature. This requirement explains the broad melting range characteristic of random copolymers and the large decrease in the equilibrium level of crystallinity with co-unit content.(33) Although the concern here is with kinetic processes, the equilibrium requirements serve as a necessary bound. In addition there is a further restriction on the fraction of sequences that can participate in forming a nucleus of critical size. This restraint, which can be severe, depends on the undercooling at which the crystallization is conducted and the co-unit concentration. This is a major factor that severely limits the level of crystallinity that can be actually attained relative to the theoretical equilibrium expectation. These considerations also contribute with rare exception to the retardation of the crystallization rate with increasing comonomer content, at a constant undercooling, when there is only one specific crystallization.(33a)

The deviations that are observed from either the free-growth or Avrami relations can be attributed in part to the general problem encountered in homopolymer crystallization, i.e. the role of chain entanglements. In addition, there is a major contribution to the deviations due to the decreasing availability of eligible sequences as the transformation proceeds. This is due to the decrease in the undercooling at constant temperature. As a consequence, in contrast to homopolymer crystallization, copolymer isotherms are not superposable. Deviations from theory are observed at much lower levels of crystallinity, although the same basic type of nucleation is involved with both homopolymers and copolymers. Nucleation catalysts influence copolymer crystallization in a similar manner to that of homopolymers.(33b)

The isotherm shapes that are characteristic of random copolymers give clear indication of a nucleation and growth type transformation taking place, despite the fact that the isotherms do not superpose. In addition, the unique temperature coefficient in the vicinity of the melting point requires the adoption of appropriate nucleation theory. However, the general concept of nucleation and growth governing the crystallization of polymers in the bulk has been seriously questioned and an alternative proposed.(34–37) The concept proposed is to be primarily motivated by electron microscopic observations that are complemented by small-angle x-ray determinations of the crystallite thicknesses upon both crystallization and fusion. It is more appropriate to discuss this concept in Volume 3, where the relation between structure, morphology and properties of copolymers will be discussed in Chapter 16.

10.2.2 Random copolymers: spherulite growth rates

Spherulites are frequently observed in random copolymers. However, as in homopolymers, spherulite development is not a universal mode of copolymer crystallization. The molecular weight, chain microstructure and crystallization temperature establish the boundary for spherulite formation in random copolymers.(38–40) Important limitations are imposed on the study of spherulite growth. Quantitative studies of spherulite growth rates, using optical microscopy, are by necessity limited to relatively low levels of crystallinity. There are, however, certain advantages in studying spherulite growth rates in random copolymers as long as the limitations are recognized.

Studies with many different random type copolymers have shown that the rate of change of the spherulite radius with time is constant.(41–43) In these cases the concentration of comonomer is low and the level of crystallinity that is attained is small. Thus, the anticipated nonlinearity, due to the changing composition and sequence distribution with the isothermal transformation, is not always observed. It is apparent from Fig. 10.6 that under these circumstances deviations from homopolymer type growth will be small. However, for higher comonomer contents, and longer crystallization times, nonlinear growth rates are observed. An example is given in Fig. 10.20 for an ethylene–vinyl acetate copolymer that contains 6.0 mol percent of comonomer and about 1 mol percent of long-chain branches.(44) Initially the growth rate of this copolymer is constant but it slowly decreases as the transformation progresses. The decrease in rate can in this case be attributed to the changing sequence distribution and composition of the melt.

The dependence of the spherulitic growth rate on co-unit content and molecular weight follows the pattern that was established by the studies of the overall crystallization rate.(3) For example, the growth rates of the ethylene–vinyl acetate copolymers decrease by several orders of magnitude, at a fixed value of the nucleation temperature function $T_{\rm m}^0/T\,\Delta T$, as the comonomer content increases from that of the homopolymer to the 6 mol percent copolymer.(44) Studies with molecular weight and composition fractions of random ethylene–1-octene copolymers are more extensive and show similar features.(8) An example is given in Fig. 10.21 where ln G is plotted against ΔT for molecular weight fractions in the range $M_{\rm w} = 17$–23×10^3

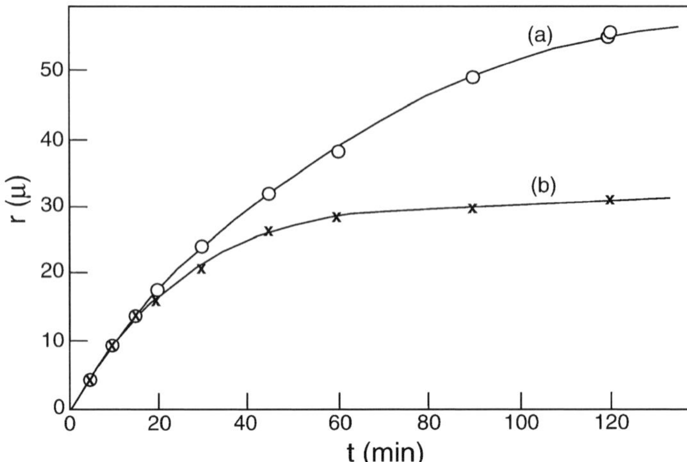

Fig. 10.20 Plot of spherulite growth rates against time for an ethylene–vinyl acetate copolymer that has 6.0 mol percent comonomer. Curves (a) and (b) represent two different directions. (From Nachtrab and Zachmann (44))

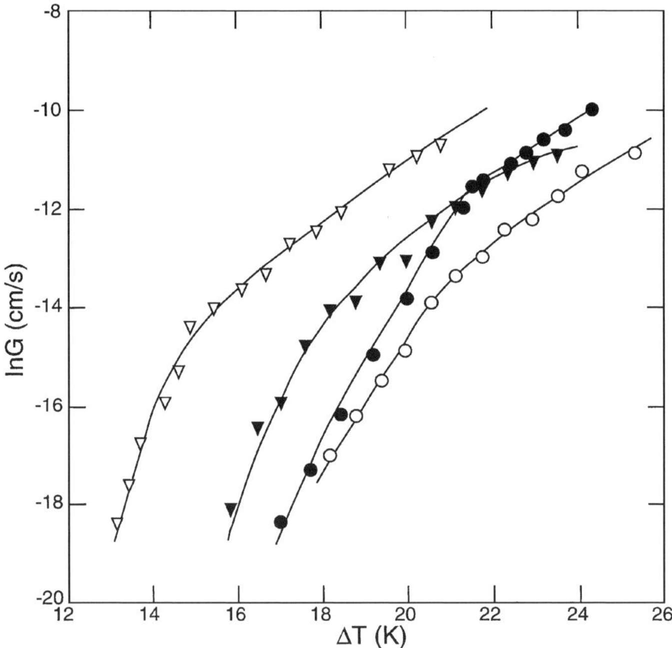

Fig. 10.21 Plot of ln G against ΔT for linear polyethylene and a set of ethylene–octene copolymers of comparable molecular weight and varying comonomer content. ● Linear polyethylene. Copolymers: ○ $M_w = 23\,600$, 0.42 mol% branches; ▼ $M_w = 18\,120$, 1.1 mol% branches; ▽ $M_w = 17\,500$, 2.2 mol% branches. (Data from Lambert and Phillips (8))

that have different comonomer contents.(8) There is a two to five order of magnitude reduction in the growth rate, depending on the value of ΔT, between the homopolymer and the copolymer with 2.2 mol percent branch points. Similar behavior is found in other molecular weight ranges.(8)

The expression that was used to analyze the temperature dependence of the growth rate of homopolymers can also be applied to copolymers. Therefore, in the infinite molecular weight approximation, well removed from the glass temperature,

$$G = G_0 \exp\left[-\frac{E_\mathrm{D}}{RT} - \frac{g_2 T_\mathrm{m}^0}{T\,\Delta T}\right] \qquad (9.173)$$

When applied to copolymers, T_m and ΔT refer to the appropriate composition of the random copolymer. The growth rates for three different molecular weight ranges of random ethylene–1-octene copolymers, each with varying comonomer content, are plotted according to Eq. (9.173) in Figs. 10.22a, b and c.(8) The T_m values are calculated from ideal equilibrium theory with the crystalline phase being pure. In each of the figures the growth rate for the complementary homopolymer is also given. The data for some of the fractions can be represented by a single straight line. In other cases, the data can be fitted to two intersecting straight lines, indicating that the simple nucleation theory being used does not apply.

The fact that different portions of the data can be represented by two straight lines of different slopes is suggestive of regime behavior involving the relations between the nucleation and spreading rates. The temperature interval for crystallization is narrow, ranging only from 6 °C to 10 °C at the most, and is similar to that of the linear polyethylenes. The plots for the low and intermediate molecular weights, Figs. 10.22a and b, are suggestive of Regimes I and II asymptotes. However, the slope ratios range from 1.5 to 2.1 for the two intersecting straight lines drawn in Fig. 10.22a. Thus, for the lowest molecular weights the slope ratios are not constant, nor is the I–II transition sharp. The data for three of the four copolymer fractions shown in Fig. 10.22b are well represented by single straight lines. The other two, in this intermediate molecular weight region, can be represented by two intersecting straight lines. These slope ratios are 2.2 and 2.3 indicating that a I–II Regime transition might be taking place. The three fractions that are represented by straight lines have either the highest co-unit content or a molecular weight that is significantly higher than the others. These results are reminiscent of the overall crystallization rate results of the high molecular weight linear polyethylene fractions. The fraction in the highest molecular weight group (Fig. 10.22c) that can be represented by a straight line also represents the larger molecular weight. A general rule appears to be developing that, insofar as growth rates are concerned, as the chain length and comonomer content increase the ethylene copolymers behavior resembles that of much higher molecular weight linear polyethylenes.

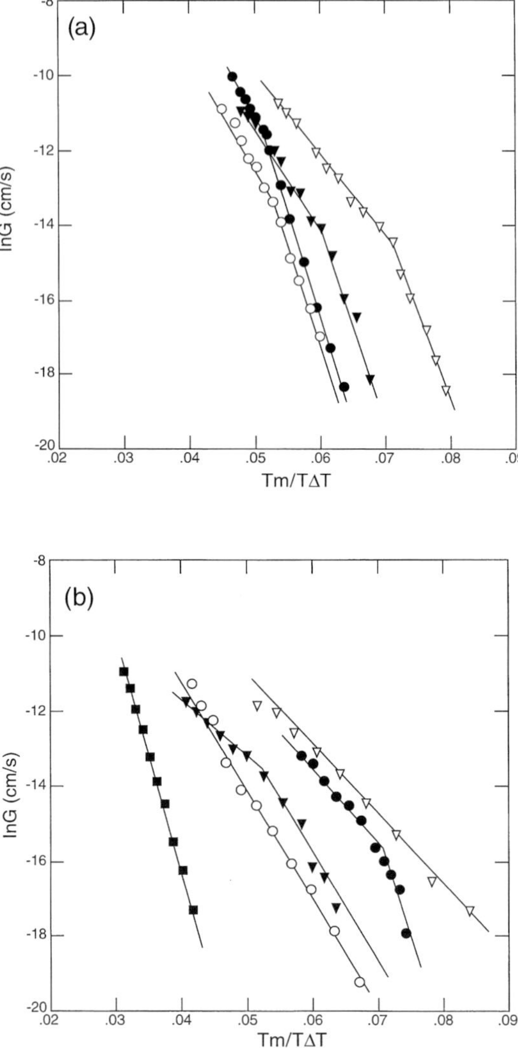

Fig. 10.22 Plot of ln G against $T_m/T\,\Delta T$ for a set of ethylene–octene copolymers of similar molecular weight and varying comonomer content. (a) ● linear polyethylene $M_w = 18\,120$. Copolymers: ○ $M_w = 23\,600$, 0.42 mol% branch points; ▼ $M_w = 18\,600$, 1.1 mol% branch points; ▽ $M_w = 17\,500$, 2.2 mol% branch points. (b) ● linear polyethylene $M_w = 51\,500$. Copolymers: ○ $M_w = 89\,900$, 0.67 mol% branch points; ▼ $M_w = 54\,700$, 1.36 mol% branch points; ▽ $M_w = 41\,800$, 1.55 mol% branch points; ■ $M_w = 65\,500$, 3.2 mol% branch points. (c) Copolymers: ● $M_w = 174\,700$, 0.29 mol% branch points; ○ $M_w = 117\,500$, 0.55 mol% branch points; ▼ $M_w = 104\,000$, 1.11 mol% branch points. (Data from Lambert and Phillips (8))

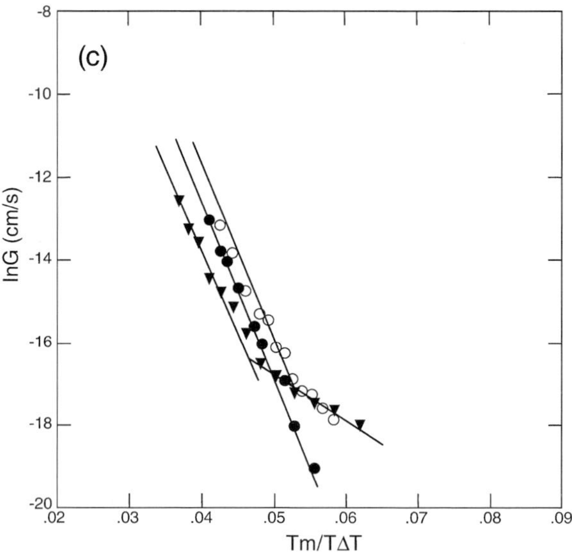

Fig. 10.22 (*cont.*)

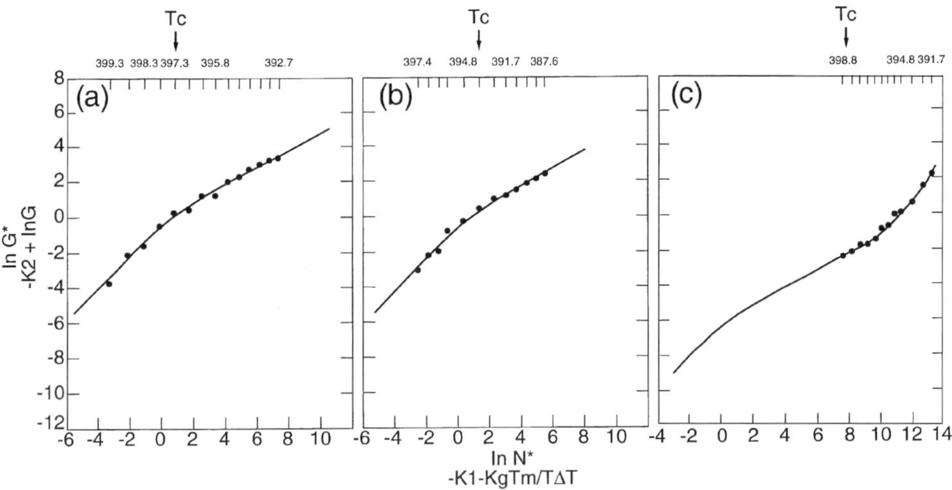

Fig. 10.23 Plot of ln G^* against ln N^*, according to Eqs. (9.203) and (9.204), for ethylene– octene copolymers. (a) $M_w = 18\,600$, 1.1 mol% branch points; (b) $M_w = 54\,700$, 1.36 mol% branch points; (c) $M_w = 117\,500$, 0.55 mol% branch points. (Data from Lambert and Phillips (8))

The data can be analyzed more exactly by applying Frank's continuum theory following the procedure used for homopolymers.(45,46) The results of such an analysis are shown in Fig. 10.23 for three typical fractions, one from each of the molecular weight groups. For the low and intermediate molecular weight fractions, curves (a) and (b), the asymptotes of regimes I and II are attained at low and high crystallization temperatures, respectively. There is a diffuse transition between the two regimes as was found in homopolymers. The plot for the highest molecular weight, curve (c), shows that crystallization occurs in Regime II at the highest crystallization temperatures. However, at the lower crystallization temperatures the theory no longer holds. At these temperatures there is probably a major retardation in the spreading rate and eventually only extremely rapid nucleation takes place. The change is very diffuse, and is the cause of the high slope ratios for the plots in Fig. 10.22c.

The product of the interfacial free energies, $\sigma_{un}\sigma_{en}$, for the copolymers can be determined from the slopes of the appropriate straight lines in the different regimes. The average value of $\sigma_{un}\sigma_{en}$ for the copolymers in the two lowest molecular weight groupings is about 1200 erg^2 cm^{-4}. This value is comparable to those found for linear polyethylene fractions. This product of the interfacial free energies is slightly greater than the 1000 erg^2 cm^{-4} found for the copolymers in the highest molecular weight group. Small changes in the equilibrium melting temperatures can account for this difference. To emphasize once again, these values are for nucleation and not for the mature crystallites.

The specific influence of copolymer composition and molecular weight on the growth rate is examined in Fig. 10.24, where different sets of copolymers are plotted,

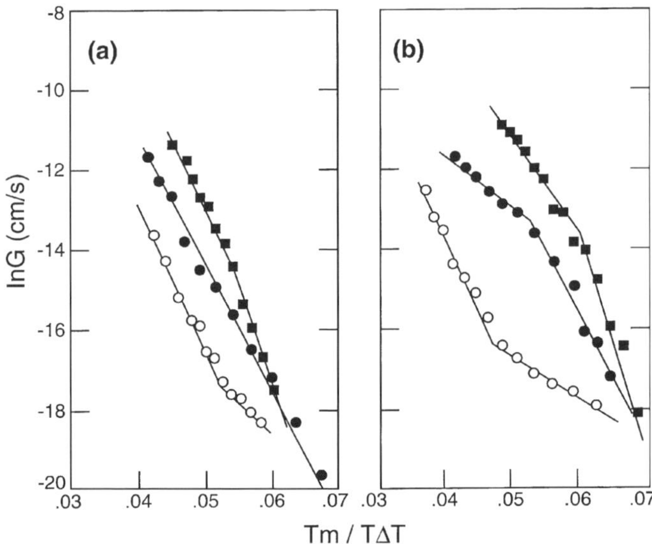

Fig. 10.24 Plot of ln G against $T_m/T\Delta T$ for ethylene–octene copolymers of comparable concentration of branch points with varying molecular weights. (a) ● M_w = 89 900, 0.67 mol% branch points, ○ M_w = 117 500, 0.55 mol% branch points; ■ M_w = 23 600, 0.42 mol% branch points. (b) ○ M_w = 104 000, 1.10 mol% branch points; ● M_w = 54 700, 1.36 mol% branch points; ■ M_w = 18 600, 1.11 mol% branch points. (Data from Lambert and Phillips (8))

so chosen that their comonomer contents are close to one another but their molecular weights vary. One such set is given in Fig. 10.24a where the fraction of branch points is about 0.5 mol percent and the molecular weights vary from 24 000 to 118 000. The branch point content is about 1.1 mol percent for the copolymer set given in Fig. 10.24b and the molecular weights are in the same range, 19 000 to 104 000. In both examples there is a significant decrease in the growth rate with molecular weight at any given value of $T_m/T\,\Delta T$. The decrease in rate is more marked for the higher co-unit content copolymers. These results are qualitatively similar to those found for the overall crystallization rates of the hydrogenated poly(butadienes) (3) and linear polyethylene above a critical molecular weight (Fig. 9.97).

In a unique and interesting set of experiments the spherulite growth rates of a linear polyethylene and a set of random type ethylene–octene copolymers, of varying comonomer content, were studied at very large undercoolings.(47) It was found that at these undercoolings the highest growth rates of the copolymers merged with one another and with that of the linear polyethylene. This result is expected, based on the plots in Fig. 10.15. At the large undercoolings in question, ζ^* is theoretically expected to be essentially independent of copolymer composition. Thus, ΔG^*, the steady-state nucleation rates, and the growth rates of the parent homopolymer and the copolymers will be close to one another. This conclusion follows from basic nucleation theory as applied to random copolymers. The convergence of growth rates at large undercooling does not represent a major breakdown of theories that describe polymer crystallization.(47)

The dependence of the spherulite growth rate on copolymer composition can be calculated from first principles. For illustrative purposes, a coherent unimolecular type nucleation is assumed. The spherulite growth rate in the vicinity of the equilibrium melting temperature can then be written as

$$G^{\mathrm{C}} = G_0^{\mathrm{C}} \exp\left\{\frac{-E_{\mathrm{D}}^{\mathrm{C}}}{RT} - \frac{\Delta G^{*,\mathrm{C}}}{RT}\right\} \tag{10.23}$$

where the superscript C refers to a random copolymer. For the parent homopolymer

$$G^{\mathrm{H}} = G_0^{\mathrm{H}} \exp\left\{\frac{-E_{\mathrm{D}}^{\mathrm{H}}}{RT} - \frac{\Delta G^{*,\mathrm{H}}}{RT}\right\} \tag{10.24}$$

where the superscript H refers to the homopolymer. It is reasonable to assume that $E_{\mathrm{D}}^{\mathrm{H}} \simeq E_{\mathrm{D}}^{\mathrm{C}}$ and $G_0^{\mathrm{H}} \simeq G_0^{\mathrm{C}}$ so that

$$\ln G^{\mathrm{C}} - \ln G^{\mathrm{H}} = \frac{4b_0\sigma_{\mathrm{en}}\sigma_{\mathrm{un}}}{RT\,\Delta G_{\mathrm{u}}^{\mathrm{H}}}\left[1 - \frac{\Delta G_{\mathrm{u}}^{\mathrm{H}}}{\Delta G_{\mathrm{u}}^{\mathrm{C}}}\right] \tag{10.25}$$

or

$$\ln G^{\mathrm{C}} - \ln G^{\mathrm{H}} = \frac{2b_0\sigma_{\mathrm{un}}\zeta^{*,\mathrm{H}}}{RT}\left[1 - \frac{\Delta G_{\mathrm{u}}^{\mathrm{H}}}{\Delta G_{\mathrm{u}}^{\mathrm{C}}}\right] \tag{10.26}$$

It has been assumed that the respective interfacial free energies for nucleation associated with the homopolymers and copolymers are the same. The product of interfacial free energies has been shown to be essentially independent of composition for a given set of copolymers. However, their identification with that of the parent homopolymers is a serious assumption. In Eq. (10.26) $\Delta G_{\mathrm{u}}^{\mathrm{H}}$ and $\Delta G_{\mathrm{u}}^{\mathrm{C}}$ are the free energies of fusion per repeating unit of the homopolymers and copolymers respectively, and $\zeta^{*,\mathrm{H}}$ is defined as ζ^* at the crystallization temperature. This development is predicated on the assumption that the crystalline phase remains pure. Recalling that

$$\Delta G_{\mathrm{u}}^{\mathrm{C}} = \Delta G_{\mathrm{u}}^{\mathrm{H}} + RT\ \ln(1 - X_{\mathrm{B}}) \tag{10.27}$$

and expanding the logarithm for small values of X_{B}, the mole fraction of noncrystallizing units, it follows that

$$\ln G^{\mathrm{C}} = \ln G^{\mathrm{H}} - \left(\frac{2b_0\sigma_{\mathrm{un}}\zeta^{*,\mathrm{H}}}{\Delta G_{\mathrm{u}}^{\mathrm{C}}}\right) X_{\mathrm{B}} \tag{10.28}$$

Thus, based on standard nucleation theory a linear relation between $\ln G^{\mathrm{C}}$ and X_{B} is to be expected at a constant value of $\Delta G_{\mathrm{u}}^{\mathrm{C}}$. Equation (10.28) can be recast in terms of $\zeta^{*,\mathrm{c}}$ as

$$\ln G^{\mathrm{C}} = \ln G^{\mathrm{H}} - (b_0\,\sigma_{\mathrm{un}}\zeta^{*,\mathrm{H}}\zeta^{*,\mathrm{c}}/\sigma_{\mathrm{en}})X_{\mathrm{B}} \tag{10.29}$$

since

$$\zeta^{*,\mathrm{C}} = \frac{2\sigma_{\mathrm{en}}}{\Delta G_{\mathrm{u}}^{\mathrm{C}}} \tag{10.30}$$

At large undercoolings Eq. (10.29) can be approximated by

$$\ln G^{\mathrm{C}} \cong \ln G^{\mathrm{H}} - K\zeta^{*,\mathrm{C}}X_{\mathrm{B}} \tag{10.31}$$

At a fixed crystallization temperature $\zeta^{*,\mathrm{C}}$ will depend on X_{B} as indicated in Fig. 10.15 for random copolymers of ethylene. At low crystallization temperatures and large undercoolings, $\zeta^{*,\mathrm{C}}$ will be essentially constant over a relatively large range in comonomer content. At intermediate crystallization temperatures ζ^* increases slowly with composition. Over a small range in co-unit content the relation is linear. At the highest crystallization temperatures, the increase in $\zeta^{*,\mathrm{C}}$ with X_{B} is more pronounced, but is still linear over a small interval in X_{B}.

Andrews *et al.* have postulated that the growth rate is attenuated in direct proportion to the probability of finding sequences of lengths $\zeta^{*,\mathrm{C}}$ or greater.(43) With this postulate, it is found that

$$\ln G^{\mathrm{C}} \cong \ln G^{\mathrm{H}} - \zeta^{*,\mathrm{C}} X_{\mathrm{B}} \tag{10.32}$$

which is of the same form as Eq. (10.31), but with $K = 1$. Experimental data for several systems (8,43) follow the linear relation expected from Eqs. (10.31) or (10.32) for small values of X_{B}. However, when the slopes are analyzed with $K = 1$, serious difficulties arise in rationalizing the apparent values of $\zeta^{*,\mathrm{C}}$ that are deduced.

10.3 Block or ordered copolymers

Since the sequence propagation parameter, p, of a block copolymer differs from that of a random type, the crystallization kinetics can be expected to be different. Since $p \rightarrow 1$ for a block copolymer, the kinetics should approach that of the homopolymer. There are, however, several important mitigating factors. The melt, from which crystallization ensues, in a block copolymer possesses unique features that could influence the kinetics. The melt can be either homogeneous, in the conventional sense, or heterogeneous. When heterogeneous, or phase separated, the melt could be comprised of one of several different domain structures, such as cylinders or spheres, that were described in Chapter 5 (Volume 1). The strength of the segregation plays an important role in the microphase separation and domain structure. During the crystallization of a block copolymer there is a competition between maintaining the domain structure, such as spheres, cylinders, lamellae and gyroids, or developing the conventional alternation of crystalline and amorphous layers. The segregation strength is important in this regard. Thus, the different possible structures in the initial melt could influence the course of the crystallization and thus the morphology and properties. When one block of a diblock copolymer does not crystallize, the question as to whether the state of the noncrystallizing block, glass or rubber-like, influences the crystallization kinetics needs to be addressed. If both blocks crystallize then the role of the already crystallized block on the crystallization of the other one is important. For triblock copolymers the location of the crystallizing block could be important. All of these possibilities, and perhaps others, must be kept in mind in the analysis of the crystallization kinetics of block copolymers.

Since block copolymers can be synthesized in a variety of chain architectures, we consider first diblock copolymers in which only one block can crystallize. An example of the overall crystallization kinetics of a diblock copolymer, atactic poly(styrene)–poly(ethylene oxide), is illustrated in Fig. 10.25.(48) Here the relative fraction transformed is plotted against log time, at different crystallization temperatures. Only the poly(ethylene oxide) block, $M_{\mathrm{n}} = 9900$, crystallizes

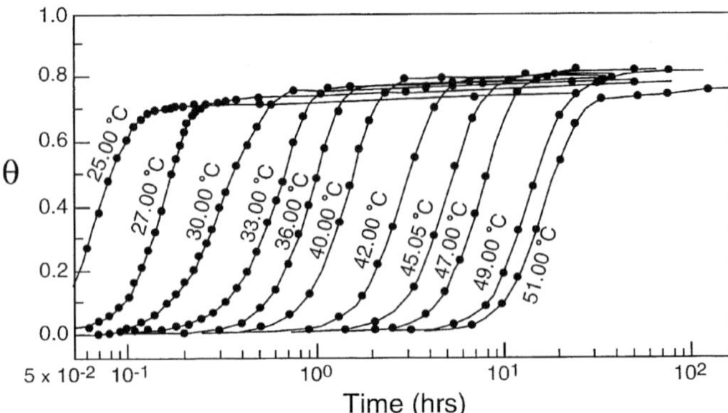

Fig. 10.25 Plot of relative fraction transformed against log time for block copolymer of poly(ethylene oxide) and atactic poly(styrene). (From Seow *et al.* (48))

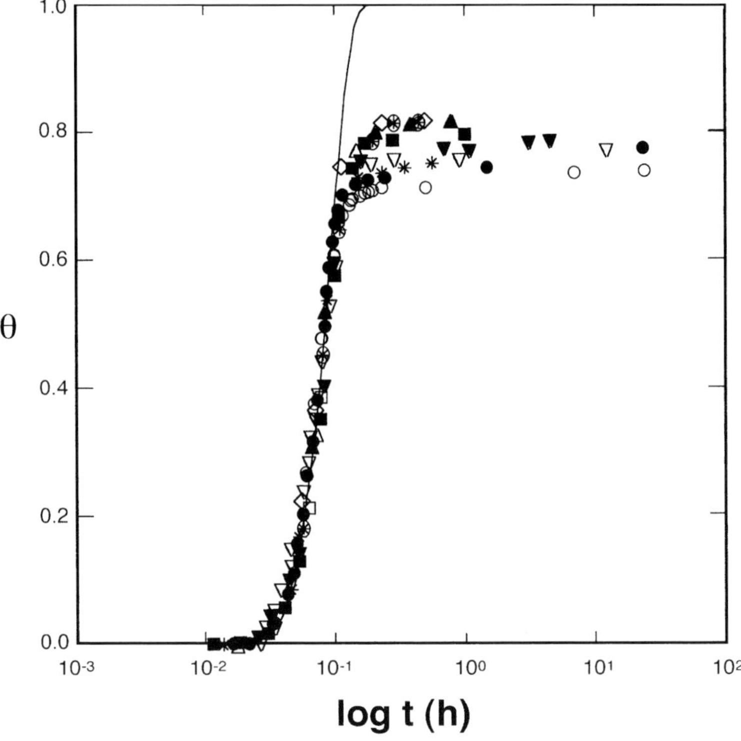

Fig. 10.26 Demonstration of superposability with isotherms of Fig. 10.25.

in this copolymer. The isotherm shapes are similar to those of the corresponding homopolymer. The superposition of the isotherms is readily apparent, as is illustrated in Fig. 10.26. The solid curve drawn in the figure represents the derived Avrami equation with $n = 3$. Good agreement between experiment and theory is attained for about 70–80% of the transformation. This behavior is virtually identical to that of the corresponding homopolymer. The fact that very long sequences of the crystallizing component are involved in both the homopolymer and block copolymer is the reason for the similarity in kinetics. A similarity in isothermal shapes between block copolymer and homopolymer is found in other copolymers as well.(49–52) This reflects the fact that the Avrami exponent is maintained. In general the crystallization rate, overall or spherulitic growth, decreases in a block copolymer relative to that of the parent homopolymer.(50,53–55) Within this generalization there are interesting specific situations. Consequently we consider in somewhat more detail the influence of the noncrystallizing block, i.e. whether it is rubber-like or glassy, on the crystallization kinetics. Also of interest is whether the microdomains in the initial melt are weakly or strongly segregated and the influence of the domains on the crystallization.

When the noncrystallizable block in a diblock copolymer is rubber-like the isotherm shapes are very similar to those of the parent homopolymers.(55,56) This situation exists even when the crystallization occurs from a well-defined melt structure.(55,57,58) However, at a fixed undercooling, there is a reduction in the overall crystallization and spherulite growth rates.(55) When the growth rates of ethylene oxide–butadiene block copolymers, and the corresponding homopolymer, are plotted against $1/\Delta T$ it is found, with the exception of the lowest content ethylene oxide polymer, that a set of parallel straight lines results irrespective of the initial melt domain structure.(55) This result implies that the products of interfacial free energies for nucleation are similar to one another.

The situation is quite different when the noncrystallizing component is a glass. When T_g of the noncrystallizable block is greater than T_m of the crystallizable block there is the potential for the crystallization to be confined to local domains. Although many interesting morphological situations can evolve, the interest at this point is focused on the crystallization kinetics in the confined space. Some typical results are given by the block copolymer tetrahydrofuran–styrene, where the poly(styrene) is the glassy component.(54) An important finding is the fact that the isotherm shapes, as manifested by the Avrami exponent n, depend on the composition of the crystallizing component. For example, when the concentration of the crystallizing block is 59% by volume, the isotherm shape, described by $n \simeq 2$, is the same as that of the parent homopolymer. In contrast, when the tetrahydrofuran is reduced to 29%, crystallinity could not be developed. At intermediate compositions, 51.5 and 38%, the n values are very small, ~0.5. It is evident that except for the highest tetrahydrofuran content, the crystallization is subject to severe constraints.

Another example of confined crystallization is illustrated by the block copolymer hydrogenated poly(butadiene)–poly(vinyl cyclohexane).(59) The glass temperature of poly(vinyl cyclohexane) is about 145 °C, well above the melting temperature of the hydrogenated poly (butadiene). Depending on the molecular weight and chain length of the crystallizing component, well-defined domain structures form in the melt. In turn these structures are maintained in the crystalline state. The final level of crystallinity that is attained does not depend on the initial mesophase geometry. However, the time scale, or crystallization rate, is dependent on the initial domain structure. When the crystalline blocks form a continuous matrix, as gyroids and cylinders of high composition, the overall crystallization rate is essentially that of the pure parent homopolymer. However, when the hydrogenated poly(butadiene) domains are not spatially continuous, as lamellar or low composition cylinders, a greater degree of supercooling is necessary in order for the copolymer to reach the same level of crystallinity as the homopolymer and aforementioned copolymers.

The differences in time scale, relative to the initial domain structures, are also reflected in the detailed kinetics, the isotherm shapes and the Avrami exponents.(59a) When there is connectivity between the domain structures, the usual sigmoidal shaped isotherm results. However, when the crystalline block is confined to a specific domain, first-order crystallization kinetics result. This corresponds to a derived Avrami equation with $n = 1$. The significance of these results will be discussed shortly in terms of other findings.

Interesting information can be obtained when the melt structure is examined in detail. In particular, it is convenient to describe the nonhomogeneous melt as two categories, weakly and strongly segregated microdomains.[4] When weakly segregated, the microdomains, or meso-phases, are destroyed by the crystallization process. Examples of this type of behavior have been found in diblocks of ethylene–ethylene *alt* propylene,(60) ethylene–ethylene ethylene,(55,57) ethylene–3-methyl-1-butene,(61) caprolactone–butadiene,(58) oxyethylene–oxybutylene (62) as well as others.[5](63) In these cases, irrespective of the initial microdomain structure, the kinetic data are well represented by Avrami type isotherms. Typically the Avrami exponent is 3.0 for an appreciable extent of the transformation as is illustrated with ethylene–ethylene ethylene diblocks.(57) Thus, when weakly segregated, the microdomains characteristic of the initial melts are overwhelmed, and have no effect on the crystallization. Conventional lamellar-like crystallites are formed even when the mesophase in the melt is not lamellar.

The situation with strongly segregated systems is quite different. In this case, the crystallization is confined to the individual domains, whose dimensions are the order of nanometers.(64,65) As a consequence, the crystallization kinetics are significantly different from the usual Avrami type. In a pioneering study, Loo *et al.*

[4] The segregation power is defined by $\chi_{12}N$ when χ_{12} is the Flory–Huggins interaction parameter and N is the overall chain length.

[5] The block that is usually termed ethylene is really a hydrogenated poly(butadiene), i.e. ethylene–butene random copolymer.

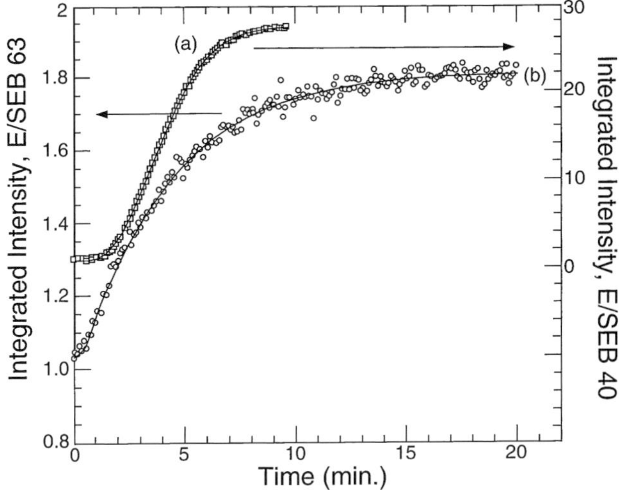

Fig. 10.27 Plot of time course of integrated intensity of small-angle x-ray scattering for (a) block copolymer hydrogenated poly(butene)–styrene-ethylene-butene crystallized at 67 °C and (b) hydrogenated poly(butadiene) crystallized at 95 °C. (Adapted from Loo *et al.* (64))

investigated the crystallization of a diblock of hydrogenated poly(butadiene) with a noncrystallizing terpolymer component of styrene-ethylene-butene, termed SEB63. In this case the noncrystalline component is rubber-like.(64) Small-angle x-ray scattering indicates that the melt of this particular copolymer is composed of spherical microdomains. Transmission electron microscopy showed that, after isothermal crystallization, the hydrogenated poly(butadiene) component crystallized as spheres of regular size and spacing. The strong segregation in the melt confined the crystallization of the copolymer to 25 nm spheres. In turn, the crystallization kinetics reflects this constraint. A subsequent study extended the molecular weight range and composition of the crystallizing block in this type of copolymer.(65) This procedure allowed for the development of spherical and cylindrical domains with different order–disorder transition temperatures. In general, confining crystallization to individual microdomains drastically affects the crystallization kinetics.

Figure 10.27 gives plots of the time course of the integrated small-angle x-ray scattering during isothermal crystallization of the diblock copolymer and of the parent crystallizing component, hydrogenated poly(butadiene).(64) The isotherm for the hydrogenated poly(butadiene) is typical of a nucleation–growth transformation. It can be represented in the customary manner by the appropriate derived Avrami relation for this random type copolymer.(3) The crystallization kinetics of the diblock copolymer is quite different. It is well described by a simple exponential

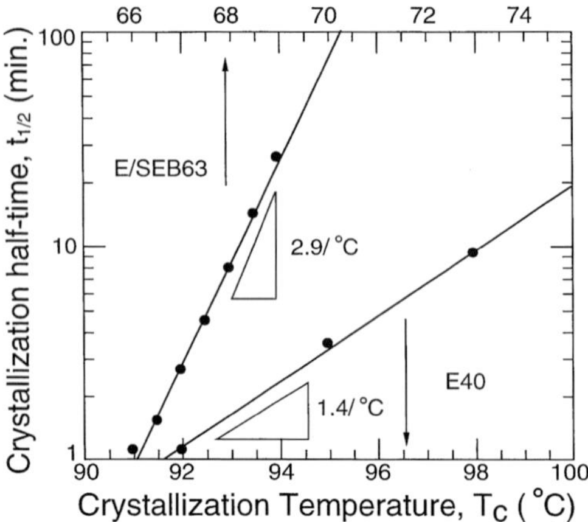

Fig. 10.28 Plot of isothermal crystallization half-time against crystallization temperature for block copolymer (E/SEB63) and hydrogenated poly(butadiene) (E40). (From Loo *et al.* (64))

decay, or first-order process. This corresponds to a derived Avrami equation with $n = 1$. Such first-order kinetics indicates that the rate of isothermal crystallization is proportional to the fraction of spheres that have yet to crystallize. This result is consistent with the crystallization being confined to individual microdomains. This result is suggestive of homogeneous nucleation and the droplet experiments discussed in Chapter 9. The reason is that the number of microdomains far exceeds the number of heterogeneities, or impurities, in the sample.(64) Thus, large undercoolings can be achieved. Because of the size of the spheres, crystal growth from the nucleation to the microdomain boundary is essentially instantaneous. Consequently only nucleation rate is determined for the crystallization. The extent of the undercooling that can be achieved in the block copolymer relative to that of the parent crystallizable polymer is illustrated in Fig. 10.28. Here, the relation between the crystallization half-time and the crystallization temperature for both the copolymer and parent polymer are given. For example, at a half-time of 10 min the undercooling is about 37 °C and 7 °C for the block copolymer and parent polymer respectively. These undercoolings are comparable to those found in homogeneous and heterogeneous droplet experiments. There is thus strong evidence that the crystallization is constrained in this system and that the nucleation process is homogeneous. In principle, experiments at lower undercooling would allow for an accurate determination of the product of interfacial free energies. These results are consistent with and can explain the kinetic results obtained with the ethylene–vinyl cyclohexane block copolymer that were described above.

Detailed studies of the ethylene–styrene–ethylene–butene copolymers has allowed for a more quantitative discrimination between weakly and strongly segregated copolymers.(65) The microdomains were spherical in one set of copolymers,

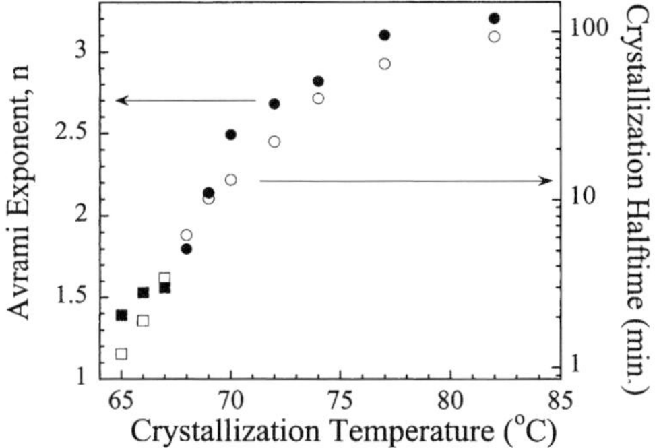

Fig. 10.29 Plot of Avrami exponent n and crystallization half-time against the isothermal crystallization temperature for the diblock copolymer ethylene–styrene-ethylene-butene $M_n = 35\,000$, weight fraction ethylene 0.14. Squares represent data acquired by synchrotron radiation, circles, data aquired by laboratory-based Kratky camera. (From Loo *et al.* (65))

and the segregation strength was varied through molecular weight control at constant copolymer composition. The order–disorder transition temperatures varied from 200 °C to greater than 300 °C for those copolymers. The crystallization kinetics of a copolymer having cylindrical domains was also investigated.

The crystallization kinetics is shown to reflect the segregation strength. An example of the key kinetic parameters is given in Fig. 10.29 for one of the blocks with spherical domains, M_n 35 000, that has an order–disorder transition temperature of 200 °C. The Avrami exponent is greater than 3 at the high crystallization temperatures and approaches 1 at the lower ones. Transmission electron micrographs indicate that the crystallization is confined to the spherical domains in this case. In contrast, at the higher temperature the domain structure is destroyed upon crystallization.(65) The higher molecular weight, strongly segregated sample, with spherical domains, exhibited confined crystallization under all conditions, even when the crystallization took several hours to complete.(64,65) The boundary between confined crystallization and so-called breakout crystallization was found to depend on the ratio of the normalized segregation strength during crystallization to that of the segregation strength at the order–disorder temperature. The ratio $(\chi N_t)_c/(\chi N_t)_{ODT} \simeq 3$ represents the boundary between confined and breakout crystallization and the concomitant first-order or sigmoidal kinetics respectively.

The block copolymers with cylindrical microdomains also show confined and breakout crystallization and the expected crystallization kinetics. There is, however, an intermediate region for this class of copolymer. In this region small-angle x-ray scattering patterns indicated that the crystallization was confined to the cylindrical domains, yet sigmoidal type kinetics resulted.(65) This intermediate rate region is defined approximately by $1.5 < (\chi N_t)_c/(\chi N_t)_{ODT}$. Here, so-called templated crystallization occurs in that the cylindrical domains in the melt guide the growing crystals but do not completely confine them. As observed by electron microscopy, so-called "rogue" crystals connect different cylinders. This allows a large volume of material to be crystallized from a single nucleus, resulting in sigmoidal type crystallization kinetics. Since the rogue crystals are infrequent the small-angle x-ray patterns are indistinguishable from the situation where the crystallization is confined to the cylinders within which the nucleus was originally formed.

There are many interesting aspects to the crystallization of diblock copolymers. The nonhomogeneity of the melt and the influence of the microdomain structures in the effects of the crystallization process on the resulting properties offer many areas for study. The ability, by means of constrained crystallization, to achieve classical homogeneous nucleation offers the possibility to further explore and expand nucleation theory as applied to long chain molecules. Strongly segregated diblock copolymers offer ideal systems with which to study homogeneous nucleation.

There have not been any reports of first-order crystallization kinetics in triblock copolymers.(59,66–72) This however does not mean that strongly segregated triblock copolymers do not exist. There is evidence for strong segregation that manifests itself in its ability to attain large supercoolings, and by implication homogenous nucleation and first-order kinetics.(66–72) As examples, in the block copolymers ethylene oxide–isoprene–ethylene oxide (66) and ethylene oxide–styrene–ethylene oxide (69) supercoolings 60–90 °C greater than that of the comparable homopolymer, poly(ethylene oxide), are easily achieved. Similar undercoolings are attained in other systems.(70–72) In all of these examples the actual isotherms were not determined. Thus, the type of kinetics, sigmoidal or first-order, is not known. However, it is implied.

When the data for the triblock vinyl cyclohexane–hydrogenated butadiene–vinyl cyclohexane is compared with the corresponding diblock copolymers, from 4 to 15 additional degrees of supercooling are needed to develop appreciable crystallinity.(50) However, when this difference in supercooling is taken into account the role of the microdomain structure in governing the isothermal crystallization is similar for the triblock copolymer as previously described for the diblock polymer.

The triblock copolymer styrene–butadiene–ϵ-caprolactone, the last being the crystallizable block, serves as example of a weakly segregated system.(67) Although both the overall crystallization and spherulite growth rates are reduced relative to the crystallizing homopolymer, the temperature ranges for isothermal crystallization are approximately the same for the copolymer and homopolymer. Consequently, the Avrami exponent is only slightly reduced from that of the parent homopolymer.

The overall crystallization and spherulite growth rates have been studied for several segmental, or multiblock copolymers.(73–76) As a rule, with but few exceptions, they crystallize at much slower rates than the corresponding crystallizing homopolymer with a molecular weight corresponding to the block length. Decreasing the concentration of the crystallizing component also reduces the rates. The spherulite growth rates of multiblock copolymers of poly(tetramethylene-*p*-silphenylene siloxane), (TMPS)–poly(dimethyl siloxane) have been investigated over an extensive range of temperatures and compositions.(74) The results are given in Fig. 10.30 and illustrate the influence of composition on the growth rate. Growth rate maxima, similar to those found in the homopolymer, are observed with the copolymers that contain smaller comonomer concentrations. Although no definite maxima are observed as the poly(dimethyl siloxane) concentrations increase above 20%, the plots give clear indication that they would be found if the studies were carried out at still lower temperatures. The crystallization range is reduced as the dimethyl siloxane content is increased. Concomitantly, the maximum growth rate decreases smoothly and monotonically with the addition of comonomers. Other segmented block copolymers behave in a similar manner in that the maximum is maintained, but at a reduced value.(76)

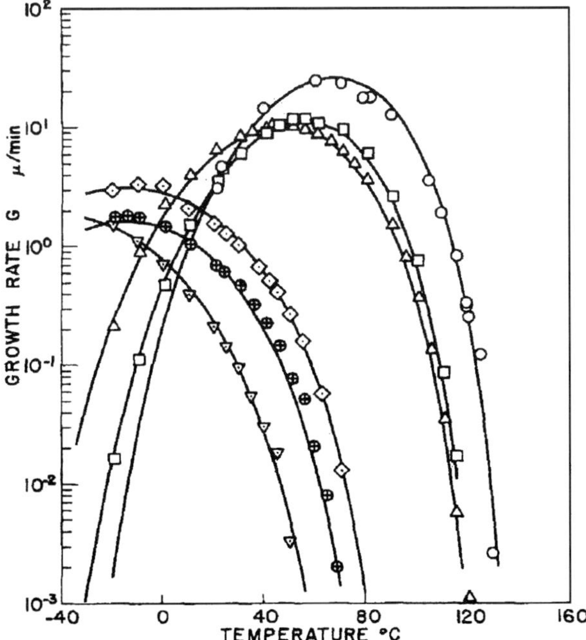

Fig. 10.30 Plot of spherulite growth rates of TMPS homopolymer and TMPS–dimethyl siloxane copolymers as a function of temperature for different copolymer compositions. ○ homopolymer; □ 90/10; △ 80/20; ◇ 50/50; ⊕ 40/100; ▽ 30/70. (From Li and Magill (74))

The glass temperatures of this set of copolymers vary from −20 °C for the homopolymer to −105 °C for the 30/70 copolymer. The extrapolated equilibrium melting temperatures range from 110 °C to 160 °C respectively over the same composition range.(74) Thus, the necessary information is available with which to compare rate data utilizing Eq. (9.209). The objective is to ascertain whether the parameters involved, U^* and C, follow a systematic pattern with copolymer composition. All of the data shown in Fig. 10.30 can be fitted quite well by Eq. (9.209) using arbitrary parameters. The best fits are illustrated in Figs. 10.31a, b, and c for the 90/10, 50/50 and 30/70 copolymers respectively. The other copolymers follow a similar pattern. Despite the good fit that is obtained there is no systematic pattern to the values of the parameters needed. For example, values of U^* are found to be in the range 5046 to 1124, while C varies from 122 to 28. There is, however, no relation between the parameters needed and copolymer composition. In another approach to fitting the data, U^* and C were fixed at 1500 and 30, parameters that were thought to be universal over time. Following this procedure respectable fits are found with the homopolymer and the 90/10 copolymer. The agreement, however, is poor for the higher poly(dimethyl siloxane) content copolymers. It is not surprising that good fits could be obtained over the complete composition and crystallization temperature range by the arbitrary selection of U^* and C values for each copolymer.

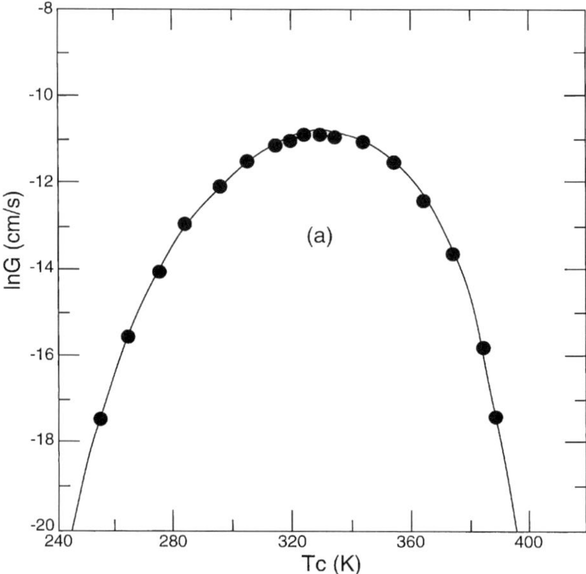

Fig. 10.31 Plot of ln G against crystallization temperature for TMPS–dimethyl siloxane copolymers. Curve calculated from Eq. (9.209) with arbitrary parameters. Solid points experimental results. Copolymer composition: (a) 90/10; (b) 50/50; (c) 30/70. (Data from Li and Magill (74))

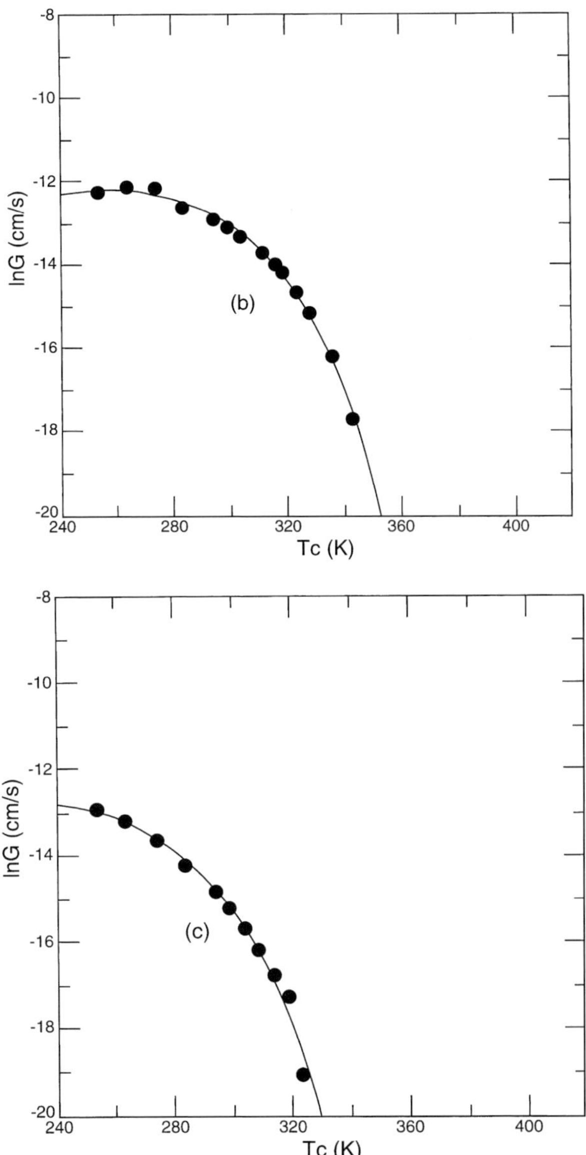

Fig. 10.31 (*cont.*)

The influence of block length on the spherulite growth rate can be found in a set of urethane linked poly(ethylene oxide) block copolymers.(75) The copolymers consisted of uniform block length of either 34, 45 or 90 repeating units with total molecular weights that varied from several thousand to 3–6 $\times$ 10^4. The 34 repeating unit blocks (M = 1500) always crystallized in extended form. In contrast the 90 repeating unit blocks (M = 3900) crystallized in a folded structure at all

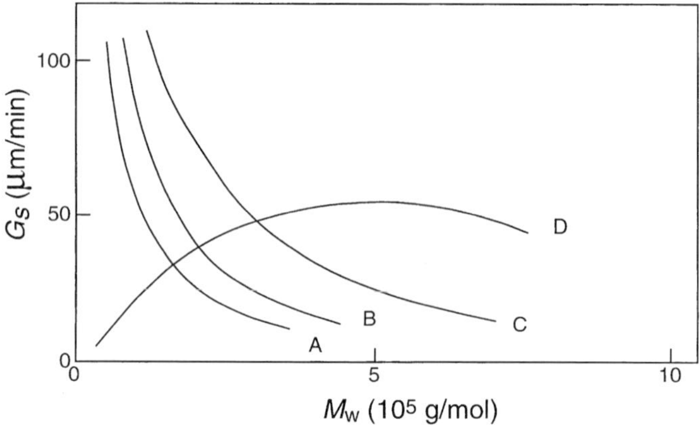

Fig. 10.32 Plot of spherulite growth rate against weight average molecular weight, $\bar{M}_w$, for poly(ethylene oxide) D and segmented block copolymers A, B, C. Block length in repeating units: (A) 34; (B) 45; (C) 90. (From Friday and Booth (75))

crystallization temperatures. The copolymers containing 45 repeating units in the block ($M = 2000$) crystallized in either an extended or folded form, depending on the crystallization temperature. The influence of polymer molecular weight on the spherulite growth rates of the block copolymers, as compared with the homopolymer crystallized under similar conditions, is illustrated in Fig. 10.32. In contrast with the homopolymer, the growth rates of the three block copolymers decrease monotonically with increasing molecular weight. The copolymers with the larger block lengths crystallize more rapidly than the others, at a fixed molecular weight. The block copolymers are of modest molecular weight, yet the dependence of their growth rate on chain length is typical of homopolymers of higher molecular weights.

A detailed analysis of the temperature coefficient of the spherulite growth of these copolymers cannot be made because of the limited range of crystallization temperatures and the uncertainty in their equilibrium melting temperatures. However, some qualitative observations can be made.(77) There is no significant change in the product $\sigma_{en}\sigma_{un}$ with molecular weight for each series of polymers of fixed block length. The interfacial free energy products of the copolymers with different block lengths are close to one another. The fact that either extended or folded mature crystallites form in one series does not manifest itself in the analysis of the temperature coefficient, i.e. in the interfacial product for nucleation.

It has already been noted that when crystallized over an extended temperature range, the crystallization rates of both random and block type copolymers display rate maxima, similar to those of homopolymers. A compilation of the ratios of the temperature of maximum growth rate, T_{max}, to the equilibrium melting temperature of the copolymer, T_m, is given in Table 10.2. For a given copolymer the ratio

Table 10.2. *Spherulite growth rate maxima in copolymers*

Polymer	Comonomer	Mol %	T_{max}/T_m	Reference
Poly(oxymethylene)	dioxolane	4	0.90	a
		7	0.90	
	1,3-dioxane	1	0.89	
		3	0.89	
		6	0.90	
	dioxepane	3	0.89	
		6	0.89	
		9	0.89	
	epichlorohydrin	2	0.89	
		5	0.90	
		7	0.90	
Poly(ethylene adipate)	hexamethyl diisocyanate	5	0.85	b
	diphenylmethane diisocyanate	5	0.87	
	1.5 naphthalene diisocyanate	5	0.88	
	2,4-toluene diisocyanate	8	0.82	
		5	0.89	
		3	0.86	
		2	0.84	
Poly(ethylene terephthalate)	azelate	6	0.77	c
		11	0.76	
		25	0.75	
		31	0.72	
Poly(tetramethylene *p*-silphenylene siloxane)	dimethyl siloxane	0	0.79	d
		10	0.77	
		20	0.76	
Copolyimides[a]		0	0.91	e
		20	0.86	
		40	0.82	
		60	0.82	
		80	0.88	
		100	0.87	
Poly(β-hydroxy butyrate)	hydroxy propionate	16	0.84	f
		18	0.87	
	hydroxy valerate	7	0.79	g
		6	0.78	
	hydroxy valerate	12	0.78	h
	hydroxy valerate	77	0.80	i

(*cont.*)

Table 10.2. (*cont.*)

Polymer	Comonomer	Mol%	T_{max}/T_m	Reference
	hydroxy valerate	8	0.83	j
		19	0.87	
		34	0.89	
		55	0.87	
		71	0.88	
		82	0.88	
		95	0.88	
	stereo isomers	92[b]	0.84	k
		88	0.87	
		76	0.89	
Poly(β-hydroxyoctanoate)[c]		—	0.82	l
Poly(L-lactide)		3[d]	0.85	m
		6	0.85	

[a] See Ref. (e) for structural formula
[b] Percent isotactic diads
[c] See Ref. (j) for composition of this terpolymer
[d] Percent D isomer

Reference

a. Inoue, M., *J. Polym. Sci.* **8**, 2225 (1964).
b. Onder, K., R. H. Peters and L. C. Spark, *Polymer*, **18**, 155 (1977).
c. Jackson, J. B. and G. W. Longman, *Polymer*, **10**, 873 (1969).
d. Li, H. M. and J. H. Magill, *J. Polym. Sci.: Polym. Phys. Ed.*, **16**, 1059 (1978).
e. Hsiao, B. S., J. A. Kreuz and S. Z. D. Cheng, *Macromolecules*, **29**, 135 (1996).
f. Cao, A., M. Ichikawa, K. Kasuya, N. Yoshie, N. Asakawa, Y. Inoue, T. Doi and H. Abe, *Polymer J.*, **28**, 1096 (1996).
g. Organ, S. J. and P. J. Barham, *J. Mater. Sci.*, **26**, 1368 (1991).
h. Akhtar, S., C. W. Pouton and L. J. Notarianni, *Polymer*, **33**, 117 (1992).
i. Pearce, R. P. and R. H. Marchessault, *Macromolecules*, **27**, 3869 (1994).
j. Scandola, M., G. Ceccorulli, M. Pizzoli and M. Gazzano, *Macromolecules*, **25**, 1405 (1992).
k. Abe, H., I. Matsubara, Y. Doi, Y. Hori and A. Yamaguchi, *Macromolecules*, **27**, 6018 (1994).
l. Gagnon, K. D., R. C. Fuller, R. W. Lenz and R. J. Farris, *Rubber World*, November 1992, p. 32.
m. Huang, J., M. S. Lisowski, J. Runt, E. S. Hall, R. T. Kean, N. Buehler and J. S. Lin, *Macromolecules*, **31**, 2593 (1998).

T_{max}/T_m is independent of the comonomer introduced. The ratio for the different copolymers ranges from about 0.75 to 0.90. These values are similar to those of the homopolymer, and for essentially the same reason.

To conclude this section we compare some aspects of the crystallization of three model polyethylenes of different chain structures.(78) The different molecular architectures are: a linear copolymer, hydrogenated poly(butadiene); a hydrogenated

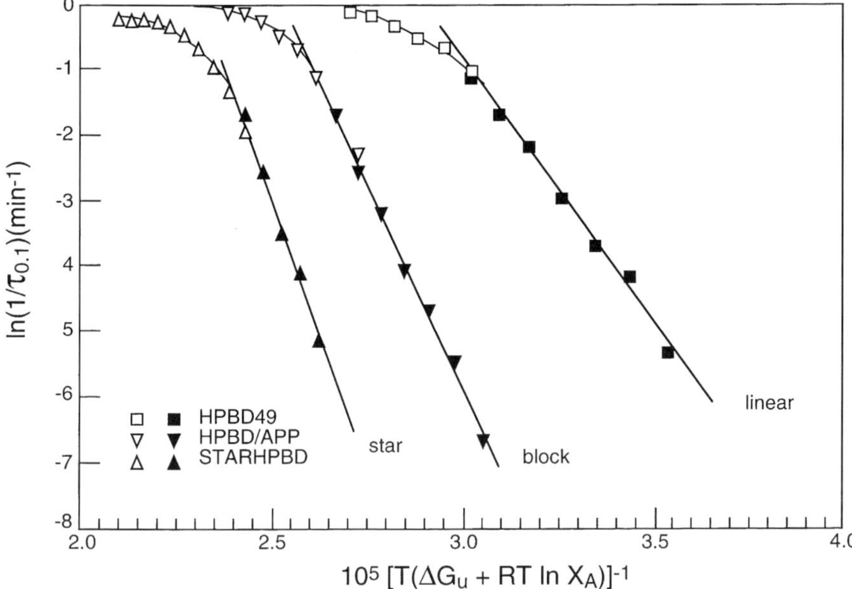

Fig. 10.33 Plot of ln of crystallization rate, $1/\tau_{0.1}$, against nucleation temperature function for indicated copolymers based on hydrogenated poly(butadiene). Open symbol rates measured from exotherms; closed symbol rates measured from endotherms.(78)

poly(butadiene)–atactic poly(propylene) diblock copolymer; and a three-arm star hydrogenated poly(butadiene). An important feature of these copolymers is that their crystallizing portions, hydrogenated poly(butadiene), all have the same molecular length and the fraction of short chain branches is the same. The isotherm shapes of these copolymers are similar to one another.(78) Figure 10.33 gives a summary of the crystallization rates in terms of $\tau_{0.1}$, the time required to reach 10% of the maximum attained crystallinity, as they depend on the nucleation temperature function for random copolymers.[6] The slopes in these plots, which reflect the product of interfacial free energies for nucleation, increase approximately two-fold from the linear copolymers to the star copolymers. The increase in the interfacial free energy can be attributed to the disorder at the crystal–liquid interfacial region. Although the three polymers have similar co-unit contents of the crystallizing species, the star and block copolymers have additional features that cannot be incorporated in the nucleus. For the block copolymer it is the junction with the noncrystallizing block, while for the star it is the region surrounding the junction point where the three chains join one another.

[6] The open symbols in this figure represent rates measured from exotherms; the solid symbols represent rates determined from endotherms.

Figure 10.33 indicates that at a fixed undercooling the copolymers have significantly different crystallization rates. Put another way, the curves in the figure are shifted along the horizontal axis. The result is that the star copolymer is the slowest crystallizer. This retardation can be attributed to the transport term in the crystallization rate. In terms of the Vogel equation, U^* varies from 1500 cal mol^{-1} for the linear copolymer for $C = 30$ to 5650 cal mol^{-1} for the star copolymer. This result is physically satisfactory since the long chain branches in the star copolymer retard the transport or diffusion of the crystallizing units to the interface. The rate of transport of the block copolymer segments would be expected to be between those of the linear and star polymers.

10.4 Both comonomers crystallize

Some interesting situations develop when both co-units are able to crystallize. The reason is that the species that crystallizes initially influences the crystallization of the other component that subsequently crystallizes at a lower temperature. In effect the crystallization of the lower melting component is constrained due to the presence of the already crystallized material. Examples of this phenomenon are found in both block, random and segmented type copolymers. A dramatic example of this effect is demonstrated by the segmented block copolymers of poly(ethylene terephthalate)–poly(butylene terephthalate).(79) The butylene terephthalate block crystallizes first on cooling, in this example. A significant enhancement in the crystallization rate of the poly(ethylene terephthalate) component is observed. For example, based on half-times, overall crystallization rate of the poly(ethylene terephthalate) block is increased about five-fold in the copolymer that only contains about 5% of the poly(butylene terephthalate). Increasing the poly(butylene terephthalate) concentration to about 20% slightly decreases the crystallization rate. The poly(butylene terephthalate) crystallites provide nucleation sites for the subsequent crystallization of the poly(ethylene terephthalate) component that results in an increase in the rate relative to that of the pure homopolymer. Superstructures that are unique to poly(butylene terephthalate) appear first in the crystallization process. They are then followed by the growth of the poly(ethylene terephthalate) spherulites. A similar enhancement in the overall crystallization rate is found in the block copolymer poly(p-dioxanone)–poly(ϵ-caprolactone).(80) In this case, the poly(p-dioxanone) crystallizes first on cooling. When this block is fully crystallized, the isothermal crystallization kinetics of the poly(ϵ-caprolactone) block is accelerated. The crystallization rate is actually faster in some of the copolymers relative to the rate of the homopolymer.

Another example of confined crystallization is found in the isothermal overall crystallization kinetics of di- and triblock copolymers of poly(ϵ-caprolactone)–

poly(ethylene oxide).(81) On cooling, the poly(ϵ-caprolactone) block crystallizes first, followed by poly(ethylene oxide). It was found that the kinetic parameters, determined from the derived Avrami equation, i.e. the half-times, rate constants and exponents for the poly(ϵ-caprolactone) component in the diblock, are very similar to those of the corresponding homopolymer. In contrast the spherulite growth rate of the p-dioxanone block in the poly(p-dioxanone)–poly(ϵ-caprolactone) copolymer is decreased by a factor of about 10 when the poly(ϵ-caprolactone) block is molten. A slightly lower crystallization rate is observed for the poly(ϵ-caprolactone) in the triblock copolymer. However, there is a significant retardation in the growth rate of the poly(ethylene oxide) block. In some cases, depending on the relative length of the two blocks, the ethylene oxide component does not crystallize at all. The Avrami exponent for the poly(ϵ-caprolactone) block is about 3 in all copolymers, comparable to that of the homopolymer. On the other hand, the value of the exponent n is about 2 for the poly(ethylene oxide) block, suggesting a different morphology from that of the corresponding homopolymer. The extrapolated equilibrium melting temperature for poly(ϵ-caprolactone) in the diblock is just slightly lower than that of the corresponding homopolymer. However, comparable melting temperatures for the poly(ethylene oxide) block are 14–18 °C lower than that of the homopolymer. These results suggest major morphological change for the poly(ethylene oxide) block, while that of the poly(ϵ-caprolactone) remains essentially unaltered. Qualitatively similar results are found in the Avrami exponent for the crystallization of the poly(ϵ-caprolactone) block in the poly(p-dioxanone)-poly(ϵ-caprolactone) copolymer.(80) Here the exponent is lower than that of the homopolymer and is reduced further with a decrease in the concentration of the poly(ϵ-caprolactone) block.

Random type copolymers, where both comonomers can crystallize, are also known. Here a distinction must be made between co-crystallization with isomorphous or isodimorphic replacement, and when the comonomers do not co-crystallize. In either case the crystallization kinetics is directly influenced by the appropriate phase diagram.

An example of isodimorphism, and the related crystallization kinetics, is typified by the random copolymers of 3-hydroxy butyrate–3-hydroxy valerate.(82) The melting temperature–composition relation of this copolymer was given in Fig. 5.17 (Volume 1). For this copolymer, depending on the composition, either of the comonomers can co-crystallize in the other's lattice. When the concentration of the 3-hydroxy butyrate is less than 40 mol percent, crystallization occurs in the poly(3-hydroxy butyrate) lattice. When its concentration is greater than 40 mol percent, crystallization takes place in the poly(3-hydroxy valerate) lattice. The copolymer that contains 41 mol percent of 3-hydroxy valerate reflects the coexistence of both crystal phases and corresponds to a pseudo-eutectic composition.

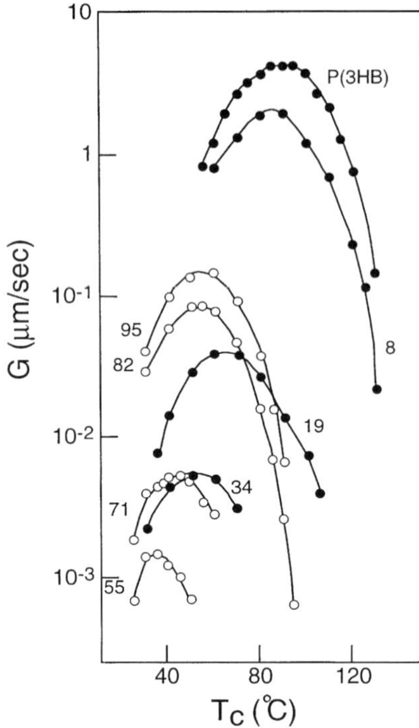

Fig. 10.34 Plot of spherulite growth rate as function of crystallization temperature for poly(3-hydroxy butyrate) and random copolymers of 3-hydroxy butyrate and 3-hydroxy valerate. Numbers on curves are mol percent 3-hydroxy butyrate. (From Scandola *et al.* (82))

The spherulite growth rates of these random type copolymers are given as functions of the crystallization temperature in Fig. 10.34. The curves in this figure follow the conventional pattern. Over the extended range of crystallization temperatures studied, a rate maximum is observed. There is, however, a striking effect of the position of these curves relative to the phase diagram. For compositions to the left of the eutectic, indicated by the closed circles in this figure, there is a marked decrease in rate with increasing concentration of the 3-hydroxy valerate component. In contrast, the opposite trend is observed for compositions to the right of the eutectic (open circles). In this composition range the growth rates increase with the 3-hydroxy valerate content. The curves shift to higher temperatures, and the growth maxima progressively increase. Thus, the change from the poly(3-hydroxy butyrate) lattice to that of poly(3-hydroxy valerate) alters the relationship between the growth rate and composition. This effect can be correlated with the maximum rate, $G^{\max}$. A plot of log $G^{\max}$ against the mol percent of 3-hydroxy butyrate is given in Fig. 10.35. There is a striking similarity between this figure and the melting

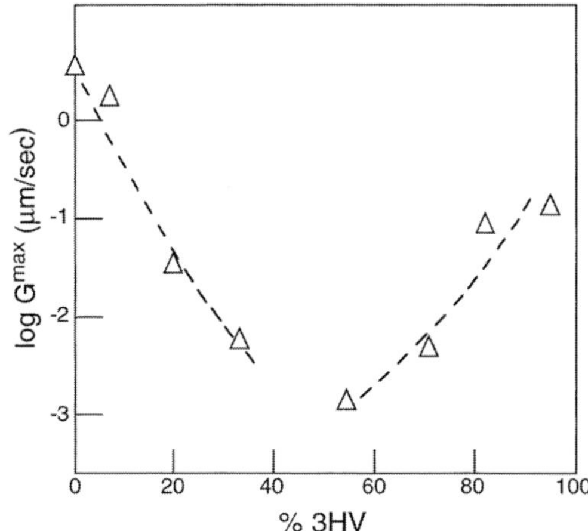

Fig. 10.35 Plot of log spherulite growth rate maxima against mol percent 3-hydroxy valerate for random copolymers of 3-hydroxy butyrate–3 hydroxy valerate. (From Scandola *et al.* (82))

temperature–composition relation (Fig. 5.17). The positions of the growth maxima in Fig. 10.35 can be directly correlated with the phase diagram.

A contrasting situation to the above is when each of the comonomers can crystallize individually. An example for random copolymers is found in the series of copolyamides containing 3,3′,4,4′-biphenyl tetracarboxylic dianhydride (BPDA), with either 1,3-bis(4-amino phenoxy) benzene (134APB) or 1,12-dodecanediamine (C12).(83,84) The spherulite growth rates of this set of copolymers, as well as the corresponding homopolymers, are plotted as functions of temperature in Fig. 10.36.(83) Except for the 20/80 composition (134APB/C12) the characters of the growth–temperature curves are very similar to one another. The usual maximum in the rate is observed when an extended range in crystallization temperatures is studied. The temperature of the maximum growth rate shifts to lower temperatures with increasing concentration of the C12 comonomer. Except for the 40/60 and 20/80 copolymers, the maximum growth rates are very close to one another. The kinetics in the composition range 100/0 to 40/60 are dominated by the crystallization of the 134APB/BPDA co-units. In the range 20/80 to 0/100 crystallization of C12/BPDA dominates. In addition to a lower growth rate, the 20/80 copolymer shows a double maximum. The double maximum is due to the growth of two different types of spherulites. This probably reflects the separate crystallization of the two comonomers. Since the equilibrium melting temperatures, as well as the glass temperatures, are known for this set of copolymers,(84) it is natural to attempt to fit the experimentally observed growth rate–temperature data to Eq. (9.209). For

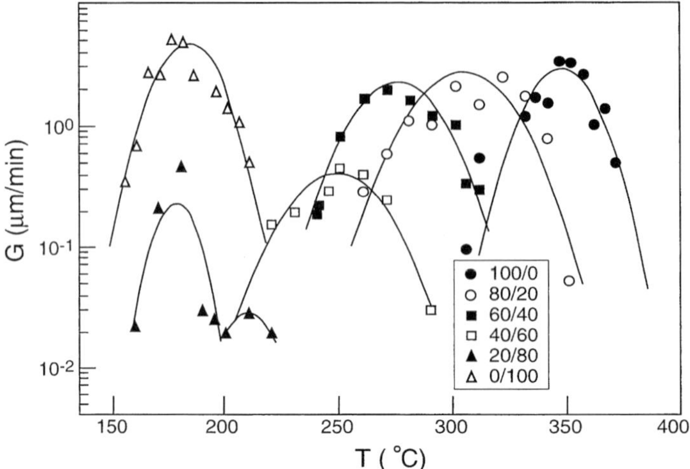

Fig. 10.36 Plot of spherulite growth rate against crystallization temperature for copolyimides described in text for indicated 134APB/C12 compositions. (From Hsaio *et al.* (83))

unexplainable reason(s), an acceptable fit cannot be attained with the copolyimide data of Fig. 10.36 for any meaningful values of U^* and C.

10.5 Long chain branches and covalent cross-links

It is convenient at this point to consider the crystallization kinetics of covalently cross-linked polymers, as well as those with long chain branches. Strictly speaking neither of these can be classified as copolymers. However, the branches and cross-links introduce structural irregularities into the chain. The isotherms illustrated in Fig. 10.2 for the crystallization of long chain branched polyethylene at low undercoolings, have the same characteristics as chemically distinct random copolymers under the same conditions. Studies of long chain branched poly(butylene terephthalate) (85) and poly(phenylene sulfide) (86) indicate that the overall crystallization rate, as well as the spherulite growth rate, decreases with increasing branching concentrations. Similar results are found with poly(ethylene terephthalate).(86a) An example of the change in crystallization rate with branching is illustrated in Fig. 10.37, where the half-time is plotted against the crystallization temperature for a series of long chain branched poly(butylene terephthalates).(85) The polymers are characterized by the number average degree of branching, which ranges from zero for pure polymer to greater than 1. The latter represents the gel point. Figure 10.37 clearly shows that the rate decreases with branching concentration at all crystallization temperatures. A substantial decrease, a factor of 5 or more, is observed at the higher crystallization temperatures. The data for both polymers can be fitted in

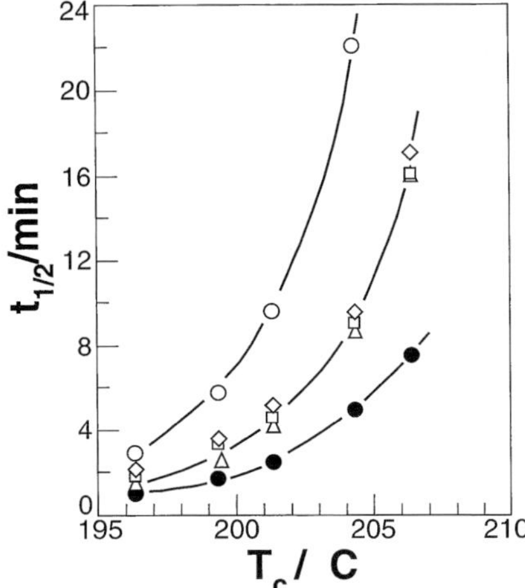

Fig. 10.37 Plot of half-time, $t_{1/2}$, against crystallization temperature for long chain branched poly(butylene terephthalate). Number average degree of branching: ○ zero; ◇ 0.5; □ 0.6; △ 0.6; ● > 1. (From Righetti and Munari (85))

the conventional manner to the Avrami theory.(85,86) Since the crystallization was conducted at relatively large undercoolings superposable isotherms were obtained as would be expected from theory. The Avrami exponent was about 3 for both the linear and branched polymers. An analysis of the temperature coefficient for the poly(butylene terephthalates), according to coherent unimolecular nucleation theory, indicates that the product of interfacial free energies, $\sigma_{un}\sigma_{en}$, increases with the long chain branching content. This conclusion is similar to that previously noted for the star hydrogenated poly(butadiene) relative to the linear copolymer.

The overall crystallization kinetics of star-branched poly(ϵ-caproamides) have also been studied.(87) The kinetics for the linear, three- and six-arm star polymers follow a similar pattern. The data are amenable to an Avrami type analysis and the isotherms are superposable. Their shapes are essentially the same for the three polymers, reflecting the similarity in the exponent n. The time scales, however, are different at the same supercoolings. This is reflected in the half-times. When compared at comparable supercoolings the half-time for the linear polymer is about twice that of the six-arm star one. The half-time for the three-arm star lies in between. The covalent cross-linking of a collection of polymer chains represents more than just the introduction of structural irregularities. As the cross-links are randomly introduced, highly ramified structures evolve from the initial set of linear chains. The molecules are also partitioned between sol and gel portions.(88–90) The details of the portioning depend on the initial molecular weight distribution and the extent of the cross-linking. Thus, the structure of the sol and gel, for a given initial molecular weight distribution, is not constant but will depend on the cross-linking level. The infinite network that is eventually formed is characterized by the molecular weight between cross-links, M_c. Such networks are rarely perfect and can contain a significant concentration of chain ends. The sol and gel portions would be expected to display different crystallization kinetics and have features that are different from linear chains with just a random distribution of defects or comonomers. For example, it was found that the isothermal crystallization rate of end-linked poly(tetrahydrofuran) networks was reduced when the sol fraction was removed.(90a)

The properties of cross-linked systems depend on the state of the polymer at the time the cross-links are introduced. It is necessary and very important to specify whether the chains are in the completely amorphous or crystalline states at the time the cross-links are introduced, or if they are cross-linked in solution.(91) When cross-linking crystalline polymers the nature of the crystalline state is also important, i.e. whether the crystallites are randomly oriented or are oriented in a specific manner. Moreover, the efficacy of cross-linking between the crystalline and noncrystalline regions is important. It is generally assumed that, irrespective of the initial polymer state, the cross-links are randomly introduced. It was shown in Chapter 7 (Volume 1) that the order initially present in a crystalline polymer is retained to some degree in the melt, if the cross-links are introduced in the crystalline state. In contrast, when the cross-links are introduced into a completely amorphous polymer the melt remains disordered even after crystallization. The melt structure in turn influences the resulting melting temperature and morphology. It can be expected that the crystallization kinetics will also be affected.

When analyzing cross-linked systems a distinction must be made between the whole polymer and the sol–gel fractions. The concentration and distribution of the cross-links will be different in each fraction. The particular method by which the cross-links are introduced could also influence the kinetics. They can be introduced by specific chemical reactions or in special cases by the action of high energy ionizing radiation. Ultimately, all these factors have to be sorted out. Most of the studies of the crystallization of cross-linked polymers have involved either natural rubber, poly(cis-1,4-isoprene), or polyethylenes with different molecular architecture.

Some general observations can be made with respect to the crystallization kinetics of cross-linked systems before specific data are analyzed.(92,93) The crystallization rates are retarded relative to the corresponding non-cross-linked polymers and the attainable level of crystallinity is significantly reduced. The influence of cross-linking in retarding the major aspects of the crystallization is much greater than their nominal concentration. Sigmodal type isotherms are typically observed, thus indicating that nucleation and growth processes are still involved.

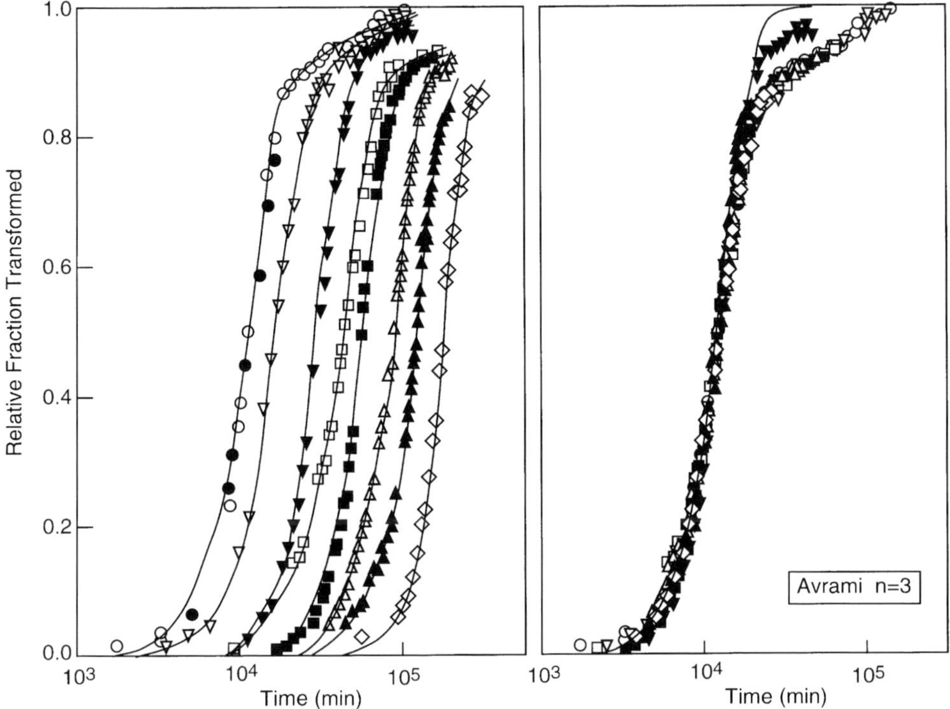

Fig. 10.38 Plot of normalized crystallinity level as a function of log time for natural rubber cross-linked to varying extents with sulfur at 2 °C. Curve derived Avrami equation with $n = 3$. Combined sulfur content in percent: ○ zero; ● 0.1; ▽ 0.2; ▼ 0.3; □ 0.35; ■ 0.40; △ 0.43; ▲ 0.46; ◇ 0.5. (Data from Bekkedahl and Wood (92))

Natural rubber is usually cross-linked in the amorphous state and the crystallization kinetics subsequently studied. Thus some of the complications that were outlined above are avoided. Figure 10.38 gives an example of the crystallization isotherms at 2 °C of natural rubber cross-linked to varying degrees with sulfur in the amorphous state.(92) The normalized fraction transformed is plotted against log time for varying amounts of combined sulfur. It can be assumed that in this system the cross-links were introduced randomly. The shapes of the isotherms are reminiscent of the crystallization of non-cross-linked polymers. As is indicated in the figure, the isotherms for the different extents of cross-linking are superposable with one another. The data can be fitted quite well with the derived Avrami equation with $n = 3$. Other methods of chemically cross-linking natural rubber lead to similar results.(92,93) The plots in Fig. 10.38 also make clear that the cross-linking severely retards the crystallization rate. For example, the time scale for the crystallization is shifted by several orders of magnitude between the non-cross-linked polymers and the one with 0.5% combined sulfur.

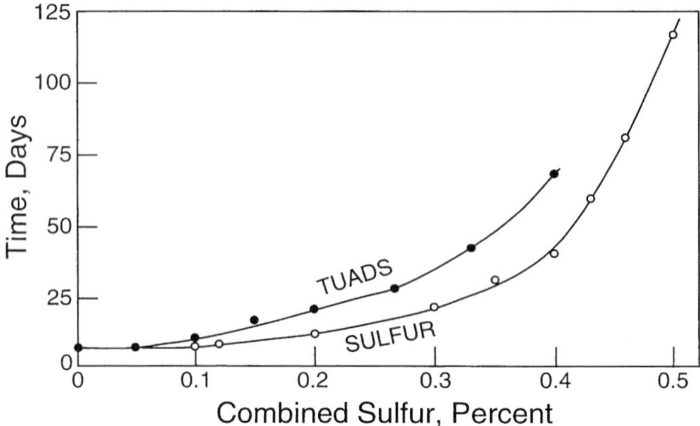

Fig. 10.39 Plot of crystallization half-time as a function of combined sulfur. (From Bekkedahl and Wood (92))

The decrease in the crystallization rate with the extent of cross-linking is illustrated definitively in Fig. 10.39.(88) Here the half-time is plotted against the amount of combined sulfur, from either just rubber–sulfur compounding or from tetramethylthiuram disulfide (Tuads). The main finding here is that although small amounts of combined sulfur only reduce the crystallization rates slightly, the retardation in rates becomes profound as the cross-linking level increases. It is likely that the decrease is caused by the influence of the cross-links on the transport term in the expression for the crystallization rate. Segmental or reptation type motion will be impeded by the intermolecular cross-links. Among other factors, the spreading rate will also be retarded. The true equilibrium melting temperature will also be slightly reduced because of the small concentration of cross-links introduced. Consequently, the nucleation should not be seriously affected.

Since the crystallization of natural rubber can be carried out over an extended temperature range, it is not surprising that rate maxima are also observed with the cross-linked polymers.(94) An example is shown in Fig. 10.40 where the crystallization half-times are plotted against the crystallization temperature. The cross-linking in this example was accomplished by Tuads, with the combined sulfur increasing from sample D_1 to D_4. The shapes of the curves in the figure are similar to one another. They are, however, displaced to longer times, at all temperatures, as the cross-link density increases. The maxima in the rate are apparent and appear at the same temperature at the different levels of cross-linking. The temperatures of the maxima are virtually identical to that of the non-cross-linked polymer.

The above examples point out that the formal aspects of the overall crystallization kinetics are unaffected by the introduction of intermolecular cross-links, when the cross-links are introduced in the amorphous state. The fanning out of the isotherms,

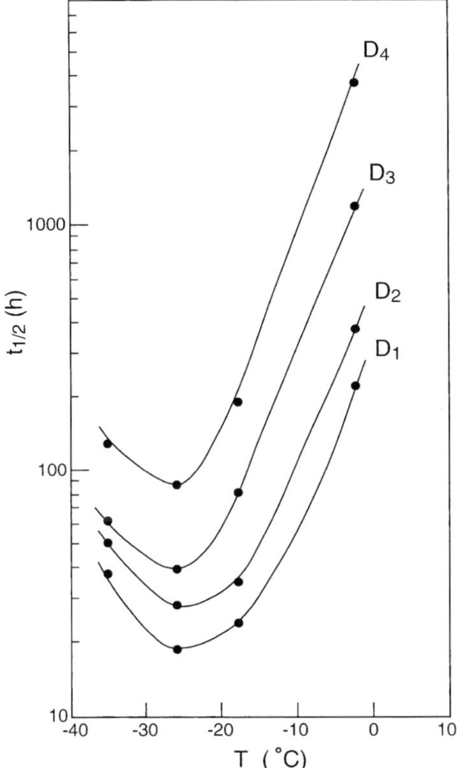

Fig. 10.40 Plot of half-time for crystallization against crystallization temperature for natural rubber crystallized with Tuads. Percent Tuads: D_1, 0.75; D_2, 1.5; D_3, 2.5; D_4, 3.5. (Data from Russell (94))

typical of random copolymers as the crystallization temperature increases, is not observed in the cross-linked natural rubber system, at least up to the level of cross-linking that has been studied. However, there is a profound effect on the time scale for crystallization at a fixed temperature.

The overall crystallization kinetics of an unfractionated linear polyethylene, cross-linked by a peroxide reaction has also been studied.[7](95,96) A special feature of this work was the study of the separated sol and gel portions at different levels of cross-linking. The overall crystallization rates, in terms of the reciprocal of the half-time, $1/t_{1/2}$, are plotted against the crystallization temperature in Figs. 10.41 and 10.42 for a set of sol and gel fractions respectively. The gel fractions are characterized by the molecular weight between cross-links, M_c, assuming ideal network formation. The sol portions are defined by their number average molecular

[7] The state of the polyethylene when the cross-linking reaction takes place is not made clear in this work.

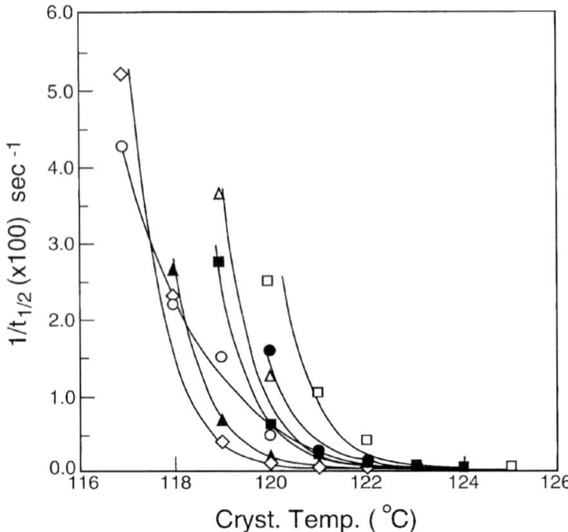

Fig. 10.41 Plot of reciprocal half-time as a function of crystallization temperature for sol fraction of cross-linked linear polyethylene. Number average molecular weight, M_n: ○ original non-cross-linked polymer; ● 7106; □ 7077; ■ 6422; △ 4561; ▲ 5634; ◇ 4150. (From Phillips and Lambert (95))

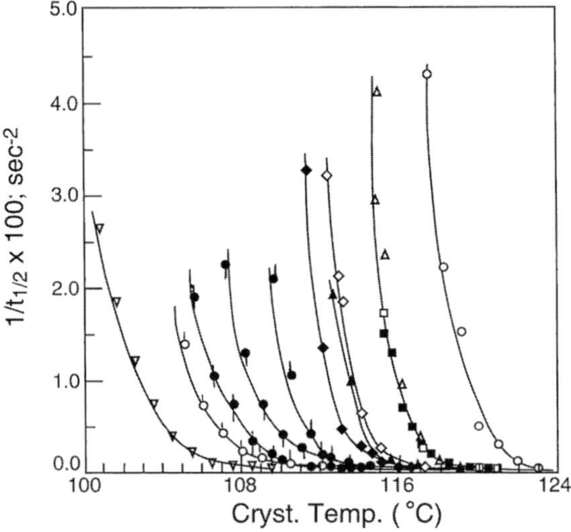

Fig. 10.42 Plot of reciprocal of half-time against crystallization temperature for the crystallization of the gel portion of cross-linked linear polyethylene. Molecular weight between cross-links, $M_c \times 10^{-3}$: ○ original polymer; □ 12.7; ■ 9.4; △ 5.7; ▲ 4.5; ◇ 3.6; ◆ 3.1; ● 2.2; ● 1.9; ○ 1.3; ● 0.96; ▽ 0.56. (From Lambert *et al.* (96))

weight. The molecular weights of these highly ramified structures were determined by gel permeation chromatography, utilizing the standard universal relation.

There is a systematic change in the crystallization rate of the gels with the concentration of cross-links (Fig. 10.42). In particular, as M_c of the network decreases, the range of isothermal crystallization moves to lower temperatures. This reflects a significant reduction in the crystallization rate at a fixed temperature as the cross-linking density increases. The data for these samples with the lowest cross-link density cannot be distinguished from one another. However, the time scale is greater than that of the uncross-linked polymers. As the concentration of cross-links increases there is a continuous decrease in the crystallization rate at fixed crystallization temperatures. For example, for a T_c of 112 °C, the crystallization rate is imperceptible for gels with molecular weights between cross-links, M_c, equal to 1300 or less. However, the rate is discernible for $M_c = 1900$ and increases about 60–70 fold for the gel with $M_c = 3600$. Conversely, at a fixed rate, for example $1/t_{1/2} = 1.5 \times 10^{-2}$ sec^{-1}, the crystallization temperatures range from 119 °C for the non-cross-linked polymer to 102 °C for the gel with $M_c = 560$. Thus, there is a significant retardation in the crystallization with cross-linking of the gel portion that is qualitatively similar to that found in natural rubber. In contrast, the crystallization rates of the sol do not give a simple pattern (Fig. 10.41). However, there does appear to be a qualitative increase in the rate with the assigned molecular weight. This may be a reflection of changes in cross-linking density and branching.

The strong negative temperature coefficients that are observed for both the sol and gel portions suggest that the kinetic data be analyzed according to nucleation theory. In principle, this will allow the question of regimes to be addressed. In order to properly perform this analysis, the equilibrium melting temperatures of both the sol and gel fractions need to be reliably known. The highly ramified, and different, structures of the sol and gel,(88,89) and the state in which the cross-links are introduced makes it extremely difficult to establish the required equilibrium melting temperatures. Thus, the required information with which to properly carry out a regime analysis is not available.

Both the sol and gel follow Avrami type kinetics of nucleation and growth. An example is given in Fig. 10.43 for a lightly cross-linked gel fraction of polyethylene with $M_c = 17\,300$.(95) Here, the relative fraction transformed is plotted against log time for different crystallization temperatures. As is indicated in the right of the diagram a set of superposable isotherms results. All of the isotherms can be fitted with an Avrami exponent $n = 3$ for about 15% of the transformation. The deviation from Avrami occurs at a much lower crystallinity level than the non-cross-linked polymers. This could be a consequence of the decrease in the sequence lengths available for crystallization because of the cross-linking. The basic kinetic data for the other gel fractions would allow for a more detailed and comprehensive

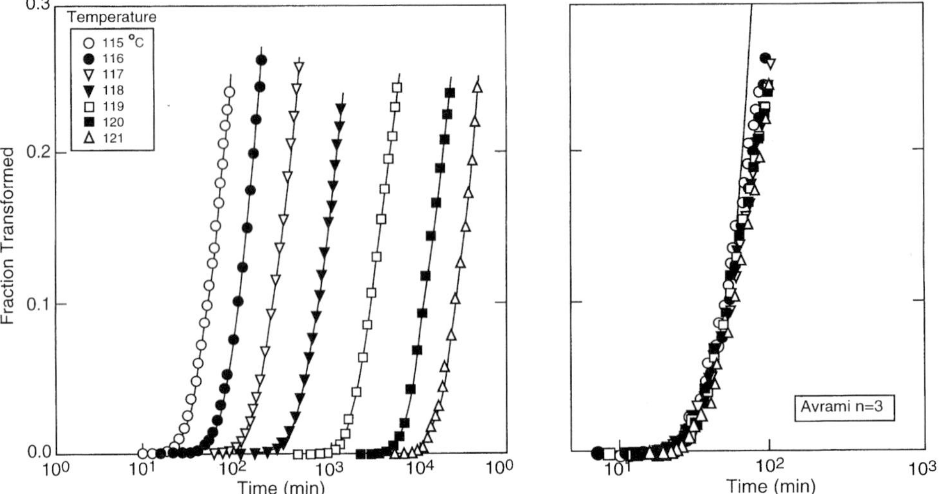

Fig. 10.43 Plot of fraction transformed against log time for gel portion of cross-linked linear polyethylene having $M_c = 17\,300$. Solid curve for superposed isotherm, derived Avrami equation with $n = 3$. Crystallization temperatures are indicated. (Data from Phillips and Lambert (95))

analysis. It is important to assess the dependence of the Avrami exponent, and the transformation level at which deviations occur, on the cross-linking level. The limited published data indicate that the sol fractions behave in a similar manner.

References

1. Mandelkern, L. in *Growth and Perfection of Crystals*, R. H. Doremus, B. W. Roberts and D. Turnbull eds, John Wiley (1958) p. 467.
2. Buchdahl, R. R., R. L. Miller and S. Newman, *J. Polym. Sci.*, **36**, 215 (1959).
3. Alamo, R. G. and L. Mandelkern, *Macromolecules*, **24**, 6480 (1991).
4. Lee, S. C., K. H. Yoon and J. H. Kim, *Polymer*, **29**, 1 (1997).
5. Reghetti, M. C., M. Pizzoli, N. Latti and A. Munari, *Macromol Chem. Phys.*, **199**, 2063 (1998).
6. Finelli, L., N. Lotti and A. Munari, *Europ. Polym. J.*, **37**, 2039 (2001).

6a. Arnold, J. R., A. Zhang, S. Z. D. Cheng, A. J. Lovinger, E. T. Hsieh, P. Chu, T. W. Johnson, K. G. Honnell, R. G. Geerts, S. J. Palachal, G. R. Hawley and M. B. Welch, *Polymer*, **35**, 1884 (1994).

7. Lambert, W. S. and P. J. Phillips, *Macromolecules*, **27**, 3537 (1994).
8. Lambert, W. S. and P. J. Phillips, *Polymer*, **37**, 3585 (1996).
9. Jackson, J. B. and G. W. Longman, *Polymer*, **10**, 873 (1969).

9a. Papageorgiou, G. Z. and G. P. Kayaryannidis, *Polymer*, **42**, 8179 (2001).

10. Gornick, F. and L. Mandelkern, *J. Appl. Phys.*, **33**, 907 (1962).
11. Akpalu, Y., L. Kielhorn, B. S. Hsiao, R. S. Stein, T. P. Russell, J. van Egmond and M. Mathukumar, *Macromolecules*, **32**, 765 (1999).

12. Goulet, B. and R. E. Prud'homme, *J. Polym. Sci: Pt. B: Polym. Phys.*, **28**, 2329 (1990).
13. Hser, J. C. and S. H. Carr, *Polym. Eng. Sci.*, **19**, 436 (1979).
14. Amelino, L. and E. Martuscelli, *Polymer*, **16**, 854 (1975).
15. Harvey, E. D. and F. J. Hybart, *Polymer*, **12**, 711 (1971).
16. Flory, P. J., *J. Chem. Phys.*, **17**, 223 (1949).
17. Fatou, J. G., C. Marco and L. Mandelkern, *Polymer*, **31**, 1685 (1990); *ibid.*, **31**, 890 (1990).
18. Isasi, J. R., J. A. Haigh, J. T. Graham, L. Mandelkern and R. G. Alamo, *Polymer*, **41**, 8813 (2000).
19. Voigt-Martin, I. G., R. Alamo and L. Mandelkern, *J. Polym. Sci.: Polym. Phys. Ed.*, **24**, 1283 (1986).
20. Bensason, S., J. Minick, J. Moet, S. Chun, A. Hiltner and E. Baer, *J. Polym. Sci.: Pt. B: Polym. Phys.*, **34**, 130 (1986).
21. Haigh, J. A., C. Nguyen, R. G. Alamo and L. Mandelkern, *J. Therm. Anal. Calorimetry*, **59**, 435 (2000).
22. Crist, B. and D. N. Williams, *J. Macromol. Sci. Phys.*, **B39**, 1 (2000).
23. Crist, B. and E. S. Claudio, *Macromolecules*, **32**, 8945 (1999).
24. Haigh, J. A. and L. Mandelkern, unpublished work (2000).
25. Okui, N. and T. Kawai, *Makromol. Chem.*, **154**, 161 (1972).
26. Kim, M. H. and P. J. Phillips, *J. Appl. Polym. Sci.*, **70**, 1893 (1998).
27. Androsch, R., *Polymer*, **40**, 2805 (1999).
28. Alizadeh, A., L. Richardson, J. Xu, S. McCartney, H. Marand, W. Cheung and S. Chum, *Macromolecules*, **32**, 6221 (1999).
29. Cheng, S. Z. D., M. Y. Cao and B. Wunderlich, *Macromolecules*, **19**, 1868 (1986).
30. Chung, J. S. and P. Cebe, *Polymer*, **33**, 2312 (1992).
31. Phillips, R. A., J. McKenna and S. L. Cooper, *J. Polym. Sci: Pt. B: Polym. Phys.*, **32**, 791 (1994).
32. Phillips, R. A. and S. L. Cooper, *Polymer*, **35**, 4146 (1994).
33. Flory, P. J., *Trans. Faraday Soc.*, **51**, 848 (1955).

33a. Kint, D. P. R., E. Rude, J. Llorens and S. Munoz-Guena, *Polymer*, **43**, 7529 (2002).

34. Strobl, G., *Acta Polym.*, **48**, 562 (1997).
35. Heck, B., T. Hugel, M. Lizima, E. Sadiku and G. Strobl, *New J. Phys.*, **1**, 17 (1999).
36. Schmidtke, J., G. Strobl and T. Thum-Albrecht, *Macromolecules*, **30**, 5804 (1997).
37. Strobl, G., *Eur. Phys. J. E*, **3**, 165 (2000).
38. Mandelkern, L., M. Glotin and R. S. Benson, *Macromolecules*, **14**, 222 (1981).
39. Glotin, M. and L. Mandelkern, *Macromolecules*, **14**, 1394 (1981).
40. Chowdhury, F., J. A. Haigh, L. Mandelkern and R. G. Alamo, *Polym. Bull.*, **41**, 463 (1998).
41. Goulet, L. and R. E. Prud'homme, *J. Polym. Sci: Pt. B: Polym. Phys.*, **28**, 2329 (1990).
42. Scandola, M., G. Ceccorulli, M. Pizzoli and M. Gazzano, *Macromolecules*, **25**, 1405 (1992).
43. Andrews, E. H., P. J. Owen and A. Singh, *Proc. Roy. Soc. (London)*, **324A**, 79 (1971).
44. Nachtrab, G. and H. G. Zachmann, *Ber. Bunsenges. Phys. Chem.*, **74**, 837 (1970).
45. Frank, C. F., *J. Cryst. Growth*, **22**, 233 (1974).
46. Haigh, J. A., L. Mandelkern and L. Howard, (unpublished results).
47. Wagner, J. and P. J. Phillips, *Polymer*, **42**, 8999 (2001).
48. Seow, P. K., Y. Gallot and A. Skoulios, *Die Makromol Chem.*, **177**, 177 (1976).
49. Nojima, S., H. Nakano, Y. Takahashi and T. Ashida, *Polymer*, **35**, 3479 (1994).

50. Heyschem, R. J. and Ph. Teyssie, *J. Polym. Sci.: Pt. B: Polym. Phys.*, **21**, 523 (1989).
51. Limtasiri, T., S. J. Grossman and J. C. Huang, *Polym. Eng. Sci.*, **29**, 493 (1989).
52. Sakurai, K., W. J. MacKnight, D. J. Lohse, D. N. Schulz and J. A. Sisson, *Macromolecules*, **27**, 4941 (1994).
53. Richardson, P. H., R. W. Richards, D. J. Blundell, W. A. MacDonald and P. Mills, *Polymer*, **36**, 3059 (1995).
54. Shiomi, T., H. Tsukada, H. Takeshita, K. Takenaka and Y. Tejuka, *Polymer*, **42**, 4997 (2001).
55. Shiomi, T., H. Takeshita, H. Kawaguchi, M. Nagai, K. Takenaka and M. Maya, *Macromolecules*, **35**, 8056 (2002).
56. Hamley, I. W., J. P. A. Fairclough, F. S. Bates and A. J. Ryan, *Polymer*, **39**, 1429 (1998).
57. Ryan, A. J., I. W. Hamley, W. Bres and F. S. Bates, *Macromolecules*, **28**, 3860 (1995).
58. Nojima, S., K. Kato, S. Yamamoto and T. Ashida, *Macromolecules*, **25**, 2237 (1992).
59. Weimann, P. A., D. A. Haiduk, C. Chu, K. A. Chaffin, J. C. Brodil and F. S. Bates, *J. Polym. Sci.: Pt. B: Polym. Phys.*, **37**, 2053 (1990).
59a. Loo, Y. L., R. A. Register, A. J. Ryan and G. T. Dee, *Macromolecules*, **34**, 8968 (2001).
60. Rangarajan, P., R. A. Register, D. H. Adamson, L. J. Fetters, S. Naylor and A. J. Ryan, *Macromolecules*, **28**, 1422 (1995).
61. Quiram, D. J., R. A. Register, G. R. Marchand and A. J. Ryan, *Macromolecules*, **30**, 8338 (1997).
62. Ryan, A. J., J. P. A. Fairclough, I. W. Hamley, S. M. Mai and C. Booth, *Macromolecules*, **30**, 1723 (1997).
63. Hamley, I. W., J. P. A. Fairclough, N. J. Terrill, A. J. Ryan, P. M. Lipis, F. S. Bates and E. Towns-Andrews, *Macromolecules*, **29,** 8835 (1996).
64. Loo, Y. L., R. A. Register and A. J. Ryan, *Phys. Rev. Lett.*, **84**, 4120 (2000).
65. Loo, Y. L., R. A. Register and A. J. Ryan, *Macromolecules*, **35**, 2365 (2002).
66. Robitaille, C. and J. Prud'homme, *Macromolecules*, **16**, 665 (1983).
67. Balsamo, V., F. vonGyldenfeldt and R. Stadler, *Macromol. Chem. Phys.*, **197**, 3317 (1996).
68. Fu, Q., B. P. Livengood, C. C. Shen, F. Lin, F. W. Harris, S. Z. D. Cheng, B. S. Hsiao and F. Yen, *Macromol. Chem. Phys.*, **199**, 1107 (1998).
69. O'Malley, J. J., *J. Polym. Sci.*, **60C**, 141 (1977).
70. Balsamo, V., A. J. Müller, F. vonGyldenfeldt and R. Stadler, *Macromol. Chem. Phys.*, **199**, 1063 (1998).
71. Arnal, M. L., V. Balsamo, F. Lopez-Carrasqueno, J. Contreras, M. Carrilla, H. Schmalz, V. Abetz, E. Laredo and A. J. Müller, *Macromolecules*, **34**, 7973 (2001).
72. Müller, A. J., V. Balsamo, M. L. Arnal, T. Jakob, H. Schmalz and V. Abetz, *Macromolecules*, **35**, 3048 (2002).
73. Risch, B. G., E. D. Rodriques, K. Lyon, J. E. McGrath and G. L. Wilkes, *Polymer*, **37**, 1229 (1996).
74. Li, H. M. and J. H. Magill, *J. Polym. Sci.: Polym. Phys. Ed.*, **16**, 1059 (1978).
75. Friday, A. and C. Booth, *Polymer*, **19**, 1035 (1978).
76. Andjelic, S., D. Jamiolkowski, J. McDivitt, J. Fischer and J. Zhou, *J. Polym. Sci.: Pt. B: Polym. Phys.*, **39**, 3075 (2001).
77. Mandelkern, L. and J. A. Haigh, unpublished calculations.
78. Haigh, J. A., C. Nguyen, R. G. Alamo and L. Mandelkern, *J. Therm. Anal. Calorimetry*, **59**, 435 (2000).

79. Misra, A. and S. N. Garg, *J. Polym. Sci.: Pt. B: Polym. Phys.*, **24**, 983 (1986); *ibid.*, **24**, 999 (1986).
80. Albuerne, J., L. Márquez, A. J. Müller, J. M. Raquez, Ph. Degée, Ph. Dubois, V. Castelletto and I. W. Hamley, *Macromolecules*, **36**, 1633, (2003).
81. Bogdonov, B., A. Vidts, E. Schacht and H. Berghman, *Macromolecules*, **32**, 726 (1999).
82. Scandola, M., G. Ceccorulli, M. Pizzoli and M. Gazzano, *Macromolecules*, **25**, 1450 (1992).
83. Hsaio, B., J. A. Kreuz and S. Z. D. Cheng, *Macromolecules*, **29**, 6926 (1995).
84. Kreuz, J. A., B. S. Hsiao, C. A. Renner and D. L. Goff, *Macromolecules*, **28**, 6926 (1995).
85. Righetti, M. C. and A. Munari, *Macromol. Chem. Phys.*, **198**, 363 (1997).
86. Lopez, L. C., G. L. Wilkes and J. F. Geibel, *Polymer*, **30**, 147 (1989).
86a. Manaresi, P., A. Munari, F. Pilati, G. C. Alfonso, S. Russo and M. L. Sartirana, *Polymer*, **27**, 955 (1986).
87. Risch, B. G., G. L. Wilkes and J. M. Warakomshi, *Polymer*, **34**, 2330 (1993).
88. Flory, P. J., *J. Amer. Chem. Soc.*, **63**, 3097 (1941).
89. Stockmayer, W. H., *J. Chem. Phys.*, **12**, 123 (1944).
90. Flory, P. J., *Principles of Polymer Chemistry*, Cornell University Press (1953) pp. 378ff.
90a. Takahashi, H., M. Shibayama, M. Hashumoto and S. Namura, *Macromolecules*, **28**, 5547 (1995).
91. Flory, P. J., *J. Amer. Chem. Soc.*, **78**, 5222 (1956).
92. Bekkedahl, N. and L. A. Wood, *Ind. Eng. Chem.*, **33**, 381 (1941).
93. Gent, A. N., *J. Polym. Sci.*, **18**, 321 (1955).
94. Russell, E. W., *Trans. Faraday Soc.*, **47**, 539 (1951).
95. Phillips, P. J. and W. S. Lambert, *Macromolecules*, **23**, 2073 (1990).
96. Lambert, W. S., P. J. Phillips and J. S. Lin, *Polymer*, **35**, 1809 (1994).

11

Crystallization kinetics of polymer mixtures

11.1 Introduction

A large number of studies concerned with the crystallization kinetics of polymer–polymer mixtures have been reported. One reason for this abundance of information is a basic scientific interest coupled with the many ramifications that are inherent to such systems. There is also a pragmatic interest since physical and mechanical properties, as well as environmental features, can be altered and controlled by the appropriate mixing of polymeric components. It is useful to follow the same methodology used in discussing the equilibrium aspects of crystalline polymer–polymer mixtures (see Chapter 4, Volume 1) when analyzing their crystallization kinetics.

The nature of the melt and the pathway followed during the course of the transformation are important factors in governing the crystallization kinetics of binary mixtures. Depending on the molecular weight, the structural regularity of the chain and the value of the interaction parameter χ_1, the blends can be either completely miscible over the complete composition range, partially miscible or completely immiscible. Phase separation may occur upon increasing or decreasing the temperature, depending on the change in χ_1. Thus a lower or upper critical solution temperature, or both, can be observed. In order for a proper analysis to be made of the crystallization kinetics in blends, it is important that the complete phase diagram be established for partially miscible systems. The crystallization kinetics will thus depend on the phase relationships and thermal history, or the pathway, that is taken.

It also has to be known whether only one, or both components can crystallize. If in fact both components are capable of crystallizing it has to be further determined whether they do so independently or if co-crystallization occurs between the two species. Another situation that needs to be considered separately is whether the two components are chemically identical. This could involve differences in stereo, regio or geometric isomers, molecular architecture or chain lengths. Each of these categories of binary blends is important and will be discussed in the sections that follow.

11.2 Components completely miscible in melt: only one component crystallizes

11.2.1 Experimental results

In this section we consider the case where both components are miscible over the complete composition range. However, only one of the components can crystallize. As a consequence of the miscibility, only one glass temperature is observed that varies continuously with composition from that of one pure homopolymer to the other. The depression of the melting equilibrium temperature is usually small, and follows the pattern described in Chapter 4. The window available for crystallization, between the equilibrium melting temperature and the glass temperature, will depend primarily on the changes in the glass temperature with composition. The crystallization rates can expect to vary accordingly. The crystallization kinetics of the blends are studied in the conventional manner by measuring either the overall crystallization or spherulite growth rates.

The spherulite growth rates of the crystallizing component in several different type blends are given in the following figures. Here, the rates are plotted as a function of the crystallization temperature for different compositions. The data for poly(vinylidene fluoride)–poly(methyl methacrylate) blends are given in Fig. 11.1,(1) where poly(methyl methacrylate) is the noncrystallizing component. Miscibility of the mixture has been demonstrated for the composition and crystallization range that was studied. In this blend, the glass temperature increases from $-50\,^{\circ}\mathrm{C}$ for pure poly(vinylidene fluoride) to $+45\,^{\circ}\mathrm{C}$ for the 50/50 mixture. In contrast the equilibrium melting temperatures only decrease by about $10\,^{\circ}\mathrm{C}$ over the same composition range. Characteristically, there is a decrease in the growth rate with dilution at any crystallization temperature. However, the relative decrease in the growth rate is much greater at the lower crystallization temperatures. The spherulite radii vary linearly with time over the composition and temperature ranges studied, indicating a constant growth rate.[1]

Figure 11.2 represents the growth rate data for blends of poly(3-hydroxybutyrate) with one of either of two cellulose acetate butylates as the noncrystallizing component.(2) There are only small changes in both the glass and melting temperatures with composition in these blends. A maximum in the growth rate is observed for all compositions. Although the shapes of the curves are qualitatively similar to one another, the decrease in the growth rate is again greater at the lower crystallization temperatures. The growth rates are linear up to 50/50 composition for both the blends as illustrated.[2] A similar behavior, including the maxima in the growth rates,

[1] The spherulite radius of the crystallizing component in binary mixtures is not always linear with time. Nonlinear growth, and reasons thereof will be discussed in Sect. 11.7.

[2] For the 50/50 and higher blends the growth rate increases with time in one of the blends. This behavior has been attributed to the crystallization of the cellulose ester component at the higher concentration.(2)

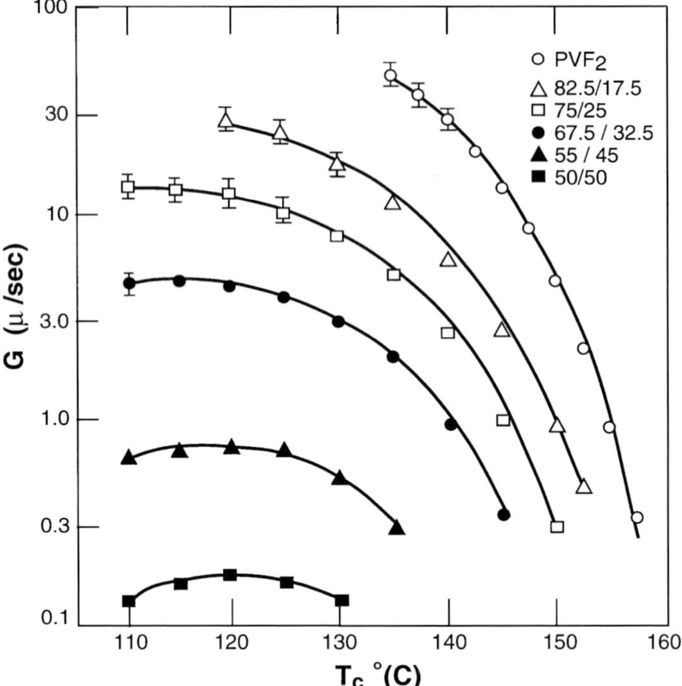

Fig. 11.1 Spherulite growth rate of poly(vinylidene fluoride) as a function of crystallization temperatures in blends with poly(methyl methacrylate) at indicated composition. (From Wang and Nishi (1))

has also been observed in other miscible blends of poly(3-hydroxybutyrate) with other noncrystallizing components.(3,4)

Spherulite growth rates, over a broad composition range, of blends of poly(pivalolactone) with poly(vinylidene fluoride) are given in Fig. 11.3.(5) In this mixture the poly(vinylidene fluoride) component is noncrystalline at all compositions over the range of crystallization temperatures studied. As the concentration of poly(vinylidene fluoride) increases in this blend the glass temperature decreases from −3 °C for pure poly(pivalolactone) to −38 °C for the pure noncrystallizing component. Concomitantly, the equilibrium melting temperature is reported to decrease by about 30 °C. This is a rather large change for a binary mixture.[3] The data in Fig. 11.3 dramatically demonstrate the decrease in growth rates with dilution. The retardation in growth rate is again much greater at the lower crystallization temperature.

A common feature of the cited examples is the decrease in growth rate with dilution. Yet, in one case the glass temperature increases with the concentration of the

[3] The equilibrium melting temperatures were obtained by extrapolative methods. The nonlinear nature of the data involved in the extrapolation allows for the distinct possibility of a large uncertainty in the T_m^0 values obtained.

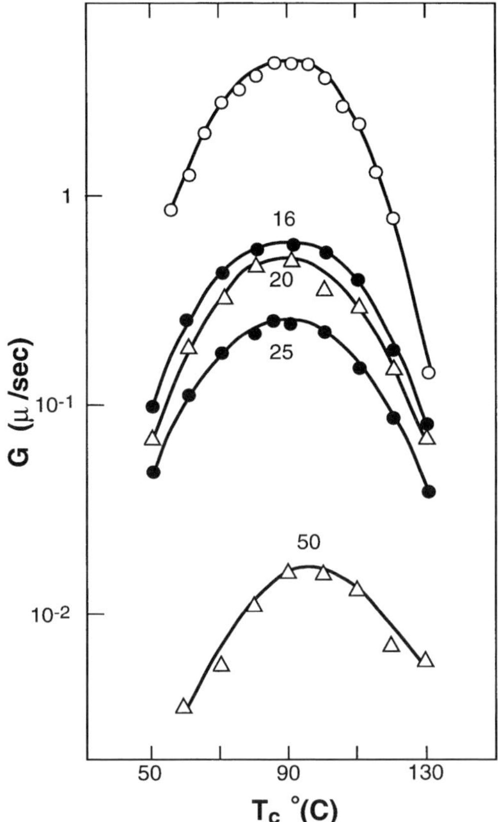

Fig. 11.2 Spherulite growth rate of poly(3-hydroxybutyrate) as a function temperature in blends with two different cellulose acetate butyrates. Numbers on curves cellulose ester weight percent. (From Pizzoli *et al.* (2))

noncrystallizing component,(1) in another it decreases,(5) and yet in the third there is not much change in the glass temperature with dilution.(2) The decrease in the growth rate, as well as in the overall crystallization with an added noncrystallizing component is a general, but not unique, phenomenon. It has been observed in blends of a wide variety of crystalline homopolymers.(6–13a) However, exceptions to this generalization have been observed. For example, the spherulite growth rate of isotactic poly(styrene) in blends with poly(vinyl methyl ether) was found to increase with the added second component.(14) In a blend with poly(1,4-butylene adipate), poly(vinylidene fluoride) crystallizes more rapidly than the pure polymer.(15) In another example, poly(ethylene-2,6-naphthalate) follows a similar pattern.(16) It crystallizes more rapidly in a blend with poly(1,4-butylene sebacate) than it does by itself. It can be concluded, however, as a general proposition, that in miscible blends the spherulite growth rate of the crystallizing component decreases with the

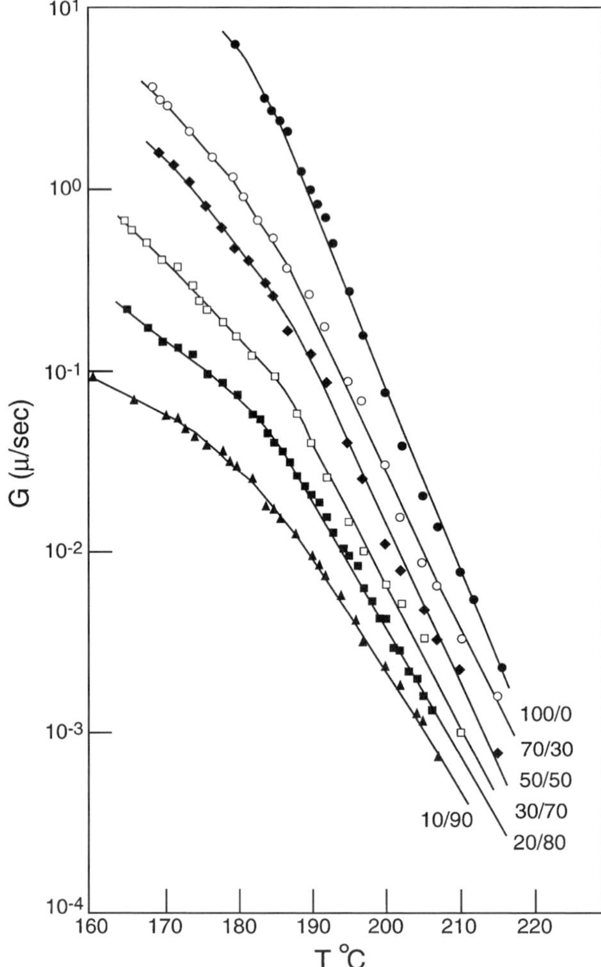

Fig. 11.3 Spherulite growth of poly(pivalolactone) as a function of temperature in blends with poly(vinylidene fluoride) at indicated compositions. (Data from Huang *et al.* (5))

addition of the non-crystallizing component. This effect is observed irrespective of how the glass temperature changes with composition. Except in a few reported cases, the equilibrium melting temperature is only depressed a small amount by the addition of the second component. Hence, there will not be any significant change in the effective undercooling during isothermal crystallization. Although the interval between glass and melting temperatures must influence the growth kinetics, other factors must be involved as well. There is an obvious dilution effect and the influence of the second component on the transport and nucleation term in the crystallization rate. These factors will be discussed in the next section.

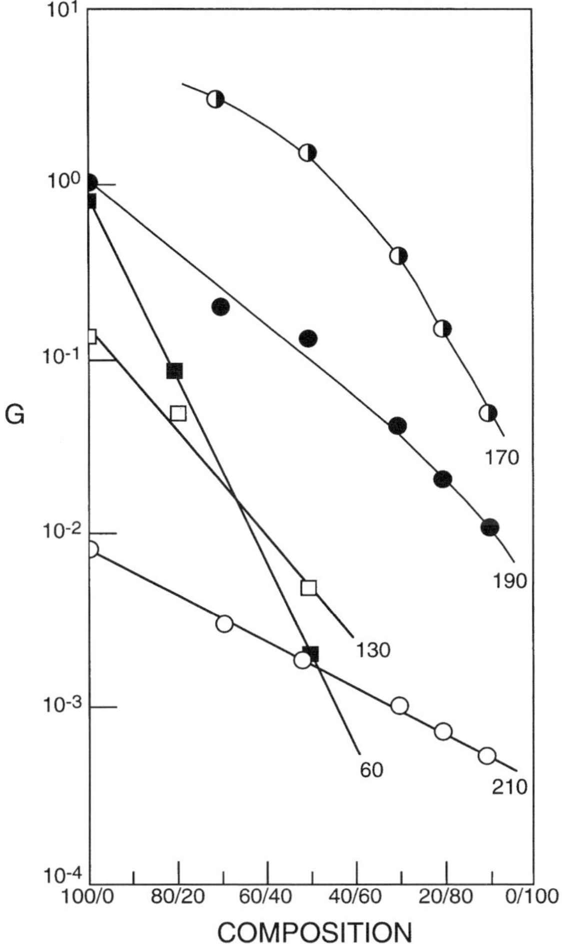

Fig. 11.4 Plot of spherulite growth rates of poly(pivalolactone) as a function of composition in blends with poly(vinylidene fluoride) (circles), pivalolactones and poly(3-hydroxybutyrate) in cellulose acetate butyrate (squares), at indicated crystallization temperatures. (Data from (2) and (5))

The varying influence of dilution at a fixed crystallization temperature is illustrated in Fig. 11.4. Here, the growth rates for the poly(pivalolactone)–poly(vinylidene fluoride) mixtures and one of the poly(3-hydroxybutyrate) blends are plotted against the composition for several different crystallization temperatures. The more extensive data for the poly(pivalolactone)–poly(vinylidene fluoride) blends emphasize the role of crystallization temperature. At the highest temperature shown, 210 °C, the growth rate changes by about an order of magnitude over the concentration range studied. In contrast, at the low crystallization temperatures the growth rates span more than two orders of magnitude over the same composition

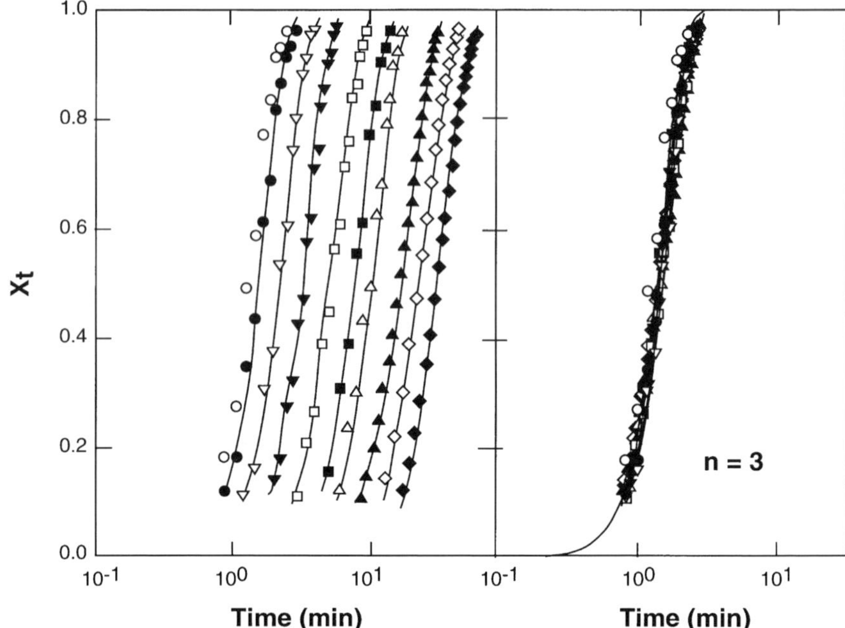

Fig. 11.5 Plot of relative fraction transformed against log time for poly(ϵ-caprolactone) in a 30/70 blend with poly(hydroxy ether) of bisphenol A phenoxy for a set of isothermal crystallization temperatures. (Data from (17))

range. The poly (3-hydroxybutyrate)–cellulose acetate butyrate blends follow a similar pattern, as do other blends that have been studied.

The spherulite growth and overall crystallization rates complement one another in blends, as they do in other systems. A typical set of isotherms of the crystallizing component in binary mixtures is shown in Fig. 11.5. The plots in Fig. 11.5 are for a 30/70 blend of poly(ϵ-caprolactone) with poly(hydroxy ether) of bisphenol A phenoxy for a range of isothermal crystallization temperatures.(17) Sigmoidal shape isotherms, typical of homopolymers, as well as other blends of this type, are observed.(6) The isotherms superpose very nicely as is indicated by the plot on the right of the figure. The data are fitted quite well with an Avrami $n = 3$ over the complete transformation range, indicative of interface control of the growth. The isotherm shapes are maintained when the composition is named at a fixed isothermal crystallization temperature. This point is illustrated in Fig. 11.6. Here a set of isotherms is given for different compositions at a fixed crystallization temperature. The isotherm shapes are maintained at all compositions so that they superpose very nicely with an Avrami exponent equal to 3. There is a significant increase in the time scale for crystallization as the concentration of the crystallizing component decreases. These results are qualitatively similar to those observed for

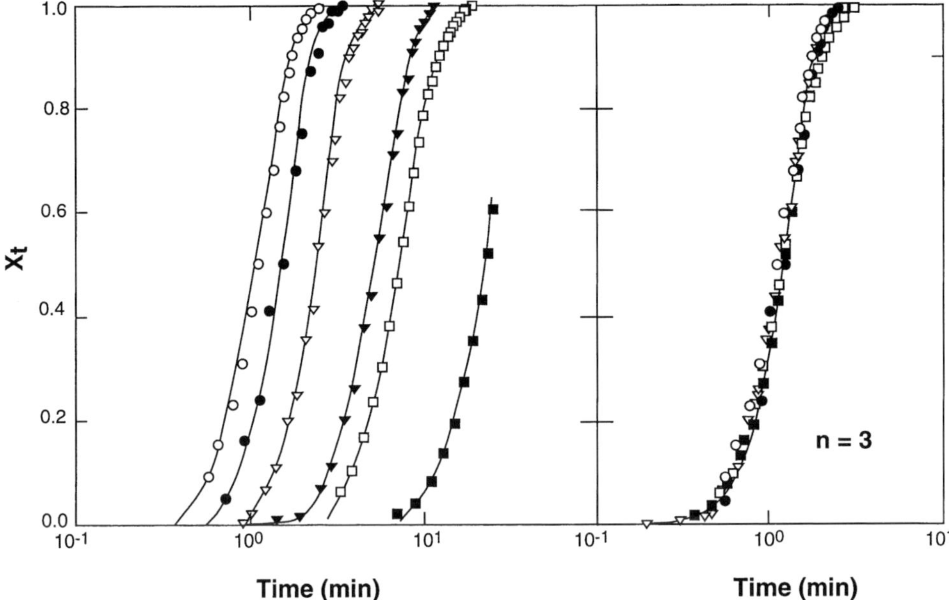

Fig. 11.6 Plot of relative fraction transformed against log time at different compositions at 309.6 K for same blend as in Fig. 11.5. Composition: ○ 100/0; ● 90/10 ▽ 80/20; ▲ 70/30; □ 65/35; ■ 60/40. (Data from (17))

spherulite growth rates. The overall crystallization kinetics of other binary blends behave in a similar manner.

In the example given in Figs. 11.5 and 11.6 the Avrami n remains constant with dilution. However, this is not always the case. There are examples where n increases with addition of the second component,(6) as well as those where it decreases.(12,22) These variations in n apparently reflect changes in the morphology with dilution.

Two typical examples of the overall crystallization rate, expressed as either $t_{0.5}$ or peak time, are given in Fig. 11.7 for poly(ethylene oxide)–poly(vinyl phenol) (18) and for poly(aryl ether ether ketone)–poly(ether imide) (19) in Fig. 11.8. The dependence of the crystallization rates on composition are similar to one another and are closely related to the results for other binary mixtures. The overall crystallization rates follow the pattern established for spherulite growth rates. At the higher crystallization temperatures only a modest decrease in the rate is observed with the addition of the noncrystallizing component. However, with a decrease in the crystallization temperature the polymeric diluent becomes more effective in reducing the rate. Because of the retardation in the rate with dilution a much wider range in isothermal crystallization temperatures can be studied. Thus, for the more dilute blends a maximum in the rates with temperature can be observed. This is

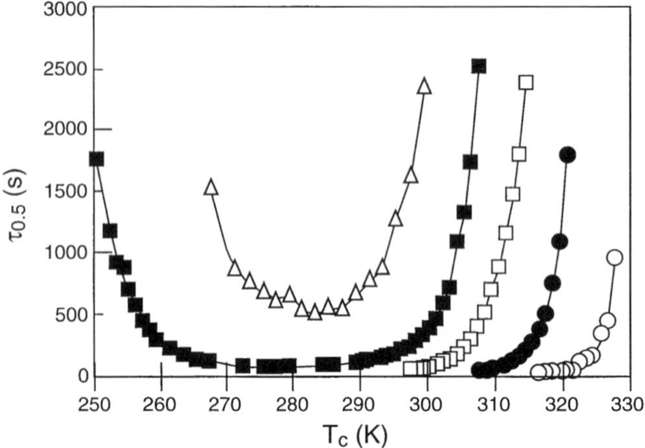

Fig. 11.7 Plot of half-time, $\tau_{0.5}$, for the crystallization of poly(ethylene oxide) against crystallization temperatures of blends of poly(ethylene oxide)–poly(*p*-vinyl phenol). Composition: ○ 100/0; ● 90/10; □ 80/20; ■ 70/30; △ 65/35. (From Pedrosa *et al.* (18))

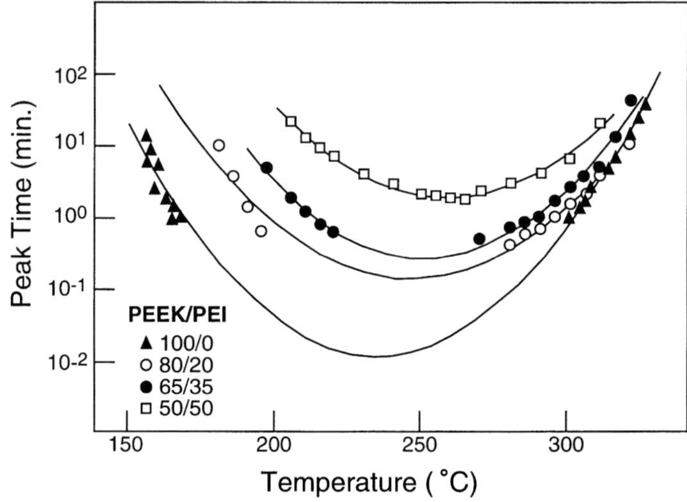

Fig. 11.8 Plot of isothermal crystallization peak time for poly(aryl ether ether ketone) in blend with poly(ether imide) for different compositions. Compositions indicated in figure. Peak time from isothermal exotherms using differential scanning calorimetry. (From Hsiao and Sauer (19))

indicative of the influence of the transport term in the crystallization process as the temperature is lowered. The ratios of $T_{\text{max}}/T_{\text{m}}$ for the two blends shown are in the usual range for pure homopolymers and copolymers discussed previously. It can be noted in Fig. 11.8 that the rate maximum moves to higher temperatures with the addition of poly(ether imide). A similar phenomenon is also observed in blends of poly(ethylene-2,6-naphthalene dicarboxylate) with poly(ether imide).(20) The homopolymer poly(ethylene-2,6-naphthalene dicarboxylate) (PEN) shows a temperature maximum in its overall crystallization rate. When blended with poly(ether imide) the crystallization rate decreases with dilution at all crystallization temperatures. Concomitantly, the temperature of the maximum is shifted to higher temperatures. The reason for the shift in both cases can be attributed to a significant increase in the glass temperature with increasing concentration of the poly(ether imide) component. This shift in the maximum shows that in fact the location of the glass temperature, and the narrowing of the window, can influence the crystallization rate in a significant manner.

Attention also has to be given to the role of molecular weight of each of the components. Of particular importance is the molecular weight of the noncrystallizing component. Some interesting results have been reported.(21) An example is given in Fig. 11.9 for blends of poly(ethylene oxide), $M = 90\,000$, and poly(methyl methacrylate) of varying molecular weight. Each blend in this example has the same composition, 70% by weight of poly(methyl methacrylate). As the molecular weight increases the spherulite growth rate decreases at all crystallization temperatures. It should be noted that the asymptotic values of the glass temperatures are

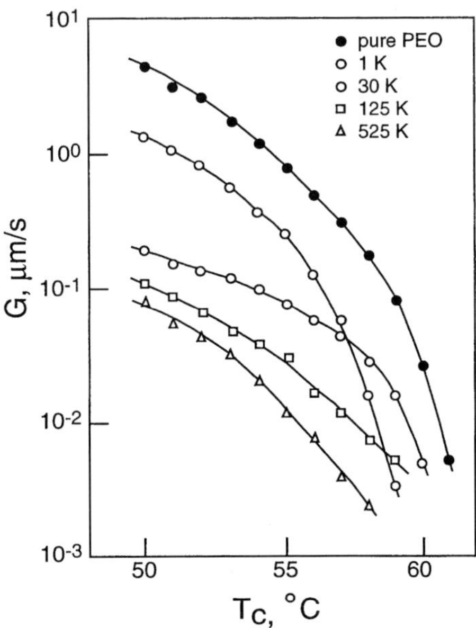

Fig. 11.9 Plot of spherulite growth rate of poly(ethylene oxide) in a blend with poly(methyl methacrylate) as a function of crystallization temperatures. Composition of blends 70/30 by weights. Molecular weights of the poly(methyl methacrylate) fractions are indicated. (From Alfonso (21))

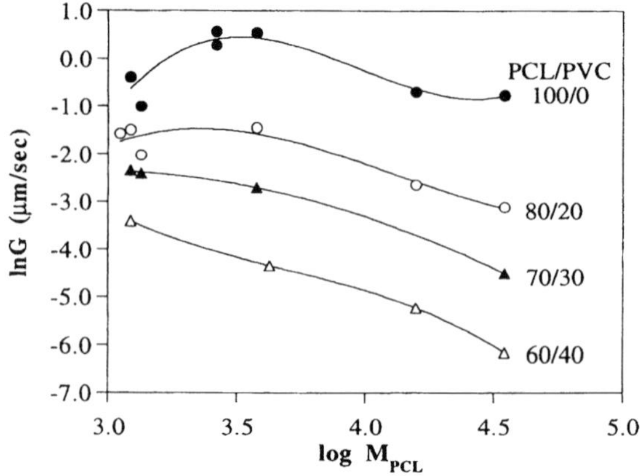

Fig. 11.9a Plot of spherulite growth rate at 30 °C of poly(ϵ-caprolactone) as a function of its molecular weight for mixtures with poly(vinyl chloride). Compositions of mixtures are indicated. (From Chen *et al.* (21a))

reached in the blends with the two highest molecular weight poly(methyl methacrylate) fractions. Thus, the decrease in growth rates cannot be attributed to variations in the glass temperature. There is also no sensible change in the equilibrium melting temperature. Thus, the temperature window for crystallization is the same in both blends. Therefore, other factors must be involved in causing the decrease in the growth rate. It has been suggested that the diffusion of the noncrystallization component away from the growth front plays a major role.(21)

Similar results have been obtained with blends of poly(vinylidene fluoride) and poly(vinyl pyrrolidene).(21) The growth rate of the crystallizing poly(vinylidene fluoride) in a 9 w/w mixture is reduced by about one order of magnitude as the molecular weight of poly(vinyl pyrrolidene) increases from 10 000 to 750 000. The results are again suggestive of the influences of diffusion processes. A more detailed consideration of diffusion of the noncrystallizing component will be given in Sect. 11.7.

Based on the experimental results that have been described, it is natural to inquire whether the basic theoretical framework that has been developed for the crystallization kinetics of one-component systems can be adapted to polymer mixtures. This question is addressed in the next section.

The molecular weight of the crystallizing component also has a strong influence on the spherulite growth rate. This is illustrated in Fig. 11.9a for the growth rate, at 30 °C, of poly(ϵ-caprolactone) as a function of its molecular weight for different mixtures with poly(vinyl chloride).(21a) A maximum in the growth rate is observed for the pure poly(ϵ-caprolactone) as was previously noted (see Fig. 9.96). The maximum persists in the 80/20 blend, although it is much broader. With the further addition of poly(vinyl chloride) the maximum disappears and the growth rates decrease monotonically with the molecular weight of the crystallizing component. In contrast, a maximum in the growth rate is observed for the 70/30 blend as the molecular weight of the noncrystallizing component, poly(vinyl chloride), increases.(21a)

11.2.2 Theory: two-component miscible melts

In order to examine the role of composition on the spherulite growth rate of the crystallizing component in a miscible blend, and indirectly on the overall crystallization rate, the effect of dilution on each of the terms in the growth rate equation for a one-component system needs to be modified. Consequently, starting with Eq. (9.205) it is found that (23–25)

$$G(v_2) = v_2 G_0 \exp\left[\frac{-U^*(v_2)}{RT - T_{\mathrm{g}}(v_2) + C(v_2)} - \frac{\Delta G^*(v_2)}{RT}\right] \tag{11.1}$$

The conventional pre-exponential factor, G_0, needs to be multiplied by v_2, the volume fraction of the crystallizing component, since the nucleation rate is proportional to the concentration of crystallizable units.

The exponential transport term is influenced by dilution in several ways. The constants C and U^* will in general depend on composition. The glass temperature, T_{g}, of a miscible system varies continuously with composition. It can be either determined experimentally, or calculated from one of several theoretical relations that are available. The influence of the constants U^* and C on the growth–temperature relation was discussed in detail in Chapter 9. It was found that a set of universal constants does not exist. These constants can be arbitrarily chosen, within broad limits. The choice has a profound effect over an extended range of crystallization temperatures.

There are several reasons why the free energy of forming a critical size nucleus depends on composition. Therefore, it is informative to derive the appropriate expressions.[4] The free energy of homogeneously forming a three-dimensional cylindrically shaped nucleus from the melt of a collection of uniform polymer chains in a binary mixture can be expressed, in the high molecular weight approximation, as (23–25)

$$\Delta G(v_2) = \zeta\rho\,\Delta G'_{\mathrm{u}} - 2\rho\sigma_{\mathrm{en}} - 2\zeta\sigma_{\mathrm{un}}(\pi\rho)^{1/2} + RT \ln v_2^{\rho_2} \tag{11.2}$$

The last term in Eq. (11.2) is an entropic contribution to the free energy that results from the probability of selecting the ρ polymer sequence from the binary mixture.

[4] The analysis that follows is very similar to that for the mixture of a polymer with a low molecular weight diluent.(22–25)

The other terms are similar to those for homopolymer nucleation. The bulk free energy of fusion per repeating unit at the specified composition is given by $\Delta G'_{\rm u}$. It represents the bulk free energy of fusion of the pure polymer plus that of mixing, or dilution. It has been tacitly assumed in this derivation that the two species are uniformly distributed throughout the system. Therefore, Eq. (11.2) will not be valid when either of the components is present at a low concentration. Analysis of the free energy surface given by Eq. (11.2) yields

$$\zeta^* = \frac{4\sigma_{\rm en} - 2RT \ln v_2}{\Delta G'_{\rm u}(v_2)} \tag{11.3}$$

and

$$\rho^* = \frac{4\pi\sigma_{\rm un}^2}{[\Delta G'_{\rm u}(v_2)]^2} \tag{11.4}$$

for the coordinates of the saddle points. It then follows that

$$\Delta G^*(v_2) = \frac{8\pi\sigma_{\rm un}^2\sigma_{\rm en} - 4\pi RT\sigma_{\rm un}^2 \ln v_2}{\Delta G'_{\rm u}(v_2)} \tag{11.5}$$

There are thus two specific places where the dilution directly influences $\Delta G^*(v_2)$; the $\ln v_2$ term and $\Delta G'_{\rm u}(v_2)$. The latter can be approximated by $\Delta G' \cong \Delta H_{\rm u}(T_{\rm m} - T)/T_{\rm m}$. $T_{\rm m}$ is now the equilibrium melting temperature of the crystallizing polymer in the mixture. There is also the distinct possibility, which cannot be ruled out, that either, or both, of the interfacial free energies, $\sigma_{\rm en}$ and $\sigma_{\rm un}$, depend on composition.

A similar analysis can also be carried out for the formation of a coherent unimolecular nucleus on an already existing crystal face. Following the procedure used for the pure polymer, it is found that (25)

$$\Delta G^* = \frac{4b_0\sigma_{\rm un}\sigma_{\rm en}}{\Delta G'_{\rm u}} - \frac{2\sigma_{\rm un}RT \ln v_2}{b_0\Delta G'_{\rm u}} \tag{11.6}$$

for this type of nucleus. In analyzing the experimental spherulite growth rates it will be assumed that a coherent, unimolecular nucleation process is involved.[5] Consequently,

$$G(v_2) = G_0 v_2 \exp\left\{\frac{-U^*}{T - T_{\rm g} + C} - \frac{4b_0\sigma_{\rm un}\sigma_{\rm en}T_{\rm m}}{RT\Delta H_{\rm u}(T_{\rm m} - T)} + \frac{2\sigma_{\rm un}T_{\rm m} \ln v_2}{b_0\Delta H_{\rm u}(T_{\rm m} - T)}\right\} \tag{11.7}$$

In writing Eq. 11.7 the tacit assumption has been made that the crystallization is taking place in Regimes I or III, i.e. only the steady-state nucleation is rate

[5] It should be recalled that the adoption of this type of nucleus involves a basic assumption that has not been given independent verification. Other types of two- and three-dimensional nucleations also fit the experimental data. Upon analysis, they lead to the same general conclusions.

determining. In Regime II, where the spreading rate has to be taken into account, ΔG^* of Eq. 11.6 needs to be divided by two. In the diffuse transition region, between the asymptote of Regimes I and II, an analysis similar to that given for pure homopolymers and copolymers needs to be employed. It is very possible that dilution will cause a shift or change in regime transitions.

The last term in Fig. (11.7) is often written as $0.2T_m \ln v_2/(T_m - T)$.(26) This expression, which is commonly used, is a consequence of the assumption that $\sigma_{un} = 0.1b_0\Delta H_u$. The basis for this assumption has been discussed in detail in Chapter 9. It is clear that severe limitations are imposed on this expression. There are serious questions as to the validity of the constant, 0.1, that is used. As has been discussed, this constant is not even on a firm basis when applied to linear polyethylenes. The ratio $\sigma_{un}/b_0\Delta H_u$ is best taken as an arbitrary parameter when analyzing experimental data.

Equation (11.7) is based on the corresponding relation for the pure polymer by accounting for the concentration dependence of each term. Although this appears to be a valid procedure, the crystallization process could in fact be more complex. Implicit in deriving Eq. (11.7) is the assumption that the spherulite radius increases linearly with time. This observation implies interface control. The growth rate is then constant at a given crystallization temperature. Although this assumption is valid for pure homopolymer crystallization, deviations have been observed in miscible blends. The possible importance of the diffusion of the noncrystallizable component away from the growth form has already been alluded to. These two factors are intimately related and will be discussed subsequently. At this time the adherence of experimental data to Eq. (11.7) will be examined. In examining the validity of Eq. (11.7) examples are selected where the spherulite growth rate is constant. Thus, the complexities of nonlinear growth and the importance of diffusion can be neglected.

As a first approximation in using Eq. (11.7) it is assumed that U^*, C, σ_{en}, σ_{eu} and G_0 are independent of composition. The melting temperature, T_m, the glass temperature, T_g, b_0 and ΔH_u will be known from independent measurements. With these assumptions Eq. (11.7) can be written

$$\ln G(v_2) - \ln v_2 - \frac{U^*}{T - T_g(v_2) + C} - \frac{2\sigma_{un} T_m(v_2) \ln v_2}{b_0\,\Delta H_u\,\Delta T(v_2)} = \ln G_0 - \frac{4b_0\sigma_{un}\sigma_{en}T_m(v_2)}{R\,\Delta H_u\,T\,\Delta T(v_2)} \tag{11.8}$$

The left-hand side of Eq. (11.8) contains the measured growth rate, the known composition dependent terms, the constants U^* and C and the ratio $\sigma_{un}/b_0\,\Delta H_u$. The latter factor is taken to be an arbitrary parameter. The values for the constant U^* and C are arbitrarily selected. It is recalled, from the analysis of pure homopolymers, that drastic alterations can result by varying U^* and C. A plot of the left-hand side of Eq. (11.8) should be linear for a given composition. A common straight line for different compositions should result if the assumptions made are valid.

Appropriate plots to assess Eq. (11.8) are given in Fig. 11.10 for blends of poly(ethylene oxide)–poly(methyl methacrylate).(27) Different values of the ratio $\sigma_{un}/b_0\Delta H_u$, defined as ψ, are used in the figure. The values $U = 1500$, $C = 30$ were arbitrarily selected for all compositions for illustrative purposes. The T_g and T_m values reported were used. The data for the different compositions, plotted according to Eq. (11.9), fall on the same straight line for each of the three values of ψ that are illustrated. Thus, the three values of ψ used give good agreement between theory and experiment although $\psi = 0.1$ gives a slightly better fit. Analysis shows that poly(vinylidene fluoride)–poly(methyl methacrylate) blends behave in a similar fashion.(1) The results are plotted in Fig. 11.11 according to Eq. (11.8). In this plot $U^* = 1500$ and $C = 30$ are again assumed and ψ is taken to be 0.05. Once again the blends of different composition delineate the same straight line. When different values of ψ are used, as is illustrated, 0.05 gives the best fit. However, when the results for other type blends are analyzed they do not always follow the pattern given in Figs. 11.10 and 11.11. In some cases, close to parallel straight lines are found with varying blend composition. In other blends, straight lines are not observed. A case in point is the spherulite growth rates of blends of poly(pivalolactone)–poly(vinylidene fluoride).(5)

The growth rate data of these blends, given in Fig. 11.12, encompass a wide range in composition. These data are plotted in Fig. 11.12 according to Eq. (11.8) for a range in ψ values. The glass and melting temperatures used in the calculations are those reported. A difference in melting temperatures of 30 °C between the pure homopolymer and the 10/90 blend was reported. The values of U^* and C were again taken to be 1500 and 30 respectively. As is indicated in Fig. 11.12 the data cannot be represented by a common straight line or even by a straight line for each composition, for any of the values of ψ. Rather, a family of curves, with gradual curvature, emanate from a common origin. For a ratio of 0.125 (Fig. 11.12c) there is a change in curvature direction with dilution. On the other hand, for a ratio of 0.05 (Fig. 11.12b), the direction of curvature is the same for all compositions. As the ratio decreases below 0.05, that nature of the curves remains the same but they are positioned closer to one another. Even if the last term in Eq. (11.8) is neglected, as in Fig. 11.12a, the data for the different compositions do not fall on a common curve. The common origin for the curves in Fig. 11.12, for each value of ψ, indicates that G_0 is independent of composition. It is thus clear that within the limits of the assumptions that were made Eq. (11.8) does not explain the dependence of the growth rate data of these blends on composition and crystallization temperature.

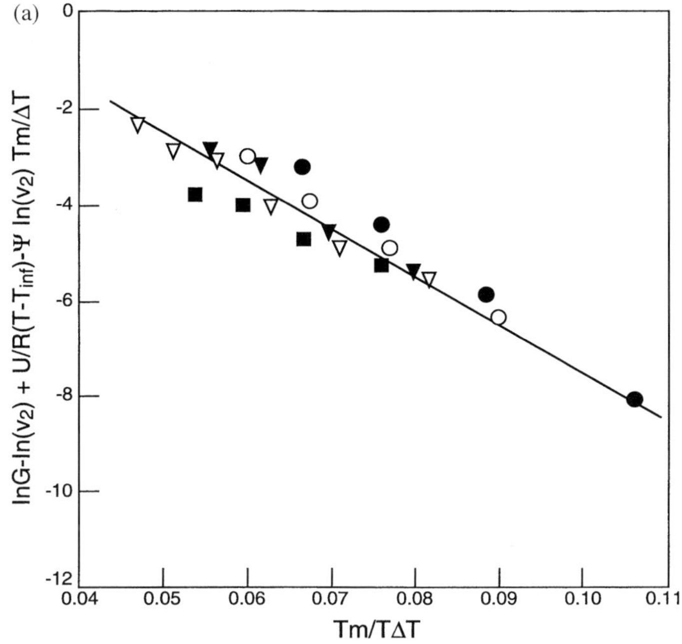

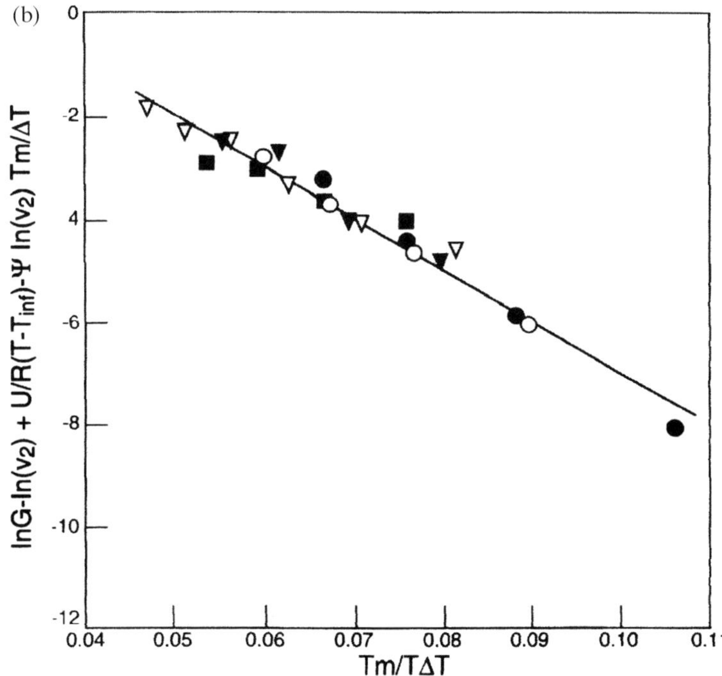

Fig. 11.10 Plots of left-hand side of Eq. (11.8) against $T_{\mathrm{m}}/T\ \Delta T$ for blends of poly(ethylene oxide)–poly(methyl methacrylate). Composition poly(ethylene oxide)/poly(methyl methacrylate): ● 100/0; ○ 90/10; ▼ 80/20; ▽ 70/30; ■ 60/40. (a) $\psi = 0.05$, (b) $\psi = 0.10$, (c) $\psi = 0.15$. (Data from (27))

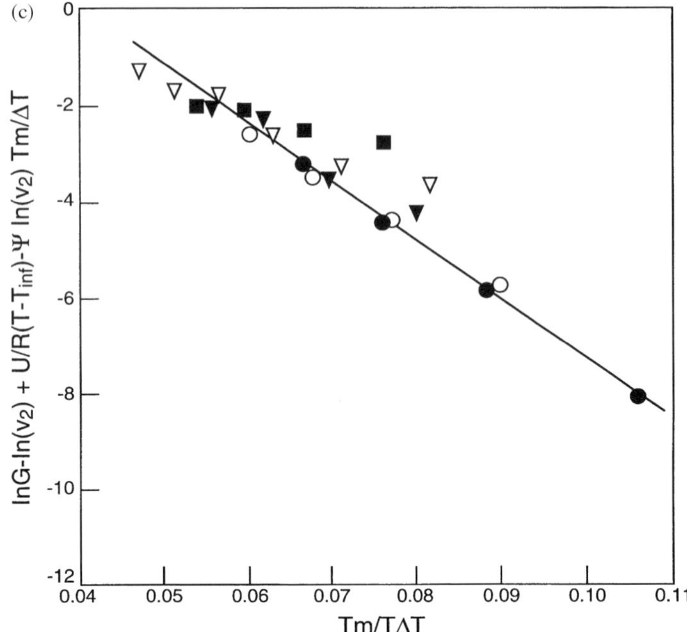

Fig. 11.10 (*cont.*)

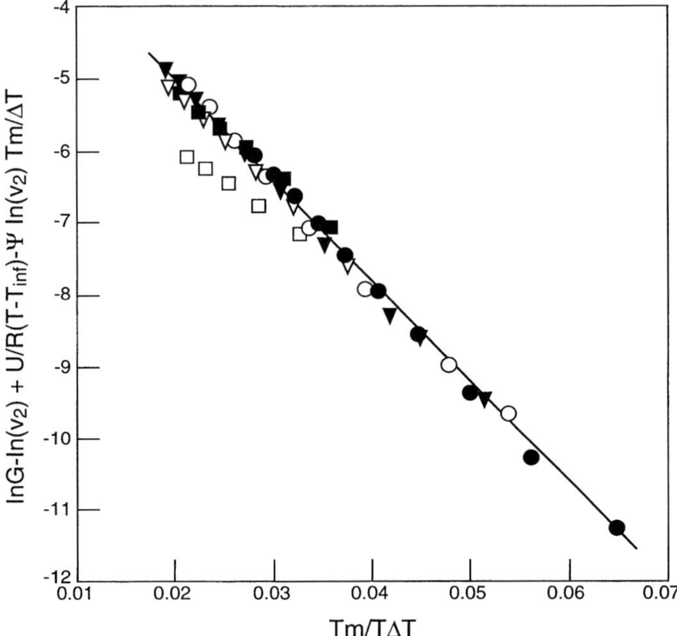

Fig. 11.11 Plot of left-hand side of Eq. (11.8) against $T_m/T\,\Delta T$ for blends of poly(vinylidene fluoride)–poly(methyl methacrylate). Composition poly(vinylidene fluoride)/poly(methyl methacrylate): ● 100/0; ○ 82.5/17.5; ▼ 75/25; ▽ 67.5/32.5; ■ 55/45; □ 50/50. $\psi = 0.05$. (Data from (1))

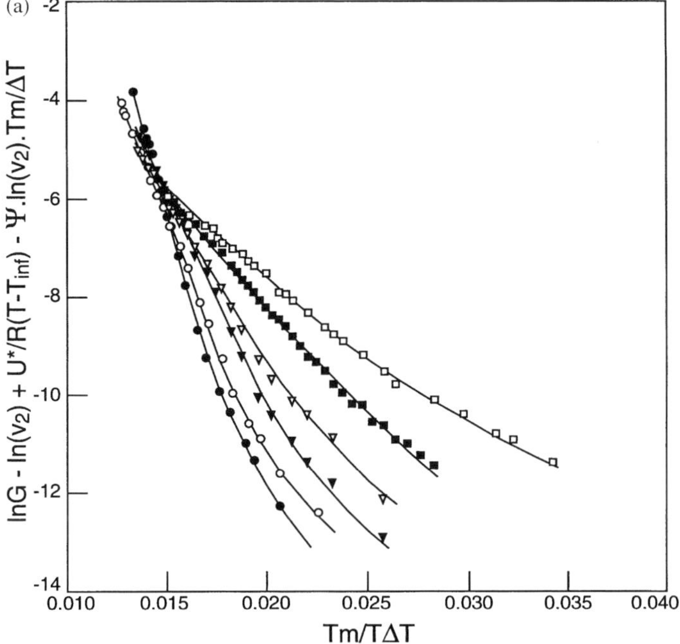

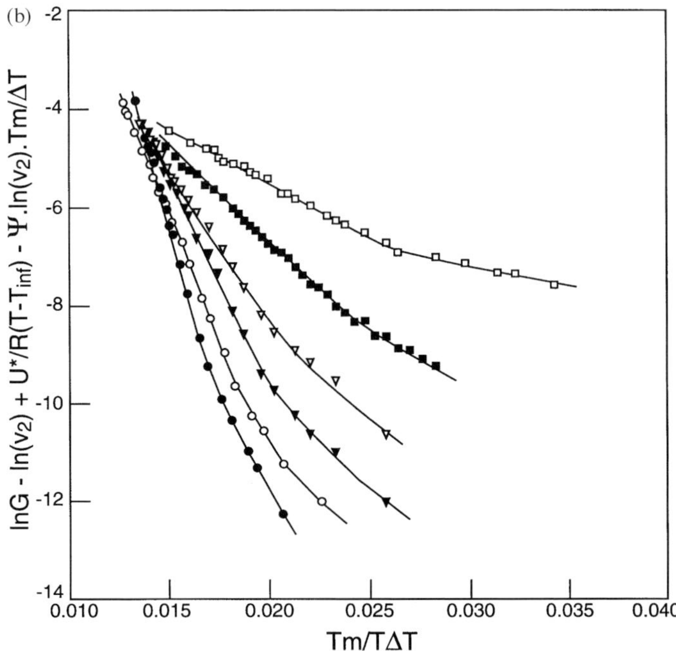

Fig. 11.12 Plot of left-hand side of Eq. (11.8) against $T_m/T\,\Delta T$ for blends of poly(pivalolactone)–poly(vinylidene fluoride). Composition poly(pivalolactone)/ poly(vinylidene fluoride): □ 100/0; ■ 70/30; ▽ 50/50; ▼ 30/70; ○ 20/80; ● 10/90. (a) $\psi = 0$; (b) $\psi = 0.05$; (c) $\psi = 0.125$. (Data from (5))

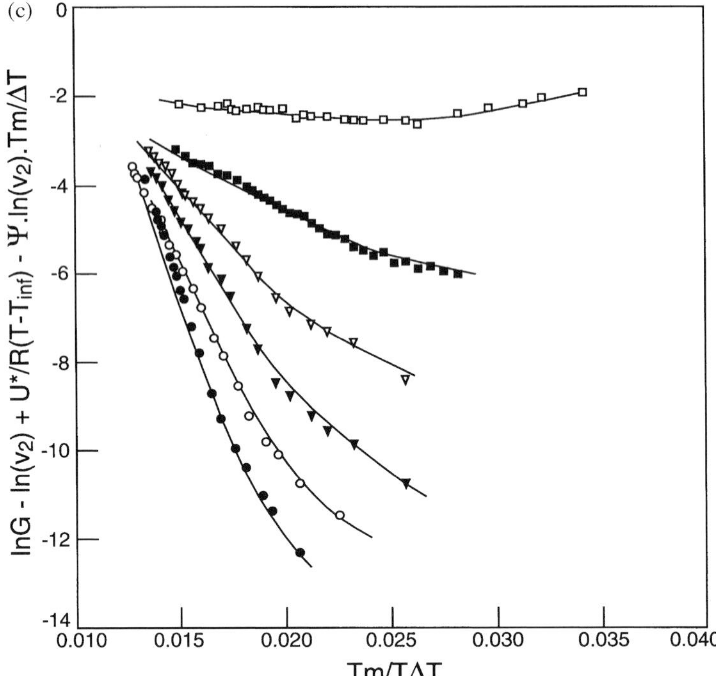

Fig. 11.12 (*cont.*)

Before inquiring as to possible reasons for the difference in behavior between these blends and those illustrated in Figs. 11.10 and 11.11 it is helpful to closely examine the nature of the curves in Fig. 11.12. Taking the plots in Fig. 11.12b as an example, each curve can, within experimental error, be r.easonably represented by two intersecting straight lines. This possibility is reminiscent of the Regime III–II transition that has been discussed in Chapter 9. However, only the straight lines for the pure homopolymers and the 70/30 mixture give slope ratios close enough to 2.0 to be consistent with such a regime transition. The slope ratio of the other blend compositions cannot be reconciled with such a regime transition. In general a discussion of regime transitions in blends is premature unless the large number of parameters involved in Eq. (11.8) can be firmly established. The different results that can be obtained by the selection of constants has been demonstrated in the discussion of pure homopolymers.

There is concern as to why this particular blend behaves so differently from those illustrated in Figs. 11.10 and 11.11. Other blends such as poly(ethylene oxide)–poly(vinyl acetate),(28) poly(3-hydroxy butyrate)–cellulose acetate butyrate,(29) poly(3-hydroxybutyrate)–poly(ethylene oxide) (30) and poly(3-hydroxybutyrate)–

poly(epichlorhydrin),(31) among others, give a straight line for each composition when the data are plotted according to Eq. (11.8). However, all of the straight lines are displaced from one another, so that a common straight line is not observed. Important in the analyses are the values taken for the equilibrium melting temperatures of the pure crystallizing homopolymers and those of the blends. As was pointed out in Chapter 4, the decrease in the melting temperature with dilution should be small, except in unusual circumstances. However, for the poly(pivalolactone)–poly(vinylidene fluoride) blends the decrease in equilibrium melting temperature is reported to be rather large.(5) Equilibrium melting temperatures have to be determined by extrapolative methods.(32) These methods will be discussed and assessed in detail in the chapters dealing with morphology and structure. For present purposes it suffices to state that a linear extrapolation is involved. However, the data for the poly(pivalolactone) blends were curved. This fact complicates the determination of the equilibrium melting temperature and results in a serious uncertainty in the values. It is therefore reasonable to examine the influence of T_m on the calculated growth rate curves.

To accomplish this task, it has been arbitrarily assumed that the decrease in T_m between the pure crystallizing polymer and the 10/90 mixture is only 10 °C. This difference was then proportioned among the blends. The results of the calculation, with this change in T_m, are shown in Fig. 11.13. Focusing first on Fig. 11.13a, where the ratio equals 0.05, the curves for each of the compositions are now much closer to one another than the corresponding plot of Fig. 11.12. In fact, the curves for the pure polymers, the 70/30 and 50/50 blends are virtually identical to one another. Moreover, the data from the more dilute blends can now be represented by straight lines that are close to one another. Increasing or decreasing the value of ψ has the same effect as was observed in the plots of Fig. 11.11. Thus, the analysis of growth data of this blend requires, at a minimum, reliable values of the equilibrium melting temperature.

In general, no definitive conclusion can be made with regard to the quantitative validity of Eq. (11.8) to any given blend without independent knowledge of the many parameters that are involved. Of particular importance is the knowledge of the equilibrium melting temperatures of each composition. This problem is no different than discussed in Chapter 9 for the crystallization of pure homopolymers. This is a fundamental problem that underlies the analysis of all aspects of the crystallization kinetics of polymers. It has been found that Eq. (11.8) qualitatively satisfies the growth rates of several different type blends. The uncertainties in U^*, C and T_m for specific cases do not allow for the specification of ψ for any mixture. The universal use of $\psi = 0.1$ is unwarranted without further supporting evidence.

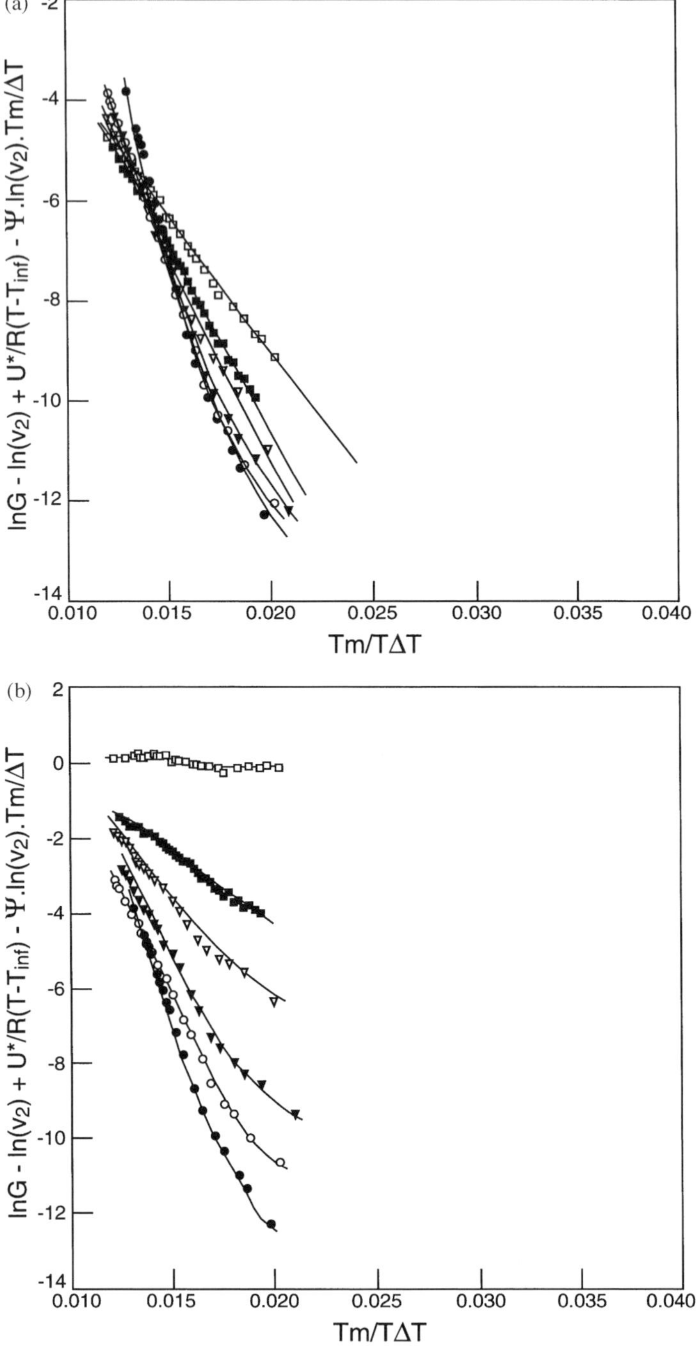

Fig. 11.13 Plot of left-hand side of Eq. (11.8) against $T_m/T\Delta T$ for blends of poly(pivalolactone)–poly(vinylidene fluoride). Equilibrium melting temperature arbitrarily proportional over a 10 °C interval (see text). Composition poly(pivalolactone)/poly(vinylidene fluoride): ● 100/0; ○ 70/30; ▼ 50/50; ▽ 30/70; ■ 20/80; □ 10/90. (a) $\psi = 0.05$, (b) $\psi = 0.25$. (Data from (5))

11.3 Miscible blend: both components crystallize

Interesting situations develop when both of the components crystallize from a miscible melt. The component that crystallizes first, usually the higher melting polymer, crystallizes from the two-component melt at the nominal composition. It would be expected to follow the same crystallization pattern that was discussed previously. However, the crystallization of the second, lower melting, component does not take place at the nominal composition. Rather, it takes place from a mixture of the residual noncrystalline portion of the high melting polymer and the second component. Moreover, it can be anticipated that the morphology that was established by the first component crystallizing will influence the crystallization of the second component. There is adequate data in the literature to analyze the crystallization kinetics of such systems and to test the expectations.

The crystallization kinetics of binary blends of poly(vinylidene fluoride)–poly(1,4-butylene adipate) provides a data set suitable for analysis.(33) To set the boundaries for the analysis, the melting temperature of the pure poly(vinylidene fluoride) is 175 °C while that of the poly(butylene adipate) 61 °C. There is, however, only a 15 °C difference between the glass temperatures of the two polymers: −45 °C for poly(vinylidene fluoride) and −60 °C for poly(butylene adipate). Poly(butylene adipate) does not form spherulites in this blend at any composition. However, poly(vinylidene fluoride) does so over a wide composition range. For blends that contain less than 60% by weight of poly(butylene adipate) the spherulite radii increase linearly with time up to impingement, indicating a constant growth rate. On the other hand, in blends that contain 60% or more of poly(butylene adipate) a non-linear spherulite growth rate develops at the advanced stages of the crystallization.

Plots of the spherulite growth rate of the poly(vinylidene fluoride) component are given in Fig. 11.14 for different compositions. The main features that were found previously in other miscible blends are again observed. It is evident that the spherulite growth rate is significantly depressed due to the addition of poly(butylene adipate). The retardation in rate is much greater at the lower crystallization temperatures than the higher ones. At the lower crystallization temperature, the growth rate is reduced by more than an order of magnitude. Another way of depicting these results is shown in Fig. 11.15. Here the growth rates are plotted against the undercooling for the different blends. At the high undercoolings there is a significant decrease in the growth rate with dilution. On the other hand, at the low undercoolings the diminution in the rate is small. These results indicate that the changes in rate with composition cannot be due solely to changes in the equilibrium melting temperatures of the blends. At the same time, the change in the glass temperature from one pure species to the other pure one is no greater than 15 °C. The observed results must, therefore, primarily be a direct consequence of the dilution itself.

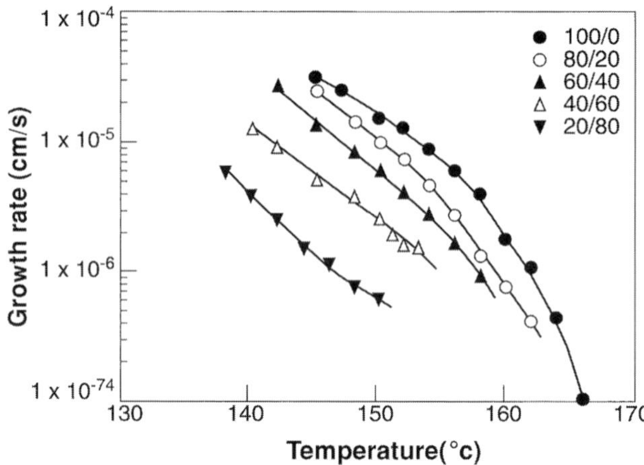

Fig. 11.14 Plot of spherulite growth rates of poly(vinylidene fluoride), in blends with poly(1, 4-butylene adipate), as a function of temperature at indicated composition. (From Pennings and Manley (33))

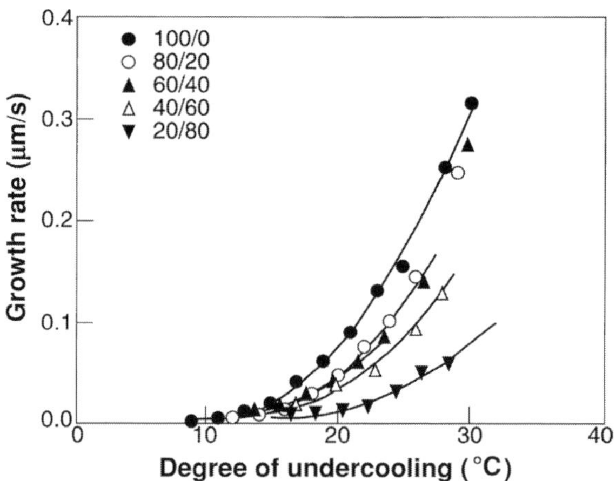

Fig. 11.15 Plot of spherulite growth rates of poly(vinylidene fluoride) in blend with poly(1,4-butylene adipate) as a function of temperature at indicated composition. (From Pennings and Manley (33))

The temperature interval for crystallization of most of the blends is small so that a quantitative analysis of the data according to Eq. (11.8) is precluded.

In contrast to the spherulite growth rates, the overall crystallization of both components can be resolved in these blends.(33) Typical isotherms are observed for the crystallization of poly(vinylidene fluoride). They can be fitted with an Avrami $n = 3$ for a significant portion of the transformation. There is a progressive shift of the isotherms to longer times with dilution. These results are thus consistent with the reduction in spherulite growth rates with the addition of poly(butylene

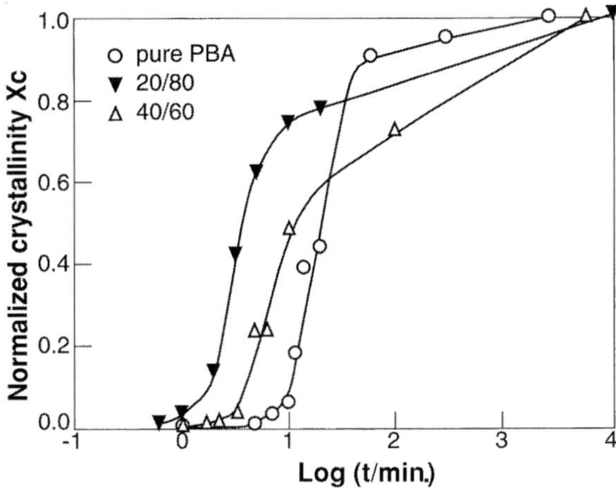

Fig. 11.16 Plot of normalized isothermal crystallization of poly(1,4-butylene adipate) in blends with poly(vinylidene fluoride) as a function of log time. Blend composition indicated. Crystallization temperature 43 °C. (From Pennings and Manley (33))

adipate). The crystallization of the poly(vinylidene fluoride) is expected because it crystallizes from a homogeneous melt.

The overall crystallization kinetics of the poly(butylene adipate) in the blends behaves quite differently. A set of isotherms, at a fixed temperature, for different compositions is given in Fig. 11.16.(33) In this case isotherms of the blends are shifted to shorter crystallization times relative to those for pure poly(butylene adipate). For this component the crystallization rates are faster in the blends illustrated. The overall crystallization rate is greatest in the 20/80 blend, indicating that there is a maximum in the rate when plotted against the concentration of poly(vinylidene fluoride). The overall crystallization rate of the blend that contains 20 wt percent of poly(vinylidene fluoride) is approximately five times faster than the pure poly(butylene adipate). The enhancements to the crystallization rate manifest themselves in the isotherm shapes that are illustrated in Fig. 11.16. The isotherms for pure poly(butylene adipate) fit an Avrami $n = 3$ quite well. However, the best fit is obtained with $n = 2$ for all blend compositions. Thus, a change in either the nucleation or growth rates, or both, is indicated. The unique crystallization of the poly(butylene adipate) component can be attributed to the influence of the already crystallized poly(vinylidene fluoride). These crystallites offer surfaces that can enhance the nucleation process. In addition, and important, is the fact that the crystallization takes place in constrained regions due to the presence of the poly(vinylidene fluoride) crystallites. The poly(butylene terephthalate) crystallizes within an ordered matrix formed by the component that crystallized first. Thus, the crystallization is probably restricted to the interlamellar region. This restriction will enhance the crystallization

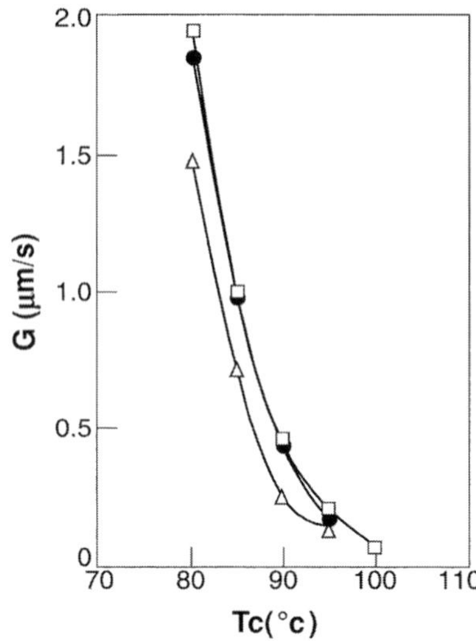

Fig. 11.17 Plot of spherulite growth rate of poly(butylene succinate) in blends with poly(vinylidene fluoride) as a function of crystallization temperature. Composition poly(butylene succinate)/poly(vinylidene fluoride): □ 100/0; ● 80/20; △ 60/40. (From Lee *et al.* (35))

rate. An enhanced crystallization rate of the lower melting component has also been observed in blends of poly(phenylene sulfide)–poly(ethylene terephthalate).(34) In blends of poly(ϵ-caprolactone)–poly(carbonate) the half-time for the crystallization of the lower melting poly(ϵ-caprolactone) increase slightly with the initial poly(carbonate) concentration. The rate then levels off as the poly(carbonate) concentration increases.(34a) This is another example of rate enhancement.

Both components in blends of poly(vinylidene fluoride)–poly(butylene succinate) also crystallize. However, in this mixture both polymers can form spherulites independent of one another so that their respective growth rates can be measured as a function of temperature and composition.(35) The growth rate of poly(vinylidene fluoride), the component that crystallizes first, follows the expected pattern with crystallization temperature and composition. The usual significant reduction in growth rate is observed with dilution. The spherulite growth rates, as a function of the crystallization temperature, of the lower melting component, poly(butylene succinate) are shown in Fig. 11.17. In this case the effect of dilution on the growth rate is quite small. The growth rate–temperature plot for the blend containing 80% poly(butylene succinate) is identical to that for the pure polymer. There is only a very small decrease in the rate, at all temperatures, for the blend containing 60% of this component. The small effect of dilution gives strong indication that the crystallite morphology established by the crystallization of the poly(vinylidene fluoride) component plays an important role in governing the crystallization of the

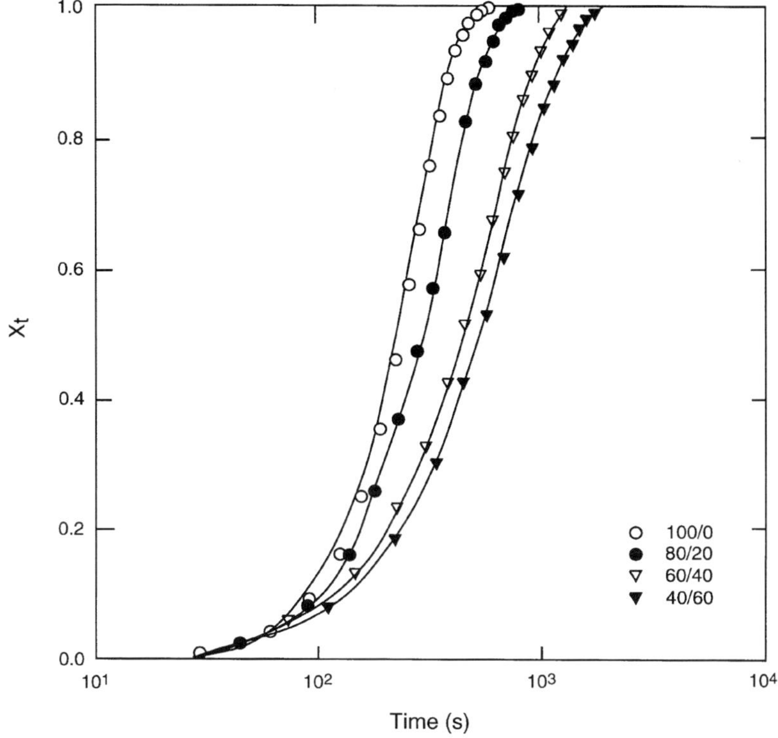

Fig. 11.18 Plot of normalized crystallinity level of poly(butylene succinate) in blends with poly(vinylidene fluoride) as a function of log time. Composition poly(butylene succinate)/poly(vinylidene fluoride): ○ 100/0; ● 80/20; ▽ 60/40; ▼ 40/60. (Data from (35))

poly(butylene succinate). This influence also manifests itself in the overall crystallization kinetics. Figure 11.18 gives the isotherms for crystallization at 190 °C for the different blends.(35) The time scale, from the pure polymer to the 40/60 blend, is limited. It is different from what would be expected if a crystalline matrix did not pre-exist. Although the initial stages of the crystallization can be fitted with an Avrami $n = 2$, deviations set in at the early stages of the transformation. The deviations become more accentuated as the blend becomes more dilute in poly(butylene succinate).

Poly(ethylene terephthalate) and poly(butylene terephthalate), which crystallize separately from one another, form miscible blends over the complete composition range.(36) A single glass temperature is observed, that varies from 300 K for pure poly(butylene terephthalate) to 340 K for pure poly(ethylene terephthalate). There are only small changes in the melting temperatures for either the poly(ethylene terephthalate) or poly(butylene terephthalate) rich mixtures and there is but a few degrees change in the melting temperature of blends dilute in one or the other

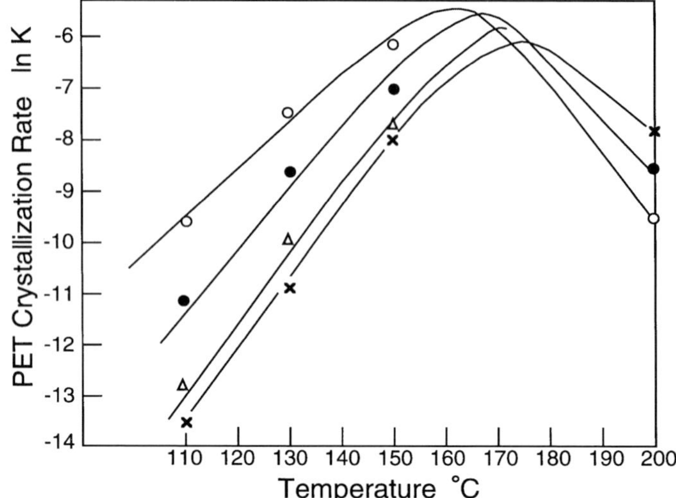

Fig. 11.19 Plot of crystallization rate constants of poly(ethylene terephthalate) in blends with poly(butylene terephthalate) as a function of crystallization temperature. Composition poly(butylene terephthalate)/poly(ethylene terephthalate): × 0/100; ▽ 10/90; ● 20/80; ○ 40/60. (From Escala and Stein (36))

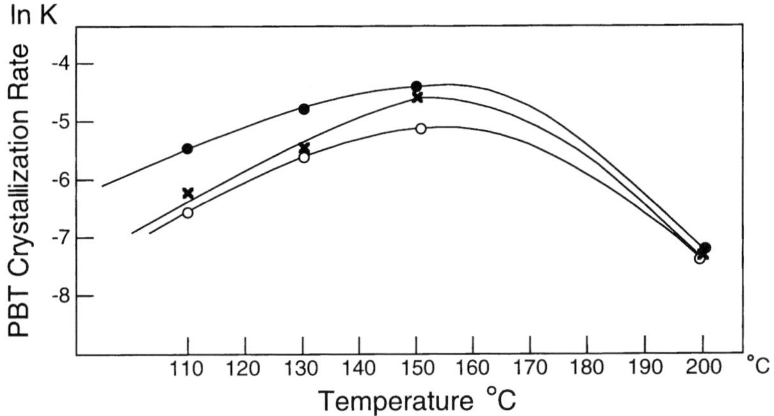

Fig. 11.20 Plot of crystallization rate constant of poly(butylene terephthalate) in blends with poly(ethylene terephthalate) as a function of crystallization temperature. Composition poly(butylene terephthalate)/poly(ethylene terephthalate): ● 100/0; × 80/20; ○ 60/40. (From Escala and Stein (36))

species. The higher melting poly(ethylene terephthalate) crystallizes first. The overall crystallization rates for each of the polymers are given in Figs. 11.19 and 11.20 as a function of the crystallization temperature for a series of compositions. The poly(ethylene terephthalate) component displays the characteristic rate maximum, typical of this polymer, at all compositions. The maxima shift to lower temperatures as the poly(ethylene terephthalate) concentration is reduced. The conventional behavior is found at the right side of the maxima. Here the overall crystallization rate, expressed in terms of the Avrami rate constant, decreases with dilution. However, the situation is quite different at the left side of the maximum. Here an inversion occurs. The crystallization rate actually increases with decreasing poly(ethylene terephthalate), which is the first component to crystallize. The differences in rate with dilution, depending on the crystallization temperature, are illustrated more clearly in Fig. 11.21. In this figure, half-times are plotted against the composition

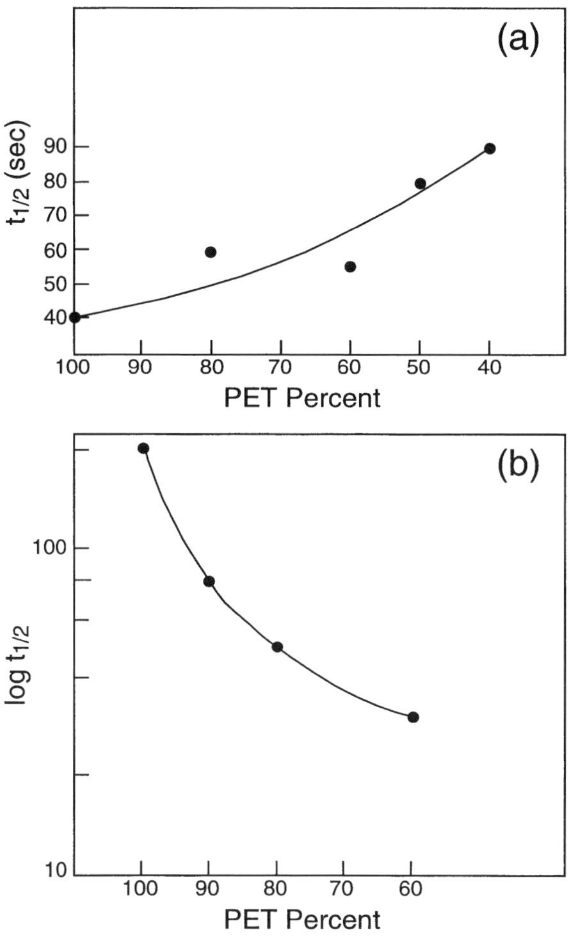

Fig. 11.21 Plot of half-time of poly(ethylene terephthalate) as a function of its concentration in blends with poly(butylene terephthalate). (a) Crystallization temperature 200 °C. (b) Crystallization temperature 130 °C. (From Escala and Stein (36))

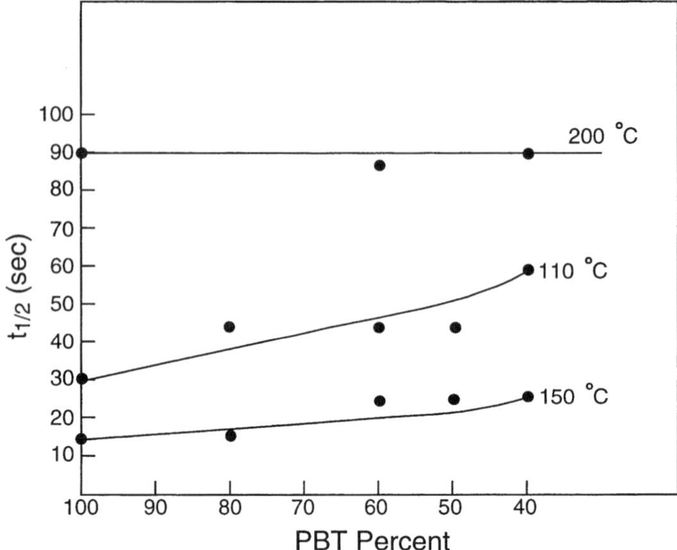

Fig. 11.22 Plot of crystallization half-time of poly(butylene terephthalate) as a function of its concentration in blends with poly(ethylene terephthalate), at the indicated crystallization temperatures. (From Escala and Stein (36))

for a temperature selected to be on either side of the maximum. In Fig. 11.21a, for crystallization at the right side of the maximum, there is a modest increase in the half-time indicating a retardation in the rate with dilution. In contrast, for crystallization at 130 °C, Fig. 11.21b, which is at the left side of the maximum, there is a decrease in $t_{1/2}$ indicating a significant increase in the crystallization rate. This is clearly opposite to what is expected solely from dilution. This enhancement of the crystallization rate can be attributed to the decrease in the glass temperature with the addition of poly(butylene terephthalate). For the composition range of interest here, pure polymer to the 60% blend, the glass temperature decreases by about 30 °C.

The overall crystallization rate of poly(butylene terephthalate), shown in Fig. 11.20, follows a different pattern. Although a maximum is still observed, it is much broader than that of poly(ethylene terephthalate). The influence of concentration on the crystallization rates in different temperature regions is summarized in Fig. 11.22, where plots of the half-time are given. At high temperatures there is virtually no change with composition. At the lower temperatures there is only a slight increase in half-time with dilution. The large change that is expected due to dilution alone is tempered by two factors. One is the increase of the glass temperature with the added poly(ethylene terephthalate). The other is the fact that the crystallization is occurring within an already existing crystallite morphology. The restraints that are imposed on the amorphous region will in general enhance the crystallization of the second crystallizing component.

The examples that have been used to illustrate the crystallization of miscible binary blends, wherein both components crystallize, allow for some generalizations to be made. The higher melting component behaves in the manner expected for miscible blends where only one component crystallizes, in that there is a continuous decrease in the crystallization rate as the noncrystallizing second component is added. However, the crystallization of the lower melting component is different. The crystallization is strongly influenced by the already existing crystallites. In turn the influence of dilution is minimal, and in some cases completely overcome. Thus, the crystallization rate is no longer reduced and in some cases enhanced as the second component is added.

11.4 Chemically identical components

11.4.1 Introduction

Blends that are composed of two polymers that have the same chemical repeating units present a special situation. There are several reasons for differences in crystallization behavior between chemically identical chains. These include stereo-differences between the chain units, as for example blends of isotactic, syndiotactic and atactic polymers. There can also be geometric differences in that chains can have predominantly cis or trans units. The presence of regio structures in one of the polymers is also a possibility. There can also be differences in chain length between the two components. Another set of examples in this category involves differences in the chain architecture. For the same repeating unit there can be linear chains, chains with either long or short chain branches, or both, as well as star type molecules. Blends where both components are chemically identical are well known and have been extensively studied. Irrespective of the chemical nature of the chains involved it is necessary to specify whether the species are miscible in all proportions, immiscible or partially miscible in order for a proper analysis to be made. Chemical identity does not necessarily ensure miscibility between the two components.

11.4.2 Blends of two molecular weight fractions and defined distributions

The dependence of the crystallization rate on molecular weights was discussed in Chapter 9. The profound influence of molecular weight fractions on the rate raises the interesting question as to the role of polydispersity in governing the crystallization kinetics. Studies of the crystallization kinetics of polydisperse homopolymers have been the norm. Consequently, there is a large amount of data in the literature that deals with such systems. However, studies with well-defined distributions, so that comparison can be made with fractions, and the role of polydispersity can be quantitatively assessed, are few in number. One approach to the problem would be to study analytically defined distributions and to compare the kinetic results with fractions that correspond to different moments of the distribution. An elementary example of a molecular weight distribution would be a binary mixture of the same species with different chain lengths. If the crystallization kinetics of the two pure species are established, then a study of the kinetics of the binary blends, as a function of composition, will give information as to the role of polydispersity. Studies of such blends, amenable to analysis, are available.(37–39) Also available for analysis are mixtures of several fractions having distributions with different moments.(40)

The half-times, $\tau_{1/2}$, for this isothermal crystallization of two different sets of binary blends of linear polyethylene are plotted in Fig. 11.23.(37) The molecular weights of the two pure components in the blend illustrated in Fig. 11.23a are 9×10^3 and 3.7×10^5 respectively, while in Fig. 11.23b the molecular weights of the pure species are 2.6×10^4 and 3.8×10^6. A family of curves is generated at different crystallization temperatures for each blend. The dependence of the half-time on composition is quite different for the two mixtures. There is a smooth, monotonic increase in $\tau_{1/2}$ as the concentration of the low molecular weight component increases in the blend illustrated in Fig. 11.23a. On the other hand, in the blends illustrated in Fig. 11.23 there is initially a monotonic decrease in $\tau_{1/2}$ as the concentration of the low molecular weight species increases until a minimum is reached at about 70%. As the concentration of the low molecular weight component increases further, $\tau_{1/2}$ monotonically increases reaching the value of the pure lower molecular weight component. The time scale is dependent on the crystallization temperature in the usual manner. The shapes of the curves in either set are displaced from one another but are independent of the crystallization temperature.

The apparent disparity in behavior between the two blends can be explained when the half-times are analyzed in terms of the weight average molecular weights of the mixtures rather than composition. The results are given in Fig.11.24a and b for the two sets of blends. Although the two plots still appear to be quite different, there is a natural explanation when the plots are compared with the results for molecular weight fractions of linear polyethylene (see Fig. 9.23a). The $\tau_{1/2}$ for the low molecular weight component in the 9×10^3–3.7×10^5 blends lies well to the left of the minimum in Fig. 9.23a, while the high molecular weight component is in the plateau region for the two crystallization temperatures of interest. Consequently, as is shown in Fig. 11.24, there is essentially a linear decrease in $\tau_{1/2}$ with the weight average molecular weight of the mixture. In contrast, in the 2.6×10^4–3.8×10^6 blends the low molecular weight species are still located to the left of the minimum in Fig. 9.23a. However, the high molecular weight component is now located well to the right of it. Therefore, as would be expected, and shown in Fig. 10.24b, $\tau_{1/2}$ goes through a minimum when plotted against the weight average molecular weight of the blend. Thus, the apparently diverse results between the two blends can be given a consistent and rational explanation by comparing $\tau_{1/2}$ of the weight average molecular weight of the binary mixtures with those for the corresponding pure species.

There is a question whether during the course of isothermal crystallization of binary blends, as well as other polydisperse systems, molecular weight fractionation

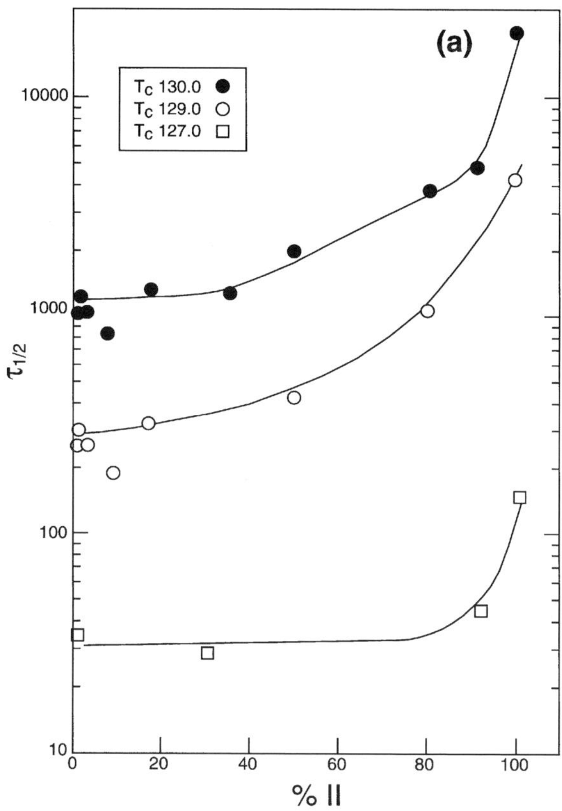

Fig. 11.23 (a) Plot of crystallization half-time, $\tau_{1/2}$, against percent $M = 9000$ for binary blends of linear polyethylene molecular weight fractions. Pure components: $M = 9000$ and 370 000. Crystallization temperature: ● 130 °C; ○ 129 °C; □ 127 °C. (b) Plot of crystallization half-time, $\tau_{1/2}$, against percent $M = 26\,000$ for binary blends of linear polyethylene. Pure components $M = 26\,000$ and 3.8×10^6. Crystallization temperature: ● 130 °C; ○ 128 °C. (From Ohno (37))

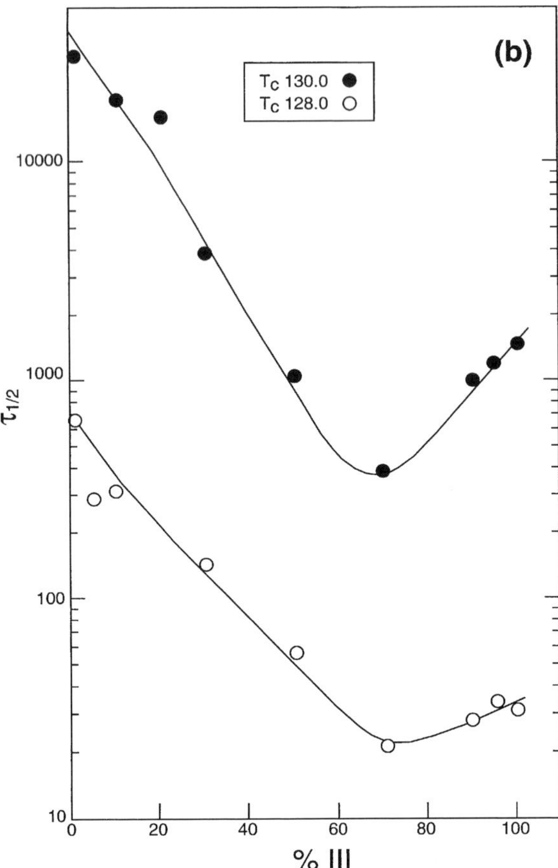

Fig. 11.23 (*cont.*)

occurs. It has been shown that for a linear polyethylene with a broad molecular weight distribution only molecular weights less than 5000 separate out, or fractionate, during isothermal crystallization.(41,42) On the other hand, molecular weights greater than about 10 000 co-crystallize and do not segregate during isothermal crystallization. The crystallization isotherms of the 9×10^3–3.7×10^5 mixtures can be examined to ascertain whether fractionation can be detected. A set of typical Avrami type plots with differing concentrations of this blend is given in Fig. 11.25. The theoretical plot, with $n = 3$, is given by the solid curve in the figure. The superposed, experimental isotherms give the same type of fit that is conventionally observed between theory and experiment. The low molecular weight pure component follows the theory over the complete extent of the transformation. As the higher molecular weight component is added, progressive deviation from theory sets in since the average molecular weight increases. There is, however, no indication in the

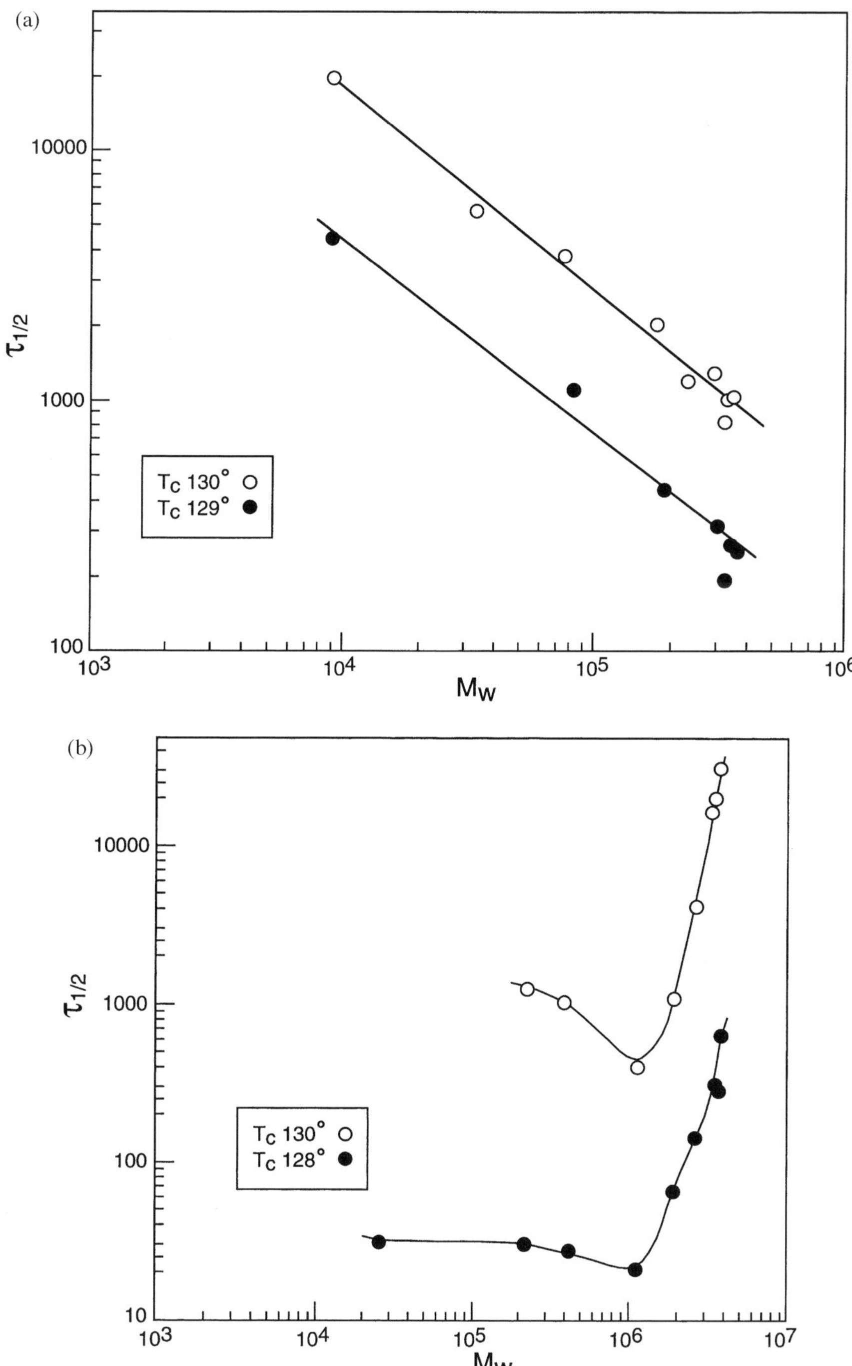

Fig. 11.24 Plots of crystallization half-time, $\tau_{1/2}$, against weight average molecular weight M_w for binary blends of linear polyethylenes. (a) Pure components $M =$ 9000 and 370 000. Crystallization temperature: ○ 130.0 °C; ● 129 °C. (b) Pure components $M = 260\,000$ and 3.8×10^6. Crystallization temperature: ○ 130 °C; ● 128 °C. (Data from Ohno (37))

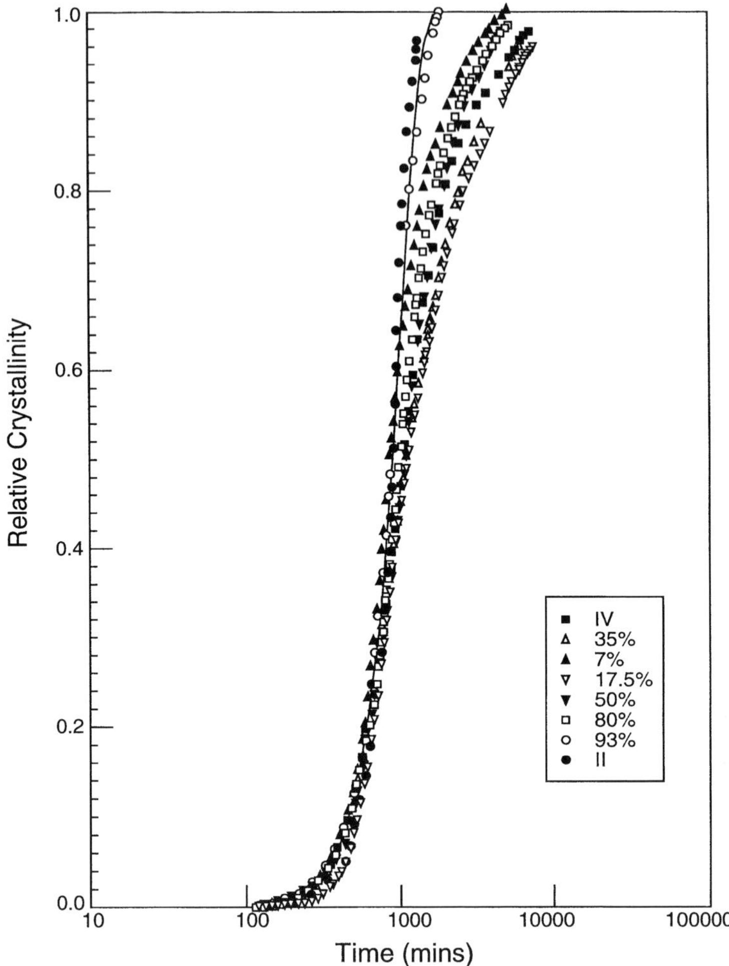

Fig. 11.25 Plot of relative crystallinity against log time for superposed linear polyethylene blends of $M = 9000$ and 370 000. Compositions indicated: IV, $M = 370\,000$; II, $M = 9000$. Crystallization temperature 130 °C. (Data from Ohno (37))

crystallization kinetics that fractionation occurs during the crystallization. Blends of molecular weight fractions of poly(ethylene oxide) behave in a similar manner.(38) It was shown in Fig. 9.27a that when plotted in the appropriate manner there is no indication from the kinetics of any fractionation in the poly(ethylene oxide) blends. Although fractionation of the low molecular weight species may very well be taking place in both type blends, it does not manifest itself in measurements of the overall kinetics.

Spherulite growth rates of blends of linear polyethylene, $M = 66\,000/M = 2500$, plotted as a function of the crystallization temperature for different

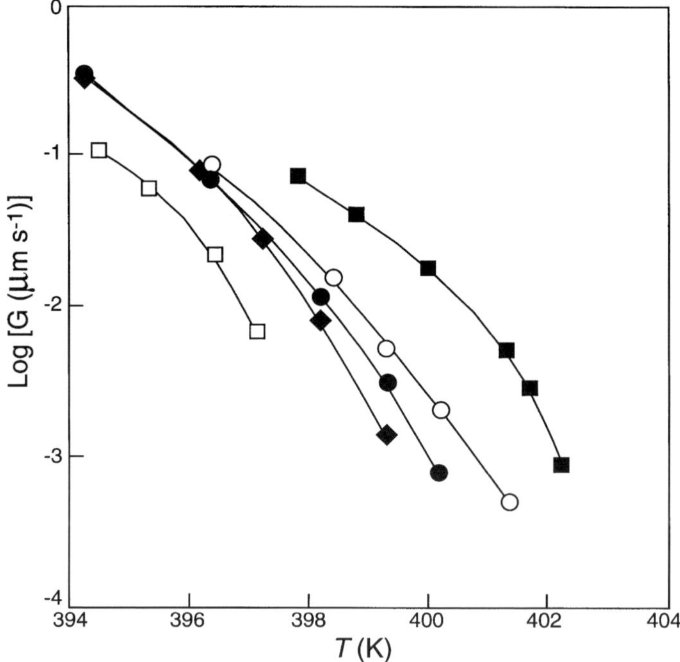

Fig. 11.26 Plot of log spherulite growth rate of a linear polyethylene fraction $M = 66\,000$ in a blend with $M = 2500$, as a function of temperature. Composition: M 66 000/M 2500: ■ 1.0/0; ○ 0.8/0.2; ● 0.6/0.5; ◆ 0.4/0.6; □ 0.2/0.8. (From Rego *et al.* (43))

compositions are given in Fig. 11.26. (43) The curves are essentially parallel but shifted along the temperature axis. Accompanying morphological studies demonstrate that fractionation occurs at these crystallization temperatures. Yet this is not expressed in the growth rate data. It was assumed that the displacement in growth rate curves is equivalent to a shift in the equilibrium melting temperature of the blends. The equilibrium melting temperatures can be calculated for the case when the low molecular weight component does not enter the crystalline lattice.(44) These are the values needed to further analyze the temperature dependence of the growth rate. We can conclude that fractionation, when it is known to occur, is not detected by either overall crystallization or spherulite growth rate studies.

The spherulite growth rates of poly(ethylene oxide) blends, $M = 5000/M = 270\,000$, display a qualitatively similar behavior.(39) In this case, both components co-crystallize in a common lattice at large undercoolings. At the low undercoolings separate phases result. There is no indication of this fractionation in the growth rate–temperature plots. Morphological studies demonstrate the fractionation.

The growth rates of molecular weight distributions composed of poly (tetramethyl-*p*-silphenylene siloxane) fractions have been studied.(40) The fractions and polydisperse samples follow the same pattern when compared on the basis of the number average molecular weight. The growth rates of the mixtures are independent of chain length at the higher molecular weights, thus following the pattern found for the fractions.

It is interesting to note that a pattern of retardation in growth rates is also observed in binary mixtures of high molecular weight *n*-alkanes.(45,46) The radial growth rate is slower in dilute blends relative to that of the major component, irrespective of whether the added component is longer or shorter.

11.4.3 Blends with different molecular architectures

The polymerization of some monomers leads to different structures or molecular architectures. The polyethylenes, with the same chemical repeating unit, represent good examples of this phenomenon. Linear polyethylene, often referred to as high density polyethylene, is a linear chain. Long chain branches can also be introduced that have short chain branches appended. This chain structure is often referred to as low density polyethylene. Polyethylene can also form copolymers that can be divided into two groups. In one grouping the comonomers are 1-alkenes, and the copolymers, with short chain alkane branches, are commonly referred to as linear low density polyethylenes. More polar comonomers, such as vinyl acetate, methacylic acid, and methyl acylate, among others, comprise another group.

Both components are usually crystallizable in blends composed of two different structural polyethylenes. Fundamental to understanding crystallization kinetics in such blends is the nature of the melt, i.e. whether or not it is homogeneous. It does not necessarily follow that although both components have the same chemical repeat that they are miscible over the complete composition range. In order for a valid comparison to be made it is also important that both components have comparable molecular weights. Small-angle neutron scattering provides direct information on the melt homogeneity of such mixtures.(47) Such studies have shown that in the melt blends of linear and low density polyethylenes are homogeneous at all compositions.(47,48) Similar studies of blends of linear polyethylene with ethylene–butene copolymers, hydrogenated poly(butadiene), have shown that they are homogeneous in the melt when the branch content is relatively low, ~4 branch points per 100 backbone carbons.(49) However, when the branch content is higher, ~8 branch points per 100 carbons, phase separation takes place in the melt. The more polar copolymers, such as ethylene–vinyl acetate, when admixed with linear polyethylene undergo phase separation over a wide range in copolymer compositions.

The properties of the blend in the solid state depend on several factors: the homogeneity of the initial melt state; the crystallization kinetics; and whether or not the two species co-crystallize. Co-crystallization depends not only on the homogeneity of the melt but also the presence of sufficiently long sequences of ethylene units in each of the components. Thus, the sequence distribution of the crystallizable sequences in a copolymer determines co-crystallization since it controls both melt homogeneity and crystallization kinetics.

A key factor in governing the extent of co-crystallization in binary polyethylene blends is the closeness of the crystallization rates of each of the components.(50–52) The difference in rate diminishes with increasing concentration of the linear component in the blend. The amount of co-crystallization is favored by lower isothermal crystallization and is maximized by rapid, quenched crystallization conditions. Blend composition and molecular structure of the components also have an influence on co-crystallization. The different results that can be obtained for different modes of crystallization have important implications for the morphology and properties in the solid state. Therefore, to gain further insight it should be fruitful to examine and analyze the crystallization kinetics of such blends in more detail.[6]

The temperature for the onset of crystallization at a constant cooling rate, T_0, is a qualitative measure of the crystallization rate of the system. This temperature reflects the behavior of the most rapidly crystallizing component. An example of such a kinetic study is given in Fig. 11.27 for blends of a linear polyethylene fraction with different random ethylene-1-alkene copolymers.(39) All the components have similar weight average molecular weights and comonomer composition. Hydrogenated poly(butadiene), HPBD, is a random ethylene–butene copolymer with a very narrow molecular weight and composition distribution. In contrast, although the copolymer designated as E-B is also an ethylene–butene copolymer, prepared by Ziegler–Natta catalysis, it has a broad molecular weight and composition distribution. The distribution in this copolymer is such that the lower molecular weight chain contains a greater concentration of comonomer units while the longest chains are very lightly branched and approach the chain structure of a linear chain. The copolymer designated E-H is an ethylene–1-hexene random copolymer with a narrow composition and most probable molecular weight distribution prepared by a metallocene type catalyst. The plots in the figure indicate that the crystallization rates, as reflected in T_0, are dominated by the linear component up to a copolymer concentration of

[6] The extensive overall crystallization kinetics that have been reported for blends of linear polyethylene with random ethylene-1-alkene copolymers cannot be analyzed in a consistent manner.(40) Hence, they are not reported here. The reason is that the crystallization temperatures were expressed in terms of an arbitrarily defined undercooling, $T_0 - T_c$, where T_0 is the peak crystallization temperature obtained by dynamic cooling.

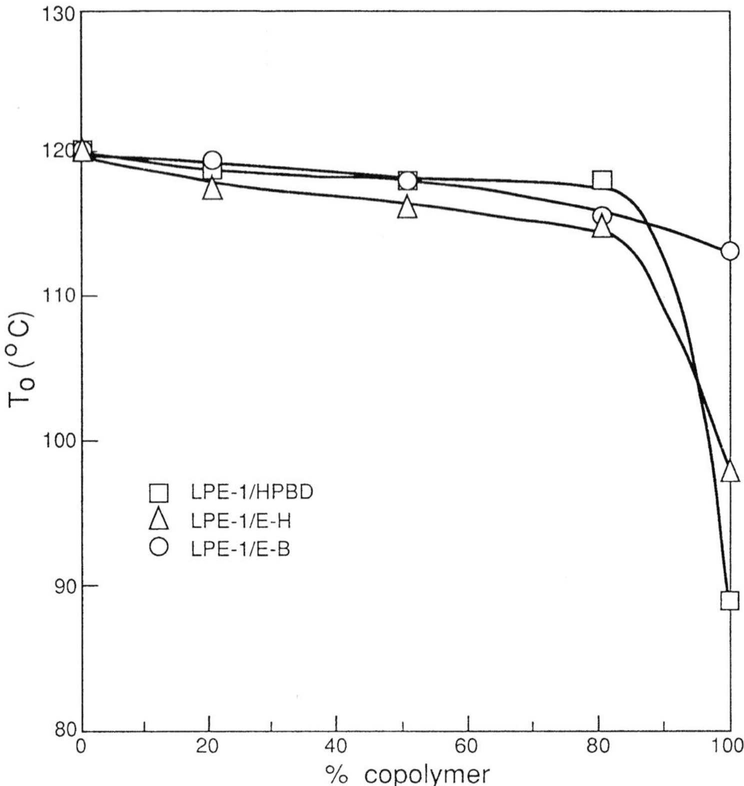

Fig. 11.27 The dependence of the crystallization temperature, T_c, for linear polyethylene blends with different copolymers as a function of composition. □ hydrogenated poly(butadiene); △ ethylene–hexene; ○ ethylene–butene. Crystallization was carried out by cooling at 10 K min^{-1} from melt. (50)

about 75%. Hydrogenated poly(butadiene) blend has the greatest difference in rates between the pure components. Following the principle that has been enunciated, calorimetric studies have shown that the extent of co-crystallization in this blend is the smallest among the three studied. In contrast, the T_0 value for the Ziegler–Natta catalyzed ethylene–butene copolymer is close to that of the linear component. The reason is that, in the distributions of this copolymer, the longer chains, which are lightly branched or unbranched, drive the crystallization. Consequently, the largest extent of co-crystallization is found. The ethylene–hexene blends, where the T_0 values for the pure components are between the other two copolymers, have an intermediate degree of co-crystallization.

In another example, the spherulite growth rates of a similar type ethylene–butene copolymer and of a low molecular weight fraction of linear polyethylene (M = 2500) are very close to one another over a range of crystallization temperatures.(53)

It is found that extensive co-crystallization occurs in blends of these two species, in accord with the principles outlined.

11.4.4 Blends with different stereoirregularities

An interesting set of binary mixtures are those between polymers of different stereo structures. Among others the crystallization kinetics of isotactic poly(styrene)–atactic poly(styrene), (25,54,55) syndiotactic poly(styrene)–atactic poly(styrene), (56) isotactic poly(propylene)–atactic poly(propylene) (55,57) and blends of isotactic and atactic poly(3-hydroxy butyrate),(58,59) and poly(D-lactic acid) with poly(L-lactic acid) (60) have been studied. In analyzing the kinetics of such blends the distinction still needs to be made as to whether the components are miscible in all proportions, partially miscible, or immiscible at all concentrations. Surprising as it may seem, this is an important consideration for these kinds of mixtures.

A typical example of the spherulite growth rates of these blends is given in Fig. 11.28 for the isotactic–atactic poly(styrene) pair as a function of temperature for different compositions.(54) The molecular weights in this mixture are 5.5×10^5 and 4.8×10^3 for the isotactic and atactic polymers, respectively. The usual rate maximum for the pure polymer is found at about 180 °C and is maintained for all the blends, even the most dilute one. There is a systematic, continuous decrease in growth rate as the atactic component is added. Similar results are found with

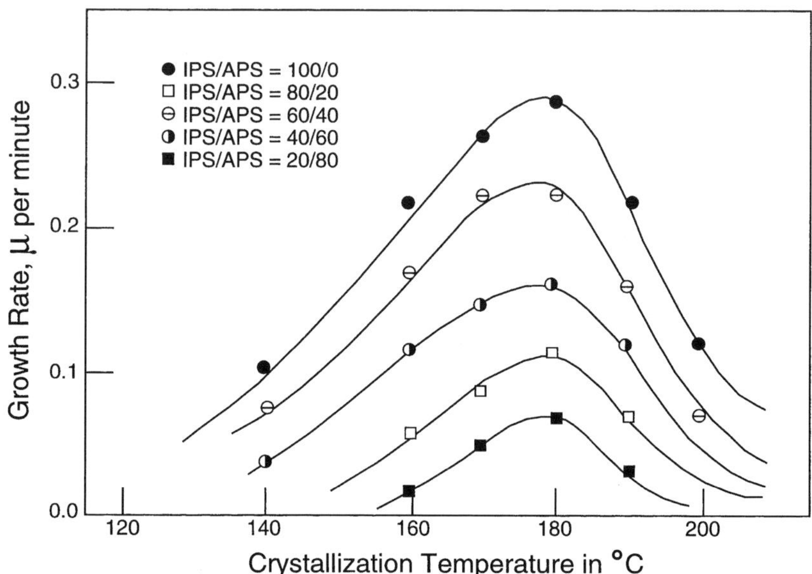

Fig. 11.28 Spherulite growth rates of isotactic poly(styrene) in blend with atactic poly(styrene) at indicated composition. (From Yeh and Lambert (54))

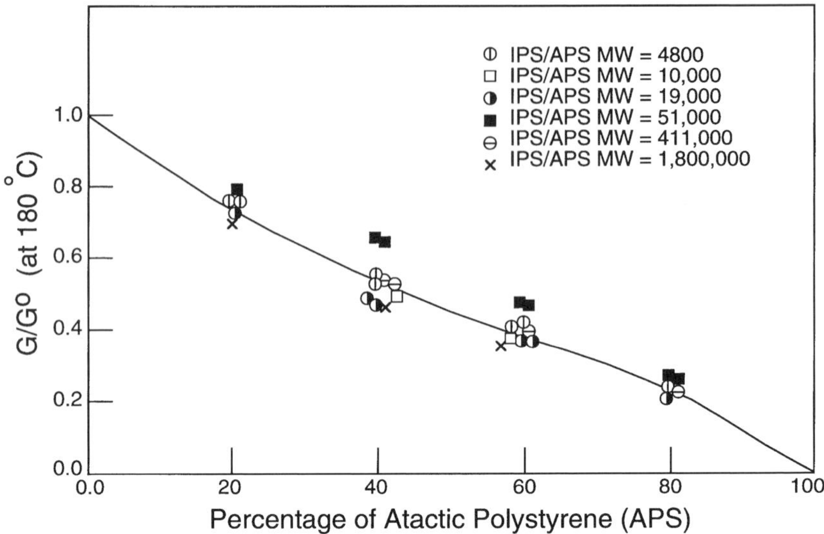

Fig. 11.29 Plot of ratio (G/G^0) of growth rate G in mixture to that of pure isotactic poly(styrene) at 180 °C against percent atactic poly(styrene). Molecular weights of atactic polymer indicated. (Adapted from Yeh and Lambert (54))

molecular weights as high as 1.8×10^6 for the atactic opponent. The growth rate maxima for the different molecular weights are in the range 178 °C to 183 °C. To illustrate the influence of the atactic polymer concentration, the ratio of the growth rate G to that of the undiluted polymer G^0 is plotted in Fig. 11.29 as a function of the concentration.(54) The systematic depression of the growth rate with the addition of the atactic component is significant. The growth rate has been reduced by a half in the 50/50 mixture. There is no sensible influence of the molecular weight of the atactic polymer on the growth rate at all blend compositions. The magnitude of the depression is typical of the crystallization of a component from a miscible melt. Close examination of the data indicates that there is a discontinuity in the growth rate, at fixed compositions, between molecular weight 1.98×10^4 and 5.1×10^4. The reason for this is not evident, but it could be due to changes in morphological features.

Essentially the same features are found in isotactic–atactic blends of poly(3-hydroxy butyrate).(58,59) The rate maximum that is observed in the pure isotactic polymer is maintained in the blends. The growth rate and maxima are systematically reduced with dilution. The maximum temperatures do not vary by more than just a few degrees.

Blends of syndiotactic poly(styrene) with the corresponding atactic polymer are miscible over the complete composition range.(56) The spherulite growth rate

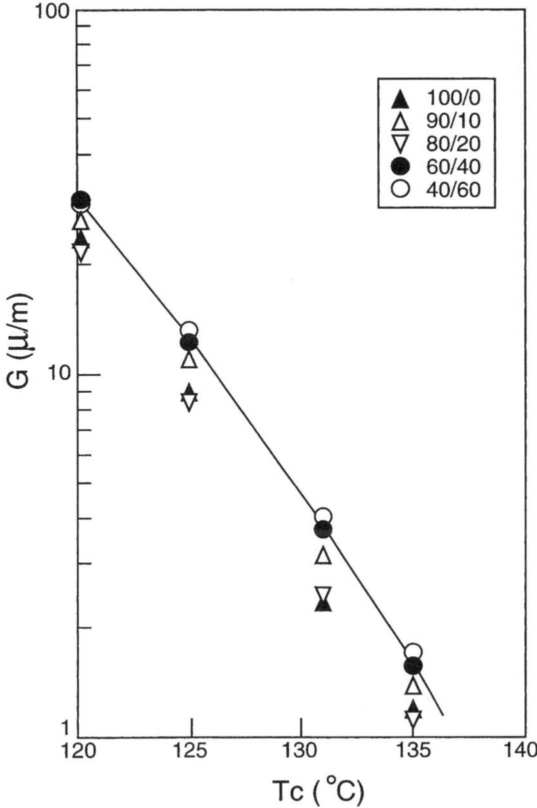

Fig. 11.30 Spherulite growth rate, G, of isotactic poly(propylene) in blend with atactic poly(propylene) as a function of crystallization temperature. Compositions of blends indicated in figure. (Data from Keith and Padden (55))

depends on both composition and crystallization temperature in the conventional manner for miscible systems. As in other miscible systems, the dilution effect is much greater at the lower crystallization temperatures than the higher ones.

Blends of isotactic–atactic poly(propylene) have been widely used to study morphological features and to develop mechanisms for spherulite formation.(55) The results of such a study are given in Fig. 11.30 for a blend of isotactic poly(propylene), $M_w = 1.78 \times 10^5$, with an atactic poly(propylene) component, $M_w = 8.7 \times 10^4$.(55) The spherulite growth rate is plotted against the crystallization temperature in this figure for compositions that range from the pure isotactic poly(propylene) to a mixture that contains 60% of the atactic component. The spherulite growth rate decreases with increasing crystallization temperature as would be expected. Most striking is the fact that at any given temperature the growth rate only changes slightly, if at all, with added atactic component. These results have been confirmed in another study,(57) and are in marked contrast to those just described and what is expected for a melt miscible system. More properly the comparison of growth rates should be made at the same undercooling using equilibrium melting temperatures for the calculation. The data define one curve when plotted against ΔT. This behavior is a characteristic of blends with immiscible components. Such systems will be discussed in more detail subsequently. However, theoretical consideration (61,62) as well as other types of experiments (63) indicate that the isotactic–atactic poly(propylene) pair is miscible in the melt. Thus, there is a dilemma between these results and the spherulite growth rates.

11.5 Partially miscible blends

In this section the crystallization kinetics of blends whose components are miscible in some proportion in the melt will be discussed. In such partially miscible systems, liquid–liquid phase separation could intervene and influence the crystallization process. Therefore, in order to properly analyze the crystallization kinetics in such systems it is necessary that the phase diagram be established. These systems represent a classical example where the melt structure plays an important role in governing the ensuing crystallization. The reason for this requirement will become clear as some typical phase diagrams, involving a crystallizable polymer, are examined.

Figure 11.31 represents the complete phase diagram for a mixture of poly(ϵ-caprolactone) with an atactic noncrystallized poly(styrene). A typical upper critical solution temperature (UCST) is exhibited.(64) The two-phase region is delineated by the binodial. The spinodial within the heterogeneous region is also indicated. The melting temperature of poly(ϵ-caprolactone) decreases slightly with added poly(styrene) until the two-phase region is reached. The melting temperature within the two-phase region (not shown) is invariant as a consequence of the phase rule. In this example, the glass temperature was calculated by the Fox equation.(65) Comparable phase diagrams can also be found for binary mixtures that display a lower critical solution temperature (LCST).(66–68) The region between the binodial and spinodial is metastable. The region at temperatures below the spinodial is unstable.

A unique situation can be observed with a UCST type phase diagram. In this particular case the blend is miscible above T_m, but displays a miscibility gap below.(69,70) There is no equilibrium basis for liquid–liquid phase separation to take place below the melting temperature–composition boundary. However, it is possible for the melt to be sufficiently supercooled below T_m that liquid–liquid phase separation does in fact occur. Therefore, in a practical sense crystallization occurs in competition with liquid–liquid phase separation.

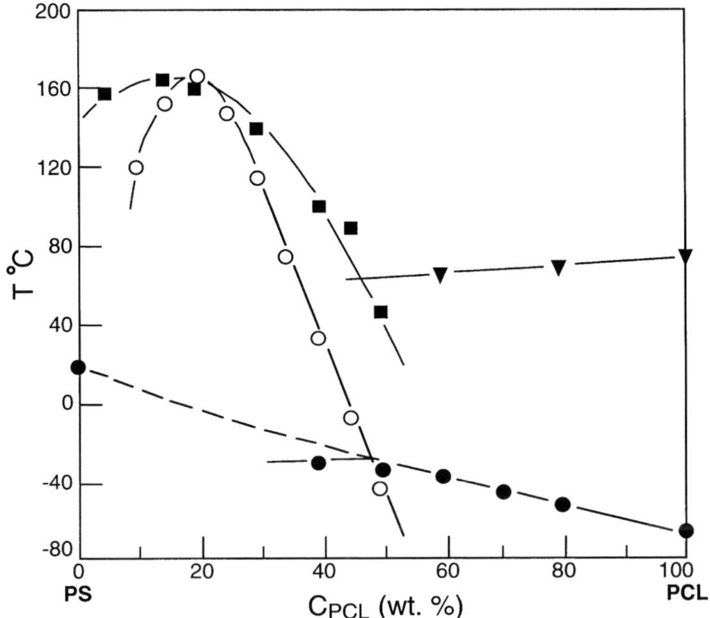

Fig. 11.31 Phase diagram of poly(ϵ-caprolactone)–atactic poly(styrene) blends. ■ binodial; ○ spinodial; ▼ melt–crystal coexistence curve; ● glass transition temperature; — — glass transition temperature calculated. (From Li and Jungnickel (64))

The role of the phase diagram in directing the crystallization kinetics and eventually the morphology and structure, can be analyzed. Crystallization of either or both components in the homogeneous region will be the same as has already been discussed. However, the situation is quite different when crystallization takes place from the two-phase region. The two phases that are in equilibrium with one another will have different compositions. Depending on the choice of temperature and nominal composition the crystallization can occur from either the metastable region or the unstable spinodial region. Crystallization from the metastable region involves the usual nucleation and growth processes. However, in the unstable region, crystallization proceeds by a spinodial decomposition mechanism.(71–74)

In phase separation that is governed by nucleation and growth a new phase is initiated by a nucleus that proceeds to grow in the conventional manner. Molecules, or polymer segments, that feed the new phase follow ordinary transport behavior. The diffusion coefficient is thus positive ("downhill diffusion") because in the metastable region the second derivative of the Helmholz free energy with respect to composition is positive. This coefficient makes the major contribution to the variation of composition. In contrast, in the unstable, spinodial region, the second derivative of the Helmholz free energy is negative. Segments move from low to high concentration in what has been termed "uphill diffusion". An important consequence of spinodial decomposition is the continuous change in composition and the interconnected phase morphology.(75,76) The connectivity of phases results in some unique morphological features as compared to the nucleation and growth. These morphological features, and the domains that are involved, could promote nucleation at their interface. However, the immediate interest at this point is to ascertain how the different paths that can be followed from the melt are reflected in the crystallization kinetics.

An excellent example of the role played by liquid–liquid phase separation in the ensuing crystallization is found in blends with syndiotactic poly(styrene).(77) Measurements of the glass temperature in mixtures with poly(2,6-dimethyl-1,4-diphenylene oxide) (PPO) indicate that the components are miscible in all proportions in the melt. However, mixtures of syndiotactic poly(styrene) with poly(vinyl methyl ether) represent partially miscible blends. When the poly(vinyl methyl ether) content exceeds 20% by weight, the melt separates into two liquid phases, one rich in syndiotactic poly(styrene), the other in poly(vinyl methyl ether). Thus, the two blends have a common crystallizing component. However, in one the crystallization takes place from a homogeneous melt; in the other from one that is phase separated. The different melt structures profoundly affect the crystallization kinetics. This can be seen when a comparison is made between the crystallization kinetics of syndiotactic poly(styrene) from a homogeneous or phase separated melt.

The spherulitic growth rates of syndiotactic poly(styrene) in both blends are given in Fig. 11.32. The addition of poly(vinyl methyl ether) causes an increase in the growth rates relative to the pure polymer that are independent of the composition at all crystallization temperatures. The growth rate of the pure polymer is represented by the solid circles in this figure. This unusual behavior is a consequence of the crystallization taking place from a phase separated melt. There could, however, be some contribution from the decrease in the glass temperature. On the other hand, for the homogeneous blend the growth rate decreases in the conventional manner with the addition of PPO. Thus, there is a very striking difference in the polystyrene growth rates in the two blends because of the different initial melt structures.

The dependence of the overall crystallization rate on the composition of these two blends differs from those of the growth rates.(77) The half-time for syndiotactic poly(styrene) crystallization is plotted against the crystallization temperature for both type blends in Fig. 11.33. The half-times for the pure syndiotactic polymer are represented again by the solid circles. For this type of measurement the half-times increase with the addition of either poly(vinyl methyl ether) or PPO. The addition of poly(vinyl methyl ether) causes a larger increase in the half-time and thus a greater reduction in the crystallization rate. The overall crystallization rate is a reflection of both initiation and growth of crystallization. Both of these processes

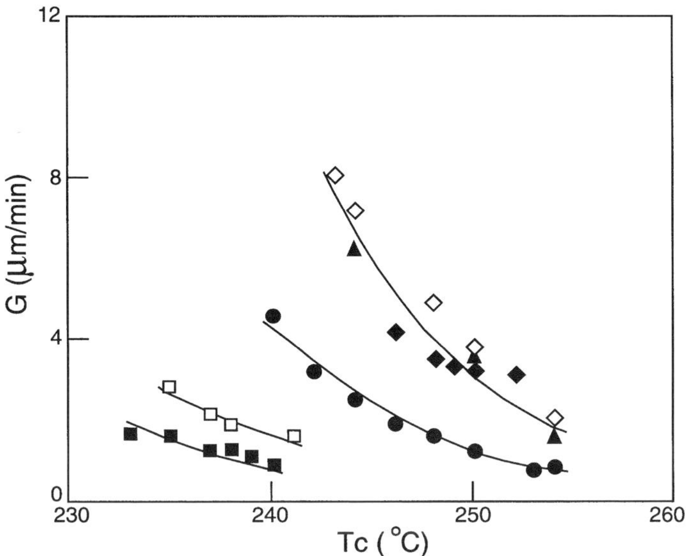

Fig. 11.32 Spherulite growth rate, G, for syndiotactic poly(styrene) in blends with poly(vinyl methyl ether) and with PPO, as a function of the crystallization temperature T_c. Composition: ● pure syndiotactic poly(styrene); poly(styrene)–PPO □ 90/10, ■ 80/20, poly(styrene)–poly(vinyl methyl ether) ▲ 80/20, ◇ 70/30, ◆ 60/40. (From Cimmino *et al.* (77))

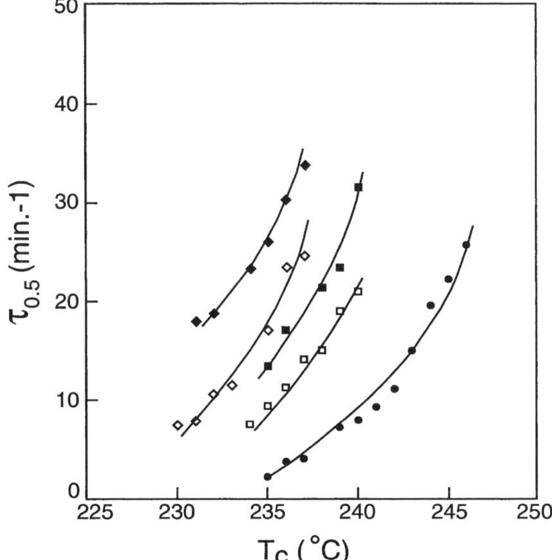

Fig. 11.33 Crystallization half-time, $\tau_{1/2}$, as a function of crystallization temperature, T_c, for different blend compositions of syndiotactic poly(styrene) with either poly(vinyl methyl ether) or PPO. Composition syndiotactic poly(styrene)–PPO: ● pure syndiotactic poly(styrene); □ 90/10; ■ 80/20. Composition syndiotactic poly(styrene)–poly(vinyl methyl ether) ◇ 90/10; ◆ 80/20. (From Cimmino *et al.* (77))

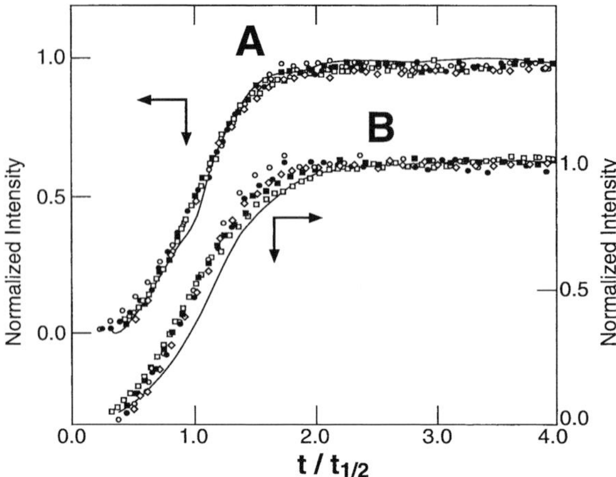

Fig. 11.34 Plot of superposed normalized intensity maximum against reduced time, $t/t_{1/2}$, for poly(ϵ-caprolactone)–poly(styrene) blend. Curve A: Fraction poly(ϵ-caprolactone) crystallized at 40 °C: ○ 1.0; ● 0.9; △ 0.8; ■ 0.7; □ 0.6. Curve B: fraction poly(ϵ-caprolactone) crystallized at T_c = ○ 30 °C, ● 35 °C, ◇ 39.5 °C, ■ 41.0 °C and □ 42.5 °C. (From Nojima *et al.* (78))

are nucleation controlled. Hence, the difference between the spherulite growth and overall crystallization rates of the blends is due to the initiating nucleation process. Analysis of spherulite densities indicates that each of the added polymers reduces the nucleation density of the syndiotactic poly(styrene); poly(vinyl methyl ether) more so than PPO. This then explains the difference between the spherulite and overall crystallization rates of the two blends.

A potentially interesting situation can develop when the crystallization is initiated in the homogeneous region of the phase diagram. As crystallization proceeds the composition will become such that the mixing of the crystallized units with the noncrystallizing component will lead to liquid–liquid phase separation. In this case the crystallization kinetics and resulting morphology will be altered. This case has been examined with poly(ϵ-caprolactone)–atactic poly(styrene) blends.(64,78) Crystallization kinetics studies have addressed this phenomenon using small-angle x-ray scattering.(78) The results of such studies are given in Fig. 11.34, where the normalized scattering intensity is plotted against the reduced time, $t/t_{1/2}$. Sets of superposable isotherms result. The plots for different initial conditions in the homogeneous regions, including that for the pure poly(ϵ-caprolactone) indicate that the kinetics are not influenced by the eventual intervention of liquid–liquid phase separation.

11.6 Blend with two completely immiscible components

The crystallization of a mixture of two polymers that are immiscible over the complete composition range presents some features that have not been previously encountered. In immiscible systems, v_2 is effectively unity so that Eq. (11.7) reduces to Eq. (9.205). One naturally has to distinguish between the situation where both polymers can crystallize or where only one can. In either case, the initial crystallization involves one of the components. The other component will be dispersed in the melt in droplet-like domains.(79–83) Details of the domain structure will depend on the composition. Therefore, one can anticipate that the initial morphology in the melt of incompatible blends could very well influence the ensuing crystallization, irrespective of whether the second component crystallizes or not.

Spherulite growth rates of blends with two immiscible components fall into two categories. Typical results for one group are illustrated in Fig. 11.35 for blends of isotactic poly(propylene)–low density polyethylene(84) and of poly(ethylene oxide)–poly(vinyl chloride).(85) Despite the presence of immiscible spherical-like domains in the melt, the growth rates in each of the blends are independent of composition over the temperature range studied. Similar results are found in immiscible blends of isotactic poly(propylene) with a polyethylene based ionomer,(86) poly(ethylene oxide)–poly(propylene oxide) blends (87) and isotactic poly(propylene)–atactic poly(styrene) blends.(88) A striking example of this invariance in growth rate is found when the crystallization of poly(3-hydroxybutyrate) in a blend with poly(vinyl acetate) is compared with one where an ethylene–propylene copolymer is the second component.(89) In the former case the two components are miscible in all proportions and the dependence of the growth rate on concentration is that expected. On the other hand, with ethylene–propylene the two components are immiscible. As a result the growth rates are independent of composition at all crystallization temperatures. In some immiscible blends the growth rates are very similar to one another, but not quite independent of composition. As an example, the growth rates are constant with composition for blends of poly(3-hydroxybutyrate)–poly(methylene oxide) at the high crystallization temperatures. However, a slight decrease in the rates takes place with dilution at lower crystallization temperatures.(80)

In contrast to the behavior described above, there are many examples where the growth rate of the crystallizing component is influenced by the dispersed droplets of the second compound. This is particularly true of blends of isotactic poly(propylene) with various elastomers.(79,83,90) An example is given in Fig. 11.36 where the second components are poly(isobutylenes) of different molecular weights.(79,83) The spherulite growth rate–composition relations for this system are different from those previously encountered and vary with the molecular weight of the poly(isobutylene). These results indicate that the dispersed phase plays a different role in each molecular weight range. In fact, morphological studies show that the dispersed poly(isobutenes) behave differently, depending on molecular weight.(79)

At low concentrations of the low molecular weight polymer, PiB(LM), all of the polymer is rejected between the spherulites at all crystallization temperatures. As

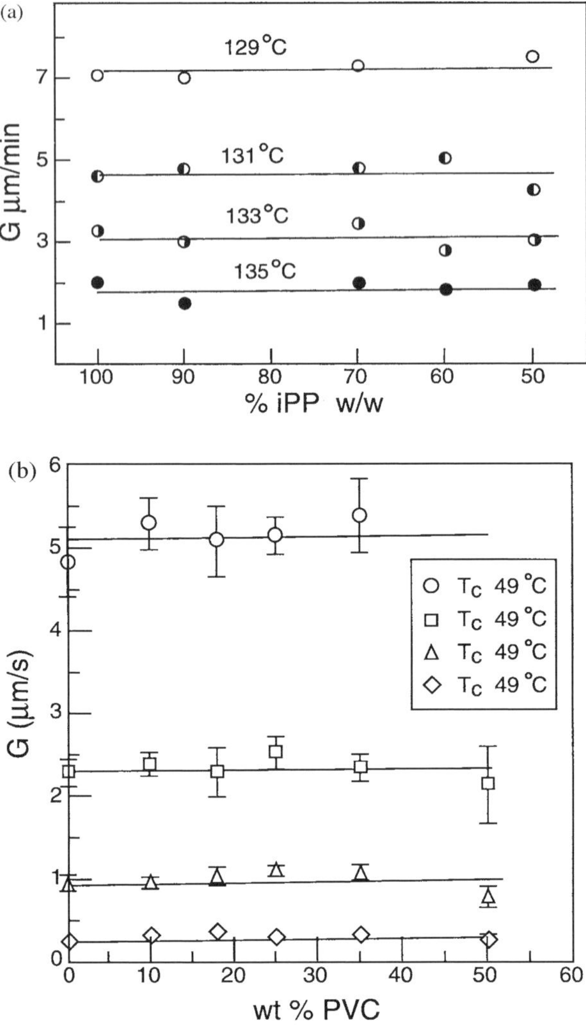

Fig. 11.35 (a) Spherulite growth rate of isotactic poly(propylene) in blends with low density polyethylene as a function of composition at indicated crystallization temperature. (From Galeski *et al.* (84)) (b) Spherulite growth rate of poly(ethylene oxide) in blends with poly(vinyl chloride) as a function of composition at indicated crystallization temperatures. (From Martenette and Brown (85))

the concentration increases, however, some of the droplets are trapped in the interior; the higher the concentration the more droplets are occluded. These changes are reflected in the kinetic data of Fig. 11.36a. Distinct droplets of the medium molecular weight poly(isobutylene), PiB(MM), are observed at all concentrations and temperatures. The droplets are rejected by the spherulite boundaries, including those generated within the spherulites. The concomitant decrease in spherulite

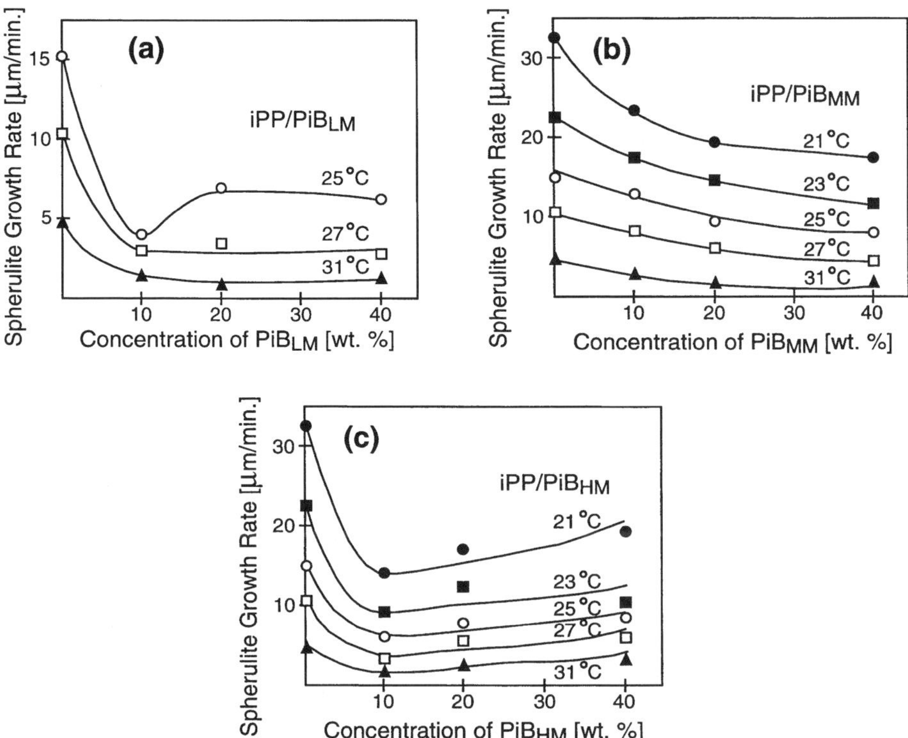

Fig. 11.36 Plots of spherulite growth rates of isotactic poly(propylene) in blends with poly(isobutylene) (PiB) as a function of composition at indicated crystallization temperatures. (a) Low molecular weight PiB; (b) medium molecular weight PiB; (c) high molecular weight PiB. (From Martuscelli (79))

growth rate, at all temperatures, for blends with PiB(MM), Fig. 11.36b, is consistent with the rejection of the dispersed particles. The plots in Fig. 11.36c for blends with the high molecular weight elastomer, PiB(HM), show some unique features. A minimum in the growth rate is observed at about 10% PIB concentration at all temperatures followed by a monotonic, small increase. Concomitant morphological studies show that the blends containing less than 10% of PiB do not have any droplet structure. However, when the PiB concentration is increased above this value a droplet structure appears in the melt and occlusion occurs during crystallization. As the crystallization temperature increases more material is rejected into the interspherulitic region and less remains inside the spherulite. The results summarized in Fig. 11.36 demonstrate a definite correlation between the melt morphology, as it involves the droplets, and the molecular weight dependence of the growth rate concentration curves.

In the initial melts of blends of two immiscible components the noncrystallizable component is segregated in droplet domains. During crystallization the domains can be rejected by the crystallizing front at the growth boundary or may be occluded or deformed. In these situations additional energy is dissipated so that Eq. (9.205) must be modified accordingly. It has been suggested that the growth rate can now be expressed as (79)

$$G = G_0 \exp - \left\{ \frac{U^*}{R(T - \mathrm{T}_\infty)} - \frac{\Delta G^*}{RT} - \frac{E}{RT} \right\} \tag{11.9}$$

In effect, an additional term has been added to the conventional transport and nucleation terms. The additional activation energy, E, for these additional processes can be expressed as

$$E = E_1 + E_2 + E_3 + E_4 + E_5 \tag{11.10}$$

Here E_1 is the energy dissipated by the growing spherulite rejecting the noncrystallizing component into the interlamellar region; E_2 is the energy required by the growth front to effect the rejection of the droplet domain in the melt; E_3 is the kinetic energy required to overcome the insertion of the droplet; E_4 is the energy required to form new interfaces between the spherulite and droplet; and E_5 is the energy dissipated when the engulfed drops are deformed by the crystallizing front. Details of the modification of the growth rate by these energy dissipation processes have been summarized.(83) In essence the growth rate in immiscible systems is influenced by the size of the dispersed phase and the interfacial energies involved. Central to the problem for rejection, engulfment or deformation is a quantity $\Delta\phi$. It is defined as the difference between the interfacial free energy of the crystallizing entity and the included material, and that of the melt and the included material, γ_{pl}. Thus

$$\Delta\phi = \gamma_{\mathrm{ps}} - \gamma_{\mathrm{pl}} \tag{11.11}$$

When $\Delta\phi < 0$ the dispersed domains in the melt are more likely to be engulfed than rejected by the growing front. The situation is more complex when $\Delta\phi > 0$. Under this condition of slow rates of crystallization the domains are pushed along by the growing front. At higher rates they will be engulfed. At some intermediate crystallization rate the domain may be displaced for short distances before being engulfed. For a given value of $\Delta\phi$ there is a critical domain size such that, at a constant growth rate, domains of this size and larger are engulfed.

With all these possibilities, modifications of the conventional growth rate can be expected in some blends.(79) It is, therefore, not surprising that the growth rate–composition relations of blends with two immiscible components are not unique and a variety of results can be expected, as is observed.

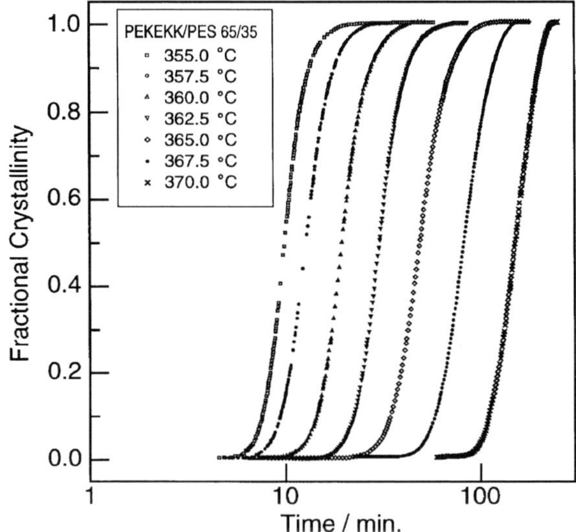

Fig. 11.37 Plot of fraction crystalline against log time for indicated crystallization temperatures for poly(ether ketone ether ketone ketone) in a 65/35 immiscible blend with poly(ether sulfone). (From Androsch *et al.* (82))

As with other systems the overall crystallization rates of immiscible blends complement those involving spherulite growth rates and provide additional insight to the problem. In general, isotherms representing the overall crystallization kinetics obey the derived Avrami formulation in a manner similar to that of the corresponding pure crystallizing components.(82,87) As an example, Fig. 11.37 is a plot of the fraction crystallinity as a function of log time of poly(ether ketone ether ketone ketone) (PEKEKK) in 65/35 immiscible blend with poly(ether sulfone) over a range of crystallization temperatures.(82) Typical superposable isotherms are observed whose major characteristics are similar to those of pure homopolymer.

An unusual feature is found in immiscible systems when the crystallization rates are examined as a function of concentration. An example is given in Fig. 11.38 for the blend of PEKEKK and poly(ether sulfone). Here the isothermal exotherm is plotted against the composition of the noncrystallizing component at various temperatures. The initial addition of poly(ether sulfone) results in a decrease in the crystallization rate. A maximum is reached at about 30% of the added component. Further additions of the diluent polymer result in an increase in the crystallization rate. These results contrast rather sharply with those of the miscible blend of the same crystallizing polymer and poly(ether imide).(82) For the miscible blend, as illustrated in Fig. 11.39, the peak times increase continuously with added poly(ether imide). Thus, the structure of the immiscible melt, both initially and as crystallization progresses, is an important factor in governing the rate–composition relation for isothermal crystallization.

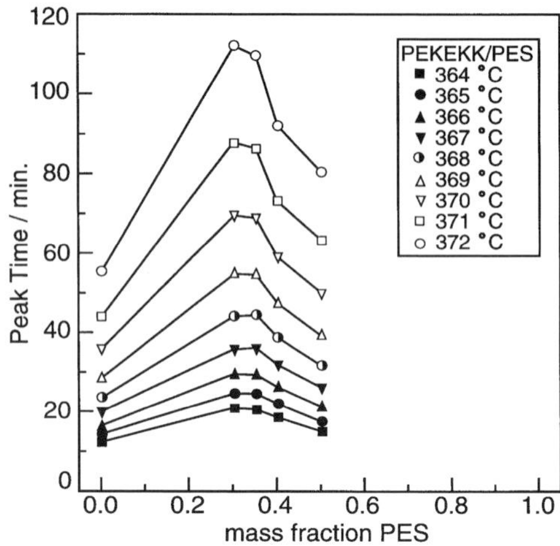

Fig. 11.38 Plot of dsc peak time dependence on concentration of poly(ether sulfone), PES, in immiscible blends with poly(ether ketone ether ketone ketone) at indicated isothermal crystallization temperatures. (From Androsch *et al.* (82))

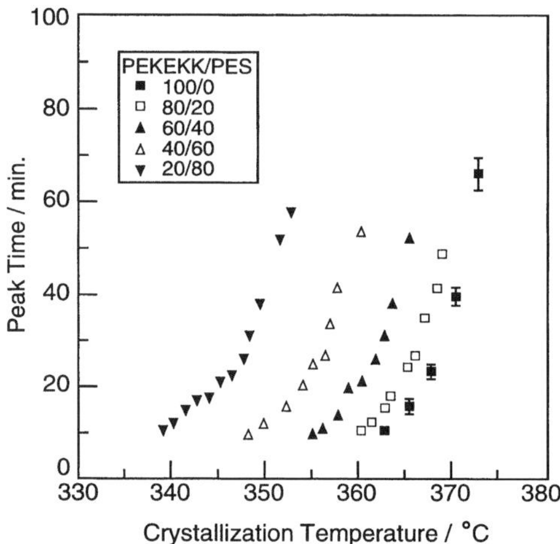

Fig. 11.39 Plot of dependence of dsc peak time on the crystallization temperature for poly(ether ketone ether ketone ketone) in immiscible blends with poly(ether imide) at indicated compositions. (From Androsch *et al.* (82))

The enhancement of the overall crystallization rate with increasing concentration of the noncrystallizing component in immiscible systems is found in many immiscible blends.(80,87,91–93) A striking example is illustrated in Fig. 11.40 for poly(ethylene terephthalate)–poly (carbonate) blends.(92) For blends containing less than 80 wt percent of the poly(carbonate) component the crystallization rate of the poly(ethylene terephthalate) is greatly enhanced. Kinetic studies for

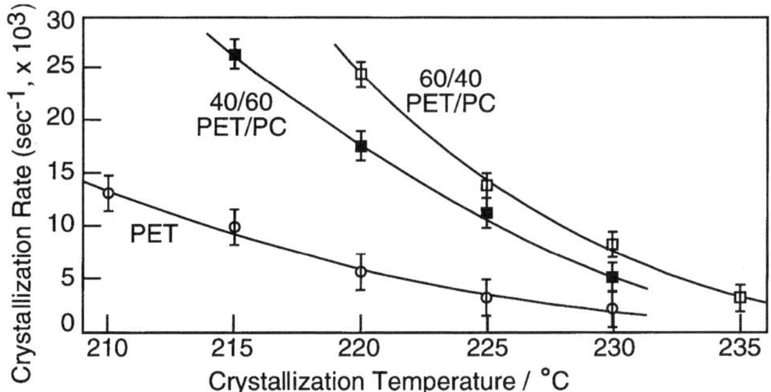

Fig. 11.40 Plot of crystallization rate, $1/\tau_{1/2}$, of poly(ethylene terephthalate) as a function of crystallization temperature in blends with poly(carbonate) for indicated compositions. (From Reinsch and Rebenfeld (92))

poly(carbonate) content greater than 80% could not be carried out. The increase in the overall crystallization rate can be attributed to the noncrystallizing surfaces that are introduced. These foster the heterogeneous initiation of stable nuclei and a more rapid overall crystallization rate. This mechanism is not involved in the spherulite growth rate and thus the difference in the influence of the concentration of the noncrystallizing component can be explained.

Morphology is not the only factor that determines the overall crystallization kinetics in immiscible binary mixtures. The nature of the noncrystallizing component can also play a significant role. We consider as an example two different immiscible blends in which poly(phenylene sulfide) is the crystallizing component.(81) In one blend, the added component is linear polyethylene, in the other poly(ethylene terephthalate). Typical melt morphology is observed in both blends. Poly(phenylene sulfide) is the continuous phase at high concentration. However, at lower poly(phenylene sulfide) concentrations both linear polyethylene and poly(ethylene terephthalate) form the continuous phase. Despite the similarities of phase morphology the overall crystallization kinetics of poly(phenylene sulfide) is different in the two blends. These differences are illustrated in Fig. 11.41 where the crystallization half-time is plotted against the crystallization temperature for different compositions of both blends. Although the general characters of all the curves are similar to one another there are important quantitative differences. In one case the crystallization rate of the poly(phenylene sulfide) is enhanced; in the other case it is retarded. The half-times are systematically reduced by the addition of linear polyethylene. In contrast, the half-time is increased by the addition of noncrystallizing poly(ethylene terephthalate). Analysis of the rate constants indicates that there is a minimum in the rates at about 50/50 composition. Therefore, there must be some type of specific interaction, or lack thereof, in the melt that leads to the opposite behavior despite the similarity in morphological features.

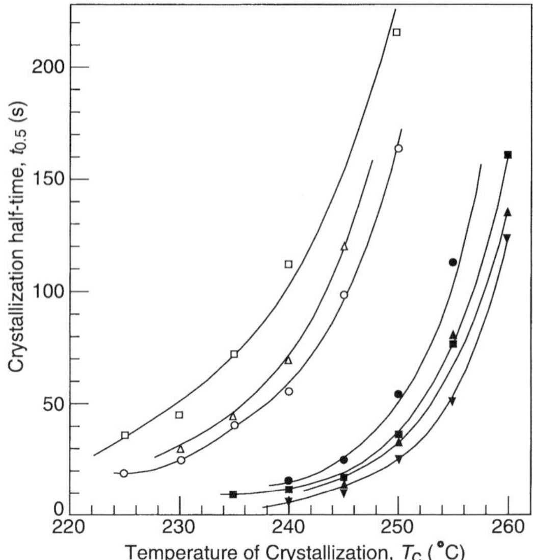

Fig. 11.41 Plot of crystallization half-time for crystallization of poly(phenylene sulfide) in blends with linear polyethylene and with poly(ethylene terephthalate) as a function of crystallization temperature. Pure poly(phenylene sulfide) ●. Composition poly(phenylene sulfide)–linear polyethylene blends: □ 50/50; △ 75/25; ○ 90/10. Composition poly(phenylene sulfide)–poly(ethylene terephthalate) blends: ■ 50/50; ▼ 75/25; ▲ 90/10. (From Jog *et al.* (81))

11.7 Nonlinear growth and diffusion

It has been tacitly assumed in the discussion up to now that the spherulite radius increases linearly with time. Thus, the growth rate is constant at a given crystallization temperature. Linear growth is most commonly observed in one-component systems, as well as in many binary blends. As has been pointed out earlier this type of behavior is indicative of interface controlled growth. However, in binary mixtures nonlinear growth of the crystallizing component is not uncommon. The nonlinearity indicates that the radial growth is no longer interface controlled. It is found in all types of blends, including completely miscible ones, those that are partially or completely immiscible, as well as some whose melt structure was not investigated.

Nonlinear growth is often observed in miscible melts that have a low molecular weight noncrystallizing component. An example of such behavior is illustrated in Fig. 11.42 for a mixture of isotactic poly(propylenes) with a low molecular weight ($M = 540$) atactic poly(propylene).(55) The nonlinearity in the growth is quite apparent in this example. It is found that when the molecular weight of the atactic component in this blend is equal to or greater than 2600 the growth is linear. Okada *et al.* found similar results in their study of a 60/40 isotactic–atactic poly(styrene) blend.

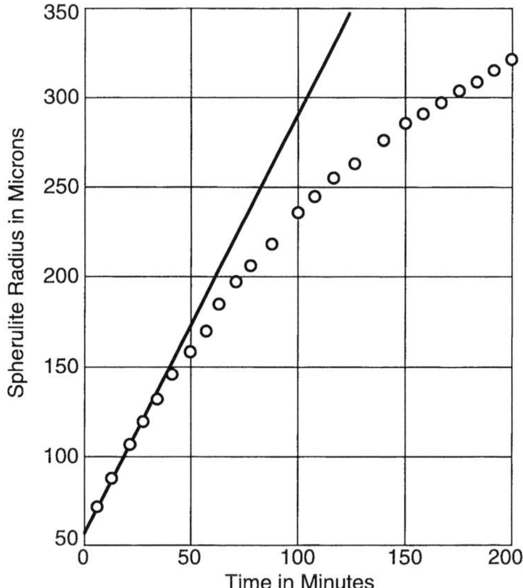

Fig. 11.42 Plot of radius of isothermally growing spherulite as a function of time, at 135 °C, in a blend of isotactic poly(propylene), $M = 178\,000$, and atactic poly(propylene), $M = 540$. (From Keith and Padden (55))

In this blend, when the molecular weight of the atactic component was greater than 1.94×10^4 linear growth was observed.(94) However, when the molecular weight was less than 5.2×10^3 the growth was nonlinear, and decreased with time. Other examples of nonlinear growth caused by a low molecular weight, noncrystallizing added species have also been found in blends of poly(ϵ-caprolactone)–atactic poly(styrene)(69,98) and isotactic poly(propylene)–liquid paraffin ($M = 338$).(96) In general, for a melt of a given composition nonlinear growth is favored by reducing the molecular weight of the added component or increasing the crystallization temperature.

Several mechanisms have been proposed to explain the underlying reason for nonlinear growth in miscible systems. Major attention has been given to the role played by the diffusion of the noncrystalline component away from the growing spherulite front, assuming that all, or a major portion, is rejected during the crystallization.(13,55,97,98) The growth rate is not only influenced by the initial melt composition but also by the changing composition at the growth front, due to the rejection of the added components. In a qualitative sense, when a steady state in the concentration of the added component is reached, the growth rate should be linear. However, if a steady-state is not reached, then commensurate with the time scale, the growth rate will be nonlinear because of the varying composition. In the extreme, for high molecular weights, where the diffusion will be slow, then a steady state is effectively reached at $t = 0$ and growth will be linear. On the other hand, for low molecular weights a steady state will not be reached at $t = 0$ and nonlinear growth will result. Thus, the relative roles of the diffusion and growth rates are important in this problem. The influence of diffusion of the second component will be explored further. This is distinct from direct diffusion controlled crystallization that can be observed in both one and two component systems. In this latter case, growth is not interface controlled.

It has been shown that for a planar growth front (97,99)

$$\frac{C(x)}{C_0} = 1 + \exp\left(-\frac{Gx}{D}\right) \tag{11.12}$$

Here $C(x)$ is the concentration of the added component at a distance x from the growth front, C_0 is the initial concentration, and that at large x, and D is the diffusion rate of the noncrystallizing species. It is assumed here, for simplicity, that all of the species is rejected by the growing spherulite. The left-hand side of Eq. (11.12) is the fractional decrease in the concentration between the value at the interphase and that far removed. Equation (11.12) can also be generalized to include other shape growth fronts.(97,98) When this fractional decrease reaches the value 1/e, a quantity δ, termed the diffusion length, is given by

$$\delta = D/G \tag{11.13}$$

This characteristic length was originally introduced by Keith and Padden (99) in their discussion of spherulite growth and texture. The diffusion length is a measure of the distance where there is an appreciable concentration gradient. It is an important concept not only in the present context but in describing spherulite texture.[7]

When δ is small growth will be linear. This corresponds to high molecular weights and low diffusion rates and reasonable values of G. In contrast, when D is large, corresponding to low molecular weights and low values of G, δ will be large. The growth rate will become nonlinear with the larger diffusion length. The growth rate can be varied over a wide range by choice of the crystallization temperature. This involves either or both the undercooling and the relation to the glass temperature. Okada *et al.* have shown that by calculating D, and using measured values of G, the variation of δ with molecular weight of the noncrystallizing component, in isotactic–atactic blends of poly(styrene) follows the pattern outlined.(94)

The increasing concentration of the noncrystallizing component at the growth front with time serves to reduce the growth rate, as is observed. There are several reasons for the decrease. One is the usual decrease in growth rate with dilution of the melt as is found in blends with constant growth rates. The other, only pertinent in

[7] A similar calculation was carried out by Kit (100) for lamellar growth.

general to low molecular weight noncrystallizing components, is the reduction of the equilibrium melting temperature. In turn, the undercooling at a fixed crystallization temperature is reduced, as is the growth rate.

Alfonso and Russell (13) have given a phenomenological expression for the spherulite growth rate in terms of the competition between the inherent capability of the crystallite to grow and the diffusion or transport away from the growth of the noncrystallizing component. The slower of the two processes will be rate determining. Their expression for the growth rate is

$$G = \frac{v_2 k_1 k_2}{k_1 + k_2} \exp\left(\frac{-\Delta G^*}{RT}\right) \tag{11.14}$$

Here, k_1 is the rate of transport of the crystallizing segments across the liquid–crystal interface and k_2 is the rate at which the noncrystallizing component diffuses away from the growth front. There are two extreme cases of interest. If $k_2 \gg k_1$, the rate of transport across the interface is rate controlling. Equation (11.7) is regenerated since k_1 can be expressed in terms of the Vogel equation. If, on the other hand, $k_1 \gg k_2$, k_2 dominates the process and the diffusion step becomes rate determining. If the maximum distance that the segments must diffuse away from the growth front is defined as d, then

$$k_2 = d/(d^2/D) = D/d \tag{11.15}$$

where D is the diffusion coefficient of the noncrystallizing component. Consequently

$$G = v_2 \frac{D}{d} \exp\left(-\frac{\Delta G^*}{RT}\right) \tag{11.16}$$

where d plays a role equivalent to δ.

The discussion of nonlinear growth caused by a low molecular weight noncrystallizing added component has focused on the quantity δ and the diffusion coefficient of the added component. With just a few exceptions, little attention has been given to the kinetics when nonlinear growth rates are involved. An exception is found in a 60/40 isotactic–atactic poly(styrene) blend.(94) The growth rates of this blend are plotted in Fig. 11.43 as a function of log M_A at the indicated temperatures. The molecular weight, M_A, is that of the atactic component. The growth rates were obtained from the initial slope of the spherulite radius as a function of time. At the highest temperature, $T_c = 195\,^\circ\mathrm{C}$, G initially increases sharply with M_A and then displays a shallow maximum as M_A increases further. In contrast, at the lower temperatures G initially decreases with increasing M_A, reaches a minimum value and then undergoes a small increase with further increase in molecular weight. Similar results have been obtained by others.(54) Of particular interest in these

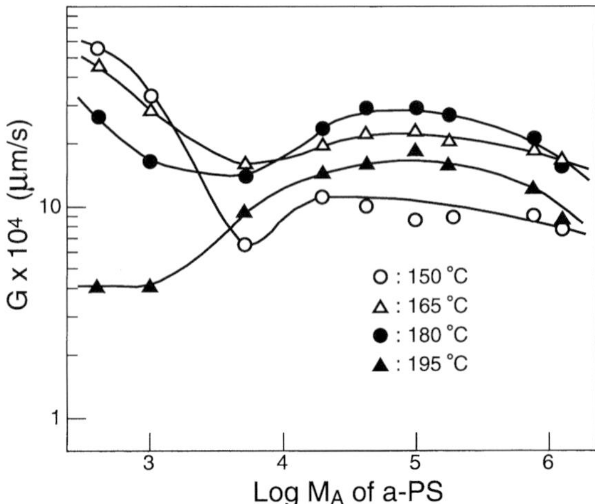

Fig. 11.43 Plot of initial growth rate G for a 60/40 isotactic–atactic poly(styrene) blend as a function of molecular weight of the atactic component, at indicated crystallization temperatures. (From Okada *et al.* (94))

data is the fact that nonlinear growth is observed when M_A is less than its value at the minimum. Conventional, linear growth is observed when M_A exceeds the minimum molecular weight. It was proposed that these results could be explained by the influence of the coefficient of the atactic component on the nucleation and spreading rates.(94) It should be noted, however, that the glass temperature of this blend increases with M_A and then levels off. The leveling off of molecular weight corresponds to the minimum observed in Fig. 11.43. The glass temperature is 0 °C at the lowest value of M_A, and levels off at about 100 °C. Thus, it is in this range of the glass temperature where the transport term is affected. Thus, the continuous increase in the glass temperature will cause a concomitant decrease in the growth rate. A minimum in the rate will be needed when the glass temperatures remain constant. This effect will be greater the lower the crystallization temperature.

In order to develop a firmer understanding of the underlying basis for nonlinear spherulite growth in immiscible systems, more extensive experimental investigations are needed. It is important that more attention be given to the kinetics of nonlinear growth. This involves studies over the complete composition range, carried out over extensive isothermal crystallization temperatures. In particular, blends that undergo a Regime I to II transition would be important since the relative nucleation and spreading rates could be assessed. It is also important to quantitatively evaluate the parameter δ and its role in governing nonlinear growth. Varying the growth rate, by crystallization temperature and molecular weight of the crystallizing component, while keeping the molecular weight of the low molecular weight species constant, would be informative.

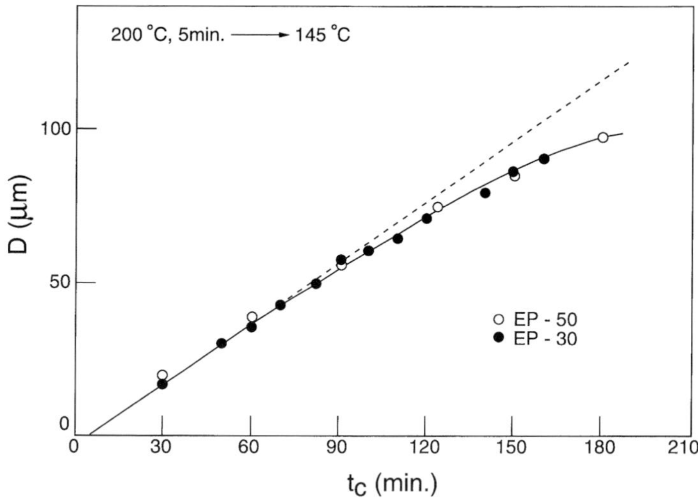

Fig. 11.44 Plot of spherulite diameter of isotactic poly(propylene) as a function of time at 145 °C, in blends with a random ethylene–propylene copolymer. Weight fraction isotactic poly(propylene) copolymers: ○ 50/50; ● 70/30. (From Inaba *et al.* (102))

Nonlinear spherulitic growth of the crystallizing component is also observed as a consequence of the concurrence of crystallization and liquid–liquid phase separation.(69,95,101–103) This case differs from the previous situation in that the noncrystallizing component need not be of low molecular weight.(102,103) Examples of nonlinear growth, caused by this process, in blends having two high molecular weight components are given in Figs. 11.44 and 11.45. Figure 11.44 is a plot of the spherulite diameter as a function of time for blends of two different compositions of isotactic poly(propylene), $M_w = 2.35 \times 10^5$, with an ethylene–propylene random copolymer, 73 mol percent ethylene, $M_w = 1.49 \times 10^5$.(102) The weight percent of the components isotactic poly(propylene)/ethylene propylene copolymer are 50/50 and 70/30. The blends were subject to demixing at 200 °C for 5 min and then isothermally crystallized at 145 °C. The phase separation occurred by a spinodal decomposition mechanism. It is clear that with time the growth becomes nonlinear. However, linear growth is observed when the crystallization temperature is reduced to 140 °C. The example in Fig. 11.45 is for a blend of poly(ethylene terephthalate)–poly(ether imide) that displays an upper critical solution temperature in supercooled melts.(103) Marked deviations from linearity of the growth rates are observed for the higher crystallization temperatures. However, at crystallization temperatures below 210 °C the growth rates are linear. All of the crystallization temperatures are within the two phase regions. Similar results are found with other compositions of the two components. The morphological features of these blends

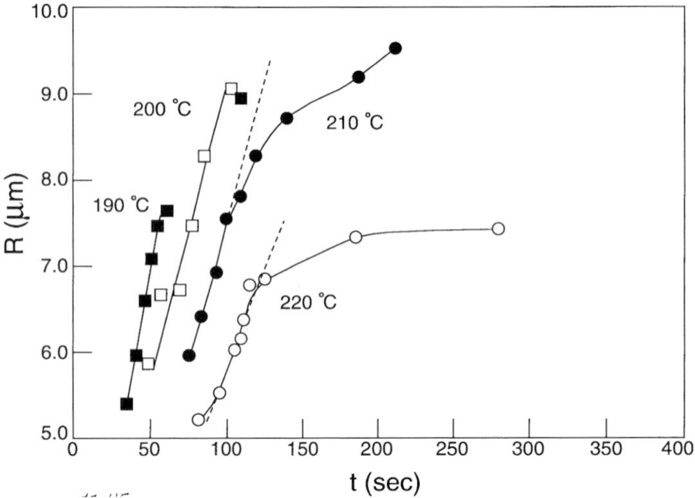

Fig. 11.45 Plot of spherulite radius of poly(ethylene terephthalate) in a blend with poly(ether imide) as a function of time at indicated crystallization temperatures. Blend composition poly(ethylene terephthalate)/poly(ether imide) 70/30. (From Chen *et al.* (103))

are similar to those that were illustrated in Fig. 11.44 since the phase separation also occurred by spinodal decomposition.[8]

When demixing occurs by spinodal decomposition, the noncrystalline component forms droplets, and a characteristic morphology develops.(102,103) These structures serve as an impediment to spherulitic growth, and rearrange with time. Depending on the growth rate, which is governed by the crystallization temperature, the crystallizing front will either encounter a fixed environment or one that is changing with time. Thus, at low crystallization temperatures, with relatively rapid growth, the spherulite radius will increase linearly with time. In contrast, at high crystallization temperatures, with relatively slow growth rates and the changing noncrystalline environment, the radius will no longer increase linearly with time.

Another example of nonlinear growth in blends with two high molecular weight components is illustrated in Fig. 11.46.(104) The blends illustrated are poly(ethylene oxide) as the crystallizing component, with either ethylene–methacrylic acid or styrene–hydroxy styrene copolymers as the noncrystallizing ones. The components in blends are completely miscible over the complete composition range. Each of the copolymers is thought to have strong intermolecular interactions with poly(ethylene oxide). As a consequence, it is claimed that there is a significant decrease in the equilibrium melting temperature with the added noncrystallizing component. If

[8] The characteristic morphology of such phase separated blends will be discussed in more detail in Volume 3.

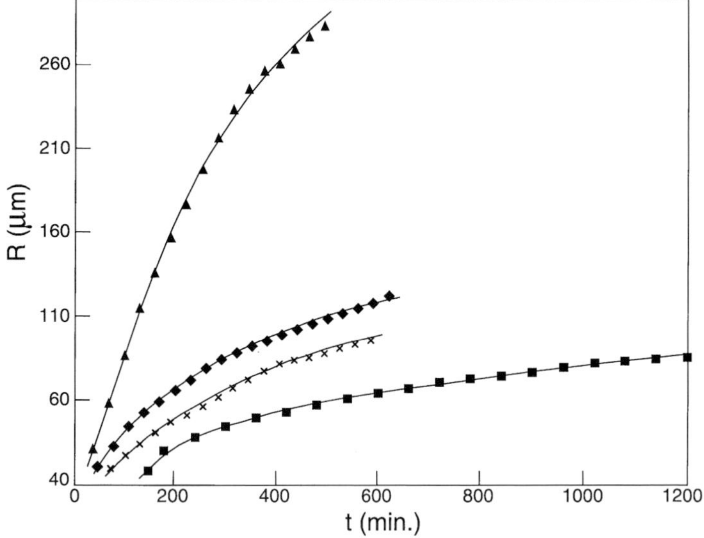

Fig. 11.46 Plot of spherulite radius as a function of time for poly(ethylene oxide) in blends with either ethylene–methacrylic acid or styrene–hydroxy styrene copolymer. With ethylene–methacrylic acid: poly(ethylene oxide)/ethylene–methacrylic acid ▲ 80/20, $T_c = 52.5\,^{\circ}\text{C}$; × 70/30, $T_c = 48\,^{\circ}\text{C}$. With styrene–hydroxy styrene: poly(ethylene oxide)/styrene–hydroxy styrene ◆ 80/20, $T_c = 55.5\,^{\circ}\text{C}$; ■ 70/30, $T_c = 46.5\,^{\circ}\text{C}$. (From Wu *et al.* (104))

correct, these blends represent an unusual situation, considering the colligative nature of melting point depressions. However, this concept can explain the nonlinear growth. The relatively large melting point depression, with the accompanying decrease in undercooling, will reduce the growth rate with time as the concentration of the added component in the noncrystalline region increases.

An unusual situation is illustrated in Fig. 11.47. Here the spherulite radius of poly(3-hydroxy butyrate) in a 50/50 blend with cellulose acetate butyrate is plotted against the crystallization time.(2) Although the growth rate is constant for isothermal crystallization at 65 °C it is not so at the higher crystallization temperature. At these temperatures the radius is not linear, and the growth rate actually increases with time. This is indeed a unique case. In all the systems studied heretofore, the growth rates were either linear or decreased with time. This behavior is only observed over a limited composition range in blends of these two components.

The conventional decrease in growth rate with time has been attributed to the rejection of the noncrystallizing component from the crystallizing species. Thus, the melt becomes richer in the noncrystallizing component with the resultant decrease in the growth rate. For a consistent interpretation of the results shown in

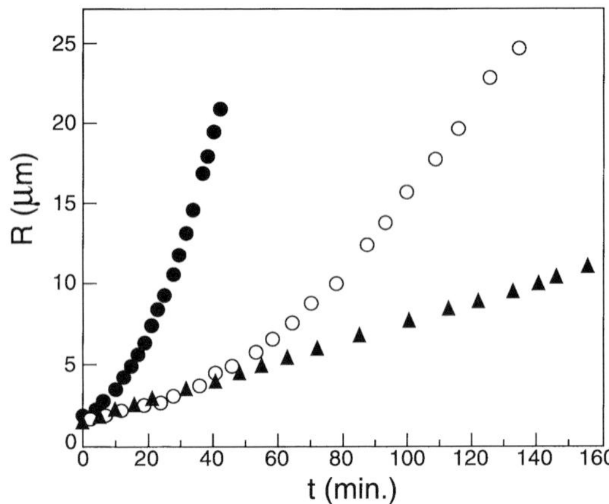

Fig. 11.47 Plot of spherulite radius of poly(3-hydroxy butyrate) as a function of time in a 50/50 blend with cellulose acetate butyrate at three crystallization temperatures: ▲ 65 °C, ● 95 °C, ○ 130 °C. (From Pizzoli *et al.* (2))

Fig. 11.47 one needs to formulate a mechanism for a decrease in the concentration of the added component in the melt. It is postulated, and demonstrated, that in the composition range of interest the cellulose acetate butyrate component also crystallizes.(2) Its concentration in the residual melt thus will be reduced, and an acceleration in the growth rate will result. The crystallization is shown to occur over a limited concentration range and is thus consistent with other aspects of the experimental results. Although unique, these results can be interpreted in a manner that is consistent with other results.

References

1. Wang, T. T. and T. Nishi, *Macromolecules*, **10**, 421 (1977).
2. Pizzoli, M., M. Scandola and G. Ceccorulli, *Macromolecules*, **27**, 4755 (1994).
3. Finelli, L., B. Sarti and M. Scandola, *J. Macromol. Sci.*, **A34**, 13 (1997).
4. An, Y., L. Dong, G. Li, Z. Mo and Z. Feng, *J. Polym. Sci.: Pt. B: Polym. Phys.*, **38**, 1860 (2000).
5. Huang, J., A. Prasad and H. Marand, *Polymer*, **35**, 1896 (1994).
6. Ong, C. J. and F. P. Price, *J. Polym. Sci.*, **63C**, 59 (1978).
7. Stein, R. S., F. B. Khambatta, F. P. Warner, T. Russell, A. Escala and E. Balizer, *J. Polym. Sci.*, **63C**, 313 (1978).
8. Dreezen, G., M. H. J. Koch, H. Reynaers and G. Groeninckx, *Polymer*, **40**, 6451 (1999).
9. Dreezen, G., Z. Fang and G. Groeninckx, *Polymer*, **40**, 5907 (1999).
10. Runt, J., D. M. Miley, X. Zhang, K. P. Gallagher, K. McFeaters and J. Fishburn, *Macromolecules*, **25**, 1929 (1992).
11. Yan, W. Y., J. Ismail, H. W. Kammer, H. Schmidt and C. Kummerlöwe, *Polymer*, **40**, 5545 (1999).
12. Rahman, M. H. and A. K. Nandi, *Polymer*, **43**, 6863 (2002).

13. Alfonso, G. C. and T. P. Russell, *Macromolecules*, **19**, 1143 (1986).
13a. Mareau, V. H. and R. E. Prud'homme, *Macromolecules*, **36**, 675 (2003).
14. Arrelino, L., E. Martuscelli, C. Selliti and C. Silvestre, *Polymer*, **31**, 1051 (1990).
15. Liu, L. Z., B. Chu, J. P. Pennings and R. St. John Manley, *J. Macromol Sci.*, **B37**, 485 (1998).
16. Gao, X., J. Manna and H. Bu, *J. Polym. Sci.: Pt. B: Polym. Phys.*, **38**, 3285 (2000).
17. de Juana, R. and M. Cort´Azar, *Macromolecules*, **26**, 1170 (1993).
18. Pedrosa, P., J. A. Pomposo, E. Calahorra and M. Cortázaŕ, *Polymer*, **36**, 3889 (1995).
19. Hsiao, B. and B. B. Sauer, *J. Polym. Sci.: Pt. B: Polym. Phys.*, **31**, 901 (1993).
20. Chen, H. L., J. C. Hwang and R. C. Wang, *Polymer*, **39**, 6067 (1998).
21. Alfonso, G. C., in *Integration of Fundamental Polymer Science and Technology*, vol. 5, P. J. Lemstra and L. A. Kleintjens eds., Elsevier (1991) p. 143.
21a. Chen, H. L., L. J. Li and T. L. Lin, *Macromolecules*, **31**, 225 (1998).
22. Jenkins, M. J., *Polymer*, **42**, 1981 (2001).
23. Mandelkern, L., *J. Appl. Phys.*, **26**, 443 (1955).
24. Mandelkern, L., *Polymer*, **5**, 637 (1964).
25. Boon, J. and J. M. Azcue, *J. Polym. Sci. Pt. A-2*, **6**, 885 (1968).
26. Hoffman, J. D., *Polymer*, **24**, 3 (1982).
27. Calahorra, E., M. Cortazar and G. M. Guzmán, *Polymer*, **23**, 1322 (1982).
28. Martuscelli, F., C. Silvestre and C. Gismondi, *Makromol. Chem.*, **186**, 2161 (1985).
29. El-Shafee, E., G. R. Saad and S. M. Fahurg, *Eur. Polym. J.*, **37**, 2091 (2001).
30. Avella, M. and E. Martuscelli, *Polymer*, **29**, 1731 (1988).
31. Paglia, E. D., P. L. Betrame, M. Caretti, A. Sevez, B. Marcandalli and E. Martuscelli, *Polymer*, **34**, 996 (1993).
32. Mandelkern, L. and R. G. Alamo, *Physical Properties of Polymers Handbook,* J. E. Mark ed., American Institute of Physics (1990) p. 119.
33. Pennings, J. P. and R. St. John Manley, *Macromolecules*, **29**, 84 (1996).
34. Shingankuli, V. L., J. P. Jog and V. M. Nadkarni, *J. Appl. Polym. Sci.*, **51**, 1463 (1994).
34a. Cheung, Y. W., R. S. Stein, B. Chu and G. Wu, *Macromolecules*,**27**, 3589 (1994).
35. Lee, J. C., H. Tazawa, T. Ikehara and T. Nishi, *Polymer*, **30**, 327 (1998).
36. Escala, A. and R. S. Stein, in *MultiPhase Polymers,* S. L. Cooper and G. M. Estes eds., Advances in Chemistry Series, American Chemical Society (1979) p. 455.
37. Ohno, K., Master of Science Thesis, Florida State University (1965).
38. Cheng, S. Z. D. and B. Wunderlich, *J. Polym. Sci.: Pt. B: Polym. Phys.*, **24**, 595 (1986).
39. Balijepalli, S., J. M. Schultz and J. S. Lin, *Macromolecules*, **29**, 6601 (1996).
40. Magill, J. H. and H. M. Li, *Polymer*, **19**, 416 (1978).
41. Glazer, R. H. and L. Mandelkern, *J. Polym. Sci.: Pt. B: Polym. Phys.*, **26**, 221 (1988).
42. Jackson, J. F. and L. Mandelkern, in *Analytical Calorimetry*, vol. 1, F. Johnson and R. Porter eds., Plenum (1968) p. 1.
43. Rego, J. M., M. T. Conde Braña, B. Terselius and U. W. Gedde, *Polymer*, **29**, 1045 (1988).
44. Mandelkern, L., F. L. Smith and E. K. Chan, *Macromolecules*, **22**, 2663 (1989).
45. Hosier, I. L., D. C. Bassett and A. S. Vaughan, *Macromolecules*, **33**, 8781 (2000).
46. Hosier, I. L. and D. C. Bassett, *J. Polym. Sci.: Pt. B: Polym. Phys.*, **39**, 2874 (2001).
47. Mandelkern, L., R. G. Alamo, G. D. Wignall and F. C. Stehling, *Trends Polym. Sci.*, **4**, 377 (1996).
48. Alamo, R. G., J. D. Londono, L. Mandelkern, F. C. Stehling and G. D. Wignall, *Macromolecules*, **27**, 411 (1998).

49. Alamo, R. G., W. W. Graessley, R. Krishnamoorti, D. J. Lohse, J. D. Londono, L. Mandelkern, F. C. Stehling and G. D. Wignall, *Macromolecules*, **30**, 561 (1997).
50. Galante, M. J., L. Mandelkern and R. G. Alamo, *Polymer*, **39**, 5105 (1998).
51. Tashiro, K., K. Imarrishi, Y. Izumi, M. Koboyashi, K. Kobayashi, M. Soto and R. S. Stein, *Polymer*, **39**, 5119 (1998).
52. Wignall, G. D., J. D. Londono, R. G. Alamo, M. J. Galante, L. Mandelkern and F. C. Stehling, *Macromolecules*, **28**, 3156 (1995).
53. Iragorri, J. I., J. M. Rego, I. Katime, M. T. Conde Braña and U. W. Gedde, *Polymer*, **33**, 461 (1992).
54. Yeh, G. S. Y. and S. L. Lambert, *J. Polym. Sci. Pt. A-2*, **10**, 1183 (1972).
55. Keith, H. D. and F. J. Padden, Jr., *J. Appl. Phys.*, **35**, 1270 (1964); *ibid.* **35**, 1286 (1964).
56. Hong, B. K., W. H. Jo and J. Kim, *Polymer*, **39**, 3753 (1998).
57. Phillips, R. A., M. D. Wolkowicz and R. L. Jones, *Soc. Plast. Eng. Antec 97*, p. 1671 (1997).
58. Pearce, R., G. R. Brown and R. H. Marchessault, *Polymer*, **35**, 3984 (1994).
59. Abe, H., Y. Doi, M. M. Satkowski and I. Noda, *Macromolecules*, **27**, 50 (1994).
60. Yamane, H. and K. Sasai, *Polymer*, **44**, 2569 (2003).
61. Haliloglu, T. and W. L. Mattice, *J. Chem. Phys.*, **111**, 4327 (1999).
62. Clancy, T., C. M. Pritz, J. D. Weinhold, J. G. Curo and W. L. Mattice, *Macromolecules*, **33**, 9452 (2000).
63. Maier, R. D., R. Thomann, J. Kressler, R. Mülhaupt and B. Rudolf, *J. Polym. Sci.: Pt. B: Polym. Phys.*, **35**, 1135 (1997).
64. Li, Y. and B. J. Jungnickel, *Polymer*, **34**, 9 (1993).
65. Fox, T. G., *Bull. Amer. Phys. Soc.*, **1**, 123 (1956).
66. Nishi, T., T. T. Wang and T. K. Kwei, *Macromolecules*, **8**, 227 (1975).
67. Martuscelli, E., C. Selliti and C. Silvestre, *Makromol. Chem. Rapid Comm.*, **6**, 125 (1985).
68. Endres, B., R. W. Garbella and J. H. Wendorff, *Colloid Polym. Sci.*, **263**, 361 (1985).
69. Chen. H. L., J. C. Hwang and R. C. Wang, *Polymer*, **39**, 6067 (1998).
70. Tanaka, H. and T. Nishi, *Phys. Rev. Lett.*, **55**, 1102 (1985).
71. Chan, J. W., *J. Chem. Phys.*, **42**, 93 (1965).
72. Chan, J. W., *Trans. Metall. Soc. AIME*, **242**, 166 (1968).
73. Nishi, T., *CRC Crit. Rev. Solid State Mater. Sci.*, **12**, 329 (1985).
74. Hilliard, J. E., in *Phase Transformations*, American Society of Metals (1970) Chapter 12.
75. Seward, T. P., D. R. Uhlmann and D. Turnbull, *J. Amer. Ceramic Soc.*, **51**, 634 (1968).
76. Inoue, T. and T. Ougizawa, *J. Macromol. Sci. Chem.*, **A26**, 147 (1989).
77. Cimmino, S., E. Di Pace, E. Martuscelli and C. Silvestre, *Polymer*, **34**, 2799 (1993).
78. Nojima, S., K. Kato, M. Ono and T. Ashida, *Macromolecules*, **25**, 1922 (1992).
79. Martuscelli, E., *Polym. Eng. Sci.*, **24**, 563 (1984).
80. Avella, M., E. Martuscelli, G. Orsello, M. Raines and B. Pascucci, *Polymer*, **38**, 6135 (1997).
81. Jog, J. P., V. L. Shingankuli and V. M. Nadkarni, *Polymer*, **34**, 1966 (1993).
82. Androsch, R., H. J. Radusch, F. Zahradnik and M. Mûrstedt, *Polymer*, **38**, 397 (1997).
83. Bartczak, Z., A. Galeski and E. Martuscelli, *Polym. Eng. Sci.*, **24**, 1155 (1984).
84. Galeski, A., M. Pracella and E. Martuscelli, *J. Polym. Sci.: Polym. Phys. Ed.*, **22**, 739 (1984).

85. Martenette, J. M. and G. R. Brown, *Polymer*, **39**, 1415 (1998).
86. Caldesi, V., G. R. Brown and J. W. Willis, *Macromolecules*, **23**, 338 (1990).
87. Morales, E., M. Salmerōn and J. L. Acosta, *J. Polym. Sci.: Pt. B: Polym. Phys.*, **34**, 2715 (1996).
88. Weng, W., H. W. Fiedel and A. Schell, *Colloid Polym. Sci.*, **268**, 528 (1990).
89. Greco, P. and E. Martuscelli, *Polymer*, **30**, 1475 (1989).
90. Martuscelli, E., C. Silvestre and L. Bianchi, *Polymer*, **24**, 1458 (1983).
91. Martuscelli, E., M. Pracella, M. Avella, R. Green and G. Ragestor, *Makromol. Chem.*, **181**, 957 (1980).
92. Reinsch, V. E. and L. Rebenfeld, in *Crystallization and Related Phenomena in Amorphous Materials*, MRS Symposium Proceedings, M. Libera, T. E. Haynes, P. Cebe and J. R. Dickinson, Jr. eds., **321**, 543 (1994).
93. Dell'Erba, R., G. Groeninckx, G. Maglio, M. Malinconico and A. Migliozzi, *Polymer*, **42**, 783 (2001).
94. Okada, T., H. Saito and T. Inoue, *Polymer*, **35**, 5699 (1994).
95. Nanbu, T., M. Ugazin, H. Tanaka and T. Nishi, *Rep. Prog. Polym. Phys. Japan*, **26**, 155 (1983).
96. Okada, T., H. Saito and T. Inoue, *Macromolecules*, **23**, 3865 (1990).
97. Schultz, J. M., *Polymer*, **32**, 3268 (1991).
98. Keith, H. D. and F. J. Padden, Jr., *J. Polym. Sci.: Pt. B: Polym. Phys.*, **25**, 229 (1987).
99. Keith, H. D. and F. J. Padden, Jr., *J. Appl. Phys.*, **34**, 2409 (1963).
100. Kit, K. M., *Polym. Commun.*, **39**, 4969 (1998).
101. Tanaka, H. and T. Nishi, *Phys. Rev. A*, **39**, 783 (1989).
102. Inaba, N., T. Yamada, S. Suzuki and T. Hashimoto, *Macromolecules*, **21**, 407 (1988).
103. Chen, H. L., J. C. Hwang, J. M. Yang and R. C. Wang, *Polymer*, **39**, 6983 (1998).
104. Wu, L., M. Lisowski, S. Talibuddin and J. Runt, *Macromolecules*, **32**, 1576 (1999).

12

Crystallization under applied force

12.1 Introduction

The discussion of crystallization kinetics, and the mechanisms involved, has so far been limited to quiescent crystallization. In essence, except for polymer structure and composition, the only variable considered was temperature. This has given a limited perspective to the overall subject, since polymers crystallize when subject to applied forces such as hydrostatic pressure, tensile, biaxial and shear deformation among others. Different morphologies and structures result from these types of crystallization. As a result, properties can be drastically altered. The analysis of the crystallization kinetics of polymeric systems when subject to such external forces is the subject of the present chapter. Although this an important area, the literature is not as rich as in some of the other areas that have been discussed. However, the results that have been obtained are interesting and hopefully will stimulate further inquiry.

12.2 Effect of hydrostatic pressure

12.2.1 Overall crystallization kinetics

Under hydrostatic pressure the force is applied uniformly in all directions. The discussion will be divided into two parts: the overall crystallization kinetics in this section and spherulite growth rates in the following. The application of hydrostatic pressure increases both the glass and equilibrium melting temperatures. However, since each increases at a different rate the window for crystallization to occur will be altered. The change in the melting temperature can be calculated by application of the Clapeyron equation. It should be recalled that in order to properly calculate the equilibrium melting temperature by this method the specific volumes of the crystal and liquid have to be known as a function of temperature. If not, significant errors can occur. The increase in the glass temperature needs to be determined experimentally.

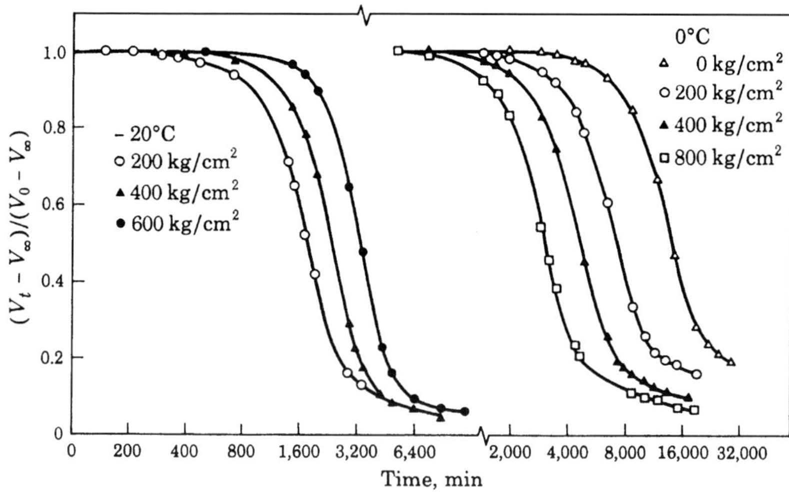

Fig. 12.1 Plot of quantity $(V_t - V_\infty)/(V_0 - V_\infty)$ against log t for poly(cis-isoprene) at indicated temperatures and pressures.(2)

An important consideration in analyzing crystallization kinetics under these conditions is the distinct possibility of the formation of polymorphs that in turn influence the crystallization process. For example, referring to the phase diagram of linear polyethylene (Fig. 6.13, Volume 1) a hexagonal form develops after crystallization from the melt at high temperatures and pressures. It is possible, in general, for the polymer to be nucleated in one form and eventually transform to the more stable form with time. It is also possible for the hexagonal form to be nucleated at temperatures and pressures where this structure does not appear in the phase diagram. It has been suggested that other polymers, particularly poly(cis-1,4-isoprene), also crystallize through a metastable form.(1)

In order to establish the basis for analyzing the experimental results for both types of measurements, it is necessary to ascertain what modifications have to be made in Eq. (9.209), and related expressions, as a consequence of the applied pressure. There is the question with respect to the nucleation term in Eq. (9.209) whether the two interfacial free energies, σ_{en} and σ_{un}, as well as ΔH_u, or ΔS_u, depend on pressure. There is also concern as to whether the parameters U^* and C in the transport term vary. Pertinent experimental results can be examined, keeping in mind these concerns.

Figure 12.1 gives a set of isotherms, obtained dilatometrically, for the crystallization of poly(cis-isoprene) at different pressures and two crystallization temperatures.(2) The isotherms have similar shapes, irrespective of the temperature and pressure of crystallization. This leads to a set of superposable isotherms, for both temperatures and pressures, as is illustrated in Fig. 12.2. The experimental points

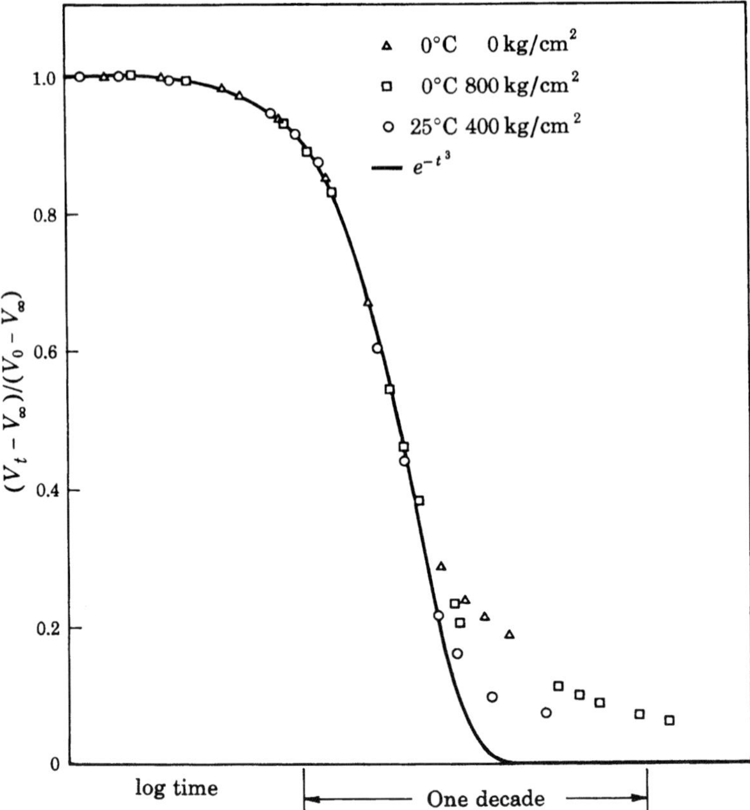

Fig. 12.2 Demonstration of superposability of isotherms of poly(cis-isoprene) for indicated temperatures and pressures. Solid line plotted according to Eq. (9.31a) with $n = 3$.(2)

represent the isotherms obtained at the indicated temperatures and pressures. They are plotted on an arbitrary time scale so as best to bring them into coincidence. The solid line is drawn in accordance with the derived Avrami equation, Eq. (9.31a), with $n = 3$. The plots demonstrate good agreement between experiment and theory for a major portion of the transformation. As in quiescent crystallization, deviations from the theory develop towards the end of the transformation. The disparity for poly(cis-isoprene), natural rubber, crystallized under pressure is observed at a crystallinity level of about 25%. Linear polyethylene,(3–5) see below, and other polymers,(6) yield superposable isotherms and adhere to the Avrami equation when crystallized under applied pressure.

The crystallization rate of poly(cis-isoprene) displays a maximum with crystallization temperature under atmospheric pressure (Fig. 9.6). As is illustrated in Fig. 12.3 this maximum is still maintained for crystallization under applied pressure.(2) Here the rate constant, from the derived Avrami expression, is plotted

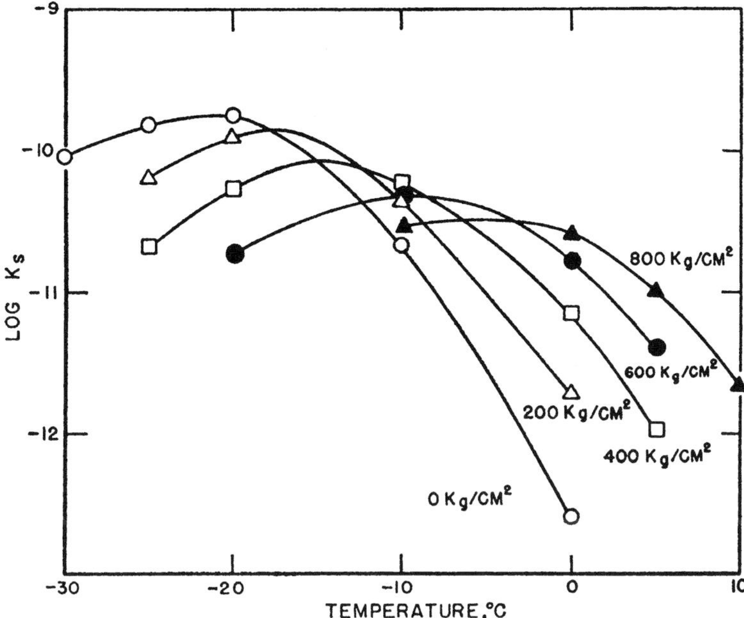

Fig. 12.3 Plot of log k_s (Avrami rate constant) for poly(cis-isoprene) against crystallization temperature at indicated applied hydrostatic pressure.(2)

against the crystallization temperature for different pressures. At one atmosphere the rate maximum is observed at −24 °C. As the pressure is increased the maximum progressively moves to higher temperatures. At a pressure of 800 kg cm^{-2} the maximum has increased to −5 °C. Concomitantly there is a decrease in the rate constant at the maximum. An inversion in the rate with applied pressure is also observed. The overall crystallization rate increases with pressure at temperatures to the right of the maximum. The opposite is observed at temperatures to the left of the maximum. In this temperature region the rate decreases with applied pressure. These results indicate that at least two mechanisms are still involved that vary differently with temperature and pressure. At high temperatures, where the nucleation term predominates, the increased undercooling with applied pressure, at a fixed crystallization temperature, is important. At the lower temperature, the change in the transport term with pressure is important as it involves the glass temperatures, U^* and C. The ratio T^*/T_m is still in the range 0.81–0.83 for this polymer over the pressure range studied.

The overall crystallization rate of linear polyethylene, as a function of temperature and pressure has been extensively studied.(3–13) The polymorphism of linear polyethylene at elevated temperatures and pressures makes the analysis of its crystallization kinetics more complex. Depending on conditions, either a hexagonal form or the conventional orthorhombic can be observed. Consequently, in principle, a true phase diagram, i.e. one that relies on equilibrium melting temperature, needs to be available in order to properly interpret the kinetic results. Such a diagram is unfortunately not available. Recourse then has to be made to a pseudo-phase diagram such as Fig. 6.13. Such a diagram is based on observed melting temperatures. The phase diagram, true or pseudo, will depend on molecular weight and polydispersity. In many instances this requirement makes it difficult to definitively interpret the results of different investigators. In general the usual orthorhombic form is observed after crystallization at low to moderate pressure. However, at high pressures and temperatures, above about 3 kbar and 220 °C, the hexagonal form develops from the melt. At these high temperatures and pressures, crystallites in the range of 1000 Å to 3 micrometers are formed.(14,15) Such crystallites have been termed extended chain crystals.[1] In analyzing the kinetic data it is important to establish which of the two polymorphs forms initially, and if any transformation occurs during the time course of crystallization. It should be recognized that with rare exceptions (5,9) the studies were carried out with samples having a broad molecular weight distribution with a significant concentration of low molecular weights.

Superposable isotherms are observed following crystallization from the melt, irrespective of whether orthorhombic or hexagonal crystals are formed. A set of isotherms for an unfractionated linear polyethylene ($M_n = 14\,000$, $M_w = 64\,000$) that is stated to be crystallizing in the hexagonal form is given in Fig. 12.4.(7) The superposition of the isotherms is quite clear. Crystallization under conditions that yield the orthorhombic phase also gives superposable isotherms that resemble those obtained at atmospheric pressure. There is, however, a major difference between the shapes of the isotherms in the two cases. At high pressure, and the formation of the hexagonal phase, the Avrami exponent is about 1.(7–11) In contrast, when the orthorhombic crystal structure is formed n is 3 or 2, depending on the particular sample being studied. The n values thus reflect the different structures that are evolving. The shapes of the isotherms are obviously quite different in the two cases.(10,12) The formation of the hexagonal structure, with $n = 1$, leads to the thick extended chain type crystallites. The value of $n = 1$ suggests one-dimensional growth. This conclusion is supported by microscopic observations.(12a) In contrast, the orthorhombic form leads to lamellar crystallites with the conventional thicknesses. Deviations from the derived Avrami expression are observed as the crystallization progresses in both cases.

[1] It does not necessarily follow that the crystallite thickness can be identified with the length of the completely ordered chain.(16) Details of the structure and morphology of such polyethylene crystallites will be discussed in detail in Volume 3.

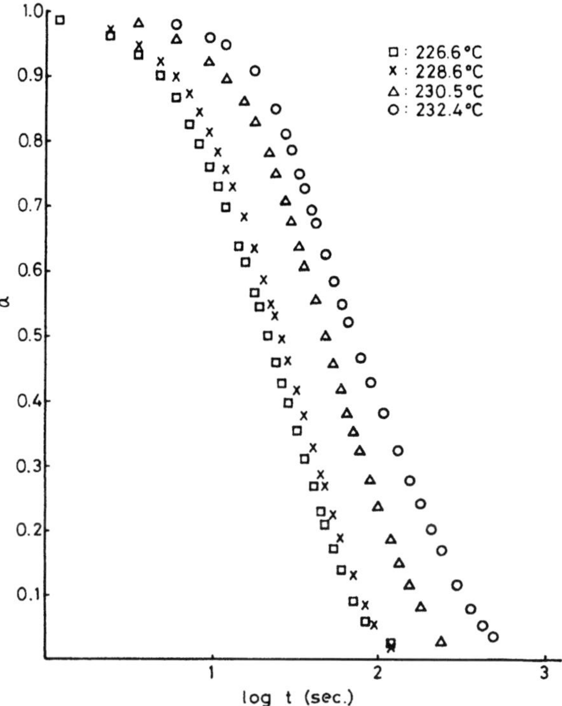

Fig. 12.4 Plot of relative extent of crystallization of an unfractionated linear polyethylene as a function of log time at 5300 bar at indicated temperatures. $A = V_\infty - V_t/V_\infty - V_0$. (From Kyotani and Kanetsuna (7))

The use of molecular weight fractions results in some interesting observations.(5) Figure 12.5 illustrates an unusual two-stage crystallization process that takes place isothermally at high pressure when fractions are used. The conditions here are such that the hexagonal phase initially forms from the pure melt. The early stages of the crystallization display conventional type isotherms. A long plateau region then develops, followed by the further development of substantial amounts of additional crystallization. Similar results are implied in the works of Matsuoka (3) and Hoehn *et al.*(17) These results are consistent with the observed double endotherms found in similar crystallized samples (18–20), as well as dilatometric studies that show two melting temperatures.(21) These results strongly suggest that each of the polymorphs is forming in turn. Direct structural studies and identification of the crystallizing species would be extremely helpful in interpreting these results.

It has been claimed, based on studies in the pressure–temperature range of 2.5 kbar and 180–310 °C, that, irrespective of the position in the pseudo-phase diagram, crystallization never occurs directly in the orthorhombic phase.(22,23) This important conclusion is based solely on morphological studies with a molecular weight

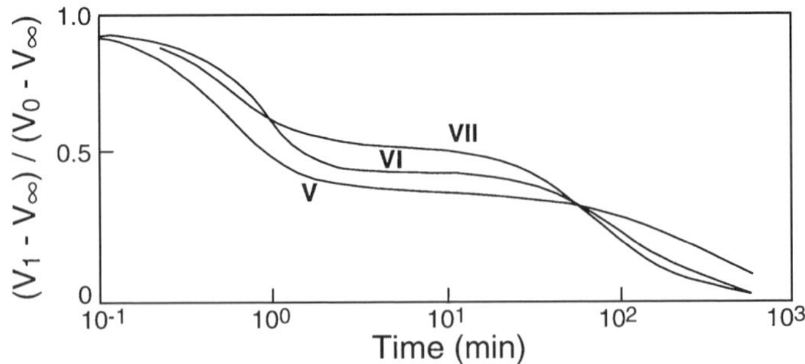

Fig. 12.5 Plot of relative extent of crystallization of molecular weight fraction of linear polyethylene as a function of log time at 5000 bar and 235 °C. Number average molecular weights: V = 17 000; VI = 50000; VII = 130000. (From Hatakeyama *et al.* (5))

fraction of linear polyethylene $M_w = 3.2 \times 10^4$, $M_w/M_n = 1.11$. Crystallization is said to proceed through the hexagonal phase, even when the orthorhombic phase is the more stable one. Put another way, it is thought that the hexagonal polymorph always initiates the crystallization, irrespective of whether it is the stable or metastable phase, or whether the crystallization is carried out above or below the triple point. When the crystallization is carried out at pressures and temperatures where the orthorhombic phase is stable, a transformation to this polymorph eventually occurs with time. No appreciable crystallite thickening takes place under these conditions. In contrast, when the hexagonal polymorph is maintained during the course of the transformation, rapid crystallite thickening to the extended form takes place.

There is a contradiction between the above concepts and the Avrami exponents. The latter are consistent with the initial formation and growth of either the orthorhombic or hexagonal forms. On the other hand, the two-stage isotherms cannot be explained by these concepts. A resolution of these complementary, but contradictory, results is clearly needed. This resolution is important not only for the understanding of the high pressure–high temperature crystallization but also the strong suggestion that has been made that crystallization of polyethylene at atmospheric pressure also proceeds from the metastable polymorph.(22–24) By implication this argument would apply to other polymers as well.

In the discussion of polymorphism in Chapter 6 several different possible equilibrium transformations from one form to another were outlined. It was pointed out that kinetic factors can also play an important role in determining which polymorph is actually observed under specific conditions. An important factor in the crystallization process is the nucleation step and the free energy barrier that needs to be overcome in order to form a nucleus of critical size. It does not necessarily follow that the barrier height, the free energy required to form a critical size nucleus, is related to the stability of the equilibrium crystallite. Taking a Gibbs nucleus as an example the free energy barrier depends on the ratio $\sigma_{en}\sigma_{un}/\Delta G_u$. Thus, depending on the relative value of this ratio between the different polymorphs, it is possible, in principle, for the less thermodynamically stable form to nucleate first. Initially this crystal structure will then be propagated. Theoretically, a transformation to the more stable form will eventually occur. This appears to be the situation for the high pressure–high temperature crystallization of linear polyethylene, if indeed the hexagonal polymorph is always formed initially. Detailed kinetic studies, as a function of temperature and pressure, that are restricted to the crystallization of the hexagonal phase, as well as the determination of its free energy of fusion, would help to better understand the problem. In contrast, when the thermodynamically most stable form is nucleated first, this structure will be maintained during the course of crystallization.

A characteristic of isothermal crystallization is that the overall crystallization rate increases with applied pressure.(5–7,9,13) This effect can in general be attributed to an increase in the undercooling. However, there is an independent influence of pressure, when the data are analyzed at constant undercooling. To appraise the situation, reliable values of the equilibrium melting temperature, and thus the undercooling, are needed as a function of pressure. This temperature can be estimated by an extrapolative method that depends on the relation between the observed melting and crystallization temperatures at different pressures,(13) or by an empirical relation based on the Clapeyron.(5) Although there is a consistent small difference between the two methods it is not significant for present purposes. Hence, we will use the equilibrium melting temperatures given by Hatakeyama *et al.* and the kinetic data they reported covering the pressure range of 1000 kg cm^{-2} to 5000 kg cm^{-2}.(5) These equilibrium melting temperatures used here are empirical in nature.

Figures 12.6 and 12.6a summarize the overall crystallization rates in terms of half-times, $t_{1/2}$, as a function of undercooling at pressures ranging from 1000 to 5000 kg cm^{-2}.(5) Linear plots are obtained for the two lowest pressures. This is the pressure region where folded chain crystallites form.(5) At the higher crystallization pressures, although the plots are not linear, the curves parallel one another. At the higher pressures, where the hexagonal structures are formed, extended chain crystals grow rapidly in one dimension in the chain direction. The plots make clear that at any undercooling there is a significant decrease in the half-time with increasing pressure. For example, at $\Delta T = 10$, $t_{1/2}$ is approximately 2000 min at 1000 kg cm^{-2}. As has been pointed out, concomitantly the Avrami exponent decreases from about 3 to 1 as the crystallization rate is enhanced.

It is appropriate at this point to interject, perhaps out of turn, results that have been reported for the lateral growth of the hexagonal form of linear polyethylene at

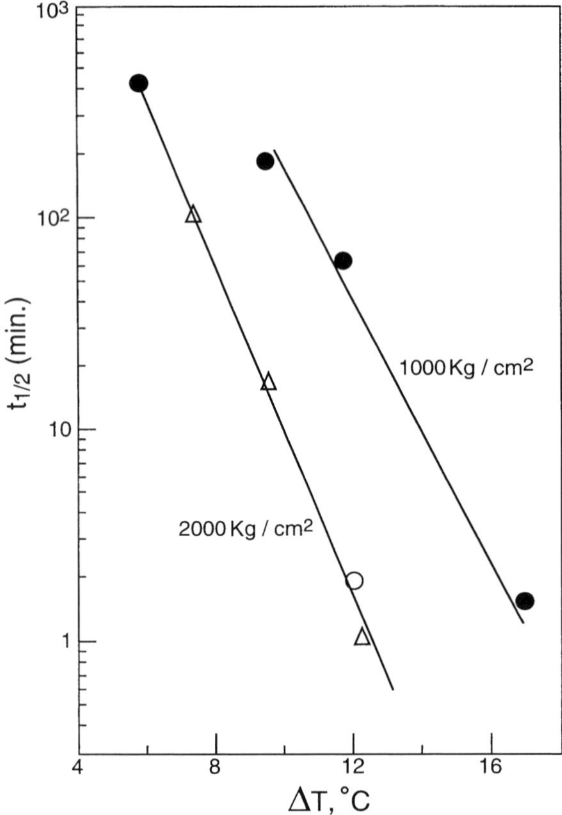

Fig. 12.6 Plot of crystallization half-time, $t_{1/2}$, against undercooling, ΔT, for linear polyethylene crystallized at 1000 and 2000 kg cm^{-2}. Unfractionated polyethylenes: △, △$M_n = 5.2 \times 10^4$, $M_w/M_n = 2.7$; ○, ● $M_n = 5.4 \times 10^4$, $M_w/M_n = 3.0$. (From Hatakeyama *et al.* (5))

pressures that are either in the stable or metastable region for this polymorph.(22) Figure 12.7 is a plot of the lateral growth rate against the reciprocal undercooling of the hexagonal form at different pressures.[2](22) In this work, a parallel set of straight lines is found for both low and high pressures. The growth rates decrease significantly with pressure, in accord with the results for the overall crystallization rates. There appears to be a discrepancy between the linearity of these results at high pressure and those of the overall crystallization rates under comparable conditions. Strictly speaking the overall crystallization and growth rates should not be compared.

[2] The plots in Fig. 12.7 are crucially dependent on the values taken for the equilibrium melting temperatures of the hexagonal polymorph.

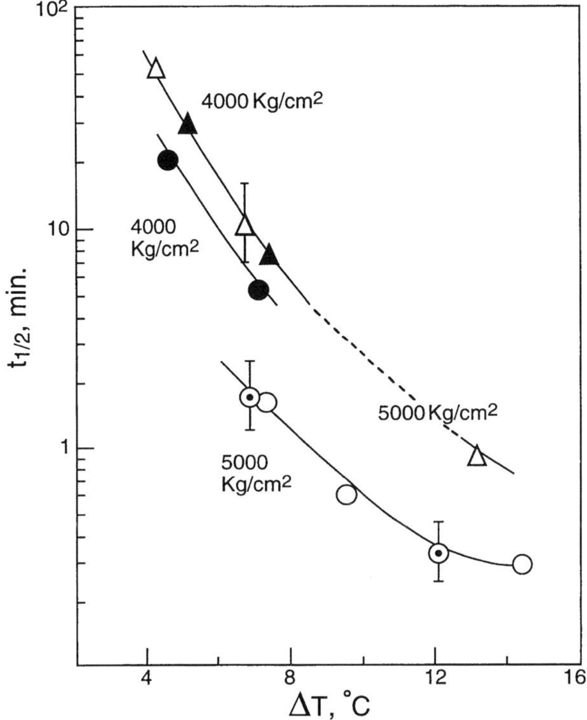

Fig. 12.6a Plot of crystallization half-time, $\tau_{1/2}$, against undercooling, ΔT, for linear polyethylene crystallized at 4000 and 5000 kg cm^{-2}. Unfractionated polyethylenes: △, ▲ $M_n = 5.2 \times 10^4$, $M_w/M_n = 2.7$; ○, ● $M_n = 5.4 \times 10^4$, $M_w/M_n = 3.0$. Fraction: ⊙ $M_n = 5.0 \times 10^4$, $M_w/M_n = 1.3$. Symbols with vertical lines represent two-stage crystallization. (From Hatakeyama *et al.* (5))

The kinetic studies suggest that at sufficiently high pressures extended chain crystals of linear polyethylene grow isothermally, directly and rapidly from the melt, at the crystallization temperature.(4,7,12,20,25,26) However, morphological studies have lead to quite different conclusions.(23,27–30) Based on morphological studies, it has been postulated that crystallization always begins with folded chain crystallites. Eventually, chain extension occurs under isothermal conditions. This conclusion is based primarily on the observation of tapered growth faces of isolated lamellae as seen by transmission electron microscopy, in samples quenched from high pressure. Other interpretations have been offered for the microscopic observation.(4,25)

Despite some differences in interpretation and discrepancies that exist between experimental results, a major feature has emerged from the high temperature–high pressure studies with linear polyethylene. This feature is related to the fact that when the hexagonal phase is maintained rapid crystallite thickening takes place in

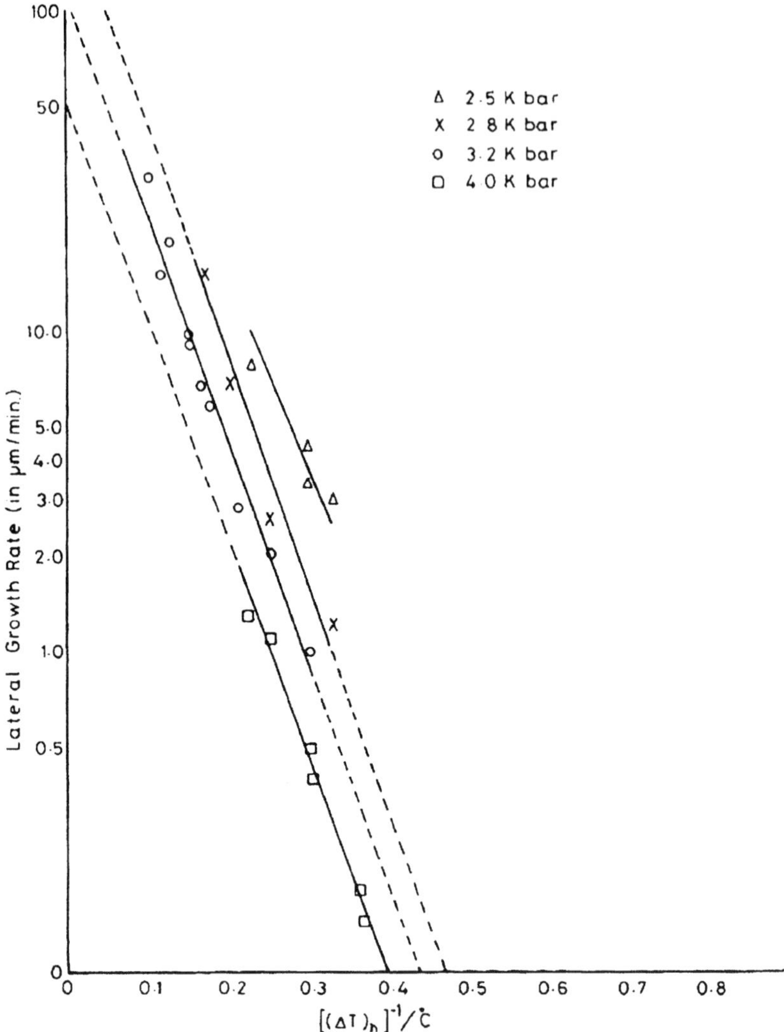

Fig. 12.7 Plots of lateral growth rates as a function of the reciprocal undercooling for linear polyethylene crystallized at the indicated pressures. (From Rastogi *et al.* (22))

the chain direction. A basic question to be addressed is why this occurs under these crystallization conditions. The hexagonal crystal has certain unique features that are consistent with the observed thickening. The crystals in this phase possess some degree of conformational disorder.(31,32) There is only order in two dimensions. The chain units in this state possess a high degree of molecular mobility and perform rotational and translational motions. NMR measurements of linear polyethylene in the hexagonal phase indicate a high rate of transactional diffusion in the chain direction.(33)

Hikosaka and coworkers have addressed some aspects of this problem by introducing the concept of sliding diffusion.(34–37) Underlying this concept is the deduction from x-ray diffraction analyses that the interaction between chains in the hexagonal form is much weaker than in the orthorhombic form.(38) This is consistent with the well-known properties of polyethylene chains in the hexagonal form that were described above. In this development attention is focused on secondary nucleation, the growth of nuclei past critical size, and the free energy region $\Delta G = \Delta G^*$ to $\Delta G = 0$. The same concepts have also been used to develop a theory for primary nucleation.(37) The chain conformations in the nuclei of critical size are assumed to be regularly folded, with adjacent re-entry in both cases.[3]

In essence, the steady-state nucleation rate of Turnbull and Fisher is modified to account for the growth of nuclei past ΔG^*. The critical free energy, ΔG^*, remains unaltered. Attention is directed to the activation energy E for transport of a repeating unit across the interfacial boundary. The activation energy is considered to consist of two terms. One is the activation energy, termed E_s, that is required to add a unit to the side surface of the nucleus. The other, E_e, is that needed to add a unit to the end surface. Consequently, E can be written as

$$E = pE_e + (1 - p)E_s \tag{12.1}$$

where p is the probability of attaching a unit to the end surface. It is assumed that E_s is related to the energy of reeling in of a chain from the melt and is estimated to be small. Equation (12.1) then becomes

$$E \simeq pE_e \tag{12.2}$$

The assertion is made that it is extremely difficult for a repeating unit to diffuse from the residual melt, or liquid, to the end surface of the growing nucleus. Therefore, in order for a nucleus to thicken, i.e. grow in the chain direction, most units must diffuse from the interior of the molecule to the end surface. This concept of nucleation past ΔG^* is what is meant by the sliding diffusion of a chain. The conventional activation energy for transport is then replaced by E_e. It will depend on the structure of the nucleus in terms of the chain conformation, lattice parameter and defects. For a close packed strongly interacting chain structure, such as the orthorhombic phase of linear polyethylene, the work required to draw a chain through the nucleus is proportional to its frictional coefficient κ, and thickness l. In contrast, in the hexagonal phase the chains are defected and loosely paired so that translational motion is facile. Thus, to include the extremes, and situations in

[3] It was pointed out in Chapter 9 that the adoption of this chain structure with the nucleus represents a major assumption. Other chain conformations within the nucleus can serve equally well. The same functional form for ΔG^* and the steady-state nucleation rate will be obtained as long as a Gibbs type nucleus is considered.

between, E can be expressed as

$$E = E_e = p\kappa l^{\upsilon} \qquad 0 < \upsilon \leq 1 \tag{12.3}$$

Following these arguments, it turns out that the quantity E effectively determines, through the nucleation rate, whether or not thickening will occur following the formation of a critical size nucleus. Parenthetically, it should be noted that for a Gibbs type nucleus to be stable at a temperature infinitesimally above the crystallization temperature thickening in the chain direction must occur, irrespective of the crystal modification involved.

The nucleation rate is calculated following the method of Frank and Tosi (39) by incorporating the assumptions outlined above. In this method a set of sequential processes, with forward and backward rates, is calculated, in this case starting at the critical size nucleus. The nucleation rate can then be written as (34)

$$N = N_0 \exp\left[-\frac{\Delta G^*}{RT} - \frac{E^*}{R(T - T_g)}\right] P_s = N_0 A P_s \tag{12.4}$$

where

$$P_s \equiv \sum_{m=0}^{\infty} \exp\left[\frac{\Delta G_m}{RT} + \frac{E_m}{R(T - T_g)} - A\right]^{-1} \tag{12.5}$$

Here the subscript m represents the stage number and E^* is the activation energy corresponding to the barrier height. In this theory E_m is not necessarily constant but is allowed to vary with m. When E is constant, Eq. (12.4) reverts to the conventional expression for the steady-state nucleation rate. As given, the steady-state nucleation rate is the product of two factors. One is overcoming the free energy barrier to form a nucleus; the other the survival probability of a critical size nucleus. In effect, what is being considered is what takes place in the region between ΔG^* and $\Delta G = 0$. The change in ΔG is known to be precipitous in this region. The key quantity in the analysis is E_m, which in turn depends on the parameters κ and ν. Analysis indicates that for large values of κ and ν nucleation will not occur, so that thickening will not be observed. However, for low values of these parameters nucleation can proceed and thickening will occur along the chain direction.

These ideas have been adapted to explain the molecular weight dependence of the primary and secondary growth nucleation of linear polyethylene in the hexagonal phase.(35–37) It is demonstrated experimentally that the free energy of forming a critical size nucleus does not depend on molecular weight. Therefore, either or both the transport term and front factors are involved. It was concluded that both of the nucleation processes involve sliding diffusion within the nucleus with the disentanglement of chains in the interfacial region.

Serious questions can be raised regarding the basic concept of sliding diffusion within the nucleus, or crystallite, and the accompanying theoretical development. Several of these merit serious consideration. Among them is the concern as to why the sliding diffusion is not invoked prior to forming a critical size nucleus. The physical reason for the assumption that E_m increases with each growth stage is crucial and needs to be justified. A related problem is that the chains within the critical size nucleus are assumed to be regularly folded, and presumably in the orthorhombic form. With such a nucleus structure how does sliding diffusion become effective for a nucleus past ΔG^*? If on the other hand the critical nucleus is composed of chains in the hexagonal form how can one conclude that folded chain crystallites are formed initially and then develop into extended form?

The discussion of the overall crystallization kinetics of linear polyethylene makes apparent that the formation of extended chain crystallites in this polymer, at high temperature and pressure, is not as well understood as would be desired. Significant theoretical and experimental problems remain to be resolved. The important fact that thickening in the chain direction takes place within the hexagonal phase is well established. However, some matters still remain to be addressed. If the complete chain exists in this mesomorphic state then facile chain extension will occur in a natural way. However, if the complete molecule is not involved in the hexagonal state, there will be some degree of entanglement among the units from different molecules. Hence, some type of chain disentanglement needs to be invoked to allow for extended chain crystallites to develop. To completely analyze crystallization involving the hexagonal phase the appropriate thermodynamic parameters need to be established. As was pointed out earlier the delineation, as closely as possible, of equilibrium phase diagrams for a range of molecular weight fractions would be highly desirable.

12.2.2 Growth kinetics

Spherulite or lamellar growth rates have been studied under high pressure and temperature for several polymers. These include poly(cis-1,4-isoprene), natural rubber,(1) the high and low melting polymorphs of poly(trans-1,4-isoprene), gutta percha,(40) linear polyethylene (13,22) and poly(ethylene terephthalate).(41) The radial growths of the two polymorphs of poly(trans-1,4-isoprene) are linear with time at all crystallization temperatures at pressures up to 3 kbar.(40) The growth rates of each of the forms are plotted in Fig. 12.8 as functions of temperature at the indicated pressures. The maxima observed at 1 bar move to successively higher temperatures with increasing pressure. The magnitude of the growth rate at the maximum decreases continuously with pressure. These results are qualitatively similar to the influence of pressure on the overall crystallization rate of

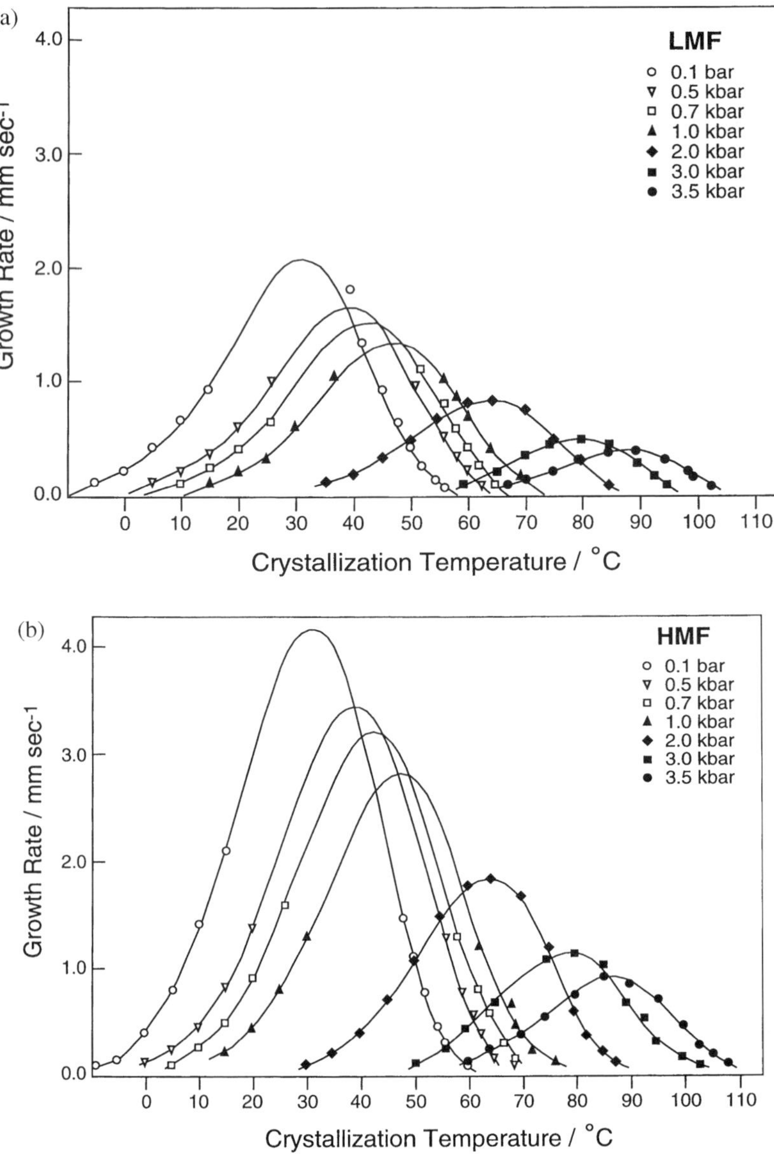

Fig. 12.8 Plot of spherulite growth rates of the low (a) and high (b) melting polymorphs of poly(trans-1,4-isoprene) as functions of temperature at the indicated pressures. (From Davies and Cucarella (40))

poly(cis-1,4-isoprene) (Fig. 12.3). In this case the growth rate increases with pressure at temperatures to the right side of the maximum and decreases at temperatures at the left of the maximum. The growth rates of poly(ethylene terephthalate) were studied up to a pressure of 2 kbar, but restricted to temperatures at the left of the maximum.(41) The growth rates decreased substantially with pressure in this temperature region.

The influence of pressure on the lamellar growth rate of poly(cis-1,4-isoprene) is quite different than what was described above.(1,42) In this case, although the maxima move to higher temperatures with increasing applied pressure, their magnitude initially increases and then levels off. As was shown in Fig. 12.7, the lateral growth rates of the hexagonal form of linear polyethylene generate a set of parallel straight lines when plotted against the reciprocal of the undercooling. The growth rates in this case decrease with pressure. It is tempting to conclude, if the undercoolings are reliable, that the product of interfacial free energies, $\sigma_{un}\sigma_{en}$ is independent of pressure. However, ΔS_u, which appears in the denominator of ΔG^*, is a function of pressure. Thus, until the relation between ΔS_u and pressure is established, no conclusion can be made with respect to the dependence of $\sigma_{un}\sigma_{en}$ on pressure.

The dependence of the spherulite and lamellar growth rates on the crystallization temperature, at each pressure studied, can be fitted by the conventional relation, Eq. (9.209), for all polymers studied. This relation, which has been successful in analyzing growth rate data at atmospheric pressure, assumes a Gibbs type nucleus and a transport term expressed by the Vogel expression. Therefore, the values of T_m^0 and T_g as a function of pressure need to be known *a priori*. In addition, the pressure dependence of the expansion terms in the free energy of fusion also need to be known, as do the parameters U^* and C. The analysis of the growth rate–temperature data, at atmospheric pressure, for many polymers has shown that there is a wide range in the values of these parameters that can satisfactorily explain the data. Thus, the conclusions reached with regard to the existence of regimes and the product of the interfacial free energies, $\sigma_{en}\sigma_{un}$, depended on the specific set of parameters, from among many, that were chosen. A similar problem exists in analyzing growth rates at elevated pressure.

The growth rate studies have not led to a definitive conclusion. This is due in part to the few polymers that have been studied and the limited pressure range covered. However, the major obstacle is the need to specify a large number of independent parameters. This presents a formidable problem to accomplishing an objective analysis.

12.3 Crystallization kinetics under uniaxial deformation

Crystallization under an applied hydrostatic pressure is uniform in all directions. In contrast, other types of deformation such as uniaxial, biaxial and shear are directional and thus anisotropic. These types of deformation have a major influence on the crystallization process and therefore on the resulting properties. In the studies involving uniaxial and biaxial deformation, which will be discussed, the sample is

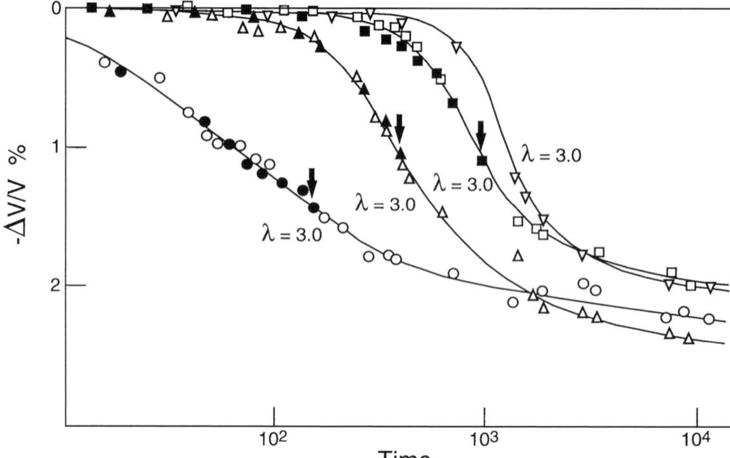

Fig. 12.9 Plot of relative volume decrease, open symbols, and relative stress decrease, closed symbols, as a function of log time for vulcanized natural rubber at −26 °C at indicated extension ratios. Stress measurements scaled appropriately. (From Gent (44))

initially deformed in the pure melt. The deformation is maintained as the temperature is lowered to allow for isothermal crystallization to ensue. The deformation is maintained during the entire course of the crystallization. It is important in these types of studies that the polymers be covalently cross-linked and the resultant networks be well characterized. If the system is not cross-linked, flow will occur with the initial deformation and continue as crystallization proceeds.[4]

Both overall and growth rate kinetics have been studied with rubber-like polymers under uniaxial deformation. The conventional techniques, described earlier, have been used to study the overall kinetics. In addition, measurement of the relaxation in the applied stress with time, while the sample is held at constant length, is an effective method by which to measure the development of crystallinity.(43) As the crystallinity progresses the stress decreases until it becomes zero. At this point the well-known spontaneous elongation occurs.(44,45) The relative decrease in the stress with time can be related to the increase in the level of crystallinity and gives results that are the same as those obtained by other methods. An example of this comparison is given in Fig. 12.9.(44) In this figure the relative decreases in volume and in stress are plotted against log time. It is evident that there is a one-to-one correspondence between the two quantities. It should also be noted that typical isotherms are observed, whose shapes vary with the elongation ratio.

[4] We shall not be concerned with the situation where crystallization takes place as the sample is deformed with time. The interest here is limited to the case where the extension ratio(s) are maintained constant during the course of the crystallization. This restraint allows for a simpler analysis of a complex problem.

The isotherms obtained for homopolymers at a constant elongation ratio, at different temperatures, can be superposed upon one another in the conventional manner. An example is given in Fig. 12.10 for natural rubber crystallized without deformation and at an extension ratio α of 5.(43) Here $1 - \lambda(t)$ is plotted against log time at the indicated crystallization temperature. Linear plots are observed over most of the transformation. The slopes of the straight lines are the same for each elongation ratio. However, they are quite different for the two cases illustrated. This difference is a reflection of the different values of the Avrami exponent n, which varies with the elongation ratio. An example of the variation in n is illustrated in Fig. 12.11 for natural rubber.(43) In this example n is about 3 for $\alpha = 1$ and monotonically decreases to $n \simeq 1$ for $\alpha = 6$. Other polymers show a similar behavior. For example, $n \simeq 1$ for cross-linked polyethylene over a range in extension ratios from 2 to 5.(46) The values of n that have been observed for poly(ethylene terephthalate) are in the vicinity of 1 at high extension ratios.(47–49a) The n value of this polymer decreases from 3 to 1 as the sample is deformed from the isotropic state to one of high deformation. A similar decrease in n with deformation has also been observed for poly(pentenamer).(50) The n value for poly(isobutylene) is also about unity at high deformation.(51) The n value for cross-linked polyethylene, swollen in xylene, is also approximately equal to unity.(52) The n values are in the range of 0.7 to 1.0 for poly(trans-1,4-butadiene) at relatively low extension ratios.(53) This could probably be attributed to the structural irregularity of the chain in this polymer.

The fact that $n \simeq 1$ at large deformations for many polymers is consistent with one-dimensional growth from nuclei that are activated at $t = 0$. The change in n values with elongation ratio suggests that there are major changes in morphology with increasing extension ratio. There are, in fact, direct morphological observations that support this conclusion. It has been demonstrated that when natural rubber is stretched in the melt, and the deformation maintained during crystallization, there is a change from spherulitic to aciform growth.(43,54) These observations give further support to the argument that the Avrami does indeed reflect the shapes of the morphological forms that evolve during crystallization.

An interesting observation associated with crystallization under uniaxial deformation is the marked enhancement of the crystallization rate at constant temperatures.(43,44,55,56) For example, the crystallization rate constant of natural rubber increases by six to nine orders of magnitude with extension ratio at constant temperatures.(43) This enhancement is also reflected in the crystallization half-time, which also increases by several orders of magnitude.(44,46,47,55,56) This enhancement of crystallization rate is not limited to rubber-like polymers. It is also observed in poly(pentenamer) (50) and in poly(ethylene terephthalate) (47–49,57,57a) when crystallized under uniaxial deformation. The change in rate can be attributed in part

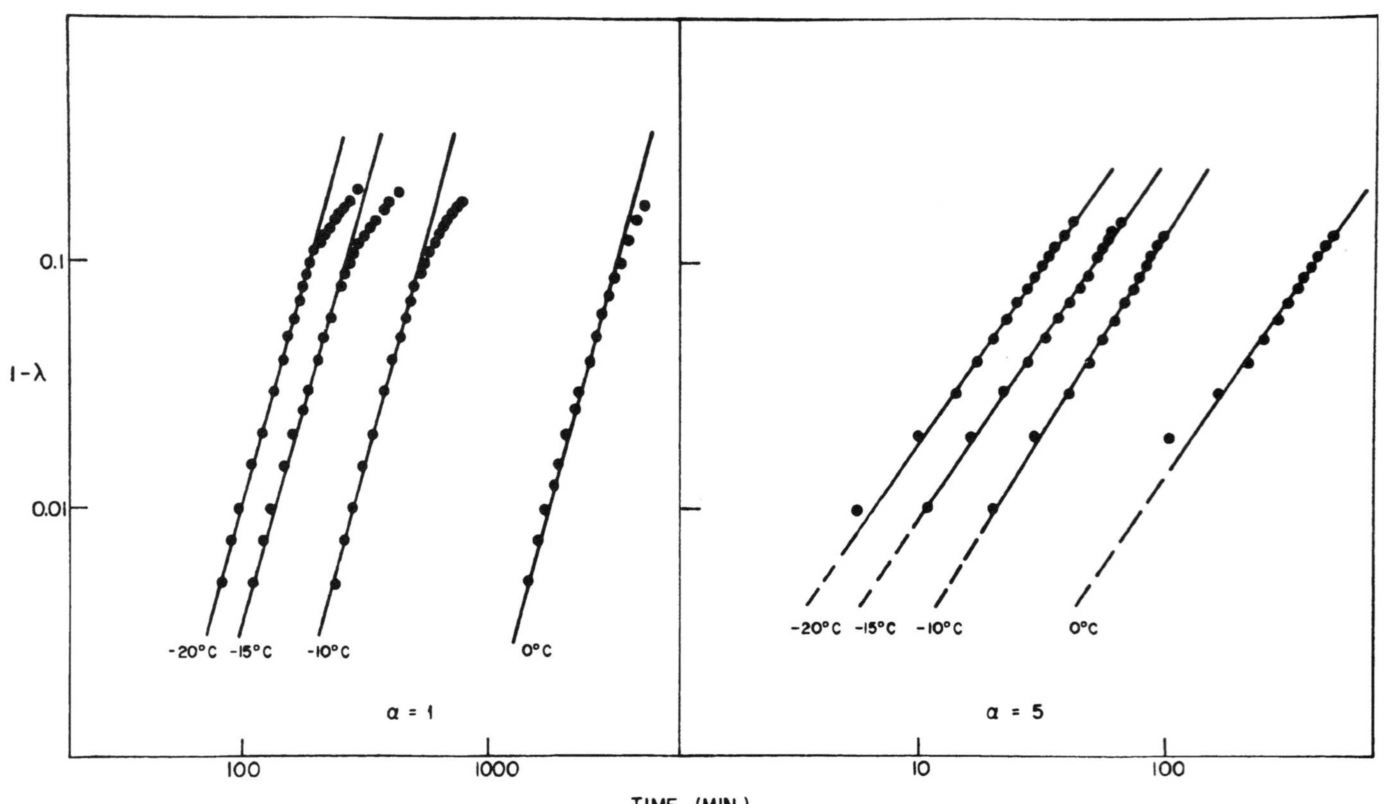

Fig. 12.10 Plot of $\ln(1-\lambda(t))$ against $\log t$ for $\alpha = 1$ and $\alpha = 5$ at the indicated crystallization temperatures.(43)

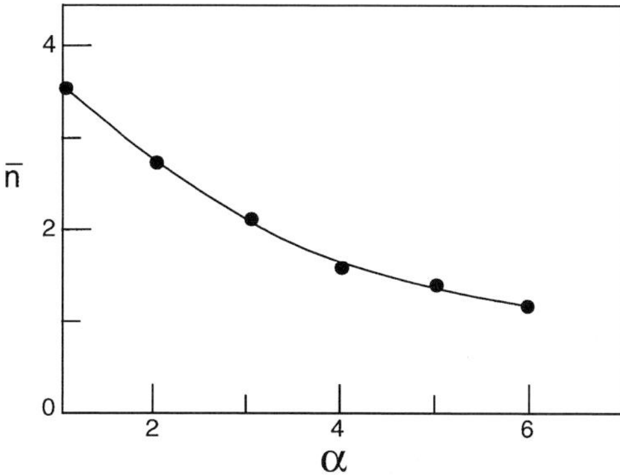

Fig. 12.11 Plot of Avrami exponent n against elongation ratio α for natural rubber.(43)

to the changing crystallization mechanism with deformation. However, the main contributor is the substantial increase that occurs in the undercooling with extension at constant temperature. It has been concluded in studies of other polymers that the crystallization rate depends solely on the undercooling and is independent of the associated changes in the crystallization mechanism.(50,55,56) In order to resolve the problem, reliable values of the equilibrium melting temperature as a function of the extension ratio are needed. The discussion in Chapter 7 (Volume 1) showed how difficult it is to reliably establish this quantity.

Figure 12.10 indicates that there is a strong negative temperature dependence of the overall crystallization rate. This behavior is found in other axially deformed systems as well.(54,55,58) This observation is usually indicative of the involvement of nucleation in the crystallization process. When studying the overall crystallization it is not possible to distinguish between either the primary or secondary nucleation processes. Even if a particular type of nucleation is selected other problems remain. These include having an accurate value of T_m^0 as a function of the extension ratio. In addition it is necessary to know the entropy of fusion per repeating unit, ΔS_u, as a function of the deformation. This quantity enters the analysis from the free energy necessary to form a critical size nucleus, which depends in part on the free energy of fusion. The temperature dependence of the latter free energy involves ΔS_u or $\Delta H_u/T_m$. There is also a fundamental question as to whether conventional nucleation theory as developed for quiescent crystallization will be directly applicable to deformed systems. The significance of classical nucleation theory in the present context needs to be examined further. Considering the complications involved prudence suggests postponing an analysis of the temperature coefficient until all these problems are resolved. However, while being cognizant of these difficulties, it is still informative to analyze the data according to conventional nucleation theory.

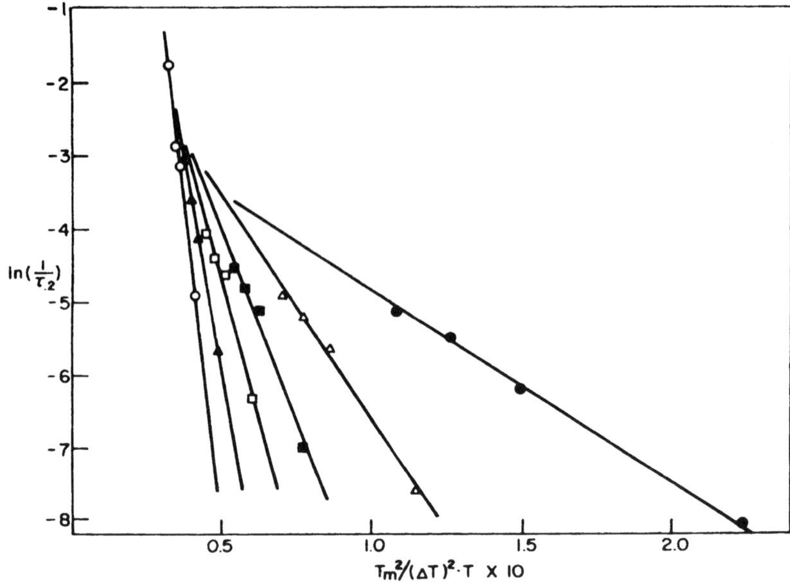

Fig. 12.12 Plot of $\ln(1/\tau_{0.2})$ against $T_m^2/(\Delta T)^2 T$ for natural rubber: ● $\alpha = 1$; Δ $\alpha = 2$; ■ $\alpha = 3$; □ $\alpha = 4$; ▲ $\alpha = 5$; ○ $\alpha = 6$.(43)

Following conventional nucleation theory, $\ln(1/\tau_{0.2})$ is plotted against either $T_m^2/T(\Delta T)^2$ or $T_m/T(\Delta T)$ for three- and two-dimensional nucleation respectively. Straight lines should result, in the absence of any regime transition. The slopes of the straight lines are proportional to $\sigma_{en}^2\sigma_u/\Delta H_u$ or $\sigma_{en}\sigma_u/\Delta H_u$. Since $\Delta H_u = T_m \Delta S_u$ the slopes, with slight error in the variation of T_m can also be taken to be inversely proportional to ΔS_u. The crystallization rates of natural rubber, $1/\tau_{0.2}$ are plotted against the appropriate temperature function for either three-or two-dimensional nucleation in Figures 12.12 and 12.13 respectively.[5](43) Straight lines are obtained for either type nucleation at each elongation ratio. However, the magnitude of the slope increases with the extent of the deformation. For example, if three-dimensional nucleation is assumed, the slope increases twelve-fold as α varies from 1 to 6. For two-dimensional nucleation, there is a ten-fold increase for the same range in α.

It is very tempting to assign the major changes in the slopes to variations in the product of the interfacial free energies. This would imply a large increase in this product with elongation. However, ΔS_u will decrease with the extent of the deformation. This factor can have a major influence in explaining the increasing magnitude of the slopes with extension ratio. The barrier height to steady-state nucleation, ΔG^*, will depend on either $1/\Delta S_u$ or $1/(\Delta S_u)^2$, according to the type

[5] In analyzing these data, the equilibrium melting temperatures were calculated from the Flory theory.(59) As discussed in Chapter 7 this theory gives good results at the higher deformations, but is lacking at the lower ones.

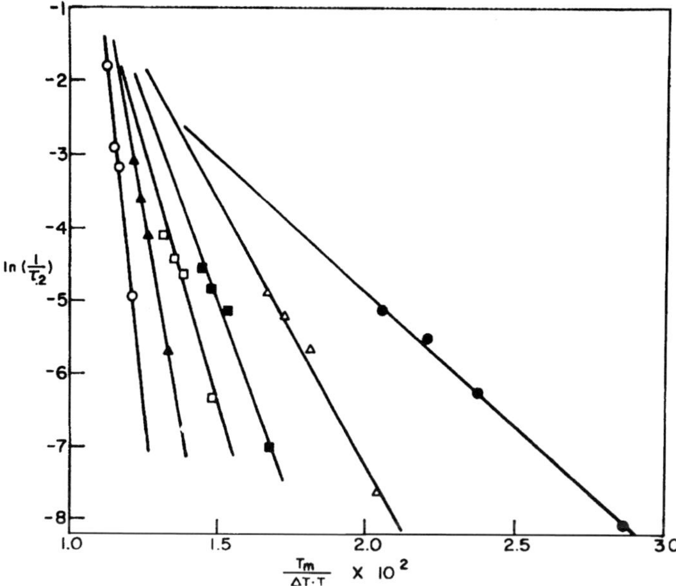

Fig. 12.13 Plot of $\ln(1/\tau_{0.2})$ against $T_m/(\Delta T)T$ for natural rubber. Symbols same as in Fig. 12.12.(43)

of nucleus suggested. At high extension ratios ΔS_u will be small. Under the circumstances ΔG^* as well as the critical dimensions will become very large. In this situation, there will be little influence, if any, of conventional nucleation on the crystallization process. There is then serious concern as to whether nucleation in the conventional sense is applicable to the crystallization of deformed systems. The development of a macroscopic crystalline phase requires the formation of a series of small crystallites, or embryos, of ever increasing size. During this process the Gibbs free energy will increase until the contributions of the surfaces of the embryos are overcome at the critical value of ΔG^*. The issue here is whether, during the crystallization of highly deformed and directionally oriented systems, the build up of small embryos is necessary in order for the new macroscopic phase to develop. The plots in Figs. 12.12 and 12.13 indicate that, in contrast to the lower values of α, at the higher ones crystallization takes place over a narrow undercooling interval. This behavior is not characteristic of conventional nucleation. There is then an important question to resolve as to whether a different type of nucleation is involved, or if nucleation is necessary at all in this situation. The problem can be summarized as follows.

Under quiescent crystallization conditions the initial melt is essentially random. There are no directing forces present that would maintain an element of preferential order. The development of order under these conditions involves concentration

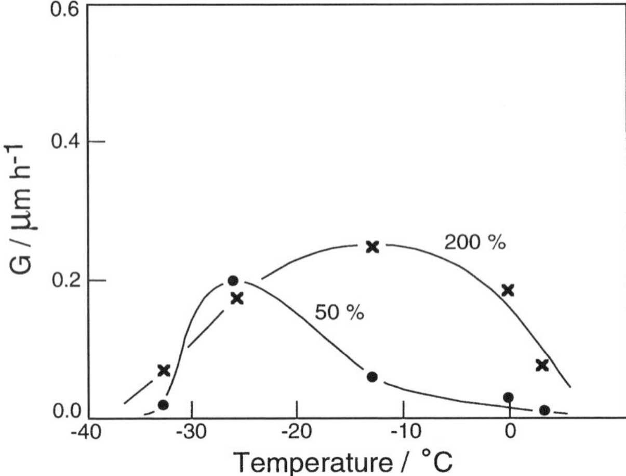

Fig. 12.14 Plot of lamellar growth rate, G, as a function of temperature for cross-linked natural rubber at indicated strains. (Adapted from Andrews *et al.* (60))

fluctuations and the formation of three-dimensional embryos, or nuclei, as a precursor to crystallization. In contrast, in a highly axially deformed melt, the chain axes are already preferentially oriented in the chain direction. Thus, under these circumstances the melt structure is biased in one direction for the development of order and crystallinity. Put another way, the chains are effectively in the "crystallization position", in that order is well-developed in the longitudinal direction in the melt at large deformations. It is only necessary, with minor adjustment, to develop lateral order and thus a three-dimensional crystalline system. Therefore, it is not clear whether under these conditions classical nucleation theory applies. Wide-angle x-ray diffraction studies of poly(ethylene terephthalate) (59a,59b) and of a random copolymer of ethylene terephthalate–ethylene napthalene-2,6-dicarboxylate (59c) suggest the formation of a mesophase precursor prior to the development of crystallinity. The subject area of the crystallization from a highly axially oriented melt presents some interesting experimental and theoretical problems.

Growth rate measurements can also be made on systems crystallizing under deformation. Andrews *et al.* have in fact carried out studies of the lamellar growth rate in cross-linked natural rubber at relatively low strains.(60) The study was limited to two extension ratios and one network.[6] A summary of the results is given in Fig. 12.14. A maximum in the growth rate with temperature is observed in each of the deformed samples, similar to that reported for undeformed natural rubber. The magnitudes of the two growth rates at the maxima are comparable to one another. However, there is an increase of about 15 °C between the two maxima.

[6] The results reported for the non-cross-linked sample are not being considered here for the reasons discussed previously.

This difference can be attributed qualitatively to the increase in the equilibrium melting temperature with strain. At the lower temperatures, $\leq 20\,^{\circ}\mathrm{C}$, the growth rates are comparable to one another. At temperatures greater than the maximum there is a precipitous drop in the growth rates in both cases. This effect is more marked with the sample under 50% strain, where the growth rate becomes severely retarded. A quantitative analysis of this data is difficult without firm values of the equilibrium melting temperatures and growth rate data for the undeformed, but cross-linked sample.

The experimental results for the crystallization kinetics of cross-linked systems that are deformed uniaxially are interesting in that new concepts are introduced. Clearly, more work is needed to explore these implications and develop a coherent understanding, particularly of the initiation process. Additional growth rate data will help in this endeavor as well as the firm establishment of the required parameters.

12.4 Crystallization kinetics under biaxial deformation and under shear

Analysis of the crystallization kinetics for polymers subject to either biaxial or shear deformation is hampered by the lack of knowledge of the equilibrium melting temperatures for these situations. This is true for both theoretical and experimental attempts to obtain this quantity. In addition, there is also a paucity of appropriate experimental data to guide efforts in obtaining these melting temperatures.

The isothermal crystallization kinetics of biaxially deformed natural rubber has been studied.(61) In this work the extension ratios in each direction are kept the same. A set of isotherms at a fixed crystallization temperature, but different extension ratios, is given in Fig. 12.15. The isotherm shapes are different from one another. There is also an enhancement of the crystallization rate with increasing extension ratio. The data could be fitted to the derived Avrami expression over a major portion of the transformation. Based on the conventional analysis, there is a significant decrease in the Avrami exponent n with the extension ratio. It decreases from 3 in the undeformed state to a little less than 1 at extension ratios 3×3. The pattern previously described for the uniaxial deformation of natural rubber (Fig. 12.11) is also followed in this case. The change in the exponent most probably reflects changes in the growth pattern, which, however, were not reported. The observed melting temperature increases by about $50\,^{\circ}\mathrm{C}$ over this deformation range. The dependence of the crystallization rate on temperature was not reported. Hence, the conventional nucleation analysis cannot be made. There should be some interesting ramifications in applying nucleation theory to crystallization under this type of deformation.

The studies of the crystallization under uniaxial and biaxial deformation involved cross-linked systems and the strain was maintained throughout the transformation.

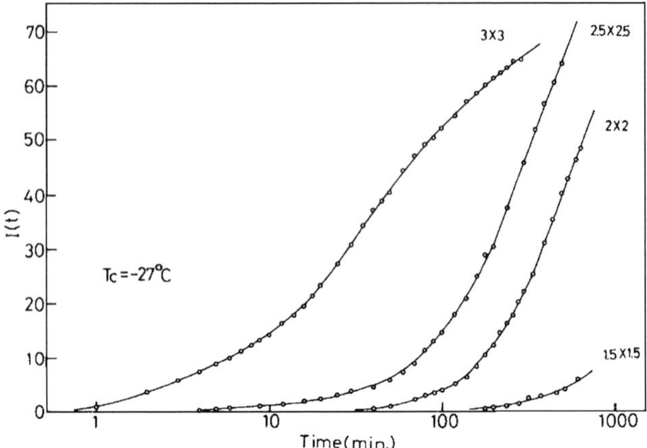

Fig. 12.15 Crystallization isotherms at $-27\,^{\circ}$C for biaxially stretched natural rubber. Plot of normalized intensity of the (120) Bragg reflection. (From Oono *et al.* (61))

Flow of the polymer was not involved. Similar studies have not been reported for systems crystallizing under shear stress. In the studies that have been reported flow was always involved. Typical instrumentation involves either a Couette type viscometer or a parallel plate rheometer, or variations thereof. Usually measured are the induction times, i.e. the growth and primary nucleation rates, as well as the overall rate of crystallization. The shear rates studied vary considerably among the different investigations. In some investigations a low and narrow range is studied; others encompass a broad range in shear rates. Two shear regimes can be recognized.(61a) Under mild conditions of applied shear there is no change in morphology. Under stronger shear conditions the problem is more complex in that a variety of morphological changes occur. It has been claimed that in the case of shear flow there is no significant change in the equilibrium melting temperature.(62)

A different type of investigation involves a brief shearing of the melt. Further crystallization is allowed to proceed with the removal of the applied stress.(62–63) The intent of this type of experiment is to separate the primary nucleation (really a precursor structure) from the crystal growth that takes place at the later times. This is obviously a very complex crystallization process to analyze. Chain relaxation dynamics will play an important role in crystallization under these conditions. It has been shown that the subsequent growth and overall crystallization rates depend on the time of pre-shearing and the shearing intensity.(63,64) It is not clear that only the primary nucleation rate is involved during the time interval that shear is applied. Utilizing this technique, the crystallization rate of a polydisperse isotactic poly(propylene) was enhanced by two orders of magnitude relative to quiescent

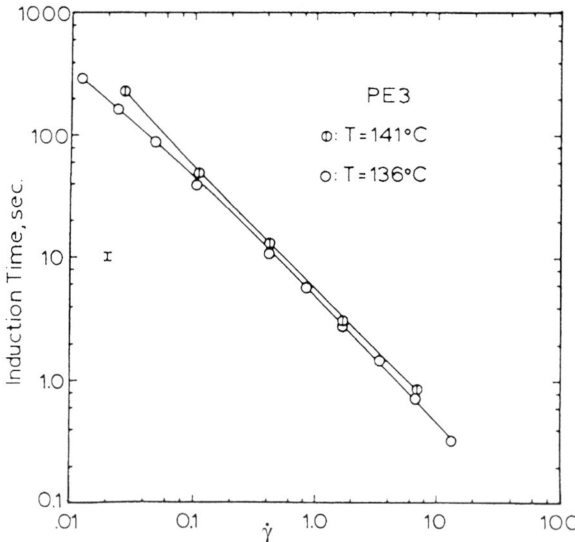

Fig. 12.16 Plot of induction time against shear rate for the crystallization of an unfractionated linear polyethylene, $M_w = 1.86 \times 10^6$ at indicated temperatures. (From Lagasse and Maxwell (66))

crystallization.(64a) However, despite these complexities there is interest in focusing attention on the precursor formation under certain conditions.(65)

A major consequence of imposing shear on the system is an enhancement of the crystallization rate. The simplest ramification of this enhancement is found in the increase in the induction time with shear rate. The onset of crystallization (the induction time) at a constant shear rate and crystallization temperature is reflected by a significant rise in the melt viscosity.[7] An example of the rate enhancement, in terms of the induction time, is shown in Fig. 12.16 (66) for a polydisperse linear polyethylene $M_w = 1.86 \times 10^6$. Crystallization is extraordinarily slow under quiescent conditions at the indicated temperature.(67) However, at a shear rate of 10 sec^{-1} the induction time is reduced to a few tenths of a second. Over the range of shear rates studied the induction times vary by four orders of magnitude.

Dilatometric or optical studies allow the degree of crystallinity to be measured as a function of time. Details of the complete crystallization process can then be analyzed. Figure 12.17 is an example of the change in the level of crystallinity as a function of time at the indicated shear rate for a poly(ethylene oxide) sample, $M = 7 \times 10^3$.(68) The enhancement of the crystallization rate with increasing shear rate is quite clear in these plots. The isotherms become much steeper with shear rate.

[7] As discussed in Chapter 9 the induction time reflects the sensitivity of the detector. However, when used for comparative purposes, as in the present case, the use of induction times to qualitatively represent the rate is adequate.

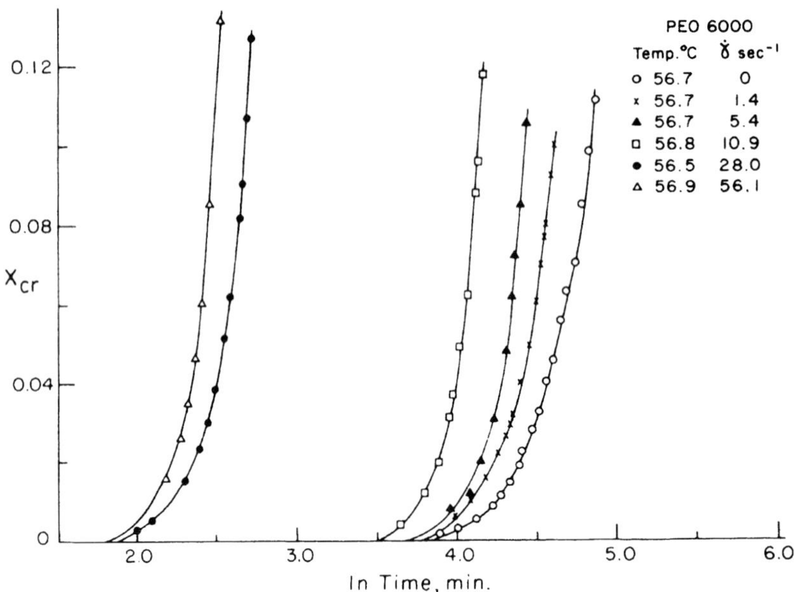

Fig. 12.17 Plot of degree of crystallinity, X_{cr}, against ln time for poly(ethylene oxide), $M = 7000$, at indicated temperatures and shear rates. (From Sherwood *et al.* (68))

As the shear rate increases the isotherms move closer together. In fact, eventually a saturation is reached. For a slightly higher molecular weight poly(ethylene oxide) the isotherms are virtually identical for shear rates in the range 28–140 $\sec^{-1}$.(69) Similar saturations are observed with poly(ϵ-caprolactone) (70) and an isotactic poly(propylene) copolymer.(64)

Isotherms such as those shown in Fig. 12.17 can be fitted formally to the derived Avrami equation. The main interest in this analysis is obtaining the value of the Avrami exponent. The results are unusual and surprising. In the shear-flow type experiment the n values are found to be relatively large and significantly increase with the shear rate. For example, n values in the range 3 to 8 have been found for poly(ϵ-caprolactone)(68,70), and 6 for poly(p-dioxanone.(61a) The value for an isotactic poly(propylene) copolymer is as high as 12.(64) Values ranging from 3 to 16 have been found for poly(ethylene oxide).[8](68,69,71) The n values that are obtained under shear are quite different from those obtained either under quiescent crystallization conditions or uniaxial and biaxial deformation. In these cases the n value can usually be interpreted in terms of the type of nucleation and the growth geometry.

[8] It was found from a study of isotactic poly(propylene), up to shear rates of 2 s^{-1}, that when the induction time was subtracted from the real time more conventional values of n were obtained.(72) In this work $n = 2$ was found for quiescent crystallization.

The physical meaning of the high n values in the shear experiment is not clear. There is a serious question as to whether the Avrami approach is even applicable for such highly deformed systems. Light microscopic observations of an unfractionated linear polyethylene have shown that above a critical shear rate, where the crystallization is accelerated, a fibrillar-like morphology is observed.(66) Below the shear rate conventional spherulitic morphology is observed. Within the Avrami framework, nucleation is indicated as the primary reason for the high n values. It has been suggested that the shear stress disrupts the growing crystallites.(69) This in turn results in the introduction of effective nuclei into the untransformed melt. If correct, this offers at least one explanation for the large n values that are observed. Other theoretical reasons within the Avrami framework have also been proposed.(62)

The nucleation rate during shear induced crystallization has been measured under the assumption that each morphological structure, such as spherulites or axialites, comes from one nucleus. Thus, by measuring the number of such structures per unit volume as a function of time the nucleation rate can be obtained. For example, by this measure there is a thirty-fold increase in the nucleation density in poly(ϵ-caprolactone) with the application of a stress of 500 Pa.(70) The nucleation density obtained by this method is not usually linear with time.(64,68,73) However, there is a linear portion that allows for a representative nucleation rate to be obtained. Another example of the increase in nucleation rate with shear rate is illustrated in Fig. 12.18

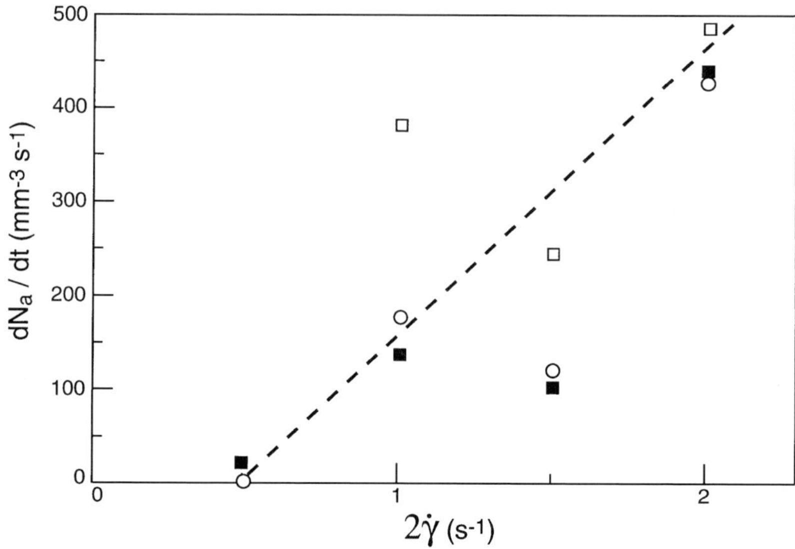

Fig. 12.18 Plot of nucleation rate against shear rate, $\dot{\gamma}$ s^{-1}, for a copolymer of isotactic poly(propylene) at indicated temperatures: □ T_c = 133.9 °C; ■ T_c = 136.4 °C; ○ T_c = 138.5 °C. (From Tribout *et al.* (64))

for a copolymer of isotactic poly(propylene).(64) There is a significant increase in the nucleation rate with shear rate in this example. Based on this and other studies (68,73,74) a major factor causing the accelerated crystallization is attributed to the increased nucleation rate.

The question can be raised whether nucleation in the conventional sense is applicable to highly deformed systems. This is similar to the problem posed in the discussion of crystallization under high axial deformation. The kind of fluctuation required for nucleation in isotropic systems may no longer be necessary in highly deformed systems. In highly deformed systems the chain axes are essentially primed for crystallization. Bearing directly on this problem is one aspect of the experiment that involves the brief shearing of the initial melt. As was pointed out earlier the structures that appear after the initial deformation are identified with nuclei on a one-to-one basis. After strongly shearing a polydisperse isotactic poly(propylene) sample the time required for the precursor to appear has been found to decrease with increasing temperature.(65) This observation is opposite to the expectation from classical nucleation theory. It would appear that, under intense initial deformation, nucleation, or effective nucleation, avoids the high activation free energy barrier characteristic of conventional nucleation. The crystallization process appears to circumvent this barrier. A similar pathway for nucleation and crystallization was suggested for systems crystallizing under large uniaxial deformation ratios. There is the suggestion from these studies that the crystallization from highly deformed polymer melts involves a unique set of initiating mechanisms.

Spherulite growth rates have also been measured as a function of shear rate.(64,73) The results for poly(butene-1) indicate that although the growth rates decrease with increasing isothermal crystallization temperature, there is very little change with shear rate up to 2 s^{-1}.(73) In contrast, the growth rate, at a shear rate of 10 s^{-1}, of a copolymer of isotactic poly(propylene) increases four to six-fold relative to that under quiescent crystallization conditions.(64) There are obviously not enough available growth rate data to reach any general conclusions with regard to the influence of shear rate.

It is to be expected that the molecular weight should have a strong influence in the shear-flow experiments. There is again a paucity of data in this regard. The results with unfractionated linear polyethylene (66) and poly(ethylene oxide) (68,69) lead to what appear to be reasonable conclusions. It is found that for a given shear rate and temperature the higher molecular weights have lower induction times. It has also been found that the higher molecular weight isotactic poly(propylene) crystallizes faster than the lower ones.(69a) The isotherms of moderate and high molecular weight poly(ethylene oxide) become saturated at surprisingly low shear rates. Decreasing the molecular weight separates these curves. Studies with a wide range of molecular weight fractions are needed to enhance the results obtained with polydisperse polymers. This would allow for a more quantitative assessment of the role of molecular weight.

The crystallization kinetic studies involving shear have involved flowing systems. Studies involving cross-linked systems, as have been reported for crystallization under uniaxial and biaxial deformation, would be extremely helpful in understanding the basic problem involved. Melting temperatures as a function of deformation and crystallization temperature could be determined. These data would help in constructing an equilibrium theory. By analogy, estimates could then be made at the undercoolings involved. A better understanding would evolve of the nucleation process and its relation to that of undeformed systems.

References

1. Phillips, P. J. and B. C. Edwards, *J. Polym. Sci.: Poly. Phys. Ed.*, **13**, 1819 (1965); *ibid.*, **14**, 377 (1976).
2. Martin, G. M. and L. Mandelkern, *J. Appl. Phys.*, **34**, 2313 (1963).
3. Matsuoka, S., *J. Polym. Sci.*, **17**, 511 (1960).
4. Calvert, P. D. and D. R. Uhlmann, *J. Polym. Sci. A2*, **10**, 1811 (1972).
5. Hatakeyama, T., H. Kametsuna, H. Kaneda and T. Hashimoto, *J. Macromol. Sci.*, **B10**, 359 (1974).
6. He, J. and P. Zoller, *J. Polym. Sci.: Pt. B: Polym. Phys.*, **32**, 1049 (1994).
7. Kyotani, M. and H. Kanetsuna, *J. Polym. Sci.: Polym. Phys. Ed.*, **12**, 2331 (1974).
8. Hoehn, H. H., R. G. Ferguson and R. R. Hebert, *Polym. Eng. Sci.*, **18**, 457 (1978).
9. Sawada, S. and T. Nose, *Polym. J.*, **11**, 477 (1979).
10. Sawada, S., K. Kato and T. Nose, *Polym. J.*, **11**, 551 (1979).
11. Brown, D. R. and J. Jonas, *J. Polym. Sci.: Polym. Phys. Ed.*, **22**, 655 (1984).
12. Kanetsuna, H., S. Mitsuhashi, M. Iguchi, T. Hatakeyama, M. Kyotani and Y. Maeda, *J. Polym. Sci.*, **42C**, 783 (1973).

12a. Hikosaka, M. and T. Seto, *Jpn. J. Appl. Phys.*, **21**, L332 (1982).

13. Tseng, H. T. and P. J. Phillips, *Macromolecules*, **18**, 1565 (1985).
14. Wunderlich, B. and T. Arakawa, *J. Polym. Sci.*, **A2**, 3697 (1964).
15. Geil, P. H., F. R. Anderson, B. Wunderlich and T. Arakawa, *J. Polym. Sci.*, **A2**, 3707 (1964).
16. Mandelkern, L., M. R. Gopalan and J. F. Jackson, *J. Polym. Sci. Polym. Lett.*, **5B**, 1 (1967).
17. Hoehn, H. H., R. G. Ferguson and R. R. Hebert, *Polym. Eng. Sci.*, **18**, 457 (1978).
18. Bassett, D. C. and B. Turner, *Nature Phys. Sci.*, **240**, 146 (1972).
19. Bassett, D. C. and B. Turner, *Phil. Mag.*, **29**, 925 (1974).
20. Yasuniwa, M., C. Nakafuku and T. Takemura, *Polym. J.*, **4**, 526 (1973).
21. Lupton, J. M. and J. W. Regester, *J. Appl. Polym. Sci.*, **18**, 2407 (1974).
22. Rastogi, S., M. Hikosaka, H. Kawabata and A. Keller, *Macromolecules*, **24**, 6384 (1991).
23. Hikoska, M., S. Rastogi, A. Keller and H. Kawabata, *J. Macromol. Chem. Phys.*, **B31**, 87 (1992).
24. Keller, A., M. Hikosaka, S. Rastogi, A. Toda and P. J. Barham, *J. Mater. Sci.*, **29**, 2579 (1994).
25. Calvert, P. D. and D. R. Uhlmann, *J. Polym. Sci. Polym. Lett.*, **8B** 165 (1970).

26. Yasuniwa, M., R. Enoshita and T. Takemura, *Jpn. J. Appl. Phys.*, **15**, 1421 (1976).
27. Wunderlich, B. and L. Melillo, *Makromol. Chem.*, **118**, 250 (1968).
28. Rees, D. V. and D. C. Bassett, *Nature*, **219**, 368 (1968).
29. Rees, D. V. and D. C. Bassett, *J. Polym. Sci. Polym. Lett.*, **7B**, 273 (1969).
30. Rees, D. V. and D. C. Bassett, *J. Polym. Sci. A-2*, **9**, 385 (1971).
31. Ungar, G., *Macromolecules*, **19**, 1317 (1986).
32. Ungar, G., *Polymer*, **34**, 2050 (1993).
33. de Langen, M., H. Lulgjes and K. O. Prins, *Polymer*, **41**, 1193 (2000).
34. Hikosaka, M., *Polymer*, **28**, 1257 (1987); *ibid.*, **31**, 458 (1990).
35. Nishi, M., M. Hikosaka, A. Toda and M. Takahashi, *Polymer*, **39**, 1591 (1998).
36. Okada, M., M. Nishi, M. Takahashi, H. Matsuda, A. Toda and M. Hikosaka, *Polymer*, **39**, 4535 (1998).
37. Nishi, M., M. Hikosaka, S. K. Ghosh, A. Toda and K. Yamada, *Polym. J.*, **31** 749 (1999).
38. Yamamota, Y., H. Miyaji and K. Asai, *Jpn J. Appl. Phys.*, **16**, 1891 (1977).
39. Frank, F. C. and M. Tosi, *Proc. R. Soc. (London)*, **A263**, 323 (1961).
40. Davies, C. K. L. and M. C. M. Cucarella, *J. Mater. Sci.*, **15**, 1557 (1980).
41. Phillips, P. J. and H. T. Tseng, *Macromolecules*, **22**, 1649 (1989).
42. Dalal, E. N. and P. J. Phillips, *Macromolecules*, **17**, 248 (1984).
43. Kim, H.-G. and L. Mandelkern, *J. Polym. Sci. Pt. A-2*, **6**, 181 (1968).
44. Gent, A. N., *Trans. Faraday. Soc.*, **50**, 521 (1954).
45. Smith, W. H. and C. P. Saylor, *J. Res. Nat. Bur. Stand.*, **21**, 257 (1938).
46. Iquchi, M., T. Matsumoto, H. Tonami, T. Kawai, H. Maeda and S. Mitsuhashi, paper presented at IUPSE Meeting Tokyo-Kyoto, Japan 1966 (Abstract 4.5.12).
47. Smith, F. S. and R. D. Steward, *Polymer*, **15**, 283 (1974).
48. Althen, G. and H. G. Zachmann, *Makromol. Chem.*, **T80**, 2723 (1979).
49. Braguto, G. and G. Gianotti, *Eur. Polym. J.*, **19**, 803 (1983).
49a. Zhang, Z., M. Ren, J. Zhao, S. Wu and H. Sun, *Polymer*, **44**, 2547 (2003).
50. Kraus, G. and J. T. Gruver, *J. Polym. Sci.: Polym. Phys. Ed.*, **10**, 2009 (1972).
51. Koch, M. J. H., J. Bordas, E. Schöla and H. Chr. Broecker, *Polym. Bull.*, **11**, 709 (1979).
52. McHugh, A. J. and W. S. Yung, *J. Polym. Sci.: Pt. B: Polym. Phys.*, **27**, 431 (1989).
53. Akana, Y. and R. S. Stein, *J. Polym. Sci.: Polym. Phys. Ed.*, **13**, 2195 (1975).
54. Andrews, E. M., *Proc. R. Soc. (London)*, **A277**, 562 (1964).
55. Gent, A. N., *J. Polym. Sci. Pt. A-2*, **4**, 447 (1966).
56. Gent, A. N., *J. Polym. Sci. Pt. A-2*, **3**, 3787 (1965).
57. Desai, P. and A. S. Abhiraman, *J. Polym. Sci.: Polym. Phys. Ed.*, **23**, 653 (1985).
57a. Alfonso, G. C., M. P. Verdona and A. Wasiak, *Polymer*, **19**, 711 (1978).
58. Riande, E., J. Guzman and J. E. Mark, *Polym. Eng. Sci.*, **26**, 297 (1986).
59. Flory, P. J., *J. Chem. Phys.*, **15**, 397 (1947).
59a. Mahendrasingam, A., C. Martin, W. Fuller, D. J. Blundell, R. J. Oldman, D. H. MacKerron, J. L. Harvie and C. Riekel, *Polymer*, **41**, 1217 (2000); *ibid.*, 7793; *ibid.*, 7803.
59b. Welsh, G. E., D. J. Blundell and A. H. Windle, *Macromolecules*, **31**, 7562 (1998).
59c. Ran, S., Z. Wang, C. Burger, B. Chu and B. S. Hsiao, *Macromolecules*, **35**, 10102 (2002).
60. Andrews, E. H., P. J. Owen and A. Singh, *Proc. R. Soc. (London)*, **A324**, 79 (1971).
61. Oono, R., K. Miyasaka and K. Ishikawa, *J. Polym. Sci.: Polym. Phys. Ed.*, **11**, 1477 (1973).

61a. Abuzaina, F. M., B. D. Fitz, S. Andjelic and D. D. Jamiolkowski, *Polymer*, **43**, 4699 (2002).
62. Eder, G., H. Janeschitz-Kriegl and S. Liedamer, *Prog. Polym. Sci.*, **15**, 629 (1990).
62a. Liedauer, S., G. Eder, H. Janeschitz-Kriegl, P. Jerschow, W. Germayer and E. Ingolic, *Int. Polym. Proc.*, **8**, 230 (1993).
62b. Eder, G., H. Janeschitz-Kriegl and G. Krobeth, *Prog. Coll. Polym. Sci.*, **80**, 1 (1989).
63. Kumaraswamy, G., A. M. Issaian and J. A. Kornfield, *Macromolecules*, **32**, 7357 (1999).
64. Tribout, C., B. Monasse and J. Haudin, *Colloid. Polym. Sci.*, **274**, 197 (1996).
64a. Somani, R. H., B. S. Hsiao, A. Nogales, S. Srinivas, A. H. Tsoa, I. Sics, F. J. Balta-Calleja and T. A. Ezquerra, *Macromolecules*, **33**, 9385 (2000).
65. Kumaraswamy, G., J. A. Kornfield, F. Yen and B. S. Hsiao, *Macromolecules*, **35**, 1762 (2002).
66. Lagasse, R. R. and B. Maxwell, *Polym. Eng. Sci.*, **16**, 189 (1970).
67. Ergoz, E., J. G. Fatou and L. Mandelkern, *Macromolecules*, **5**, 47 (1972).
68. Sherwood, C. H., F. P. Price and R. S. Stein, *J. Polym. Sci.*, **63C**, 77 (1978).
69. Fritzsche, A. and F. P. Price, *Polym. Eng. Sci.*, **14**, 401 (1974).
69a. Nogales, A., B. S. Hsiao, R. H. Somani, S. Srinivas, A. Tsoa, F. J. Balta-Calleja and T. A. Ezquerra, *Polymer*, **42**, 5247 (2001).
70. Floudas, G., L. Hilliou, D. Lellinger and I. Alig, *Macromolecules*, **33**, 6466 (2000).
71. Theil, M. H., *J. Polym. Sci.: Polym. Phys. Ed.*, **13**, 1097 (1975).
72. Masubuchi, Y., K. Watnabe, W. Nagatake, J.-I. Takimoto and K. Kogama, *Polymer*, **42**, 5023 (2001).
73. Wolkowicz, M. D., *J. Polym. Sci.*, **63C**, 365 (1978).
74. Ulrich, R. D. and F. P. Price, *J. Appl. Polym. Sci.*, **20**, 1095 (1976).

13

Polymer–diluent mixtures

13.1 High molecular weight *n*-alkanes

The crystallization kinetics of the monodisperse, high molecular weight *n*-alkanes from solution follows the unique pattern observed in the melt crystallization of such species.(1–4) The characteristic dependence of the observed rate on the crystallization temperature measured using differential scanning calorimetry is illustrated in Fig. 13.1 for the crystallization of $C_{198}H_{398}$ from a 3.85% toluene solution.(5) Typically, in crystallization from solution, as well as from the pure melt, the crystallization rate reaches a maximum several degrees below the melting temperature of the extended crystals. A minimum follows, as the temperature is lowered just a few more degrees. The further lowering of the temperature results in a steep increase in the rate. It is easily demonstrated that folded chain crystallites are initially formed at crystallization temperatures in the vicinity of the minimum.[1] Similar inversions in the crystallization rates have been reported for the *n*-alkanes $C_{246}H_{494}$ (5) and $C_{294}H_{590}$ using the same method.(6) Wide-angle x-ray scattering studies, utilizing synchrotron radiation, yield similar results for $C_{162}H_{326}$ and $C_{246}H_{494}$ crystallizing from solution.(6a) Growth rate studies with $C_{198}H_{398}$ and higher *n*-alkanes also show maxima and minima with crystallization temperature, as does the initiation and spreading of a given layer.(6b–6e)

In the crystallization of $C_{246}H_{494}$ from a dilute solution in toluene (0.14% w/w) two minima are observed in the crystallization rate–temperature plot.(7) One occurs in the vicinity of the transition from the once-folded to extended form. The other, observed at a lower temperature, occurs in the vicinity in the transition from the twice-folded form to the once-folded structure. It has been reported that the primary nucleation and growth rates also follow a pattern similar to that illustrated

[1] The detailed nature of the folding, and the associated interfacial structure, is not pertinent to the present discussion. The folded structure and the question of whether precise sharp integral folding takes place (3,4) will be discussed in Volume 3.

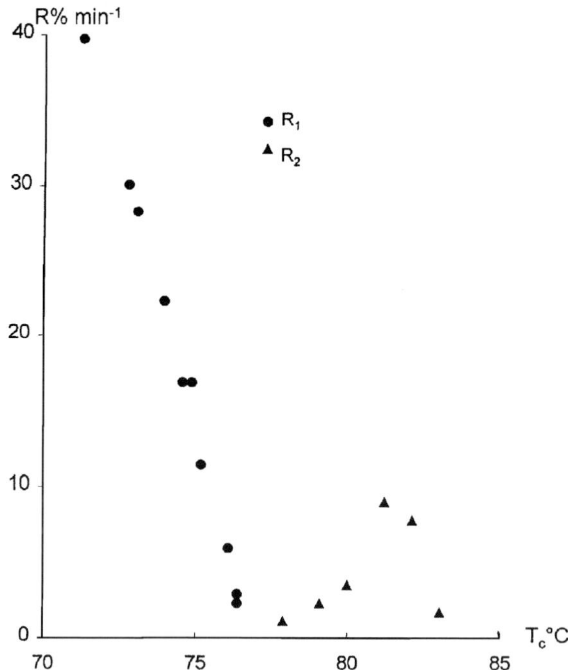

Fig. 13.1 Plot of rate of crystallization of $C_{198}H_{398}$ from a 3.85% toluene solution against the crystallization temperature, T_c. ▲ extended chain crystallites; ● folded chain crystallites. (From Organ *et al.* (5))

in Fig. 13.1.(8,9) The observed rate–crystallization temperature dependence of the high molecular weight *n*-alkanes is quite unique. They are anomalous in the realm of polymer crystallization as well as in other systems. It is, therefore, of interest and importance to explain the basis of these well substantiated results.

A poisoning, or self-poisoning, effect, as it has been termed, has been postulated to explain the crystallization rate–temperature relation typified by Fig. 13.1.(4,6c,6d,8) Particular attention has been focused on the temperature region of the minimum. Here, it has been established that a transition from folded to extended chain crystallites takes place. The minimum in the rate has been attributed to the deposition of folded chain crystallites on the lateral surface of the already growing extended chain crystals. Thus, the growth of extended chain crystallites is retarded and a minimum in the rate results.(11) This postulate fits the concept of rough surface growth that has been advanced by Sadler and Gilmer.(12–14) Calculations made on this basis have replicated diagrams such as that in Fig. 13.1.(14,15)

In analyzing and interpreting the results shown in Fig. 13.1 it is important that the structures that form initially from solution, or the melt, be clearly identified. In addition, it is essential that any changes in structure that occur during the course of the crystallization be accounted for in the analysis. If these concerns are not heeded major problems develop in the analysis of the data. A complicating feature is the fact that isothermal crystallite thickening occurs during the course of the crystallization of the high molecular weight n-alkanes.(7,10,16–19) Depending on the length of the molecule, and the mode of crystallization, isothermal thickening may proceed from what is initially close to a once-folded crystallite to the extended form. For the longer chains isothermal thickening can proceed through multiple refolding stages. In contrast, although the thickening of lamellar crystallites of polymers is known to occur during isothermal crystallization from the pure melt,(20) such thickening has not been observed during the crystallization of polymers from dilute solution.(21–24) There has been the fundamental question of how, if at all, the isothermal thickening influences the observed crystallization rate and the interpretation of the experimental observations. As a consequence of these concerns it is necessary to examine the thickening process in detail and assess its influence in the analysis of the data and the conclusions drawn.

Differential scanning calorimetry is both a convenient and a useful method with which to study isothermal thickening.[2] Both endothermic and exothermic type measurements complement each other and are helpful in understanding this problem. Particular attention will be given to the temperature region where the transition from the once-folded to extended form occurs because of the interest in the rate inversions. The DSC exotherms for the crystallization of $C_{168}H_{338}$ from a 4% toluene solution are shown in Fig. 13.2.(10) Only a narrow temperature interval, 70–76 °C, is practical for the study of the exothermic processes. The major feature in this figure is the appearance of double exothermic peaks. The first exotherm, the one that appears earliest, is found at longer times with increasing crystallization temperatures. This is characteristic of a decreasing crystallization rate with increasing temperature. On the other hand, the behavior of the second exotherm is quite the opposite. The process associated with this exotherm becomes faster with increasing temperature. At $T_c = 76$ °C both peaks overlap, resulting in a broad exotherm.

To obtain some insight into the process or processes that lead to these exotherms, the crystallization at a given temperature was interrupted at different times. For example, the crystallization was stopped at a time between the two exotherms, at a time corresponding to the maximum of the second exotherm, and after the second exotherm was completed. After each of these time intervals the crystals were rapidly melted from the crystallization temperature, T_c. The melting endotherms resulting from these interrupted experiments are given in Fig. 13.3.(18) This procedure also allows for the construction of a crystallization isotherm and an analysis of the kinetics, see below. The dissolution endotherms of Fig. 13.3 have been identified

[2] The interest at this point is limited to the influence of isothermal thickening on the kinetics. The structural and morphological changes that occur on thickening, as well as theoretical interpretations of the process, will be discussed in Volume 3.

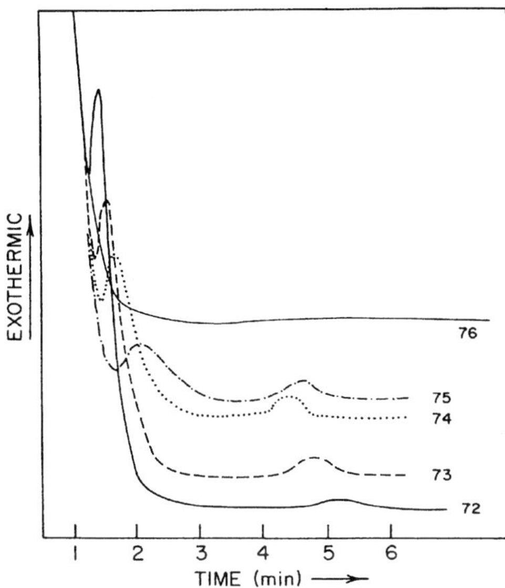

Fig. 13.2 DSC exotherms for the crystallization of $C_{168}H_{338}$ from 4 w/v% toluene solution at the indicated temperatures.(10)

with a specific chain structure. The higher temperature endotherm can be identified with the melting of the extended crystallites; the lower temperature endotherm with once-folded crystallites. For this alkane and concentration only, extended crystals were obtained at temperatures equal to greater than 77 °C. At low temperature, the lower endothermic peaks form initially and are maintained for long time periods.

Associated with the thermograms of Fig. 13.3 is the area under the endotherms. At each temperature, the total area under the endotherm remains constant with time, within experimental error. This result indicates a conservation of the level of crystallinity throughout the transformation. Concomitantly, the dissolution temperature of each structure remains constant with time. This important factor, the conservation of crystallinity, is illustrated more vividly in Fig. 13.4.(18) It is clear from the plots that the area corresponding to the extended crystals increases at the expense of the folded crystals. At 65 °C more than 1000 min is needed for this isothermal transformation to be completed. However, when the crystallization temperature is raised to 72 °C the folded crystallites disappear after about 5.5 min. The conservation of the level of crystallinity throughout the transformation in this temperature region requires that no additional crystallites be formed with increasing crystallization time. Therefore, it can be concluded that the crystallites having the higher dissolution temperature develop at the expense of the lower dissolution crystallites that are formed initially. In effect, therefore, what is taking place is the isothermal thickening of once-folded crystallites to extended ones. This behavior

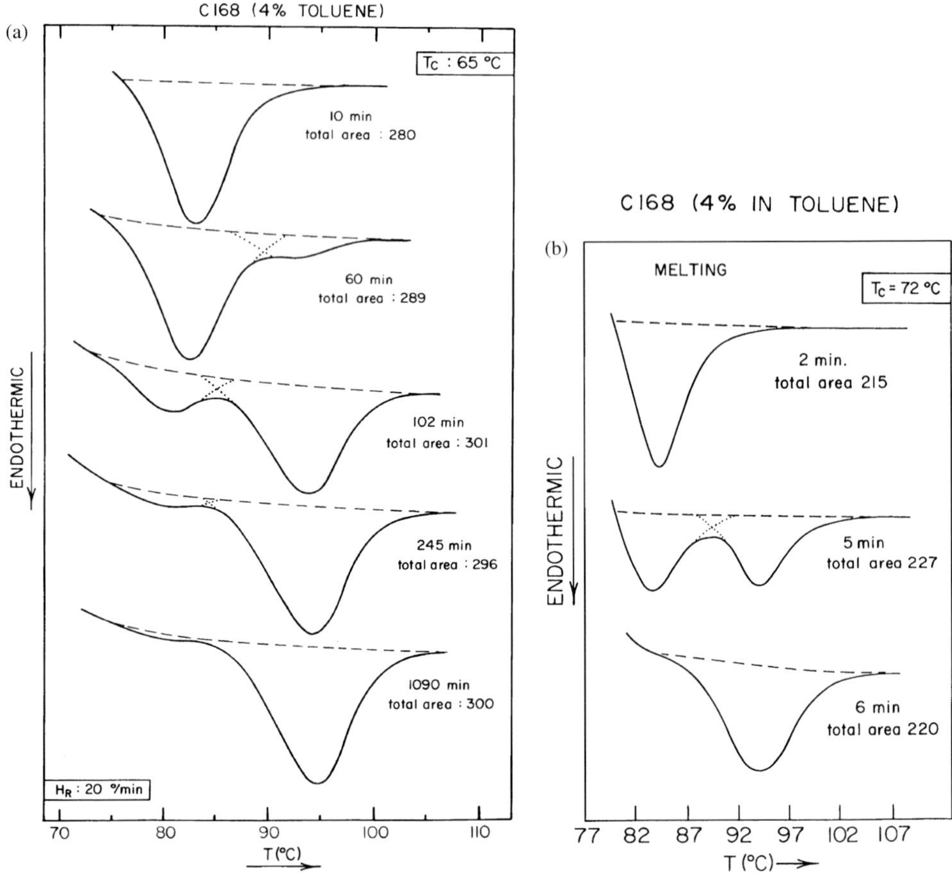

Fig. 13.3 DSC dissolution endotherms after the crystallization of $C_{168}H_{338}$ from a 4 w/v% solution in toluene for the indicated times. (a) Crystallization at 65 °C; (b) Crystallization at 72 °C. The integrated areas of the endotherms, in arbitrary units, are also indicated.(18)

is observed at all temperatures in the vicinity of the rate minima. At both high and low crystallization temperatures only a single endotherm is observed representing the extended and folded forms respectively. The transformation illustrated in Figs. 13.3 and 13.4 was observed to be independent of dilution at concentration as low as 0.15%.(18)

A concise summary of the time required for the complete transformation from the folded to extended forms, as a function of crystallization temperatures, is given in Fig. 13.4a.(18a) Here the data are for $C_{168}H_{338}$ crystallizing from a 4% toluene solution. It is evident that the transformation is very rapid at crystallization temperatures greater than 72 °C. At the highest temperatures the formation of folded crystals and their transformation to extended form is almost simultaneous. In general, it is

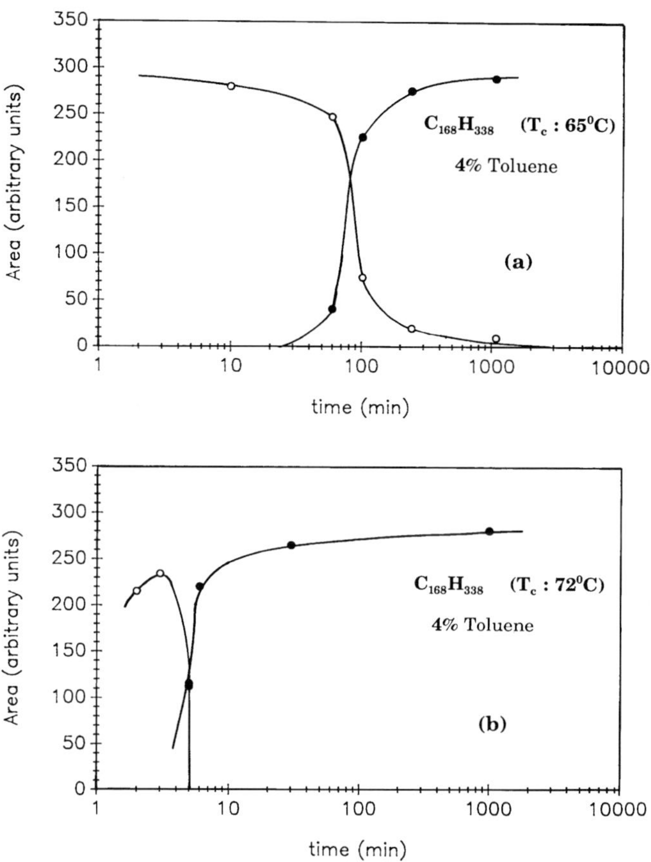

Fig. 13.4 Integrated area under dissolution (endothermic peaks) against crystallization time for $C_{168}H_{338}$ crystallizing from a 4 w/v% toluene solution at the indicated crystallization temperatures. (a) 65 °C; (b) 72 °C. ● low temperature endotherms, ○ high temperature endotherms.(18)

important for the kinetic analysis to establish the crystallization temperature range where the rate of transformation of the folded structure is comparable to its formation. This allows for the delineation of the temperature interval where the direct crystallization of the pure form cannot be quantitatively analyzed.

The isothermal thickening is observed in other high molecular weight n-alkanes as well.(5,7,10,18,19) A similar set of thermograms for the crystallization of $C_{240}H_{482}$ from dilute solution is illustrated in Fig. 13.5.(18) The characteristics of the thermograms are very similar to those found with $C_{168}H_{338}$. In the example shown, the low temperature endotherm is consistent with the melting of the twice-folded crystallites, while the endotherm centered at about 91.5 °C represents the melting of the once-folded crystallite. Studies with $C_{246}H_{494}$ show similar multiple endotherms in different temperature ranges.(7) In one range the transition from

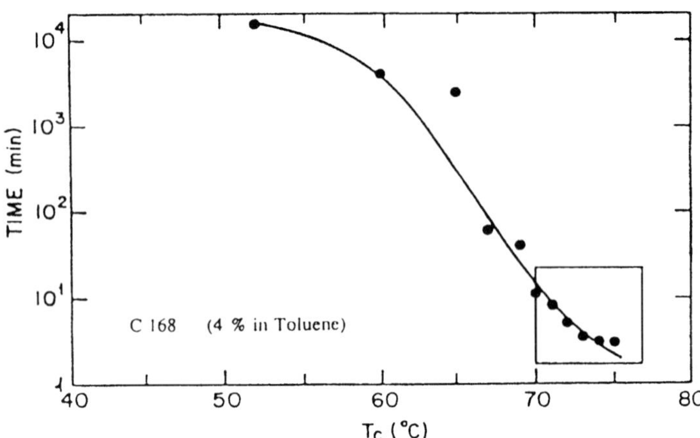

Fig. 13.4a Plot of time for complete transformation from once-folded to extended form for $C_{168}H_{338}$ crystallizing from a 4% toluene solution. (From Alamo and Chi (18a))

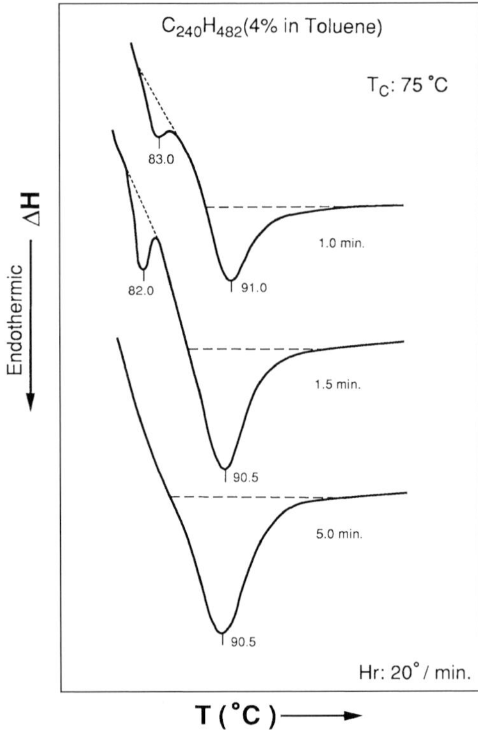

Fig. 13.5 DSC dissolution temperature after the crystallization of $C_{240}H_{482}$ from a 4 w/v% solution of toluene at 75 °C. Crystallization times are indicated.(18)

twice-folded to once-folded was observed. The transition from a once-folded to extended structure was observed in a higher temperature range. In an earlier study with $C_{246}H_{494}$ only broad endotherms were observed.(5)

With this background with regard to the origin of the endothermic peaks in the thermogram the crystallization kinetics can be analyzed. It needs to be recognized that within the data to be analyzed there are two distinctly different, and separate, kinetic processes involved. One is the formation from the melt, or solution, of a distinct crystallite structure, such as the extended form, once-folded, twice-folded, etc. The other involves the isothermal transformation from one form to another. When the two are intermixed a very complex situation results. At this point the kinetics of the formation of the distinct forms from the melt (without any further structural changes) will be treated. The consequences when the two kinetic processes are intermixed will also be considered.

The crystallization kinetics can be analyzed by taking the inverse of the time to reach 10% of crystallization as a measure of the crystallization rate. The crystallization rates of $C_{168}H_{338}$ from various concentrations in toluene are given in Fig. 13.6. Most of the data were obtained for a 4% solution. The sparser set of data that were obtained from 1% and 0.15% solutions are also included in the figure. All the data represent the crystallization rates of the structures that were initially formed. The open symbols represent the initial formation of folded chain crystallites. The closed symbols represent the extended chain crystals. The rate of crystallization decreases with increasing crystallization temperature. The data from the 4% solution show a distinct break at the crystallization temperature where folded crystals are no longer formed. Similar trends are observed for the 1 and 0.15% solutions. The data indicate that decreasing concentration decreases the crystallization rate of both the folded and extended structural forms.

Although a discontinuity is observed in the plot in Fig. 13.6, where there is a change in the structure crystallizing from the melt, either an extended crystallite or a once-folded one, no minimum is observed in the rate. There is then the question of why the minimum, or minima, is reported. In the temperature range of interest the isothermal transformation (thickening) of a folded to extended structure is very rapid.(18) If one counts this contribution to the high temperature endotherms as coming from the melt then, as indicated in Fig. 13.7, a minimum is observed in the rate. In this figure the open and closed circles represent the crystallization rates of the structures that form initially either folded or extended; the closed triangles represent the rate of the isothermal transformation. Therefore, it is important that the origin of and contribution to the endothermic peaks be clearly identified. At the expense of some redundancy it needs to be recognized that there are two distinctly different kinetic processes involved. Their intermixing leads to a minimum in the rate plot, as is illustrated in Fig. 13.7 for $C_{168}H_{338}$. This is the reason why minima

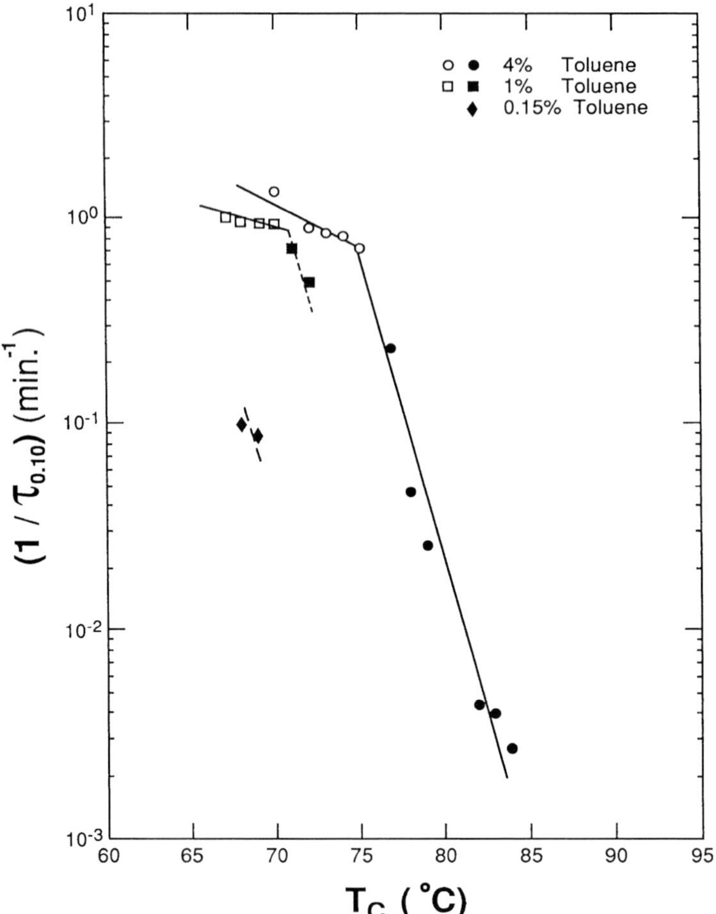

Fig. 13.6 Rate of crystallization, $1/\tau_{0.101}$ as a function of crystallization temperature of $C_{168}H_{338}$ crystallizing from toluene solution at indicated concentrations. Open symbols represent data for once-folded crystals. Closed symbols are for extended crystals.(10)

have been reported in many alkanes.(3,5–8,11) The reason for the discontinuity observed in Fig. 13.6 remains to be resolved.

The higher molecular weight *n*-alkanes such as $C_{198}H_{398}$, $C_{240}H_{482}$, $C_{246}H_{494}$ and $C_{294}H_{590}$ display, in addition to the transition from once-folded to extended, the transformation from three-extended to twice-ended as well as twice to once-extended.(2,6,7,10) Minima in the rates are observed in each of the transition regions. It is shown that the minima are a result of including the isothermal transformation of one form to another with the structures that initially crystallize from the melt. Minima in the rate of crystallization of the high molecular weight *n*-alkane

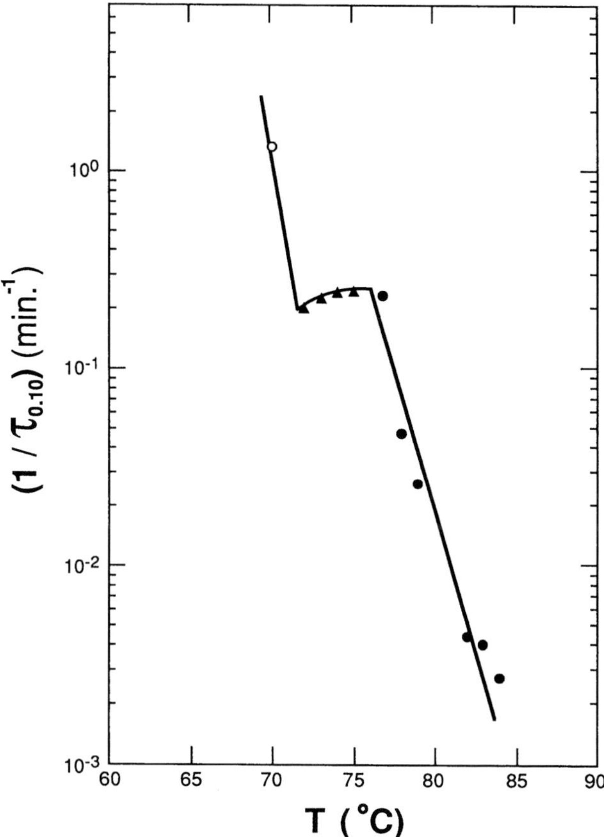

Fig. 13.7 Rate of crystallization versus temperature for $C_{168}H_{338}$ crystallizing from a 4 w/v% toluene solution. ○ data for the initial formation of once-folded crystals. ● data for extended crystals. ▲ related to the isothermal transformation of once-folded to extended crystals.(10)

from the pure melt have also been reported.(2,4,8,9,25,26) A similar explanation can be offered for these observations.

In summary, the experimental observations of minima in the kinetic studies are not in question. These are clearly the result of the type of experiment involved. In interpreting the results, however, it needs to be recognized that these observations are the consequence of the intermixing of two distinctly different kinetic processes. This intermixing leads to the observed minima. The role, if any, of "self-poisoning" in influencing the kinetics, either overall or growth, must be superposed on the rapid transformation of a folded crystallite to an extended one.(6c,6d)

It is also of interest to examine the temperature coefficient of the crystallization rate in terms of appropriate nucleation theory.(27) Accordingly, the ln of the crystallization rates of $C_{168}H_{338}$ and $C_{240}H_{482}$ from a 4% solution in toluene are plotted

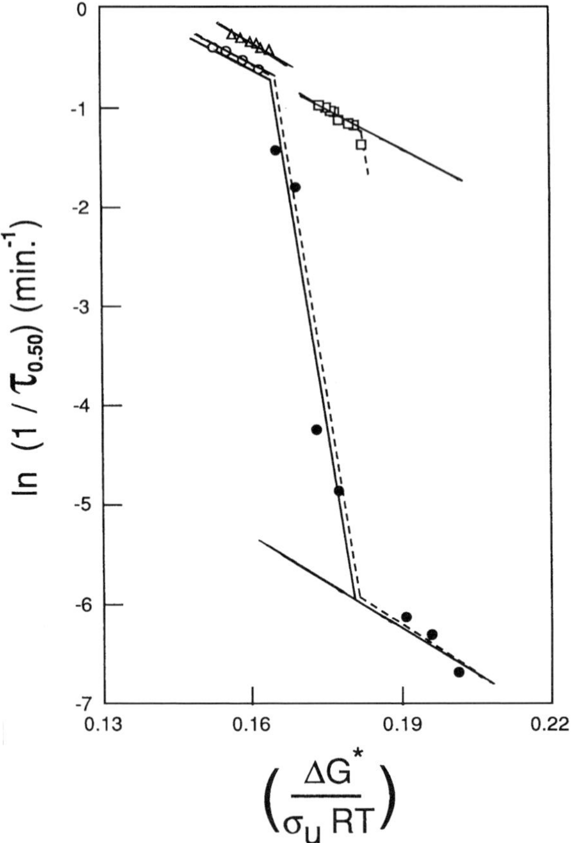

Fig. 13.8 Plot of ln crystallization rate against nucleation temperature function for coherent unimolecular surface nucleation for $C_{168}H_{338}$ and $C_{240}H_{482}$ crystallizing from a 4 w/v% toluene solution. Open and closed symbols represent folded and extended crystals respectively. ○, ● for $C_{168}H_{338}$; △, □ for $C_{240}H_{482}$.(10)

in Fig. 13.8 against the nucleation temperature function for coherent unimolecular nucleation (Gibbs type), taking into account the finite chain length. Only data for the crystallite structures that were initially formed in the original solution are utilized. Thus, for $C_{168}H_{338}$ the rates of crystallization of nearly once-folded crystallites (represented by open circles) and extended crystals (closed circles) are given in the figure. Similarly, the data for three-times-folded $C_{240}H_{482}$ crystallites (open triangles) and once-folded ones (open squares) are also included. The complications in analysis, when the transition from one form to the other, i.e. isothermal thickening, is taking place, are thus avoided. Striking similarities are found in the slope of the crystallization rate data corresponding to all of the folded forms (open symbols). Thus, three parallel straight lines, which are only slightly displaced from one another, can be drawn through the data and thus reflect almost identical temperature coefficients. This result indicates quite clearly that the same type of nucleation process is involved irrespective of the different kinds of crystallites (close to three-times or once-folded) that are developed.

The interpretation of the data that correspond to the initial formation of extended $C_{168}H_{338}$ crystallites (closed circles) is more complex. A line parallel to those of the folded crystals can be drawn for the three data points obtained at the highest temperatures. This would be consistent with the analyses of the melt-crystallized system (see Fig. 9.77). This description of the data leaves an intermediate temperature region, indicated by the dashed line, where the rate decreases very rapidly with temperature. The experimental point obtained at the highest crystallization temperature with $C_{240}H_{482}$ seems to indicate a similar rapid decrease of the rate with increasing temperature. However, this representation cannot be interpreted according to conventional regime theory. The ratio of the slopes in this case will be inverted. Alternatively, if the three highest data points are neglected only the two intersecting straight lines that are drawn need to be considered. In this case, the change in slope can be related to a Regime I to II transition in analogy to the results obtained for the rates of $C_{168}H_{338}$ from the melt. A more consistent interpretation would be obtained. However, there is no justification for neglecting the three high temperature data points.

In summary, the dependence of the crystallization rate on the temperature of the high molecular weight *n*-alkanes, for crystallization from either the melt or solution, can be divided into three regions. There is a low temperature region where folded structures are initially found. Here the thickening or transformation process is sufficiently slow so that the rates obtained from either endotherms, exotherms or direct growth measurements are not affected. Secondly, there is an intermediate crystallization temperature region where the isothermal thickening rate is rapid and the crystallization rate will be affected. The third high temperature crystallization region corresponds to the case when only the extended form crystallizes.

13.2 Crystallization from dilute and moderately dilute solutions

Investigations of the crystallization kinetics of polymers from dilute solution are of particular interest. Well-defined lamellar-like crystallites are formed in this concentration region. Moreover, in contrast to crystallization from the pure melt, or concentrated solution, the chains are relatively isolated from one another. Both the overall crystallization and lamellar growth rates have been studied in this concentration range. Attention will initially be given here to the overall crystallization studies and the analysis of the data obtained.

Figures 13.9 to 13.12 give sets of isotherms, plotted as the relative fraction transformed against the log time, for a series of molecular weight fractions of

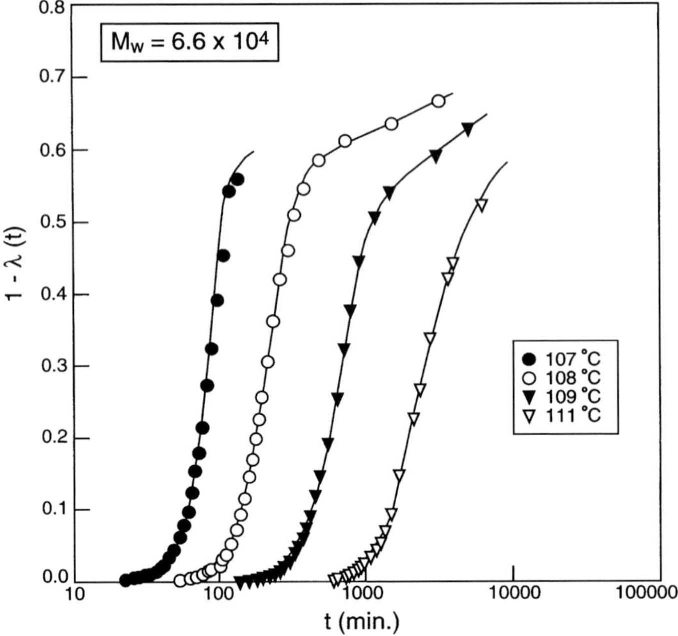

Fig. 13.9 Plot of relative extent of transformation against log time for a molecular weight fraction of linear polyethylene, $M_w = 6.6 \times 10^4$, at a weight fraction 0.025 in n-hexadecane. The crystallization temperatures are indicated. (From Chu (29))

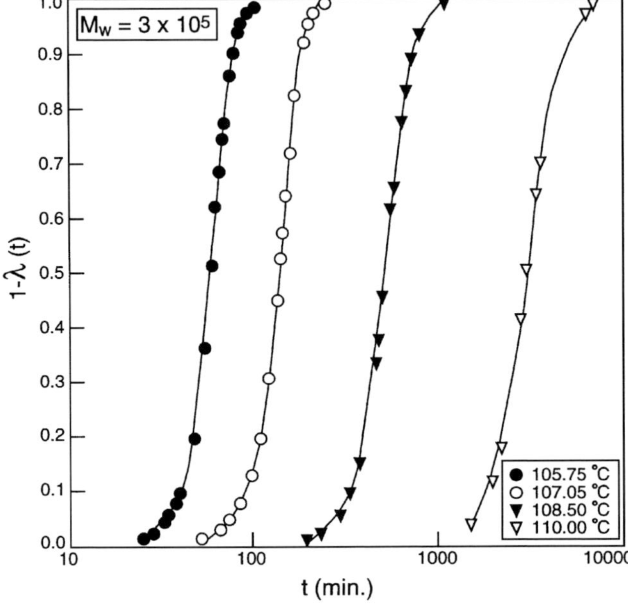

Fig. 13.10 Plot of relative extent of transformation against log time for a molecular weight fraction of linear polyethylene, $M_n = 3 \times 10^5$, at a 0.55% solution in n-hexadecane. The crystallization temperatures are indicated.(28)

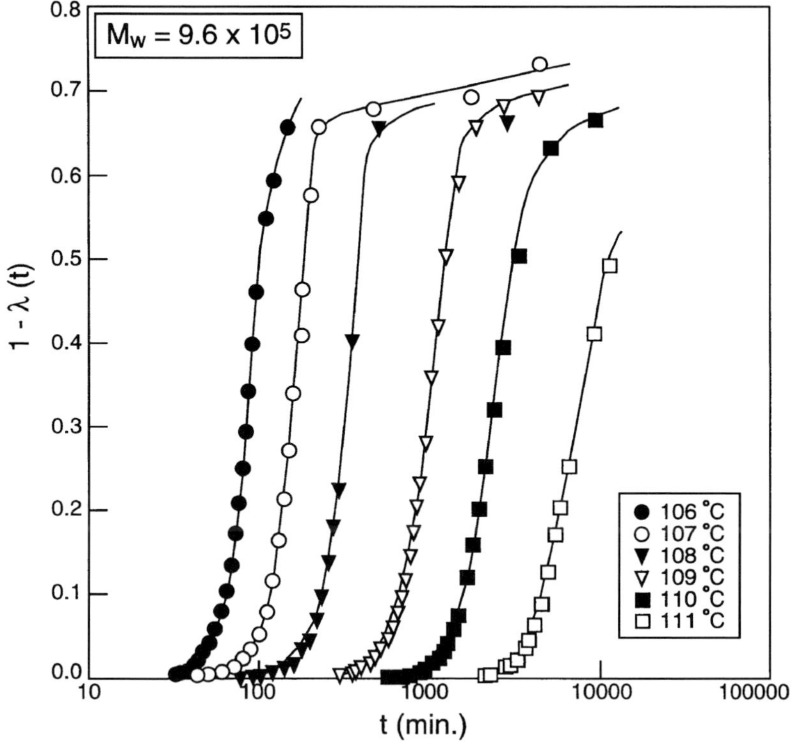

Fig. 13.11 Plot of relative extent of transformation against log time for a molecular weight fraction of linear polyethylene, $M_w = 9.6 \times 10^5$, at a weight fraction 0.025 in n-hexadecane. The crystallization temperatures are indicated. (From Chu (29))

linear polyethylene crystallized from dilute solutions of n-hexadecane.(28,29) The molecular weights range from 6.6×10^4 to 3.1×10^6. Characteristic sigmoidal shaped isotherms are again observed and are typical of nucleation and growth processes. The autocatalytic nature of crystallization from dilute solution is qualitatively similar to that observed for the crystallization from the pure melt. There is a difference, however, between the two cases. The termination of the crystallization from dilute solution is relatively abrupt. On the other hand, bulk crystallization is characterized by a small, but continuous, increase in the level of crystallinity, over an extended time period. There is also a strong negative temperature coefficient for all the molecular weights indicating, as in bulk crystallization, the importance of nucleation processes.

The isotherm shapes for a given molecular weight fraction are quite similar, suggesting that they can be superposed one onto the other. Figure 13.13 demonstrates the superposition of the data for two of the fractions. The isotherms for the other fractions behave in a similar manner. It is clear from these plots that the

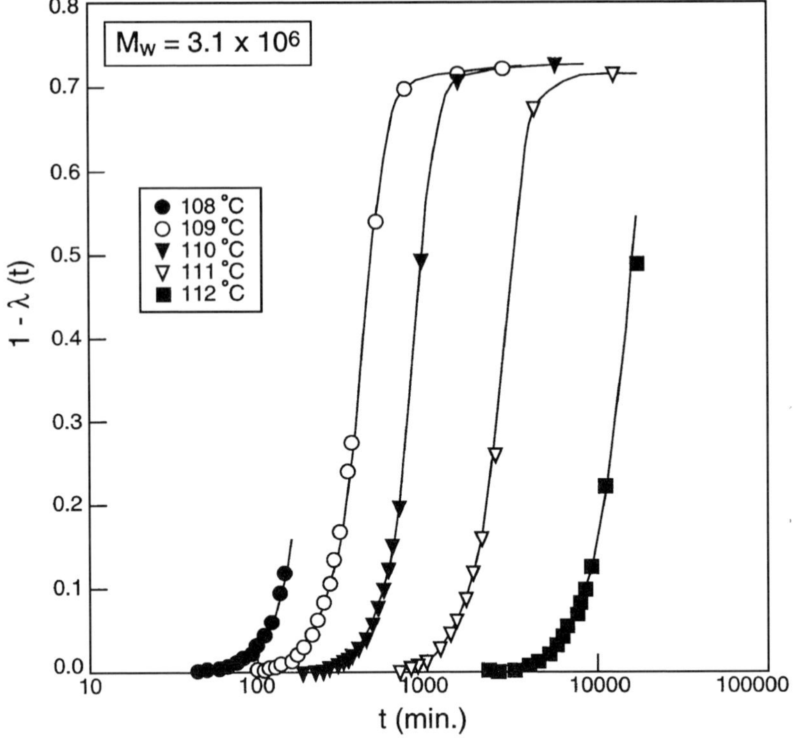

Fig. 13.12 Plot of relative extent of transformation against log time for a molecular weight fraction of linear polyethylene, $M_w = 3.1 \times 10^6$, at a weight fraction 0.025 in n-hexadecane. The crystallization temperatures are indicated. (From Chu (29))

superposition principle is obeyed for crystallization from dilute solution. The solid curve that is drawn in each of the figures represents the derived Avrami equation for $n = 4$. In contrast to bulk crystallization, there is extraordinarily good adherence between the Avrami formulation and the experimental data over almost the complete transformation. This is true even for the very high molecular weight fraction, $M = 3.1 \times 10^6$. Almost exact fits are obtained, with only small deviations at the very end of the transformations. With just a few exceptions, similar results have been reported for linear polyethylene in other solvents.(28,30–32) For example, good agreement with the derived Avrami, with $n = 4$, has been found in p-xylene and decalin.(28,30,32) An unfractionated sample, $M = 3.0 \times 10^6$, gives an n value of 3 at low undercoolings and $n = 4$ at higher undercooling (32), as do lower molecular weights in n-paraffins.(31) Two solvents α-chloronaphthalene and tetralin give $n = 3$ at all crystallization temperatures.(30)

Similar results have been found with other polymers crystallizing from dilute solutions. The crystallization of molecular weight fractions of poly(ethylene oxide),

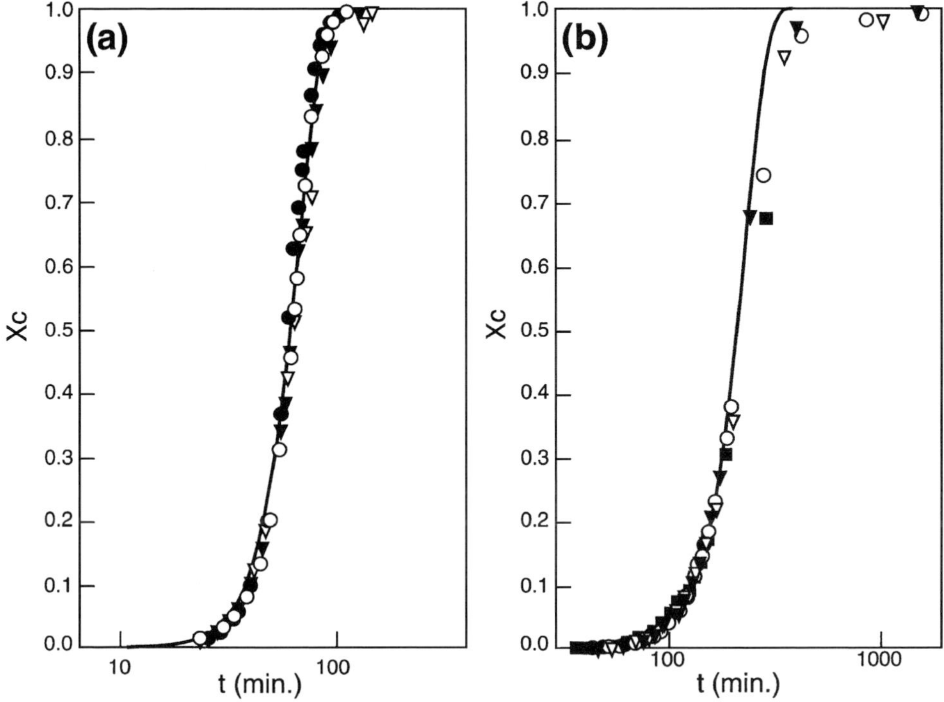

Fig. 13.13 Examples of superposition of isotherms. (a) Data from Fig. 13.10, $M_n = 3.0 \times 10^5$. (b) Data from Fig. 13.12, $M_n = 3.1 \times 10^6$. Solid curves derived Avrami equation with $n = 4$.

$M = 2 \times 10^4$ to 2×10^6, from dilute solutions of either p-xylene or ethanol showed good adherence to the derived Avrami equation with $n = 4$, over an appreciable extent of the transformation.(33,34) An unfractionated isotactic poly(propylene), $M_w \simeq 3.6 \times 10^5$, yields a set of superposable isotherms, when crystallized from a 0.3% solution of either ether tetralin or decalin.(35) A plot of the data according to Eq. (9.32) that is given in Fig. 13.14 makes clear both the superposition and adherence to a derived Avrami expression. In this case the exponent n is again found to be 4. An exponent of 4 has also been found for poly(ethylene terephthalate) crystallizing from several different solvents.(36)

There are several noteworthy and important features that are characteristic of polymer crystallization from dilute solution. A derived Avrami equation fits the isotherms over almost the complete transformation. Except for very low molecular weights, this is contrary to the results for bulk crystallization.(37) The reason for this difference and the crystallinity level that is attained will be discussed shortly. The crystallization of polymer from dilute solution can be considered to be ideal, since the kinetics resemble the crystallization of low molecular weight substances.

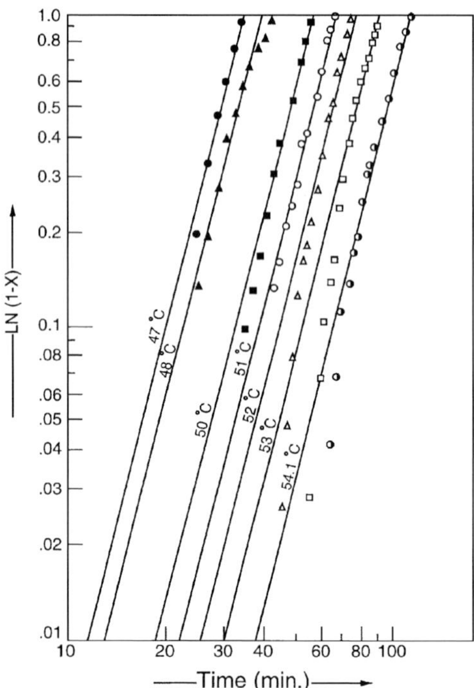

Fig. 13.14 Log–log plot of ln(1 − X) against time for isotactic poly(propylene), $M_w \simeq 3.6 \times 10^5$, crystallizing from a 0.3% decalin solution. X represents the fraction transformed. (From Shah and Lahti (35))

The value of 4 for the exponent n for crystallization from dilute solution is difficult at first to reconcile with the well-established lamellar crystallites that are observed.(28,30) However, studies have shown that the basic morphological form that crystallizes under these conditions consists of elongated three-dimensional crystallites.(32) Rather than single crystals, the crystallite structures are highly branched, three-dimensional lamellar networks, that are supported by the solvent during growth. This geometry remains the same over a wide range in molecular weights and independent of whether the Avrami exponent is 3 or 4.(32) Thus, an exponent of 4 represents steady-state nucleation and three-dimensional growth. For nucleation initiated at $t = 0$, the exponent is reduced to 3 for the same growth geometry.

It has been noted that the transformation is almost complete for linear polyethylene crystallized from dilute solution. For example, the crystallinity level that is attained at the isothermal crystallization temperature varies from about 85–90% at the lower molecular weights, $\leq 1 \times 10^5$, to about 75–80% at higher molecular weights, including $M = 3.1 \times 10^6$.(28,38) These results can be contrasted with those for bulk crystallization. At the isothermal crystallization temperatures the

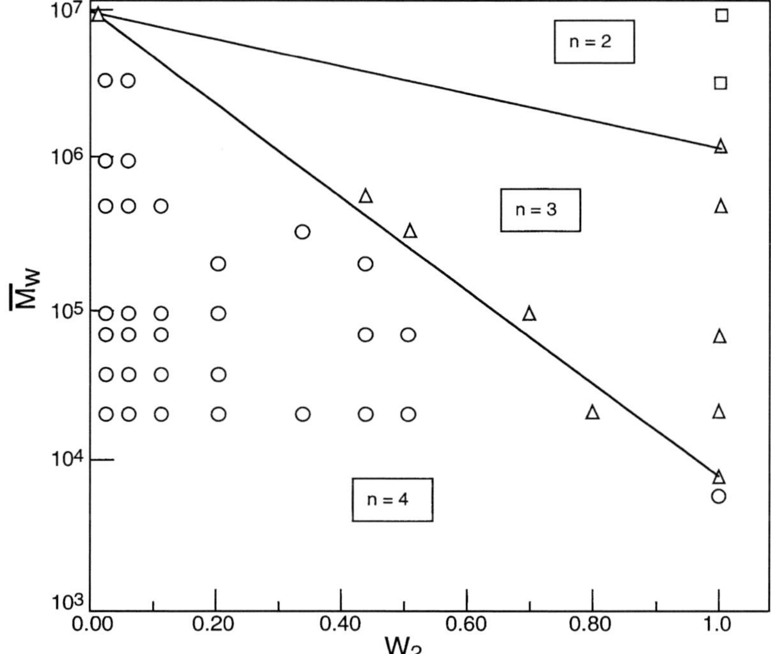

Fig. 13.15 Plot of Avrami n value for molecular weight fractions of linear polyethylene crystallizing from the pure melt and polymer–n-hexadecane mixtures. W_2 is weight fraction of polymer: $n = 4$ ○; $n = 3$ △; $n = 2$ □. (From Chu (29))

attained crystallinity level is much less. It is only about 0.40 for the high molecular weights.

There are also major differences in the Avrami n values between crystallization in the bulk and from dilute solution. These differences are illustrated in Fig. 13.15, where a map is given showing the relation between the n value and concentration for linear polyethylene fractions in n-hexadecane.(29) The interest at present is comparing the dilute range with the pure polymer, $W_2 = 1$. The results for the intermediate fractions, which are important, will be discussed in the next section. The n values equal 4 for all molecular weights, $M_w = 2.0 \times 10^4$ to 3.1×10^6, in the dilute range. In contrast, for bulk crystallization n only equals 4 at the very low molecular weights. There is a gradual reduction in n from 4 to 2 as the molecular weight is increased.(37)

The decrease in attained crystallinity level with molecular weight as a result of isothermal crystallization of the pure polymer has been attributed to the increasing entanglements along the chain with molecular weight. The entangled units, and nearby neighbors, are rejected into the residual amorphous melt by the growing crystallites. Hence, the final crystallinity level is systematically reduced and the kinetics altered accordingly. In dilute solutions the chain molecules are relatively isolated from one another so that entanglements do not present a significant impediment to crystallization. With the removal of this obstacle to crystal growth the growing polymer crystallite resembles low molecular weight systems in many respects. Hence, the ideal derived Avrami equation is obeyed with very high levels of crystallinity being achieved.

Further evidence for the influence of chain entanglement in the crystallization process is found in the crystallization of an isotactic poly(styrene) sample that was prepared from a freeze-dried dilute benzene solution.(38a) Entanglements in such a sample will be minimal. The overall crystallization rate in such samples, in terms of $t_{1/2}$, is enhanced by a factor of seven to eight relative to conventional crystallization from the pure melt.(38a) Experiments with isotactic poly(propylene), freeze-dried from *n*-octane, showed a similar enhancement in the crystallization rate.(38b) This type of experiment complements the overall crystallization rate of polymers from dilute solution.

The fact that the entanglement problem is no longer a factor has a major influence on the relation between crystallization rate in dilute solution and molecular weight. McHugh *et al.* noted that a linear polyethylene fraction, $M = 3.8 \times 10^5$, crystallized slower than a higher molecular weight, $M = 3.0 \times 10^6$, from a 0.1% xylene solution.(32) A more detailed picture of the influence of molecular weight on the crystallization rate in the dilute range is given in Fig. 13.16.(29) Here $\tau_{0.1}$ is plotted against log M_n for the crystallization at a weight fraction of 0.025, from a hexadecane solution. The quantity $\tau_{0.1}$ represent the time, in minutes, for 10% of the transformation to occur. It is the inverse of the crystallization rate $1/\tau_{0.1}$. In the lower molecular weight range, the crystallization rate increases with chain length at all the crystallization temperatures. This result is similar to that found for crystallization from the pure melt. The same explanation can be given in that there is a change in the effective undercooling because of the correction for chain length. A plateau region is reached with molecular weight and then the crystallization rate increases with chain length. In this molecular weight range the crystallization from dilute solutions is faster with increasing chain length. This result is the opposite to that observed for crystallization from the melt. However, the molecular weight dependence of the overall crystallization rate depends on the nature of the solvent. For example, when crystallizing from dilute solutions of α-chloronaphthalene the opposite effect is found.(38) The rate decreases with increasing chain length in this solvent. It will be found shortly that the variation of the growth rate with molecular weight in the dilute range depends on both concentration of the polymer and its thermodynamic interaction with the solvent. The basis for the solvent effect needs to be investigated in more detail.

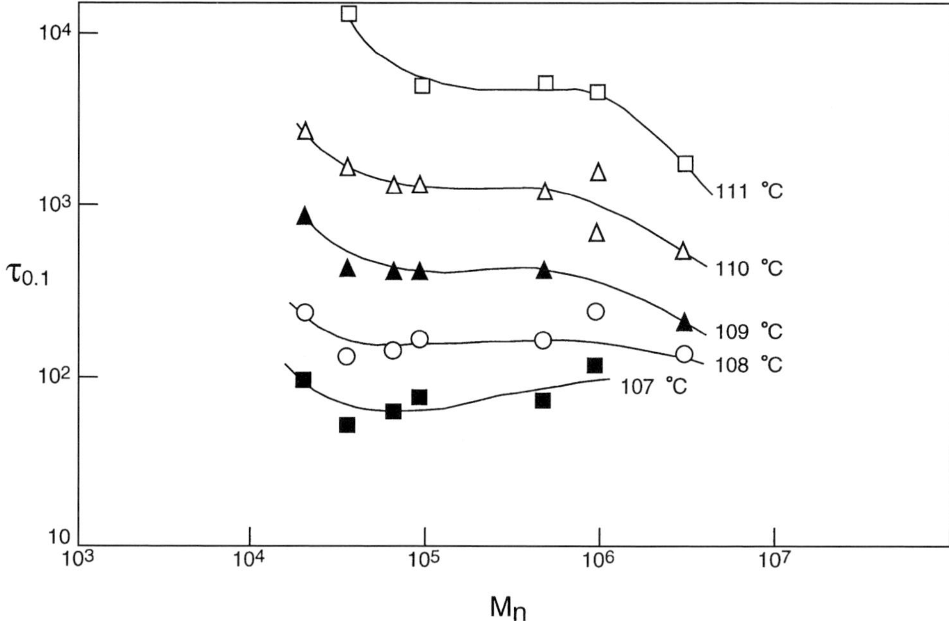

Fig. 13.16 Plot of time in minutes for 10% of the transformation to occur, $\tau_{0.1}$, as a function of molecular weight for fraction of linear polyethylene crystallizing from dilute n-hexadecane solution. Crystallization temperatures are indicated. (From Chu (29))

In dilute solution there is also a strong, negative temperature coefficient to the crystallization rate. This indicates that nucleation is still an important factor in the crystallization process. To develop a nucleation theory for polymer–diluent mixtures, the complete Flory expression for the free energy of fusion (39) is introduced into the free energy equation for forming either a cylindrical (three-dimensional) nucleus or a Gibbs type that involves the coherent unimolecular deposition of chains (two-dimensional).(40–42a) In analogy to the results for the pure polymer, each of the surfaces that result contains a saddle point that defines the critical dimensions of the nucleus, ρ^* and ζ^*, and consequently the critical free energy, ΔG^*, necessary to form such a nucleus. Then, for a three-dimensional nucleus in the high molecular weight approximation (40–42a)

$$\rho^* = \frac{4\pi\sigma_{\text{un}}^2}{(\Delta G_{\text{u}}')^2} \tag{13.1}$$

$$\zeta^* = \frac{4\sigma_{\text{en}} - 2RT\ln v_2}{\Delta G_{\text{u}}'} \tag{13.2}$$

$$\Delta G^* = \frac{8\pi\sigma_{\text{un}}2\sigma_{\text{en}} - 4\pi RT\sigma_{\text{un}}^2\ln v_2}{(\Delta G_{\text{u}}')^2} \tag{13.3}$$

where

$$\Delta G'_{\mathrm{u}} = \Delta G_{\mathrm{u}} - RT\frac{V_{\mathrm{u}}}{V_1}\left\{(1 - v_2) - \chi_1\left(1 - v_2^2\right)\right\} \tag{13.4}$$

Similarly, for a Gibbs type nucleus (40)

$$\rho^* = \frac{2\sigma_{\mathrm{un}}}{\Delta G'_{\mathrm{u}}} \tag{13.5}$$

$$\zeta^* = 2\sigma_{\mathrm{un}} - RT \ln v_2 \tag{13.6}$$

$$\Delta G^* = \frac{4\sigma_{\mathrm{un}}\sigma_{\mathrm{en}} - 2\sigma_{\mathrm{un}}RT \ln v_2}{\Delta G'_{\mathrm{u}}} \tag{13.7}$$

The free energy expression that is used here involves the Flory–Huggins mixing relation.(39) This relation involves a uniform distribution of polymer segments in the melt. Although valid for concentrated and moderately concentrated solutions it is not appropriate for dilute solutions. Hence Eqs. (13.1) to (13.7) must be used with caution. The ln v_2 term that is added to ΔG^* is in effect an entropic contribution. It represents the probability of selecting the ρ crystalline sequences from a mixture whose composition is v_2 in the vicinity of the polymer chain. For mixtures of monomeric substances and for polymer–diluent mixtures, where the segment density is uniform throughout the mixture, v_2 can be identified with the nominal concentration. However, for dilute polymer systems, where the segment distribution is nonuniform, v_2 represents the effective polymer concentration within the swollen coil. The probability of selecting the required number of sequences, whether from a single chain or a plurality of chains, will depend on the local concentrations and not the nominal one.

With this understanding of the limitation of present theory, some typical temperature coefficient data can be examined. In the usual type plot, the log of the rate is plotted against the nucleation temperature function. For three-dimensional nucleation this quantity is expressed as $(T_{\mathrm{s}}^0/\Delta T)^2(1/T_{\mathrm{c}})$; while for a Gibbs type nucleus it is given as $(T_{\mathrm{s}}^0/\Delta T)(1/T_{\mathrm{c}})$. Here T_{s}^0 is the equilibrium dissolution temperature, i.e. the equilibrium melting temperature at the concentration in question. For a pure system a linear plot should result whose slope can be related to the product of either $\sigma_{\mathrm{en}}\sigma_{\mathrm{un}}$ or $\sigma_{\mathrm{un}}^2\sigma_{\mathrm{en}}$, depending on the type of nucleation assumed. However, according to Eqs. (13.3) and (13.7), the value of the slope will also depend on ln v_2, for polymer–diluent mixtures. Therefore, for dilute solution the effective volume fraction, v_2, needs to be known in order to extract the product of interfacial free energies from the data.

The influence of dilution on the crystallization rate is illustrated in Fig. 13.17 for an unfractionated linear polyethylene crystallizing from α-chloronaphthalene.(42) Here the crystallization rate is plotted according to three-dimensional nucleation

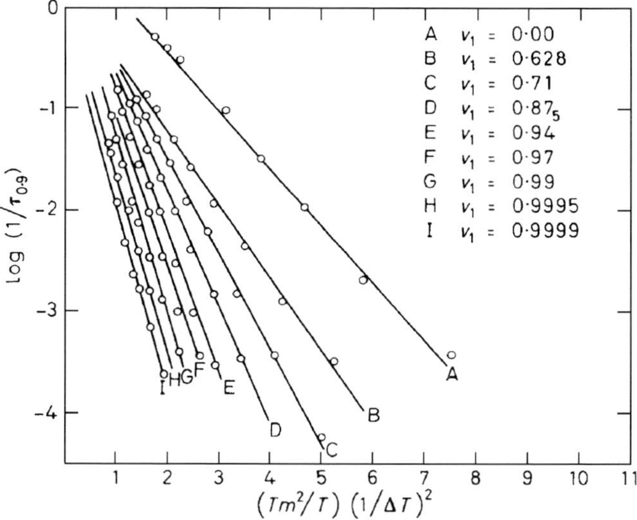

Fig. 13.17 Plot of $\log(1/\tau_{0.9})$ against temperature variable $(T_m^2/T)(1/\Delta T)^2$ for linear polyethylene crystallizing from α-chloronaphthalene mixture. Volume fraction of diluent in each mixture, v_1, is indicated.(42)

theory. The trends are the same if the data are analyzed according to the formation of a Gibbs type nucleus. The data encompass the complete composition range, from pure polymer to a 0.01% solution. A family of straight lines results, whose slopes steadily increase as the polymer concentration decreases from the pure system to the 1% mixture. However, for the hundred-fold change, of 1% to 0.01%, in concentration the slope remains essentially constant. The results for the higher concentration range will be discussed in the next section. The interest at present is in the dilute range. In this range, the results indicate that the effective polymer concentration is constant, independent of the nominal concentration, if it is assumed that the interfacial free energies are independent of dilution.

Although the temperature coefficient of the crystallization rate is independent of the concentration for concentrations less than 1%, the time scale of crystallization is not. At a fixed value of the temperature function, $(T_m/\Delta T)^2(1/T)$, the crystallization rate steadily decreases with decreasing polymer concentration. Thus, while at a fixed undercooling, the overall rate of crystallization is dependent on concentration, its temperature dependence is independent of the concentration.

Figure 13.18 illustrates the role of specific solvents in mediating the crystallization kinetics in dilute solutions. Here, the ln of the rate of a linear polyethylene fraction is plotted against the nucleation temperature function, $(T_s/\Delta T)(1/T_c)$, for a Gibbs type two-dimensional nucleus, for four different solvents.(30) In this example the molecular weight of the fraction is 1×10^5 crystallizing from a $v_2 = 0.003$

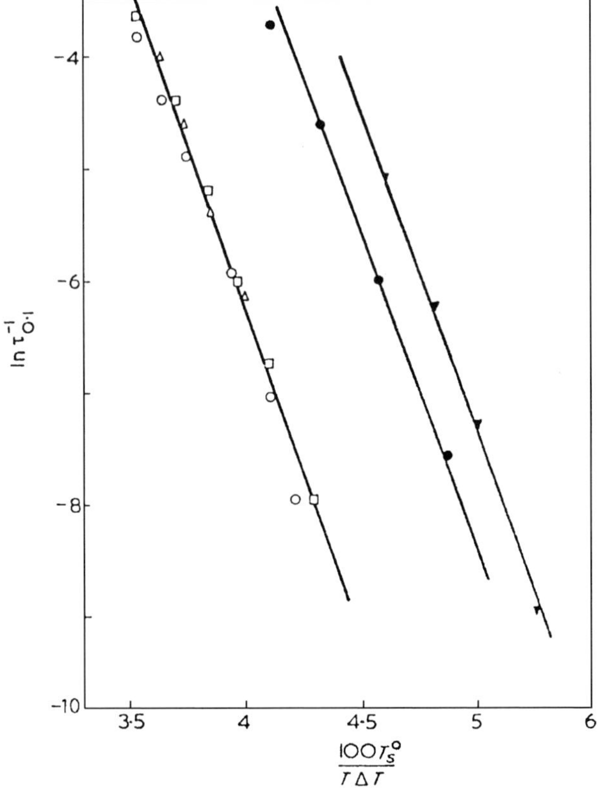

Fig. 13.18 Plot of $\ln(1/\tau_{0.1})$ against $(T_m^0/T)(1/\Delta T)$ for a linear polyethylene fraction, $M = 1 \times 10^5$, crystallizing from a $v_2 = 0.003$ solution of the following diluents: ○ decalin; □ tetralin; △ p-xylene; ● n-hexadecane; ▼ α-chloronaphthalene. (From Riande and Fatou (30))

solution. Characteristically, a set of parallel straight lines is obtained. This result is consistent with products of interfacial free energies and the effective volume fraction being independent of the nature of the diluent. However, the crystallization rates, at a given undercooling, are not necessarily the same. Decalin, tetralin and p-xylene fall on the same straight line. The diluents n-hexadecane and α-chloronaphthalene differ. The polymer crystallizes the fastest in the latter solvent. It is tempting to relate these differences in rate to the intensity of the polymer–solvent interaction as reflected in the parameter χ_1. However, the values reported for χ_1 do not appear to support this postulate.(30) Decalin has the lowest value for χ_1, xylene the highest, yet their rates are the same. The most rapid crystallization takes place in α-chloronaphthalene whose χ_1 value is close to that of p-xylene. Based on the values given for χ_1, there does not appear to be any correlation between it and the crystallization rate.

The growth rate of polymer crystallization from dilute solution has also been studied. Major emphasis has been on the crystallization of linear polyethylene from a variety of solvents, although poly(ethylene oxide) has also been studied in this regard. Usually, the growth rates of specific faces, such as the (110) and (100), in linear polyethylene have been recorded, although the light scattering technique has also been used. Of particular interest is the influence of molecular weight, polymer concentration and the thermodynamic nature of the solvent in governing the crystallization. These factors will be discussed in terms of experimental results and theoretical interpretations. Molecular weight fractionation is well established to take place during the crystallization of polymers from dilute solution.(43–45) Therefore, attention will be focused primarily on studies involving molecular weight fractions.

In a pioneering work, Holland and Lindenmeyer found that the growth rate of linear polyethylene crystallizing from a 0.01% *p*-xylene solution decreased with decreasing molecular weight at a constant crystallization temperature.(46) A similar effect has also been reported by others.(47–49) This dependence on chain length is opposite to what is observed in the same molecular weight range for crystallization from the pure melt. However, detailed studies indicate that the molecular weight dependence of the growth rate of linear polyethylene depends on both the polymer concentration and on the nature of the solvent.(50–53) An example is illustrated in Fig. 13.19 for the crystallization of linear polyethylene fractions from *p*-xylene solutions at different polymer concentrations.(50) At the lowest concentrations studied, 0.001 wt percent, the growth rate increases continuously with molecular weight at all crystallization temperatures. However, when the polymer concentration is increased to 0.01% the growth rate becomes constant at a molecular weight of about $M = 10^5$. The growth rate goes through a maximum at about $M = 10^5$ for

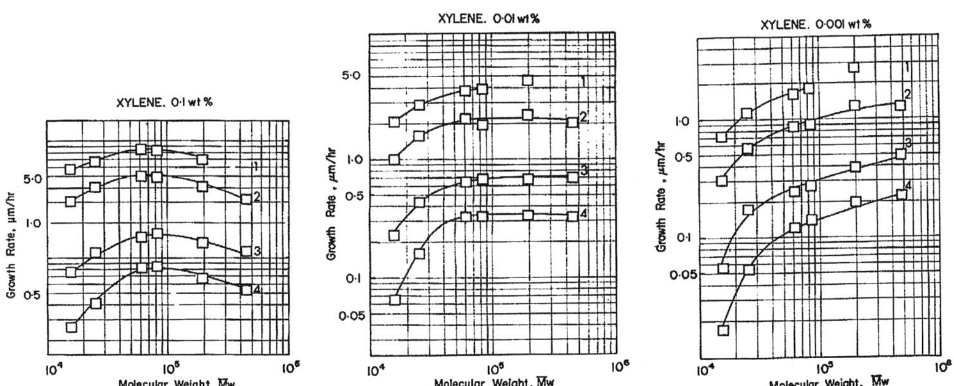

Fig. 13.19 Plot of log *G* against log *M* for molecular weight fractions of linear polyethylene crystallizing from dilute solution of *p*-xylene. (a) 0.001 wt%; (b) 0.01 wt%; (c) 0.1 wt%. Isothermal crystallization temperatures are (1) 86.60 °C; (2) 87.90 °C; (3) 89.90 °C; (4) 90.70 °C. (From Cooper and Manley (50))

the more concentrated solution, 0.1%. The growth rate–molecular weight relation at this composition is qualitatively similar to that observed for crystallization from the pure melt. A minimum in the growth rate is observed at concentration 0.001% to 0.1% when molecular weights in the range 3.1×10^3 to 1.2×10^4 are studied.(52)

Growth rate studies of linear polyethylene from other solvents have also been reported.(51,53) The results obtained with decalin, a relatively good solvent, are very similar, at all concentrations, to those obtained with *p*-xylene.(51) In contrast, in a poorer solvent, *n*-octane, a maximum in the growth rate is observed at all concentrations studied, 0.001% to 0.1 wt%.(48) However, the rate of decrease is much steeper in this case. Growth rates of linear polyethylene crystallizing from tetradeconol (0.05%), over the molecular weight range $M = 1.4 \times 10^4$ to 1.2×10^5, are essentially constant at the lower crystallization temperatures.(53) However, as the temperature is raised the crystallization rate of the lowest molecular fractions decreases.

No simple pattern has as yet emerged for the role of solvent in mediating the growth rate of linear polyethylene in dilute solution. Both the concentration and thermodynamic interaction with the solvent play an influential role in the growth rate–molecular weight relation. There is the suggestion that segment–segment interactions, both intra- and intermolecular are important.

Growth rate studies, utilizing a light scattering method, have also been reported for molecular weight fractions of poly(ethylene oxide) crystallizing from a 0.01 wt percent solution of toluene. The role of chain length is illustrated in Fig. 13.20 as a

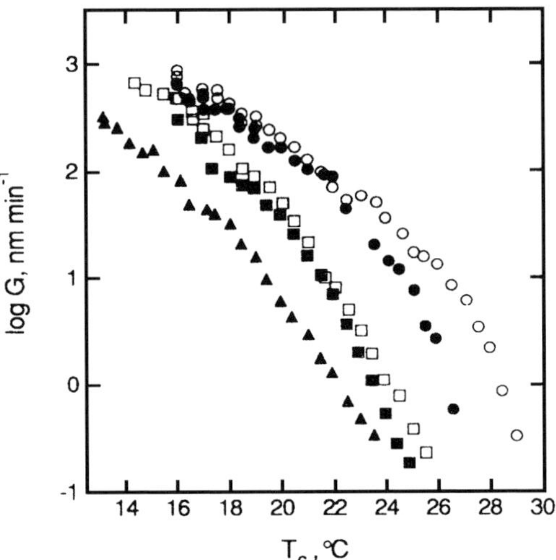

Fig. 13.20 Plot of log *G* against crystallization temperatures for molecular weight fractions of poly(ethylene oxide) crystallizing from 0.01 wt% toluene solution. Molecular weights: ▲ 5.6×10^4; ■ 1.05×10^5; □ 1.60×10^5; ● 3.25×10^5; ○ 7.70×10^5. (From Ding and Amis (54))

plot of log G against crystallization temperature for a set of molecular weights that range from $M = 5.63\times10^4$ to 7.70×10^5.(54,55) It is clear from this figure that at the same crystallization temperature longer chains crystallize much faster than shorter ones. This effect is greater at the higher crystallization temperatures than at the lower ones. Data for the crystallization of poly(ethylene oxide) from other solvents and concentrations, in the dilute range, are not available. Molecular weight fractionation can be accomplished, both for linear polyethylene and poly(ethylene oxide), based on crystallization kinetics by appropriate choice of solvent and temperature.

The dependence of the growth rate on polymer concentration, C, is usually expressed as

$$G = C^{\alpha} \tag{13.8}$$

With a few exceptions, the exponent α is less than 1.(56) Examples of the relation between the growth rate and concentration of two linear polyethylene fractions in p-xylene at several different temperatures are given in Fig. 13.21.(50) Over the concentration range studied here, C approximately 10^{-2} to 10^{-4} wt percent, α is a constant that only varies slightly with crystallization temperature. However, the value of α depends on the molecular weight as can be discerned from the plots in Fig. 13.22.(50) However, when the molecular weight is lowered to $M_w = 4050$ and $M_w = 3100$ the plots of log G and log C are no longer linear.(52) The slight increase with temperature is seen to be a property of the low molecular weight fractions. The change with temperature is negligible at the higher molecular weight. However, at

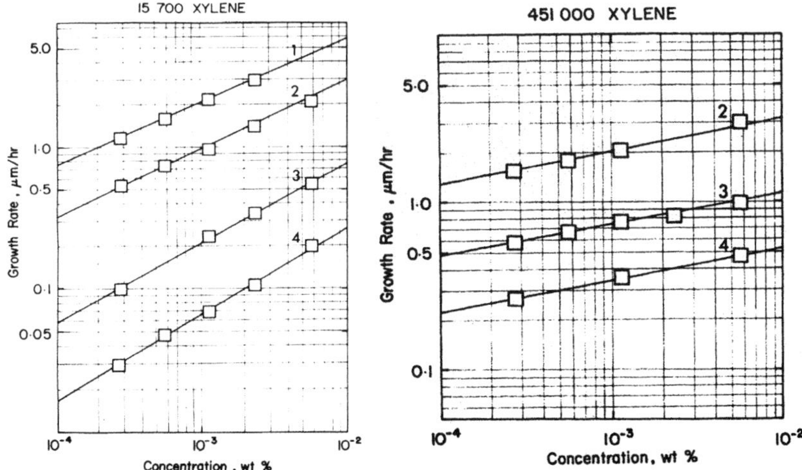

Fig. 13.21 Logarithmic plots of the growth rate of the (110) face of linear polyethylene fractions in p-xylene as a function of the concentration in weight percent. (a) $M_w = 15\,700$; (b) $M_w = 451\,000$. Crystallization temperatures: (1) 86.60 °C; (2) 87.90 °C; (3) 89.90 °C; (4) 90.70 °C. (From Cooper and Manley (50))

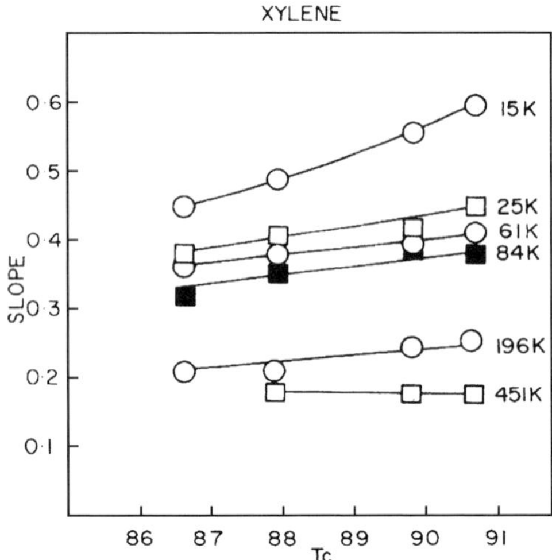

Fig. 13.22 Plot of the value of α against the crystallization temperature for the indicated molecular weight fractions. 15 K ≡ 15 700; 25 K ≡ 25 200; 61 K ≡ 61 600; 84 K ≡ 83 900, 196 K ≡ 195 900; 451 K ≡ 451 000. (From Cooper and Manley (50))

still lower weights, $M_w = 3100$ to 11 600, the α dependence on the crystallization temperature is more complex.(52) For molecular weights 6750 and 11 600 α first increases with crystallization temperature, passes through a maximum and then decreases. In sharp contrast to the results for high molecular weight fractions, α decreases monotonically with increasing crystallization temperature for the two lowest molecular weights studied. More important in Fig. 13.22 is the change in α with chain length, at a fixed crystallization temperature. At the highest crystallization temperature studied, 90.68 °C, α decreases from about 0.6 at $M_w = 15\,700$ to about 0.1 at $M_w = 451\,000$. The change in α with chain length is similar at the lower crystallization temperature. Thus, at the highest molecular weight studied, $M = 451\,000$, the growth rate is only very slightly dependent on concentration. It can be presumed that at still higher molecular weights the growth rate will become independent of concentration.

The growth rates in decalin are similar to those in p-xylene over a molecular weight range of 15 700 to 451 000.(51) In contrast, the α values obtained from crystallization from n-octane are higher than those formed with either decalin or p-xylene. In this solvent the growth rate–molecular weight plots show a maximum at concentrations 10^{-3} to 10^{-1}. Most interesting is the fact that in n-octane α initially decreases with molecular weight, similar to the observation with p-xylene and decalin. However, at $M_w \sim 70\,000$, at all crystallization temperatures, α reaches a minimum value and then increases. This is a rather unique behavior.

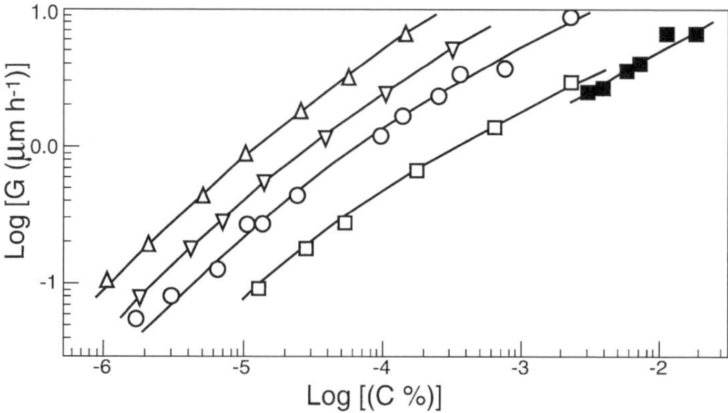

Fig. 13.23 Plot of log G (100 face) against log C for a linear polyethylene fraction, $M = 1.1 \times 10^4$, crystallizing from an n-octane solution at different crystallization temperatures: △ 88.8 °C; ▽ 90.7 °C; ○ 92.2 °C; □ 94.2 °C; ■ 94.8 °C. (From Toda *et al.* (59))

These results give further indication that the thermodynamic nature of the solvent plays an important role in the polymer crystallization from dilute solution.

A striking example of the role of the solvent is found in the difference in growth rates of isotactic poly(styrene) in dilute solution of either mesitylene, a good solvent, or a mixture of n-tetradecane and decalin, a poor solvent.(59) The growth rates are about three orders of magnitude greater in the poor solvent than in the good one, over a wide range in crystallization temperatures.

Toda and coworkers have studied the growth rate of linear polyethylene fractions over a much wider concentration range, $C = 10^{-1}$–10^{-6} wt percent, than was discussed above.[3] (58,59) The molecular weights used in their work ranged from 6.5×10^3 to 3.2×10^4. Since these studies were restricted to relatively low molecular weights, caution needs to be exercised in generalizing the conclusions reached to higher chain lengths. One of the interesting results found from the study of very dilute solutions is illustrated in Fig. 13.23 for the growth rate of a fraction $M = 1.1 \times 10^4$ in n-octane.(58) Here log G is plotted against log C for a family of crystallization temperatures. The data are now represented by gentle curves at each temperature. They are similar to those of low molecular weight fractions in p-xylene at higher concentrations.(52) The high concentration data in Fig. 13.23 can be represented by a straight line that yields an α value of about 0.5. This value is comparable to that reported by Cooper and Manley for a similar molecular weight in the same solvent.(51) It is important to recall that in this concentration range α is a decreasing function of molecular weight in xylene and in the lower

[3] The investigators expressed some doubt about the reliability of the concentration cited in the very dilute range, since the solutions were prepared by dilution.(58)

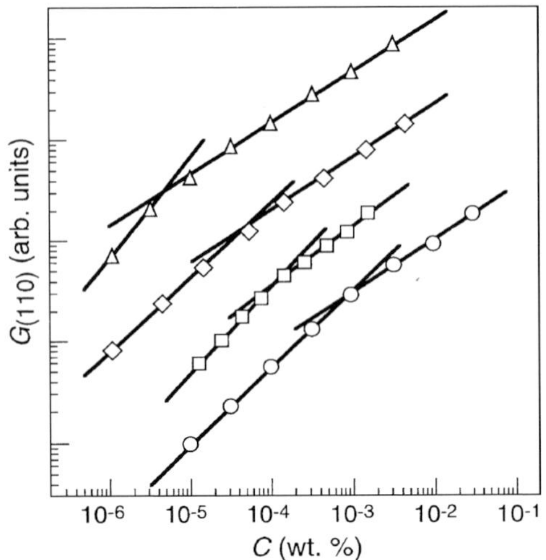

Fig. 13.24 Plot of log G, of (110) face, against log C for linear polyethylene fractions crystallizing from tetrachloroethylene solution. Molecular weight and crystallization temperature: ○ $M_w = 6.5 \times 10^3$, $T_c = 69.7\,°C$; □ $M_w = 1.1 \times 10^4$, $T_c = 69.0\,°C$; ◇ $M_w = 1.6 \times 10^4$, $T_c = 71.5\,°C$; △ M_w 3.2×10^4, $T_c = 73.5\,°C$. The growth rate data are displaced along the vertical axis for clarity. (From Toda (59))

molecular weight range for n-octane.(50,51) Hence, $\alpha = 0.5$ is not representative of all polyethylene molecular weights, in that it is only valid for molecular weights of about 1×10^4. A qualitatively similar relation between α and molecular weight has been reported for poly(ethylene oxide) in toluene.(55) The asymptotes for the low concentration in Fig. 13.26 yield an α value of 1.0 for $M = 1.1 \times 10^4$. The concentration corresponding to the intersection of the two asymptotes has been defined as C_0 and given theoretical significance.

The role of molecular weight, within the restricted range studied, is shown in Fig. 13.24.(59) Here, log G is plotted against log C for a set of molecular weight fractions covering the interval $M = 6.5 \times 10^3$ to 3.2×10^4, crystallizing from tetrachloroethylene. The data can be represented by a straight line at the higher concentrations, for all the molecular weights studied. However, the extent of the linearity is dependent on the chain length. For $M = 6.5 \times 10^3$ the linearity only extends to $C \simeq 10^{-3}$. However, for $M = 3.2 \times 10^4$ the linearity extends to solutions as dilute as $C \simeq 10^{-5}$. Following this trend, it is not difficult to presume that for slightly higher molecular weights the plot would be linear over the complete composition range studied. Under those circumstances the quantity $C = C_0$ would not exist. A similar conclusion can be reached from growth rate of a fraction $M = 1.1 \times 10^4$ from p-xylene solution over an extended concentration range at several crystallization temperatures.(59) Linearity is observed at some crystallization temperatures but not others. The actual existence of the parameter C_0 is very limited.

Plots of log G against log C of poly(ethylene oxide) fractions from dilute toluene solution are linear over the concentration range of $C = 10^{-3}$ to greater than 10^{-1} for molecular weights ranging from 5.63×10^4 to 3.25×10^5. At a fixed crystallization temperature α increases slightly with a decrease in molecular weight. This effect is qualitatively similar, although not as marked, to that observed in linear polyethylene crystals grown from dilute p-xylene solutions.(50) The poly(ethylene oxide) system is weakly dependent on temperature at a fixed molecular weight. This is again similar to the linear polyethylene–p-xylene system. In general the α values for the poly(ethylene oxide systems) are larger than those for linear polyethylene at comparable molecular weights.

Before attempting any theoretical interpretation it is useful to summarize the experimental results. There are several important factors that need to be considered. The dependence of the growth rate on concentration, in the dilute range of present interest, can be expressed by the parameter α. For low molecular weights, $\leq 3.2 \times 10^4$, α gradually increases with the concentration in both good and poor solvents. However, as the molecular weight increases, α becomes a constant. It decreases with chain length and tends toward zero at the higher molecular weights. Thus, at the higher molecular weights, $M = 2.3 \times 10^5$, the growth rate becomes independent of concentration.

The variation of the growth rate with molecular weight depends on the polymer concentration and thermodynamic nature of the solvent. In a poor solvent, like n-octane, the growth rate initially increases with molecular weight, reaches a maximum and then decreases. In the better solvent, like p-xylene, there is a strong influence of polymer concentration. In the most dilute solution studied, 0.001 wt percent, there is a continual increase in growth rate with chain length. At the highest polymer concentration the behavior is similar to that found in n-octane. There is no coherent and complete set of data that encompasses a large range in polymer concentration, molecular weight, different solvents and a variety of polymers with which to develop a general theory. It would be interesting, for example, to perform this type of growth measurements in a Θ solvent. However, it is worthwhile to consider the interpretations that have been made of the available data.

Several theories have been proposed to explain the available experimental data. These are based on the presumed role played by cilia during polymer crystallization from dilute solution. Two different approaches have been made to explain how cilia influence the crystallization process. One set of ideas has been put forth by Sanchez and DiMarzio.(60,61) Their underlying proposal is that when in solution a chain of finite length attaches itself to a crystal growth front, the crystallization (nucleation) need not start from the end of the chain. Most likely a central portion of the chain would be involved. Two ends of the chain are then left dangling in solution. The longer one crystallizes more rapidly on the growth strip by regular chain folding. It is presumed that the shorter one does not crystallize on this layer but dangles in solution, and is a cilium. These dangling end sequences are allowed to nucleate the following growth step by regular chain folding. At this point one might inquire as to what prevents the shorter sequence of chain units from crystallizing on the original growth front. If this were to happen then, depending on the length of the shorter sequence, either the growth front would be completed, leaving behind a sequence of still shorter dangling chains, or the initial cilium would be exhausted leaving a defect on the growth front. It is not clear why the proposed mechanism is limited to dilute solution crystallization and not also advocated for crystallization from the pure melt.

This proposal involves two distinctly different nucleation processes. One only involves the cilia with a steady-state nucleation rate N^{c}. The other, with a steady-state nucleation rate N^{s}, is the conventional segmental nucleation from dilute solution. The net steady-state nucleation rate, N, is expressed by

$$N = w_1 N^{\mathrm{c}} + w_2 N^{\mathrm{s}} \tag{13.9}$$

where w_1 and w_2 are weighting factors that are difficult to evaluate. Although clearly defined they are in effect arbitrary parameters. It can be expected that w_1 would be small at low molecular weights, but would become more important at the higher molecular weights. It then remains to examine the dependence of both N^{c} and N^{s} on concentration, solvent type, molecular weight and crystallization temperature.

The problem was formulated in terms of forward and backward rate constants for the two nucleating types following the conventional procedure for nucleation theory. The individual behavior of N^{c} and N^{s} was investigated. Both nucleation rates behaved in a similar qualitative manner with respect to molecular weight and crystallization temperature. However, a significant difference was found between the two with respect to their dependence on the polymer concentration. According to this theory the nucleation rate N^{s} goes through a broad maximum with molecular weight. The maximum is shifted to lower molecular weights as the crystallization temperature is lowered. The nucleation rate for cilia, N^{c}, does not show the broad maximum. However, there is a leveling off in the $\log N^{\mathrm{c}}$–$\log M_{\mathrm{w}}$ plot. If, however, the weighting factor w_1 is assumed to be inversely proportional to molecular weight, an arbitrary assumption, then a broad maximum appears in a plot of $\log(w_1 N^{\mathrm{c}})$ against $\log M_{\mathrm{w}}$. The temperature dependences of N^{c} and N^{s} are taken to be approximately the same and are represented by conventional nucleation theory.

The concentration dependences of N^{c} and N^{s} are quite different from one another. Therefore, the actual nucleation rate, or growth rate will depend on the weighting factors that can be arbitrarily selected. If only solution molecules are involved in the nucleation, the growth rate is found to be proportional to the concentration, i.e. $\alpha = 1$, except for very low molecular weights.(60) For pure cilia nucleation α is less than 1 and decreases with chain length at a given crystallization temperature. At a given molecular weight α increases with crystallization temperature. According to this theory α would not change from unity if nucleation of polymer molecules from solution were the only nucleation process involved. Therefore some other mechanism for nucleation and growth must be involved. Cilia nucleation has been proposed to fill the gap.

The plots in Fig. 13.19 allow for one comparison to be made between theory and experiment. In the linear polyethylene–xylene system, the dependence of the growth rate on molecular weight, over the range $M = 1.5 \times 10^4$ to 4.5×10^5, depends strongly on the polymer concentration. This result is not apparent from the theory. Qualitative agreement could probably be attained by appropriate choice of the weighting parameters. With respect to the temperature dependence, α increases with temperature at the lower molecular weights in agreement with theory. However, at the higher molecular weights there is a negligible dependence of α on the crystallization temperature. In accordance with theoretical expectations α decreases with molecular weight. However, it appears to be approaching zero at high molecular weights, $M \simeq 10^6$. Thus, the experimental results indicate that for very long chains the growth rate will not depend on concentration in the dilute range. This conclusion was not anticipated.(60) In contrast, α for the poly(ethylene oxide)–toluene system only shows a weak molecular weight dependence.(55) Definite conclusions cannot be reached with respect to the agreement between the theory outlined and the available experimental data, because of conflicting expectations.

Another mechanism for the role of cilia, in the crystallization from dilute solution, has been postulated by Toda and coworkers.(62) This theory is based on the works of Seto (47,63) and on Frank's theory,(64) which was discussed in Chapter 9. In essence, two steps traveling in opposite directions approach one another and are allowed to collide. As a consequence of the contact a pair of cilia is generated. Each can nucleate on the next layer and travel in both directions. The major conclusions from the analysis of this model are as follows. When the chain length is relatively small, nucleation by cilia can be neglected. For linear polyethylenes the cut-off molecular weight is approximately 3.0×10^4. Below this molecular weight cilia nucleation is not important. In Regime I, irrespective of whether or not cilia nucleation is taken into account

$$G \simeq C \exp(-K/T_c \, \Delta T) \tag{13.10}$$

In Regime II, when growth is controlled by the solute molecules

$$G \simeq C^{1/2} \exp(-K/2T_c \, \Delta T) \tag{13.11}$$

When cilia nucleation is the controlling factor in this regime

$$G \simeq C^{1/3} \exp(-2K/3T_c \Delta T) \qquad (13.12)$$

Striking in these equations is the fact that there is no explicit indication of the role of molecular weight. All experimental results show that, at a fixed polymer concentration and crystallization temperature, the growth rate depends on the molecular weight. It is also established that α varies with chain length. In particular, for the high molecular weights, where cilia nucleation would dominate, the α values for poly(ethylene oxide)–toluene are well above the 1/3 value predicted above. In the case of linear polyethylene in p-xylene α is less than 1/3 for the highest molecular weight studied and is expected to decrease further for longer chain lengths. Ding and Amis have concluded that cilia nucleation is unimportant in the crystallization of poly(ethylene oxide) from dilute toluene solution.(54,55)

It has been suggested that the adsorption of polymer chains on the growth face complements or even replaces in importance nucleation by cilia.(59,64,65) It was argued that the unusual dependence of growth rate on concentration, as illustrated in Fig. 13.24, results from the surface adsorption of the polymer. The change in the surface nucleation, which is directly related to the growth rate, is due to the extent of coverage of the growth face by the adsorbed polymer. By studying the growth kinetics of [100] twins, formed by linear polyethylene, the nucleation rate was found to be proportional to C at the lower concentrations and independent at the higher ones. Concomitantly, the spreading rate increased linearly with concentration over the whole range studied.(64,65) The surface adsorption of polymers will then cause a fractional power dependence of the growth rate on concentration. The conclusion of the importance of polymer adsorption on the growth process is based solely on an analysis of data that is limited to low molecular weights.

The interpretation of experimental results has focused mainly on the role of cilia. The conclusions have been mixed. The importance of cilia, and their influence on the crystallization from dilute solutions, cannot be adequately analyzed without a complete set of experimental data. However, some features of solvent influence become clearer when adequate data are available. This point is illustrated in Figure 13.25 where the values of α for linear polyethylene crystallizing from p-xylene, decalin and n-octane are plotted against the weight average molecular weight.(50–52) Decalin and p-xylene are considered to be relatively good solvents for linear polyethylene, while n-octane is a poorer one. There are some important features in these plots that are not apparent when the molecular weight range is limited to $1\text{–}3 \times 10^4$.

When crystallizing from p-xylene, linear polyethylene displays a maximum in α at about $M = 4 \times 10^3$ followed by a slow monotonic decrease to about 0.2 at $M = 4.5 \times 10^5$. Indications are that α will decrease further as the chain length

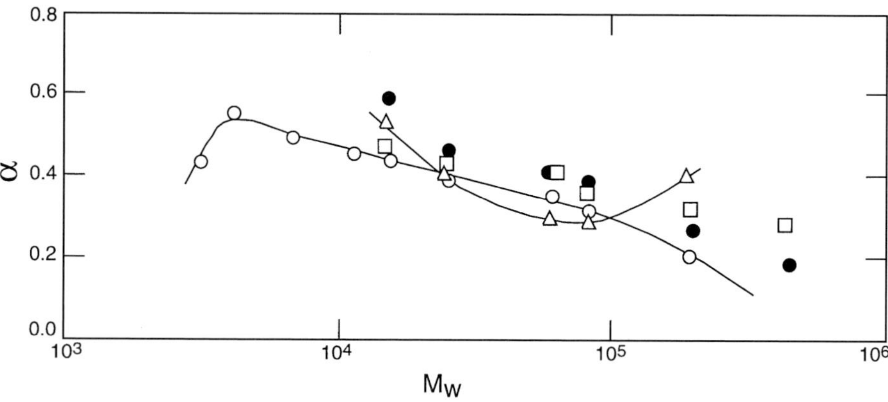

Fig. 13.25 Plot of α against the weight average molecular weight for linear polyethylene crystallizing from different solvents. *p*-xylene: ○ 86.0 °C and 86.6 °C; ● 90.7 °C. Decalin □ 85.3 °C; *n*-octane △ 98.0 °C.

increases. Thus, in this solvent the decrease in α suggests that the contribution of intermolecular interaction to growth decreases with molecular weight. A contribution from cilia nucleation to the growth is not expected for $M \leq 3 \times 10^4$.(62) Hence, the maximum must result from some other cause.

The variation of α with molecular weight in the poorer solvent, *n*-octane, is different. The lower molecular weight range, $<1 \times 10^4$, was not studied. Therefore, it is not known whether there would be a maximum also in this range. However, in this solvent there is a broad minimum centered about 7–8 $\times 10^4$. The growth rate shows a maximum in the same molecular weight range for all concentration studies 0.1%–0.001%.(51) The maximum becomes more pronounced as the nominal concentration decreases.

The summary given in Fig. 13.25 indicates that the thermodynamic interaction between polymer and solvent plays an important role in governing growth from dilute solution. A better solvent, with a lower χ_1 value, will expand the coil and reduce intramolecular segmental interaction. In contrast, the coil will contract in a poorer solvent and intramolecular interactions will become enhanced. The relation between the nominal concentration and the effective concentration, the quantity needed for nucleation theory, will also be affected by the nature of the solvent. There has not been very much attention paid to this important aspect of the problem. Studies with poorer solvents, in the lower molecular weight range, could be illuminating. The influence of cilia is minimal under these circumstances so that the role of solvent could be assessed independently. The source of the chain units that form the nucleus, i.e. from either within a molecule or between molecules, or proportional between them, needs to be elucidated.

13.3 Temperature dependence: crystallization from dilute solution

The temperature dependence of the growth rate from dilute solution follows the pattern established by the crystallization from the pure melt. An example is given in Fig. 13.26 of the growth rate of isotactic poly(styrene) from a 0.1% dimethyl phthalate solution over an extended temperature range.(66) Following the crystallization of the pure polymer, a maximum in the rate is observed and there is a severe retardation of the rate in the vicinity of the melting and glass temperatures. Both the nucleation and transport terms are operative in the dilute solution crystallization. However, the growth rate is reduced by about three orders of magnitude relative to that of the pure melt and the temperature for maximum growth is reduced about 100 °C.(67)

At the right side of the maximum, in the vicinity of the melting temperature there is strong evidence for nucleation controlled growth. This was suggested by studies of overall crystallization (30) and confirmed by many growth rate–temperature studies. The plots in Fig. 13.23 for poly(ethylene oxide) crystallizing from toluene show a three to four orders of magnitude increase in the growth rate with a

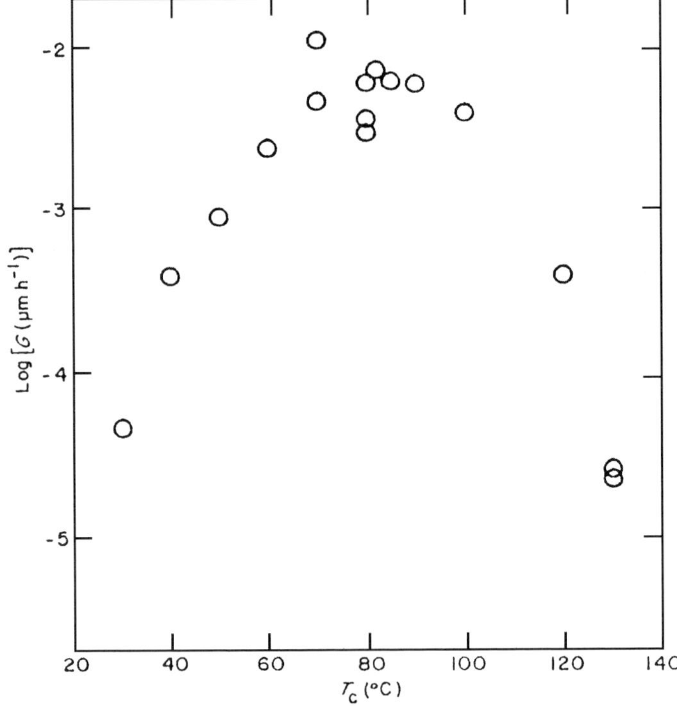

Fig. 13.26 Plot of log G against crystallization temperature T_c for isotactic poly (styrene) crystallizing from a 0.1% dimethyl phthalate solution. (From Tanzawa (66))

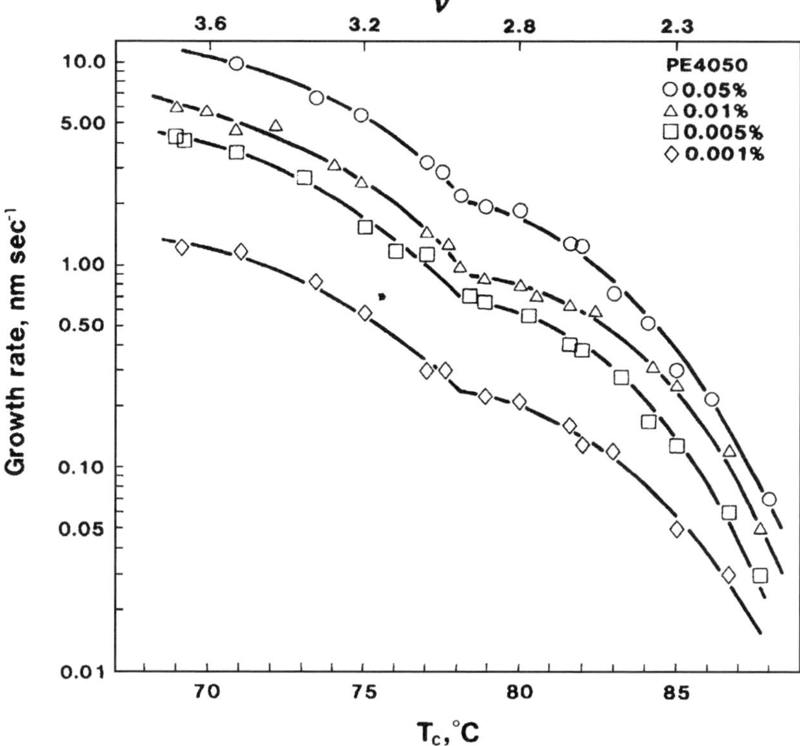

Fig. 13.27 Plot of growth rate against crystallization temperature for a linear polyethylene fraction, $M_w = 4050$, crystallizing from *p*-xylene at indicated concentrations. (From Leung *et al.* (52))

10–12 °C decrease in the crystallization temperature. Similar results have been reported for linear polyethylene crystallizing from *p*-xylene.(52) An example is shown in Fig. 13.27 for the crystallization of a low molecular weight polyethylene fraction, $M_w = 4050$. Here, the strong negative temperature coefficient is clear as is the discontinuity in the plot. A higher molecular weight, $M_w = 11\,600$, shows a similar temperature coefficient, but the data give a continuous curve.(52)

The application of nucleation theory to the data, irrespective of the type selected, requires an accurate value of the equilibrium dissolution temperature, T_s^0, because of the crucial role played by the undercooling in the analysis. The problem is that major differences in T_s^0 values have been reported for the same polymer–solvent system. For example, in the same investigation, depending on the extrapolation method used, T_s^0 for *p*-xylene is given as 118.6 ± 2 °C, for *n*-octane it ranges from 127 ± 1 °C to 124 ± 2 °C, while in decalin it varies from 113.0 ± 1.5 °C to 116.5 ± 2 °C.(68) Among different investigations T_s^0 values for *p*-xylene have been reported as 108.2 °C,(65), 110.5 ± 1 °C (69) and 118.6 ± 2 °C.(68) The reported values of T_s^0 for n-octane have varied from 116.4 ± 1 °C,(70) 120.2 ± 1.4 °C (69) and 124 °C to 127 °C.(68) A similar disparity is found with other solvents. An extreme situation is represented by decalin, where reported T_s^0 values range from 105.4 °C (72) to 116.5 °C.(68) Consequently, in the conventional plot of the growth rate against the nucleation temperature function, in the vicinity of the equilibrium melting temperature, either a straight line or a curve could result depending on the choice of the value for T_s^0. It is also possible to develop regime or pseudo regime transitions. Because of the uncertainties that are involved it is prudent not to focus on the analysis according to nucleation theory.[4]

Comparisons have been made between the product of interfacial free energies determined from the growth rate–temperature relation for bulk and dilute solution crystallization. The problem in this case is again selecting the proper values of T_m^0 and T_s^0. Rather drastic changes in the ratio of the product of interfacial free energies can be obtained by varying these two equilibrium temperatures.(49) This concern is true both for linear polyethylene and poly(ethylene oxide). By appropriate selection of T_s^0 and T_m^0 the slopes of the straight lines obtained from the data plotted according to nucleation theory can be made to be identical.(71–73) Considering the range in values reported for T_s^0 it cannot be concluded that the products of interfacial free energies, and thus the crystallization mechanisms, are the same in both cases.(72,73)

13.4 Crystallization from concentrated mixtures

The concentration range being considered in this section is one where the molecules overlap with one another. The Flory–Huggins relation for the free energy of mixing will hold in this concentration range.(74,75) Consequently, in contrast to dilute solutions, the effective concentration that is involved in nucleation theory can be closely identified with the nominal concentrations. With molecular overlap, chain entanglement becomes an important factor to consider in the analysis of crystallization kinetics. The number of chain entanglements per molecule, E, can be expressed as (76)

$$E = M_N v_2 / M_E - 1 \tag{13.13}$$

where M_E is the number of entanglements per molecule in the pure state. The final level of crystallinity attained and the crystallization rate will be strongly influenced by chain entanglements, as well as by other kinds of topological defects. The overall crystallization and spherulite growth rates from concentrated solutions will be analyzed in turn.

[4] The plots in Fig. 13.18 suffer from the same deficiency. However, in this case efforts were made to use a consistent set of T_s^0 values.(30)

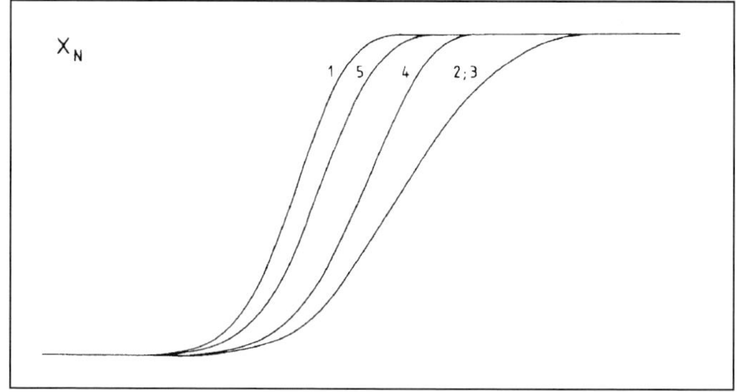

Fig. 13.28 Schematic representation of superposed isotherms of poly(dimethyl siloxane) and its mixtures with toluene. The normalized crystallinity level is plotted against log time. The curves are arbitrarily shifted along the log t axis for clarity. Curve (1), $\upsilon_2 = 1.00$; curve (2), $\upsilon_2 = 0.79$; curve (3), $\upsilon_2 = 0.59$; curve (4), $\upsilon_2 = 0.42$; curve (5), $\upsilon_2 = 0.32$. (From Feio *et al.* (78))

Large negative temperature coefficients are characteristic of the overall crystallization rates at all compositions. This is again indicative of the importance of nucleation, which is a dominant factor for the crystallization of pure polymer to dilute solution. The details of nucleation in concentrated mixtures will be discussed shortly. Focusing attention first on the character of the isotherms it is found quite generally that at a given composition their shapes are similar at different crystallization temperatures so that the superposition principle still holds. However, the shape and character of the isotherms change in a systematic manner with dilution.(29,31,34,40,77,78) A schematic representation of the isotherm shapes, typical of polymer–diluent mixtures is given in Fig. 13.28 for poly(dimethyl siloxane)–toluene mixtures.(78) The curves in this figure are arbitrarily displaced from the time origin in order of increasing steepness. There is clearly a change in isotherm shape with composition. In this example for concentrated and moderately concentrated mixtures, $0.79 > \upsilon_2 > 0.59$, there is a retardation in the overall crystallization rate in comparison with that of the pure polymer. However, with further dilution, the shapes of the isotherms become progressively closer to that of the pure polymer.

A major reason for the isotherm shape changes can be attributed to the changing composition of the residual melt as crystallization progresses.(79) The situation is similar to that discussed earlier with regard to copolymer crystallization. As crystallization proceeds, the melt becomes more dilute in polymer. The equilibrium melting temperature is thus reduced, as is the effective undercooling at the constant isothermal crystallization temperature. Concomitantly, the crystallization rate becomes progressively protracted with time. However, at high dilution, the order of

$v_2 = 0.10$ or less, the change in the effective undercooling with time, at a constant crystallization temperature, is usually small, since the equilibrium melting temperature is not as sensitive to polymer concentration in this range. The change is of the order of 1 °C or less. This is consistent with the isotherm shapes observed in the dilute range as was illustrated earlier in Figs. 13.10 and 13.11. The isotherm shapes are also tempered by chain entanglement, which plays a role similar to that in polymer crystallization from the pure melt.

The dependence of the n values in the derived Avrami expression on composition is also of interest. Because of the nature of the isotherms, and the deviations from theory, the determination of n needs to be limited to the early stages of the transformation. Figure 13.15, previously discussed for dilute mixtures of linear polyethylene and n-hexadecane, also contains n values for the complete concentration range for molecular weight fractions from 2.0×10^4 to 8.0×10^6. Referring to this figure, it is clear that n depends on both chain length and composition. At a fixed molecular weight, the integral value of n increases with dilution. At low to moderate molecular weights n increases from 3 to 4 as the polymer concentration decreases. Eventually, at high dilution theoretical isotherms are obtained. The range in composition for which $n = 4$ increases as the molecular weight decreases. For the very high molecular weights n increases from 2 for the pure polymer to 3 for the dilute systems.

An analysis of the overall crystallization rate with composition requires that the comparison be made either at constant undercooling or at one of the nucleation temperature quantities, $T_m/T\Delta T$ or $T_m^2/T(\Delta T)^2$. This requirement is essential because of the importance of nucleation to the crystallization process. The overall crystallization kinetics of a variety of polymer–diluent systems have been reported. Many different relations between the overall crystallization rate and composition have been observed. For example, as is shown in Fig. 13.17 there is a continuous decrease in the crystallization rate with dilution for linear polyethylene–α-chloronaphthalene mixtures.(42) The results for poly(trans-1,4-isoprene) in methyl oleate follow a similar pattern.(80) In contrast, the rates for poly(dimethyl siloxane) crystallizing from toluene, at compositions $v_2 = 0.32$ to 0.79, are the same at all undercoolings, but are faster than that of the pure polymer.(78) Another example is found with poly(ethylene oxide)–diphenyl ether mixtures.(77) In this case the crystallization rates for the pure polymer and composition $v_2 = 0.92$ to 0.51 are the same. However, the rates for the more dilute mixtures, $v_2 = 0.04$ and 0.30 are lower. For poly(decamethylene adipate)–dimethyl formamide mixture the rates for the pure polymer and $v_2 = 0.80$ are the same.(77) The mixture of isotactic poly(propylene) with dotricontane shows interesting behavior.(81) At all undercoolings studied, the crystallization rate initially decreases with dilution, reaches a minimum in the range $v_2 \simeq 0.7$ (a maximum in $t_{1/2}$) and then slowly increases with further dilution, up to $v_2 = 0.10$.

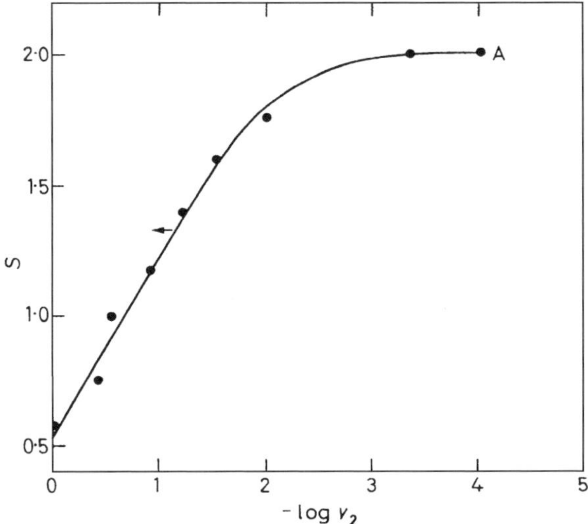

Fig. 13.29 Plot of slope, S, of straight lines in Fig. 13.17 against $-\log v_2$. (Data from Ref. (42))

The examples cited above indicate a wide diversity in the overall crystallization rate–composition relations among the different mixtures studied. A general pattern has not emerged from these studies. These results may reflect either specific differences between the different systems studied or may be a consequence of inadequate values being used for the equilibrium melting temperatures at the different compositions.

The slopes of the straight lines in Fig. 13.17 increase with dilution. According to Eq. (13.3) the slopes in this plot should be a linear function of $-\log v_2$. When plotted according to the dictates of a Gibbs type nucleus the slopes should also be a linear function of $-\ln v_2$. The volume fraction v_2 needed is the effective volume fraction of diluent. The slopes of the straight lines in Fig. 13.17 are plotted against $-\ln v_2$ in Fig. 13.29. In this example the nominal polymer volume fraction is used. The plot in Fig. 13.29 adheres quite well to a linear relation up to a polymer concentration of $v_2 = 0.03$. This result gives further verification of the need of the log v_2 term in ΔG^*. With further dilution, deviations from linearity set in. A polymer concentration higher than the nominal one is necessary in order to satisfy the linearity requirement. At these compositions the solution is no longer uniform with respect to the concentration of polymer segments. The effective volume fraction of the polymer then needs to be used. To satisfy the linearity requirement in this case v_2 would have to be of the order 0.02 to 0.03. If these values are taken then the plot in Fig. 13.29 would be linear over the complete composition range. It is not unreasonable that a concentration of this magnitude represents the polymer segments within the domain of the swollen polymer.

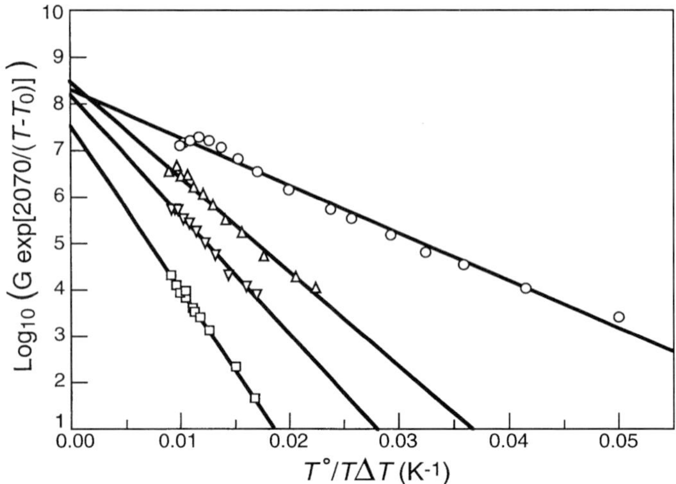

Fig. 13.30 Plot of $\log\left[G\exp\left(\frac{-U^*}{T-T_0}\right)\right]$ against $T_s^0/T\Delta T$ for isotactic poly (styrene) crystallizing from its mixture with dimethyl phthalate. Percent polymer: ○ pure polymer; △ 50%; ▽ 30%; □ 0.1%. (From Miyamoto *et al.* (67))

The radial growth rates of spherulites that develop in this composition range are usually linear with time for the early stages of the transformation.(67,82–85) However, the rates become retarded with time as a consequence of the dilution of the residual melt and the resulting decreases in the undercooling. The composition change is aided by diffusion in the vicinity of the phase boundary. When the linear portion of the growth rate is used, the rate decreases with polymer concentration at constant undercooling or nucleation temperature parameter for most of the reported systems.(67,82,86,87) In some mixtures the growth rates become close to one another at high dilutions. The measured primary nucleation rate of poly(ethylene oxide) in tripropionin has also been found to decrease with decreasing polymer concentration.(85) This conclusion depends on the values used for the equilibrium melting temperature for the pure polymer and the diluent mixtures.

Plots of $\log\left[G\exp\left(\frac{-U^*}{T-T_0}\right)\right]$ against $T_s^0/T\Delta T$ for isotactic poly(styrene) crystallizing from its mixtures in dimethyl phthalate are given in Fig. 13.30.(67) The slope continuously decreases with dilution, indicating the influence of the $\ln v_2$ term in ΔG^*. A similar effect is seen in Fig. 13.31 where the nucleation rate for the initiation of spherulite formation is plotted against $1/(\Delta T)^2$ for poly(ethylene oxide) crystallizing from tripropionin.(87) The increase in the slopes of the straight lines in the figure with polymer concentration is again consistent with the $\ln v_2$ term in Eq. (13.7). Thus, there is further experimental evidence of the importance of $\ln v_2$ in the expressions for the nucleation and growth rates.

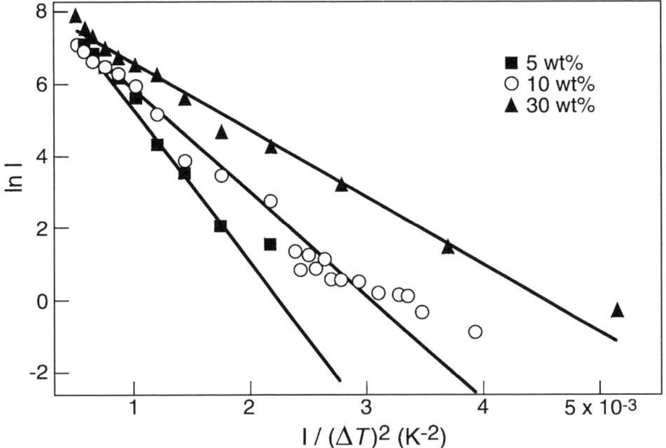

Fig. 13.31 Plot of log nucleation rate against $1/(\Delta T)^2$ for poly(ethylene oxide) crystallizing from tripropionin solutions. Weight percent polymer: ▲ 30%; ○ 10%; ■ 5%. (From Sasaki *et al.* (87))

In order to obtain the product of the interfacial free energies for nucleation from the slopes of the straight lines in Fig. 13.31 (as well as for other systems) both the effective volume fraction and reliable equilibrium melting temperature need to be known. For these reasons values of the product $\sigma_{en}\sigma_{un}$ cannot be obtained in a routine manner even for crystallization from dilute solutions. Even if the proper value of $\sigma_{en}\sigma_{un}$ could be obtained from experimental data there is a problem that has been discussed previously. It is the matter of obtaining σ_{en}. To obtain this important quantity the value of $\sigma_{un}/\Delta H_u$ needs to be independently known. As has been pointed out earlier this ratio does not have a universal value as has been implied.(73) Even the value of 0.1 given for linear polyethylene can be seriously questioned. Thus, there are several major problems that need to be overcome in order to obtain the product $\sigma_{en}\sigma_{un}$ from kinetic data and from it, the key quantity σ_{en}.

Polymers that show a rate maximum with respect to temperature in the pure state do so also when crystallizing from diluent mixtures.(42a,67,88) Two examples are shown in Figs. 13.32 and 13.33 for isotactic poly(styrene) crystallizing from ether benzophenone or dimethyl phthalate respectively.(42a,67) Characteristically, the addition of the diluent causes a shift of the crystallization range to lower temperatures. A similar effect was observed with bisphenol-A poly(carbonate).(88) In addition, the growth rate maximum increases with the initial addition of diluent. This phenomenon is observed up to about 20% diluent in the case of benzophenone (Fig. 13.32) and about 50% with dimethyl phthalate (Fig. 13.33). A similar pattern is also indicated for the poly(carbonate)–diluent mixture.(89) With further additions of diluent there is a continuous decrease in the growth maxima up to very dilute

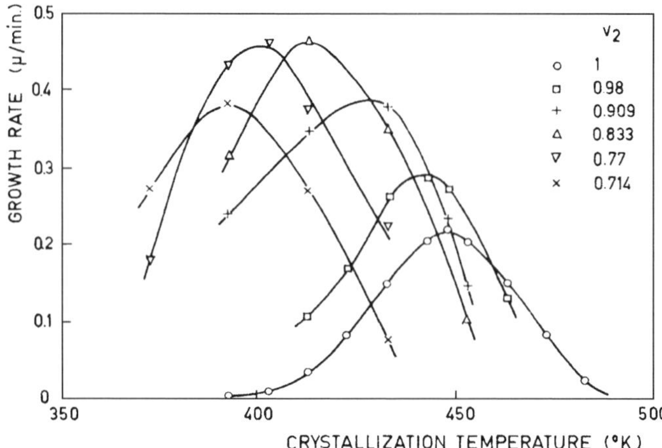

Fig. 13.32 Plot of spherulite growth rates of isotactic poly(styrene) crystallizing from its mixture with benzophenone against the crystallization temperature for indicated volume fraction of polymer. (From Boon and Azcue (42a))

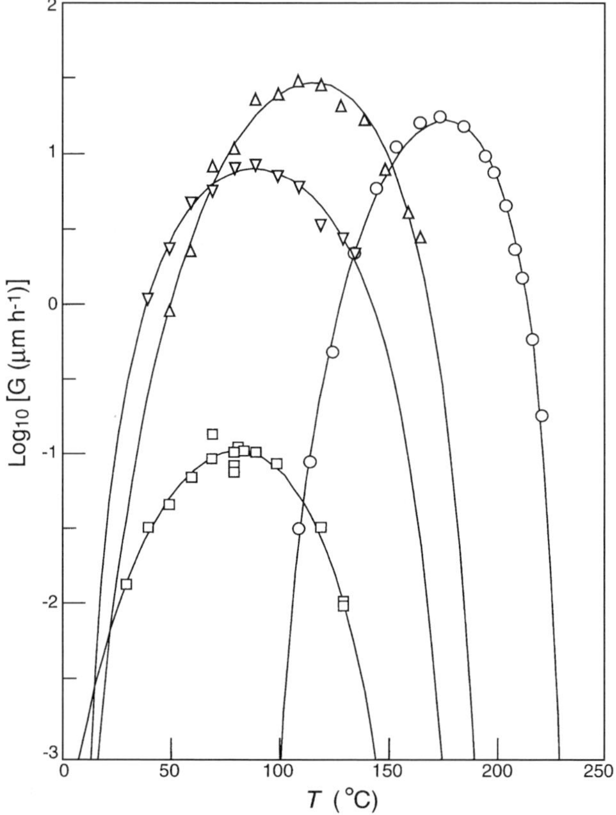

Fig. 13.33 Plot of spherulite growth rates of isotactic poly(styrene) crystallizing from its mixtures with dimethyl phthalate against the crystallization temperature. Weight percent polymer: ○ pure polymer; △ 50%; ▽ 30%; □ 0.1%. (From Miyamoto *et al.* (67))

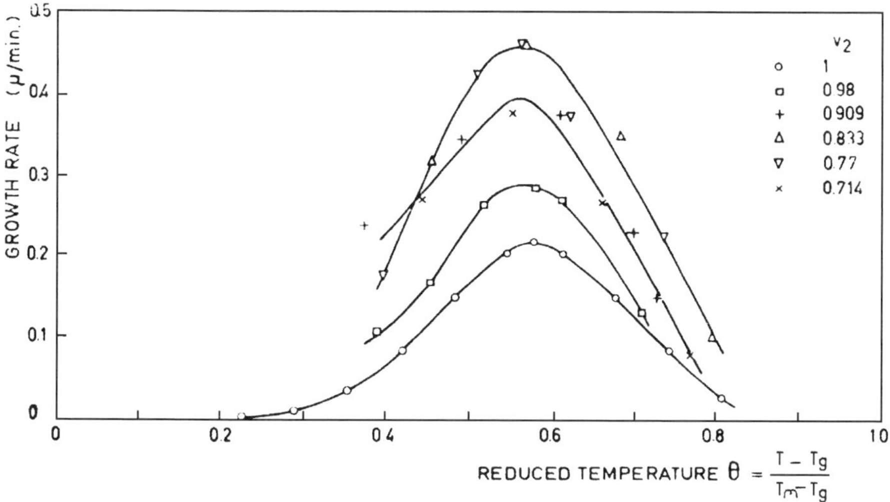

Fig. 13.34 Plot of spherulite growth rates of isotactic poly(styrene) crystallizing from mixtures with benzophenone against the reduced temperature $\Theta = (T - T_g)/(T_m - T_g)$. Data from Fig. 13.32. (From Boon and Azcue (42a))

systems as is illustrated in Fig. 13.33. It is stated that the behavior in the benzophenone system follows a similar pattern.(42a) In the examples cited an expansion of the crystallization range is observed with dilution. This can be accounted for by a larger depression of the glass temperature relative to the melting temperature. For the benzophenone mixtures the glass temperature decreases by about 150 K over the complete composition range. The melting temperature decreases by only about 75–100 K over the same range. The influence of the glass and melting temperatures in shifting the crystallization range can be seen by introducing a reduced temperature variable Θ defined as (42a)

$$\Theta = (T - T_g/T_m - T_g) \tag{13.14}$$

The experimental growth rate data of Fig. 11.32 are plotted against Θ in Fig. 13.34. The shapes of the curves are now similar to one another and the data for several of the mixtures fall on the same curve. The maxima for all compositions occurs at Θ = 0.57. From the data given, the ratio of T_{max}/T_m is about 0.85 for the isotactic poly(styrene)–benzophenone system at all compositions. The ratios for the dimethyl phthalate mixtures appear to be slightly lower.

13.5 Solvent induced crystallization

The crystallization process can be initiated, or accelerated, by interaction with appropriate solvents. Two different situations can be distinguished. In one case, crystallization of the polymer can occur from either the pure melt or the glassy state. Crystallization is induced by interaction with a solvent in this state. In the other situation the polymer, either homopolymer or copolymer, possesses sufficient structural regularity to be potentially crystallizable, but crystallization is not kinetically favored. Interaction with an appropriate solvent rectifies this situation and allows for crystallinity to develop.

The latter situation is found in some homopolymers and copolymers. It is quite commonly used to induce crystallization in stereo irregular polymers.(89–92) The reason for the difficulty is that the melting temperature and glass temperature are so close to one another that the kinetics for crystallization is unfavorable. Put another way, the window of opportunity for crystallization is quite small. This problem can be alleviated by treating the polymer with a diluent that would depress the glass temperature a greater amount than the melting temperature. According to Eq. (3.2) the addition of a diluent with a large molar volume with a poorer solvent power (positive χ_1) will result in a minimal melting point depression at a given polymer concentration. On the other hand, the depression of the glass temperature depends primarily on composition and not on the nature of the diluent. Thus, by utilizing the appropriate diluent, the temperature interval between the melting and glass temperatures can be substantially increased over that of the undiluted polymer. By expanding the temperature window in this manner the kinetics for crystallization becomes more favorable. This method has been successfully used to develop crystallinity in poly(styrene) synthesized by alfin catalysts,(89) poly(methyl methacrylate) prepared by free radical or ionic methods (90–92) and poly(ethylene isophthalate).(93,94)

The solvent induced crystallization from the glassy state differs from the conventional crystallization of polymer–diluent mixtures. In the latter case there is a uniform mixture of polymer and diluent prior to crystallization. In solvent induced crystallization the diluent has to penetrate and diffuse through the glassy polymer prior to the onset of crystallization. The diffusion of the solvent can play a crucial role in the overall crystallization process. The same requirements for a diluent to be effective still hold. Although a diluent that depresses the glass temperature can be very effective, it is important that at the same time the melting temperature depression be minimized. It is theoretically possible to narrow the crystallization window rather than expand it. By doing so crystallizations will be repressed. This method is particularly effective and useful with polymers that crystallize slowly when pure and can be quenched, without crystallization, to the glassy state. Crystallization can be carried out below the glass temperature of the undiluted polymer. For example, the crystallization temperature has to be 30–50 °C higher than the glass temperature for crystallization to occur in undiluted, nonoriented poly(ethylene

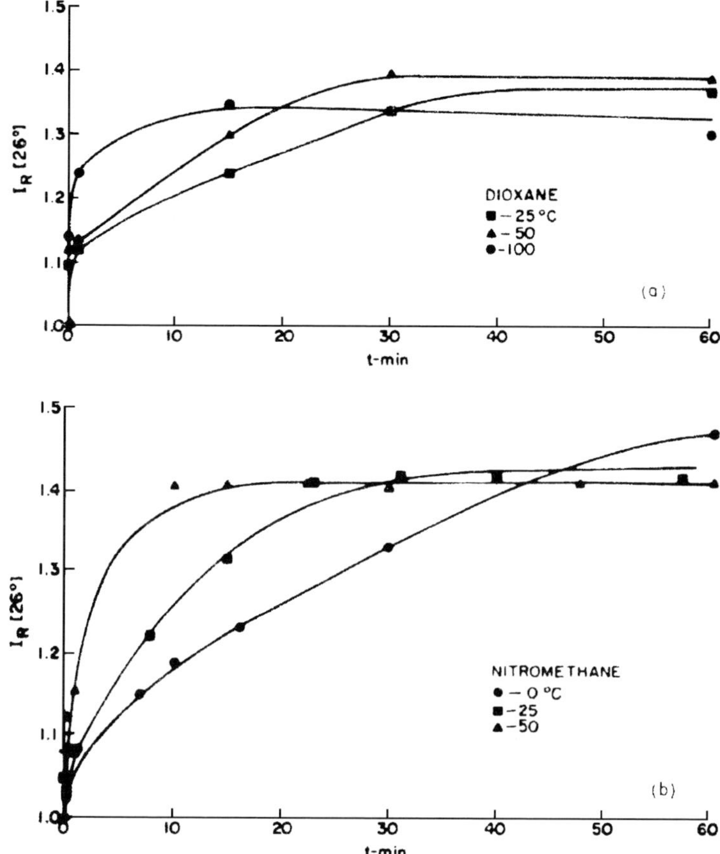

Fig. 13.35 Plot of wide-angle x-ray diffraction intensity of the (100) plane of poly(ethylene terephthalate) as a function of immersion time in either dioxane or nitromethane. Crystallization temperatures are indicated. (From Desai and Wilkes (96))

terephthalate).(95) In contrast, the crystallization of this polymer can take place at least 70 °C below T_g by immersion in an appropriate solvent.(96) Similar effects have been observed by the solvent induced crystallization of syndiotactic and isotactic poly(styrene).(97,97a)

The isotherms that represent crystallization that is solvent induced do not always give the typical sigmoidal shape found in conventional crystallization. (95,96,98–104) Examples are given in Fig. 13.35 for the solvent induced crystallization of poly(ethylene terephthalate) by either dioxane or nitromethane.(96) The isotherms in this figure do not resemble those of the Avrami type. It can be suspected that the penetration of the solvent and its diffusion through the sample is rate determining. This suspicion is confirmed by the plots in Fig. 13.36.(104) Here the crystallinity is

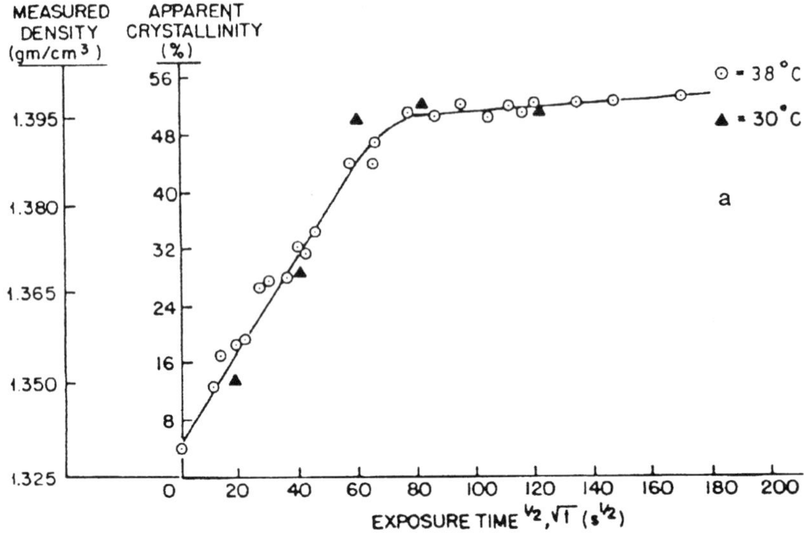

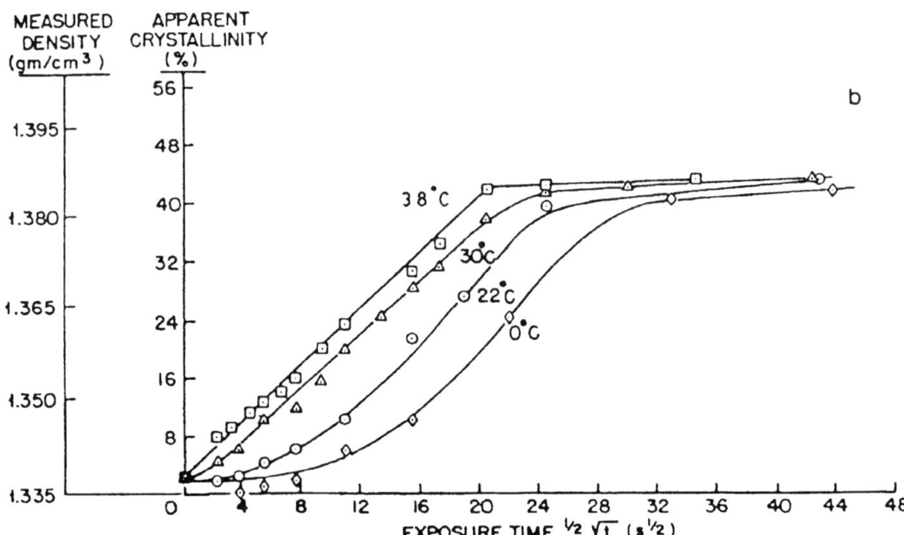

Fig. 13.36 Plot of percent crystallinity against the square-root of time for initially quenched, amorphous films of poly(ethylene terephthalate) exposed to saturated methylene chloride vapor at indicated temperatures. Film thickness: (a) 33.8 mil, 0.086 cm; (b) 12.0 mil, 0.03 cm; (c) 1 mil, 0.0025 cm. (From Durning *et al.* (104))

plotted against the square-root of time for initially amorphous films of poly(ethylene terephthalate) subject to the action of methylene chloride at the indicated temperatures. Figure 13.36a represents the results with the thickest film used (33.8 mil, 0.086 cm). In this case the straight line that represents the data is typical of Fickian

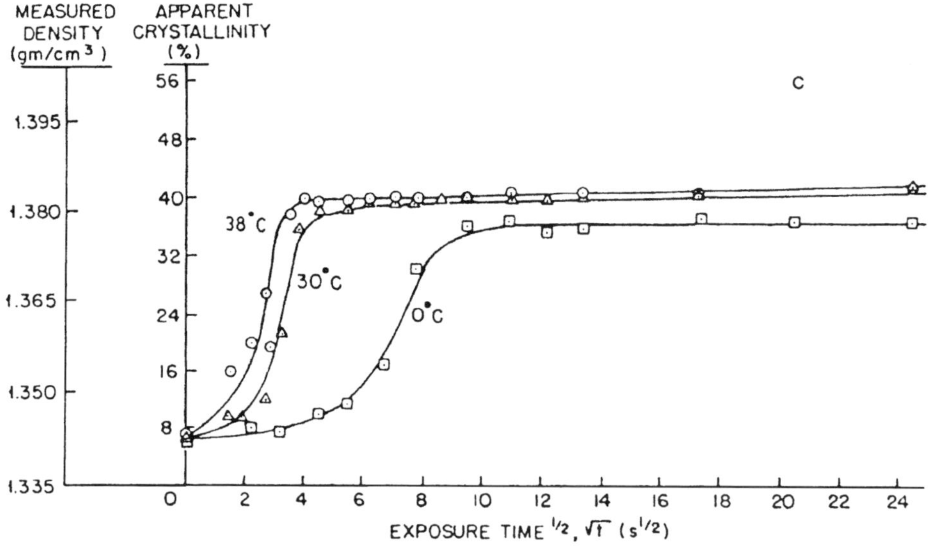

Fig. 13.36 (*cont.*)

type diffusion.[5] Thus, for films of this thickness, diffusion is rate controlling in the crystallization process. For the film of intermediate thickness (12.0 mil, 0.03 cm), diffusion dominates at the higher crystallization temperature, but not at the lower one. The plots in Fig. 13.36c, which represent the thinnest film (1.0 mil, 0.0025 cm), show that there is no longer any indication of diffusion influence on the crystallization process. In fact, the data fit the derived Avrami in the conventional manner with $n = 3$.(104) Figure 13.37 summarizes the influence of film thickness for the poly(ethylene terephthalate)–methylene chloride system.(104) At a fixed temperature of 38 °C the two thicker films show control by Fickian type diffusion, the thinner film clearly does not. Thus diffusion, as the rate determining step of solvent induced crystallization, is not always involved. Each case must be independently investigated.

The experimental results indicate that in solvent induced crystallization there is an interplay between the conventional Avrami type kinetics and the diffusion of the solvent through the polymer sample. There are many different factors involved. These include the type of diffusion, whether or not Fick's law is obeyed, the geometry and dimensions of the sample and the role of the solvent in reducing the glass and equilibrium melting temperature, and the solubility parameter among others.(104a) Complete analysis of this crystallization process is obviously

[5] In Fickian diffusion, the diffusion coefficient is a constant independent of the composition. Under these conditions the square-root of time rate law holds. However, the diffusion of vapors or liquids in a polymer film does not necessarily follow Fick's law.(105,106)

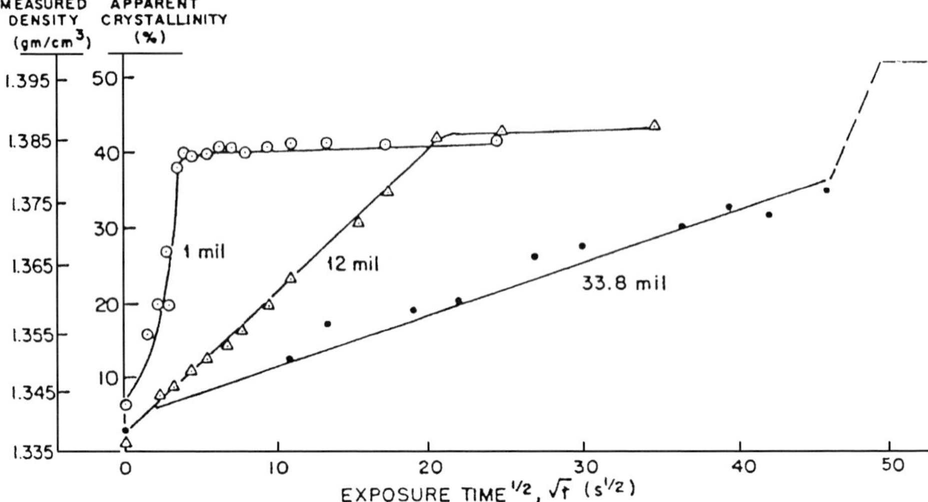

Fig. 13.37 Plot of percent crystallinity against the square-root of time for initially quenched, amorphous films of poly(ethylene terephthalate) of indicated thicknesses exposed to saturated methylene chloride vapors at 38 °C. (From Durning *et al.* (104))

complex. There is a series of papers that develop the theoretical framework for this problem.(104,107–109)

References

1. Boda, E., G. Ungar, G. M. Brooks, S. Burnett, S. Mohammed, D. Proctor and M. C. Whiting, *Macromolecules*, **30**, 4674 (1997).
2. Organ, S. J., G. Ungar and A. Keller, *Macromolecules*, **22**, 1995 (1989).
3. Ungar, G., J. Stejny, A. Keller, I. Bidd and M. C. Whiting, *Science*, **229**, 386 (1985).
4. Ungar, G., in *Integration of Fundamental Polymer Science and Technology*, P. J. Lemstra and L. A. Kleitjens eds., Elsevier (1988) p. 346.
5. Organ, S. J., P. J. Barham, M. J. Hill, A. Keller and R. L. Morgan, *J. Polym. Sci. Pt. B: Polym. Phys.*, **35**, 1775 (1997).
6. Morgan, R. L., P. J. Barham, M. J. Hill, A. Keller and S. J. Organ, *J. Macromol. Sci. Phys.*, **B37**, 319 (1998).

6a. Terry, A. E., J. K. Hobbs, S. J. Organ and P. J. Barham, *Polymer*, **44**, 3001 (2003).

6b. Ungar, G., P. Mandal, P. G. Higgs, D. S. M. De Silva, E. Boda and C. M. Chen, *Phys. Rev. Lett.*, **85**, 4397 (2000).

6c. Ungar, G. and X. Zeng, *Chem. Rev.*, **101**, 4157 (2001).

6d. Putra, E. G. R. and G. Ungar, *Macromolecules*, **36**, 3812 (2003).

6e. Putra, E. G. R. and G. Ungar, *Macromolecules*, **36**, 4214 (2003).

7. Hobbs, J. K., M. J. Hill and P. J. Barham, *Polymer*, **42**, 2167 (2001).
8. Ungar, G. and A. Keller, *Polymer*, **28**, 1899 (1987).
9. Organ, S. J., A. Keller, M. Hikosaka and G. Ungar, *Polymer*, **37**, 2517 (1996).

10. Alamo, R. G., L. Mandelkern, G. M. Stack, C. Kröhnke and G. Wegner, *Macromolecules*, **27**, 147 (1994).
11. Ungar, G., in *Crystallization of Polymers*, M. Dosiéré ed., Kluwer Academic Publishers (1993) p. 63.
12. Sadler, D. M. and G. H. Gilmer, *Polymer*, **25**, 1446 (1984).
13. Sadler, D. M. and G. H. Gilmer, *Phys. Rev. Lett.*, **56**, 2708 (1986).
14. Sadler, D. M. and G. H. Gilmer, *Polym. Commun.*, **28**, 243 (1987).
15. Higgs, P. G. and G. Ungar, *J. Chem. Phys.*, **100**, 640 (1994); *ibid.*, **114**, 6958 (2001).
16. Ungar, G. and S. J. Organ, *J. Polym. Sci.: Polym. Phys. Ed.*, **28**, 2353 (1990).
17. Organ, S. G., G. Ungar and A. Keller, *J. Polym. Sci.: Polym. Phys. Ed.*, **28**, 2365 (1990)
18. Alamo, R. G., L. Mandelkern, G. M. Stack, C. Kröhnke and G. Wegner, *Macromolecules*, **26**, 2743 (1993).

18a. Alamo, R. G. and C. Chi, in *Molecular Interaction and Time-Space Organization in Macromolecular Systems*, Y. Horishima, T. Norisaye and K. Tashiro eds., Springer (1999), p. 29.

19. Hobbs, J. K., M. J. Hill and P. J. Barham, *Polymer*, **41**, 8761 (2000).
20. Stack, G. M., L. Mandelkern and J. G. Voigt-Martin, *Polym. Bull.*, **8**, 421 (1982).
21. Jackson, J. F. and L. Mandelkern, *J. Polym. Sci. Polym. Lett. Ed.*, **5B**, 557 (1967).
22. Nakajima, A. and S. Hayashi, *Koll. Z. Z. Polym.*, **2**, 116 (1968).
23. Weaver, T. J. and I. R. Harrison, *Polymer*, **22**, 1590 (1981).
24. Organ, S. J. and A. Keller, *J. Mater. Sci.*, **20**, 1602 (1985).
25. Sutton, S. J., A. S. Vaughan and D. C. Bassett, *Polym. Commun.*, **37**, 5735 (1996).
26. Hosier, I. L. and D. C. Bassett, *Polymer*, **41**, 8801 (2000).
27. Fatou, J. G., E. Riande and V. Garcia, *J. Polym. Sci.: Polym. Phys. Ed.*, **13**, 2103 (1975).
28. Devoy, C., L. Mandelkern and L. Bourland, *J. Polym. Sci. Pt. A-2*, **8**, 869 (1970).
29. Chu, C. C. Ph.D. Dissertation, Florida State University (1976).
30. Riande, E. and J. G. Fatou, *Polymer*, **17**, 795 (1976).
31. Riande, E. and J. G. Fatou, *Polymer*, **19**, 1295 (1978).
32. McHugh, A. J., W. R. Burghardt and D. A. Holland, *Polymer*, **27**, 1585 (1986).
33. Beech, D. R. and C. Booth, *Polymer*, **13**, 355 (1972).
34. Riande, E., J. G. Fatou and J. C. Alcazar, *Polymer*, **18**, 1095 (1977).
35. Shah, J. K. and L. E. Lahti, *Ind. Eng. Chem. Prod. Res. Dev.*, **12**, 304 (1973).
36. Vane, L. M. and F. Rodriguez, *J. Appl. Polym. Sci.*, **49**, 765 (1993).
37. Ergoz, E., J. G. Fatou and L. Mandelkern, *Macromolecules*, **5**, 147 (1972).
38. Riande, E. and J. G. Fatou, *Polymer*, **17**, 99 (1976).

38a. Bu, H., F. Gu, L. Bao and M. Chen, *Macromolecules*, **31**, 7108 (1998).

38b. Xiao, Z., Q. Sun, G. Zue, Z. Yuan, Q. Dai and Y. Hu, *Eur. Polym. J.*, **39**, 927 (2003).

39. Flory, P. J., *J. Chem. Phys.*, **17**, 223 (1949).
40. Fatou, J. G., E. Riande and R. G. Valdecasas, *J. Polym. Sci.: Polym. Phys. Ed.*, **13**, 2103 (1975).
41. Mandelkern, L., *J. Appl. Phys.*, **26**, 443 (1955).
42. Mandelkern, L., *Polymer*, **5**, 637 (1964).

42a. Boon, J. and J. M. Azcue, *J. Polym. Sci. Pt. A-2*, **6**, 885 (1968).

43. Sadler, D. M., *J. Polym. Sci. Pt. A-2*, **9**, 779 (1971).
44. Koningsveld, R. and A. J. Pennings, *Rec. Trav. Chim.*, **83**, 552 (1964).
45. Pennings, A. J., *J. Polym. Sci.*, **16c**, 1799 (1967).
46. Holland, V. F. and P. H. Lindenmeyer, *J. Polym. Sci.*, **57**, 589 (1962).
47. Seto, T. and V. Mori, *Rep. Prog. Polym. Phys. Japan*, **12**, 157 (1969).

48. Blundell, D. J. and A. Keller, *J. Polym. Sci. Polym. Lett.*, **6B**, 433 (1968).
49. Jackson, J. F. and L. Mandelkern, *Macromolecules*, **1**, 546 (1968).
50. Cooper, M. and R. St. J. Manley, *Macromolecules*, **8**, 219 (1975).
51. Cooper, M. and R. St. J. Manley, *Coll. Polym. Sci.*, **254**, 542 (1976).
52. Leung, W. M., R. St. J. Manley and A. N. Panaras, *Macromolecules*, **18**, 760 (1985).
53. Organ, S. J. and A. Keller, *J. Polym. Sci. Polym. Lett.*, **25B**, 67 (1987).
54. Ding, N. and E. J. Amis, *Macromolecules*, **24**, 3906 (1991).
55. Ding. N. and E. J. Amis, *Macromolecules*, **24**, 6964 (1991).
56. Keller, A. and E. Pedmonte, *J. Cryst. Growth*, **18**, 111 (1973).
57. Keith, H. D., R. G. Vadimsky and F. J. Padden, Jr., *J. Polym. Sci. A-2*, **8**, 1687 (1970).
58. Toda, A., H. Miyaji and H. Kiho, *Polymer*, **27**, 1505 (1986).
59. Toda, A., *J. Chem. Soc. Faraday Trans.*, **91**, 2581 (1995).
60. Sanchez, I. C. and E. A. DiMarzio, *Macromolecules*, **4**, 677 (1971).
61. Sanchez, I. C. and E. A. DiMarzio, *J. Chem. Phys.*, **55**, 893 (1971).
62. Toda, A., H. Kiho, H. Miyazi and K. Asai, *J. Phys. Soc. Japan*, **54**, 1411 (1985).
63. Seto, T., *Rep. Prog. Polym. Phys. Japan*, **7**, 67 (1964).
64. Toda, A. and H. Koho, *J. Polym. Sci.: Pt. B: Polym. Phys.*, **27**, 53 (1989).
65. Toda, A., *J. Polym. Sci.: Pt. B: Polym. Phys.*, **27**, 1721 (1989).
66. Tanzawa, J., *Polymer*, **33**, 266 (1992).
67. Miyamoto, Y., Y. Tanzawa, H. Miyoji and H. Kiho, *Polymer*, **33**, 2497 (1992).
68. Mandelkern, L., *J. Phys. Chem.*, **75**, 3909 (1971).
69. Organ, S. J. and A. Keller, *J. Mater. Sci.*, **20**, 1602 (1985).
70. Toda, A., *Polymer*, **28**, 1645 (1987).
71. Nardini, M. J. and F. P. Price, *Crystal Growth. Proceedings of an International Conference*, June 1966, Pergamon Press (1967) p. 395.
72. Hoffman, J. D., *Soc. Plast. Eng. J.*, **4**, 315 (1964).
73. Hoffman, J. D., G. T. Davis and J. I. Lauritzen, Jr., in *Treatise on Solid State Chemistry*, vol. 3, N. B. Hannay ed., Plenum Press (1976) Chapter 7.
74. Huggins, M. L., *J. Phys. Chem.*, **46**, 151 (1942); *Ann. N. Y. Acad. Sci.*, **41**, 1 (1942); *J. Am. Chem. Soc.*, **64**, 1712 (1942).
75. Flory, P. J., *J. Chem. Phys.*, **10**, 51 (1942).
76. Graessley, W. W., *Adv. Polym. Sci.*, **16**, 1 (1974).
77. Mandelkern, L., *J. Appl. Phys.*, **26**, 443 (1955).
78. Feio, G., G. Buntinx and J. P. Cohen-Addad, *J. Polym. Sci.: Pt. B: Polym. Phys.*, **27**, 1 (1989).
79. Gornick, F. and L. Mandelkern, *J. Appl. Phys.*, **33**, 907 (1962).
80. Chaturvedi, P. N., *Makromol. Chem.*, **188**, 433 (1987).
81. Wang, Y. F. and D. R. Lloyd, *Polymer*, **34**, 4740 (1993).
82. Lim, G. B. A. and D. R. Lloyd, *Polym. Eng. Sci.*, **33**, 522 (1993).
83. Sasaki, T., Y. Yamamoto and T. Takahashi, *Polym. J.*, **30**, 868 (1998).
84. Keith, H. D. and F. J. Padden, Jr., *J. Appl. Phys.*, **34**, 2409 (1963); *ibid.*, **35**, 1270 (1964).
85. Okada, T., H. Saito and T. Inoue, *Macromolecules*, **23**, 3865 (1990).
86. Wang, Y. F. and D. R. Lloyd, *Polymer*, **34**, 2325 (1993).
87. Sasaki, T., Y. Yamamoto and T. Takahashi, *Polym. J.*, **32**, 263 (2000).
88. Legras, R. and J. P. Mercier, *J. Polym. Sci.: Polym. Phys. Ed.*, **17**, 1171 (1979).
89. Williams, J. L. R., J. Van Den Berghe, W. J. Dulmage and K. R. Dunham, *J. Am. Chem. Soc.*, **78**, 1660 (1956); *ibid.*, **79**, 1716 (1957).

90. Fox, T. G., W. F. Goode, S. Gratch, C. M. Huggett, J. F. Kincaid, A. Spell and J. D. Stroupe, *J. Polym. Sci.*, **31**, 173 (1958).
91. Fox, T. G., B. S. Garrett, W. F. Goode, S. Gratch, J. F. Kincaid, A. Spell and J. D. Stroupe, *J. Am. Chem. Soc.*, **80**, 1768 (1958).
92. Korotkov, A. A., S. P. Mitsengendeleu, U. N. Krausline and L. A. Volkova, *High Molecular Weight Compounds*, **1**, 1319 (1959).
93. Conix, A. and R. Van Kerpel, *J. Polym. Sci.*, **40**, 521 (1959).
94. Yamadera R. and C. Sonoda, *J. Polym. Sci. Polym. Lett.*, **3B**, 411 (1965).
95. Kolb, J. and E. F. Izard, *J. Appl. Phys.*, **20**, 571 (1949).
96. Desai, A. and G. L. Wilkes, *J. Polym. Sci.*, **46C**, 291 (1974).
97. Tashiro, K., Y. Veno, A. Yoshioka and M. Kobayashi, *Macromolecules*, **34**, 310 (2001).

97a. Overbergh, N., H. Berghmans and G. Smets, *Polymer*, **16**, 703 (1975).

98. Moore, W. R. and R. P. Sheldon, *Polymer*, **2**, 315 (1961).
99. Sheldon, R. P., *Polymer*, **3**, 27 (1962).
100. Cottam, L. and R. P. Sheldon, *Adv. Polym. Sci. Technol.*, **26**, 65 (1966).
101. Kishimiri, M. I. and R. P. Sheldon, *Brit. Polym. J.*, **1**, 65 (1969).
102. Mercier, J. P., G. Groeninckx and M. Lesne, *J. Polym. Sci.*, **16C**, 2059 (1967).
103. Lin, S. B. and J. L. Koenig, *J. Polym. Sci.: Polym. Phys. Ed.*, **21**, 1539 (1983).
104. Durning, C. J., L. Rebenfeld, W. B. Russell and H. D. Weigmann, *J. Polym. Sci.: Pt. B: Polym. Phys.*, **24**, 1321 (1986).

104a. Lawton, E. L. and D. M. Cates, *J. Appl. Polym. Sci.*, **13**, 899 (1969).

105. Mandelkern, L. and F. A. Long, *J. Polym. Sci.*, **6**, 457 (1951).
106. Kokes, R. J., F. A. Long and J. L. Hoard, *J. Chem. Phys.*, **20**, 1711 (1952).
107. Zachmann, H. G. and G. Conrad, *Makromol. Chem.*, **118**, 189 (1968).
108. Makarewicz, P. J. and G. L. Wilkes, *J. Polym. Sci.: Polym. Phys. Ed.*, **16**, 1559 (1978).
109. Durning, C. J. and W. B. Russell, *Polymer*, **26**, 119 (1985); *ibid.*, **26**, 131 (1985).

Author index

Numbers in parentheses indicate the number of mentions/citations on that page.

Subject index

Page numbers in italics, e.g. *381*, indicate references to figures. Page numbers in bold, e.g. **42**, denote entries in tables.